Unravelling Alzheimer's: Innovations in Pathogenesis, Diagnosis, and Therapeutics

Edited by

Prashant Kumar Dhakad
Gyan Vihar School of Pharmacy
Suresh Gyan Vihar University
Jaipur, Rajasthan 302017
India

Raghav Mishra
Lloyd Institute of Management and Technology
Greater Noida, Uttar Pradesh 201306
India

Shweta Sharma
Department of Pharmaceutical Chemistry
IIMT College of Pharmacy, Greater Noida
Uttar Pradesh 201306
India

Sudhanshu Mishra
Faculty of Pharmaceutical Science
Mahayogi Gorakhnath University
Gorakhpur, Uttar Pradesh 273007
India

&

Nishant Gaur
Lloyd Institute of Management and Technology
Greater Noida, Uttar Pradesh 201306
India

Unravelling Alzheimer's: Innovations in Pathogenesis, Diagnosis, and Therapeutics

Editors: Prashant Kumar Dhakad, Raghav Mishra, Shweta Sharma, Sudhanshu Mishra and Nishant Gaur

ISBN (Online): 979-8-89881-495-3

ISBN (Print): 979-8-89881-496-0

ISBN (Paperback): 979-8-89881-497-7

First published in 2026.

Bentham Science Publishers Pte. Ltd.
No. 9 Raffles Place
Office No. 26-01
Singapore 048619
Singapore
Email: subscriptions@benthamscience.net

BENTHAM SCIENCE

CONTENTS

FOREWORD

Unravelling Alzheimer's: Innovations in Pathogenesis, Diagnosis, and Therapeutics

Alzheimer's disease stands as one of the most formidable challenges to global health in the 21st century. As populations age and life expectancy increases, the burden of neurodegenerative diseases, particularly Alzheimer's, has grown exponentially, touching countless lives and posing significant emotional, social, and economic tolls. The urgency to understand, diagnose, and treat this devastating disorder has never been more critical.

It is against this backdrop that Unravelling Alzheimer's: Innovations in Pathogenesis, Diagnosis, and Therapeutics emerges as a timely and impactful scholarly contribution. Edited by Mr. Sudhanshu Mishra and Dr. Raghav Mishra, along with a dedicated team of experts, this volume brings together a diverse and interdisciplinary exploration of the many facets of Alzheimer's disease from its complex pathogenesis to state-of-the-art diagnostic approaches, and from natural therapeutics to gene and immunotherapy innovations.

The book thoughtfully integrates traditional knowledge and cutting-edge science. It explores the potential of natural compounds and phytochemicals while simultaneously embracing the future of diagnostics through artificial intelligence, neuroimaging, and biomarker discovery. The inclusion of gene therapy, RNA-based solutions, and emerging pharmacological treatments reflects the depth and foresight of the contributors. Equally valuable are the discussions surrounding comorbidities, intranasal drug delivery systems, and real-world insights gleaned from patents and case studies.

I commend the editors and contributors for their vision, dedication, and scholarly excellence in assembling this important body of work. This book is not just a repository of knowledge; it is a call to action for deeper inquiry, innovation, and hope in the fight against Alzheimer's disease.

Vandana Arora Sethi
Lloyd Institute of Management and Technology,
Greater Noida, Uttar Pradesh 201306, India

PREFACE

Alzheimer's Disease (AD), a progressive neurodegenerative disorder, represents one of the most formidable challenges in contemporary medicine and neuroscience. As global populations age, the incidence of Alzheimer's continues to rise, exerting profound social, economic, and healthcare burdens across the world. This book, Unravelling Alzheimer's: Innovations in Pathogenesis, Diagnosis, and Therapeutics, provides a comprehensive and multidisciplinary overview of the latest advancements in our understanding, diagnosis, and treatment of Alzheimer's disease. The book begins by exploring the historical background, epidemiology, and global impact of Alzheimer's Disease, laying the foundation for a deeper understanding of its pathophysiology. A thorough examination of the molecular and cellular mechanisms underpinning Alzheimer's progression is presented, followed by insights into biomarkers and their emerging role in early detection and therapeutic targeting.

Technological progress has redefined diagnostic approaches, and the book highlights the growing utility of neuroimaging techniques and digital tools, including artificial intelligence, in enhancing diagnostic precision. In addition, a significant portion is dedicated to natural products, phytochemicals, immunotherapies, gene therapy, and RNA-based innovations, showcasing novel therapeutic frontiers. Recognizing the multifactorial nature of Alzheimer's, chapters also delve into comorbid conditions that influence disease progression, as well as the latest developments in pharmacological treatments and intranasal drug delivery strategies. The inclusion of case studies and patents further connects theoretical advancements with real-world applications, offering a well-rounded perspective on ongoing innovation in the field.

Technological progress has redefined diagnostic approaches, and the book highlights the growing utility of neuroimaging techniques and digital tools, including artificial intelligence, in enhancing diagnostic precision. In addition, a significant portion is dedicated to natural products, phytochemicals, immunotherapies, gene therapy, and RNA-based innovations, showcasing novel therapeutic frontiers. Recognizing the multifactorial nature of Alzheimer's, chapters also delve into comorbid conditions that influence disease progression, as well as the latest developments in pharmacological treatments and intranasal drug delivery strategies. The inclusion of case studies and patents further connects theoretical advancements with real-world applications, offering a well-rounded perspective on ongoing innovation in the field.

This volume brings together contributions from experts across diverse disciplines, making it a valuable resource for clinicians, researchers, students, and pharmaceutical professionals. It aims not only to consolidate current knowledge but also to inspire future research directions and therapeutic breakthroughs.

We trust that this book will serve as a valuable guide in your exploration of Alzheimer's disease and stimulate continued inquiry into innovative solutions for this pressing global health issue.

Prashant Kumar Dhakad
Gyan Vihar School of Pharmacy, Suresh Gyan Vihar University,
Jaipur, Rajasthan 302017
India

Shweta Sharma
Department of Pharmaceutical Chemistry, IIMT College of Pharmacy,
Greater Noida, Uttar Pradesh 201306
India

Sudhanshu Mishra
Faculty of Pharmaceutical Science, Mahayogi Gorakhnath University,
Gorakhpur, Uttar Pradesh 273007
India

&

Nishant Gaur
Lloyd Institute of Management and Technology,
Greater Noida, Uttar Pradesh 201306
India

List of Contributors

Abhishek Kumar Singh	Faculty of Pharmaceutical Sciences, Mahayogi Gorakhnath University, Gorakhpur, Uttar Pradesh 273007, India
Amit Upadhyay	Faculty of Pharmaceutical Sciences, Mahayogi Gorakhnath University, Gorakhpur, Uttar Pradesh 273007, India
Anjali Singh	Innovative College of Pharmacy, Greater Noida, Uttar Pradesh 201310, India
Anubhav Tyagi	Department of Management Studies, Indian Institute of Technology, New Delhi, Delhi 110016, India
Arvind Kumar Patel	Narayan Institute of Pharmacy, Gopal Narayan Singh University, Jamuhar, Bihar 821305, India
Ayushi Tyagi	I.T.S College of Pharmacy, I.T.S College of Pharmacy, Ghaziabad, Uttar Pradesh 201206, India
Baby Ilma	School of Pharmacy, Sharda University, Greater Noida, Uttar Pradesh 201310, India
Chandragiri Siva Sai	Department of Pathology, The University of Oklahoma Health Sciences Center, Oklahoma, OK 73104, USA
Chandrashekhar Patil	Divine College of Pharmacy, Satana, Nashik, Maharashtra 423301, India
Deepak Kumar	Faculty of Pharmaceutical Sciences, Mahayogi Gorakhnath University, Gorakhpur, Uttar Pradesh 273007, India
Deepak Sonawane	Divine College of Pharmacy, Satana, Nashik, Maharashtra 423301, India
Ganesh Sonawane	Divine College of Pharmacy, Satana, Nashik, Maharashtra 423301, India
Harendar Kumar Nivatya	Faculty of Pharmacy, IFTM University, Moradabad, Uttar Pradesh 244102, India
Juhi Tiwari	Faculty of Pharmaceutical Sciences, Mahayogi Gorakhnath University, Gorakhpur, Uttar Pradesh 273007, India
Kajal Pansare	Divine College of Pharmacy, Satana, Nashik, Maharashtra 423301, India
Khushboo Bansal	Sunder Deep Pharmacy College, Ghaziabad, Uttar Pradesh 201002, India
Lovy Sharma	Department of Pharmaceutical Chemistry, Parmarth College of Pharmacy, Hapur, Uttar Pradesh 245101, India
Manish Mathur	Department of Academic Affairs(E), Amity University, Lucknow, Uttar Pradesh 226010, India
Manish Singh	Narayan Institute of Pharmacy, Gopal Narayan Singh University, Jamuhar, Bihar 821305, India
Mansi Verma	Division of Pharmacology, Institute of Pharmaceutical Research, GLA University, Mathura, Uttar Pradesh 281406, India
Nandani Jayaswal	Faculty of Pharmaceutical Sciences, Mahayogi Gorakhnath University, Gorakhpur, Uttar Pradesh 273007, India
Neetu Sachan	Department of Pharmaceutical Sciences, School of Pharmaceutical & Biological Sciences, Harcourt Butler Technical University, Kanpur, Uttar Pradesh, 208002, India

Neha Mathur	Amity Institute of Pharmacy, Amity University, Lucknow, Uttar Pradesh 226010, India
Neha Singh	Department of Pharmaceutical Sciences, School of Pharmaceutical & Biological Sciences, Harcourt Butler Technical University, Kanpur, Uttar Pradesh, 208002, India
Nikhila Shekhar	Neuropharmacology Research Laboratory, School of Pharmaceutical Sciences, Delhi Pharmaceutical Sciences and Research University, New Delhi, Delhi 110016, India
Niraj Kumar Singh	Division of Pharmacology, Institute of Pharmaceutical Research, GLA University, Mathura, Uttar Pradesh 281406, India
Nitin Kumar	Department of Pharmacognosy, College of Pharmacy, Shri Venkateshwara University, Gajraula, Uttar Pradesh 244236, India
Phool Chandra	Department of Pharmaceutical Sciences, School of Pharmaceutical & Biological Sciences, Harcourt Butler Technical University, Kanpur, Uttar Pradesh, 208002, India
Piyush Anand	Faculty of Pharmaceutical Sciences, Mahayogi Gorakhnath University, Gorakhpur, Uttar Pradesh 273007, India
Pooja Jaiswal	Faculty of Pharmaceutical Sciences, Mahayogi Gorakhnath University, Gorakhpur, Uttar Pradesh 273007, India
Rajan Kumar Kurmi	Metro College of Health Sciences and Research, Metro College of Health Sciences and Research, Greater Noida, Uttar Pradesh 201308, India
Sakshi Tyagi	School of Pharmacy, Sharda University, Greater Noida, Uttar Pradesh 201310, India
Sana Ahmad	School of Pharmacy and Research, People's University, Bhopal, Madhya Pradesh 462037, India
Shafiurrahman	Department of Pharmacy, Integral University, Dasauli, Lucknow, Uttar Pradesh 226026, India
Shankul Kumar	Narayan Institute of Pharmacy, Gopal Narayan Singh University, Jamuhar, Bihar 821305, India
Shashank Tewari	Department of Pharmaceutical Chemistry, Manipal College of Pharmaceutical Sciences, Manipal Academy of Higher Education (MAHE), Manipal, Karnataka 576104, India
Shashi Kant Singh	Faculty of Pharmaceutical Sciences, Mahayogi Gorakhnath University, Gorakhpur, Uttar Pradesh 273007, India
Shiva Mishra	Lloyd Institute of Management and Technology, Lloyd Institute of Management and Technology, Greater Noida, Uttar Pradesh 201306, India
Shivam Kumar	Narayan Institute of Pharmacy, Gopal Narayan Singh University, Jamuhar, Bihar 821305, India
Shubh Deep Yadav	I.T.S College of Pharmacy, I.T.S College of Pharmacy, Ghaziabad, Uttar Pradesh 201206, India
Shubhanshu Goel	I.T.S College of Pharmacy, I.T.S College of Pharmacy, Ghaziabad, Uttar Pradesh 201206, India

Snigdha Srivastava	Galgotias College of Pharmacy, Galgotias College of Pharmacy, Greater Noida, Uttar Pradesh 201306, India
Sohaib Farooqui	School of Pharmacy and Research, People's University, Bhopal, Madhya Pradesh 462037, India
Sonam Sonam	Faculty of Pharmacy, IFTM University, Moradabad, Uttar Pradesh 244102, India
Sunil Mahajan	Divine College of Pharmacy, Satana, Nashik, Maharashtra 423301, India
Supriya Gupta	Faculty of Nursing & Paramedical, Mahayogi Gorakhnath University, Gorakhpur, Uttar Pradesh 273007, India
Vishal Singh	Department of Pharmaceutics, NKBR College of Pharmacy, Meerut, Uttar Pradesh 245206, India

CHAPTER 1

Introduction to Alzheimer's Disease: History, Epidemiology, and Global Impact

Sana Ahmad[1,*], Sohaib Farooqui[1] and Shafiurrahman[2]

[1] *School of Pharmacy and Research, People's University, Bhopal, Madhya Pradesh 462037, India*

[2] *Department of Pharmacy, Integral University, Dasauli, Lucknow, Uttar Pradesh 226026, India*

Abstract: Alzheimer's Disease (AD) is a progressive neurodegenerative condition and the leading cause of dementia, marked by memory impairment, cognitive deterioration, and alterations in behaviour. Initially identified by Alois Alzheimer in 1906, the disorder has since garnered significant attention within the field of neuroscience. The defining pathological characteristics include the presence of extracellular amyloid-beta plaques, intracellular neurofibrillary tangles formed by hyperphosphorylated tau protein, and extensive neuronal degeneration. On a global scale, AD poses a considerable public health issue, currently affecting over 55 million individuals with dementia, a figure anticipated to rise beyond 139 million by 2050 due to the aging population. The disease predominantly impacts those aged 65 and older, with a higher incidence observed in women. Contributing factors to the development of AD encompass genetic vulnerabilities, such as mutations in the allele, alongside modifiable elements like cardiovascular health, educational attainment, and lifestyle choices. The economic and social ramifications of AD are significant, with an estimated annual global expenditure exceeding $1 trillion, which includes healthcare costs, caregiving expenses, and lost productivity. Furthermore, the emotional toll on families and caregivers is considerable, often resulting in psychological distress and financial hardship. Epidemiological research indicates variations in the prevalence and incidence of AD across different regions, shaped by demographic, cultural, and healthcare-related factors. Initiatives to combat AD include progress in early diagnostic methods utilizing biomarkers and neuroimaging, the creation of disease-modifying treatments, and public health strategies focused on prevention through the management of risk factors. This review seeks to present a comprehensive overview of the historical context, epidemiology, and global implications of AD, highlighting the pressing necessity for ongoing research and international cooperation to address this escalating crisis.

Keywords: Alzheimer's disease, Dementia, Epidemiology, Neuronal degeneration, Neuroscience.

[*] **Corresponding author Sana Ahmad:** School of Pharmacy and Research, People's University, Bhopal, Madhya Pradesh 462037, India; E-mail: syedsanaahmad@gmail.com

Prashant Kumar Dhakad, Raghav Mishra, Shweta Sharma, Sudhanshu Mishra & Nishant Gaur (Eds.)

INTRODUCTION

History of Alzheimer's Disease

Alzheimer's disease (AD) is a relentless, advanced neurodegenerative condition that affects widespread regions of the cerebral cortex and hippocampus. Initial deviations are typically observed in the brain tissue of the temporal as well as frontal lobes, gradually spreading to other zones of the neocortex at varying rates among individuals. This disease is defined by the accumulation of insoluble amyloid-β in plaques within extracellular spaces and blood vessel walls, along with the clustering of the microtubule protein tau into neurofibrillary tangles inside neurons. Amyloid-β originates from the proteolytic cleavage of amyloid precursor protein by a complex array of enzymes, including γ-secretases and β-secretases, which involve presenilin 1 and presenilin 2 [1]. Several stages of AD with their symptoms are stated in Table **1**.

Table 1. Phases of Alzheimer's disease with their symptoms/characteristics.

STAGE	SYMPTOMS/CHARACTERISTICS	REFERENCES
Preclinical AD	Categorized by identifiable biomarkers and observable alterations in the brain, blood, and cerebrospinal fluid, despite the nonappearance of symptoms like impairment of memory.	[5, 6]
Cognitive Impairment [Mild]	Mild Cognitive Impairment due to AD presents assessable biomarkers and brain variations associated with AD pathology, alongside a moderate decline in cognitive function. This decline is primarily evident in the implementation of daily routine activities, like managing bills or preparing meals, which require more time and exhibit reduced efficiency.	[7, 8]
Dementia	It is marked by measurable biomarkers and brain changes linked to the pathology of AD, alterations in behaviour, and severe difficulties in performing daily responsibilities.	[9]

The average lifespan of the illness ranges from 8-10 years, but the characteristic phases are headed by preclinical stages that can last up to two decades. The most prevalent form is sporadic AD, with a typical onset around the age of 80. The primary issue lies in the brain's inability to eliminate the amyloid-β peptide. Additionally, co-existing conditions such as cerebrovascular disease and sclerosis are common at this age, complicating both diagnosis and treatment. While it is not uncommon for individuals with sporadic Alzheimer's to have a family history of the disease, a small fraction (<1%) of patients experience autosomal dominant inherited Alzheimer's disease, which typically manifests much earlier, around the age of 45. In this group, PS1 and PS2 lead to either excessive production or the creation of an abnormal form of amyloid-β. In many clinical aspects, sporadic and familial AD are similar, including the pace of disease evolution and biomarker characteristics. AD, as a distinct medical condition, exhibits numerous similarities with other neurodegenerative disorders that are defined at the molecular level, including Parkinson's disease and frontotemporal dementias [2]. The stages of Alzheimer's disease, with their severity range, occur in the form of amyloid plaque and neurofibrillary tangles, are shown in Fig. (**1**).

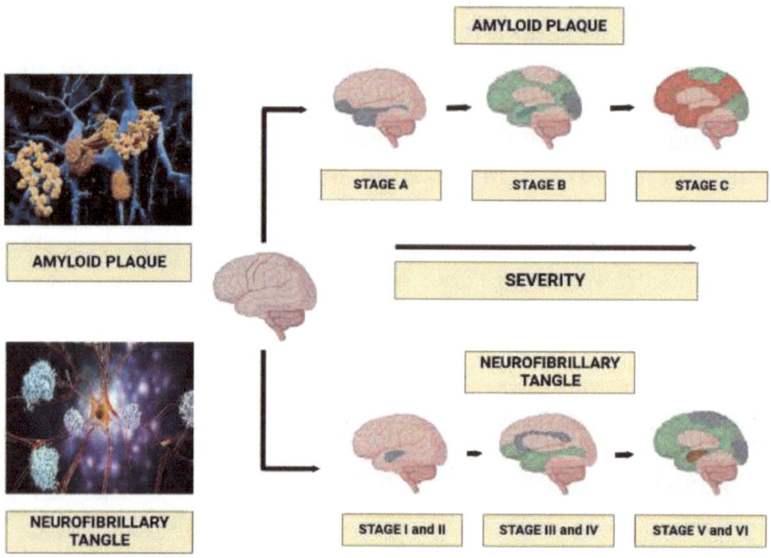

Fig. (1). Stages of Alzheimer's disease.

AD—recognized as the utmost prevalent kind of dementia—was first described in 1906 by the German psychiatrist Alois Alzheimer. He presented the case of Auguste Deter, a 51-year-old woman suffering from significant memory impairment and cognitive deterioration. After her death, Alzheimer examined her brain and identified amyloid plaques and neurofibrillary tangles, now recognized as key markers of the disease [3].

Initially considered rare and affecting younger individuals, Alzheimer's was later found to be a leading cause of dementia in older adults as life expectancy increased. The mid-20th century saw major progressions in neuropathology and biochemistry, including the discovery of amyloid-beta in 1984 and tau protein in 1985, both crucial to the disease's development. Further research in the late 20th and early 21st centuries uncovered genetic factors, such as APP, PSEN1, PSEN2, and the APOE ε4 allele, highlighting its hereditary aspects. Developments in neuroimaging and biomarkers have improved early diagnosis and disease monitoring. Despite these discoveries, Alzheimer's remains a major global health concern, affecting millions worldwide. Research continues to focus on latent treatments, including anti-amyloid therapies and approaches targeting tau pathology, as scientists strive to unravel its complexities and develop effective interventions [4].

Pathophysiology of Alzheimer's disease

In Alzheimer's disease, the most significant neuronal degeneration and pathological alterations occur in areas such as the hippocampus, amygdala, entorhinal cortex, and the association regions of the frontal, temporal, and parietal lobes. Furthermore, subcortical structures, including the serotonergic dorsal raphe, noradrenergic locus coeruleus, and cholinergic basal nucleus, are also notably impacted. The progression of neurofibrillary tangles follows a specific trajectory, beginning in the trans-entorhinal cortex and advancing to the entorhinal cortex, the CA1 region of the hippocampus, and eventually the cortical association areas, with the frontal, parietal, and temporal lobes being particularly vulnerable. The severity and distribution of these tangles are closely associated with the degree of dementia, showing a stronger correlation with cognitive decline than the accumulation of amyloid plaques. The aggregation of tau protein is especially linked to neurodegeneration, brain atrophy, and the reduction of hippocampal volume [10]. In the context of Alzheimer's pathology, substantial neuron loss and cortical atrophy in the superfrontal regions instigate inflammation, the deposition of amyloid plaques, and the abnormal clustering of protein fragments, resulting in an increased presence of monocytes and macrophages in the cerebral cortex, alongside the activation of microglial cells within the brain tissue [11]. A detailed representation of this pathophysiological process can be found in Fig. (**2**).

Hyperphosphorylated tau protein and the amyloid β hypothesis

A key pathological characteristic of AD is the development of senile plaques [SP], primarily caused by the build-up of amyloid-beta [Aβ]. Under normal circumstances, Aβ consists of soluble small peptides generated through the cleavage of the amyloid precursor protein by enzymes known as α-secretase, β-secretase, and γ-secretase. An unevenness between the production and clearance of β-amyloid leads to the development of various toxic oligomers, including fibrils, protofibrils, and plaques, depending on the degree of oligomerization. Although the exact cause of Aβ formation remains elusive, factors such as its arrangement, concentration, and stability situations play a crucial role. The pathophysiology of AD is attributed to multiple influences, including cholinergic dysfunction, the toxicity of amyloid and tau proteins, as well as mitochondrial dysfunction and oxidative stress [12].

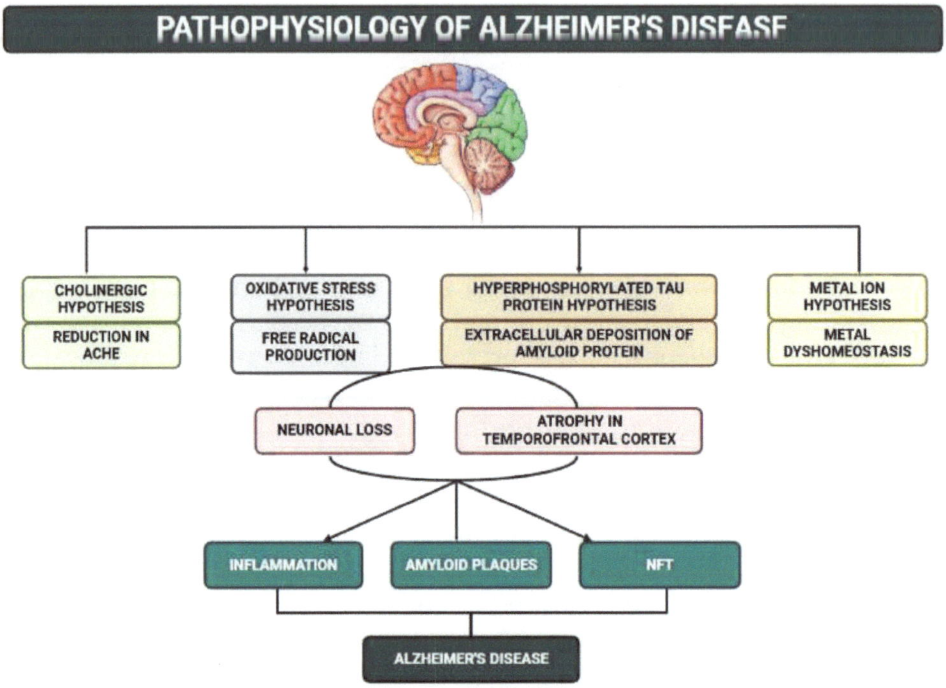

Fig. (2). Pathophysiology of Alzheimer's disease.

Oxidative Stress Hypothesis

The human body produces reactive oxygen species and reactive nitrogen species through both normal physiological functions and pathological conditions. These compounds provide a dual purpose; while they are essential for certain cellular signaling pathways, they can also trigger harmful effects that cause injury to several cellular components, like membranes, deoxyribonucleic acid, lipids, and proteins. The brain, which consumes 20-22% more oxygen than the other tissues reliant on mitochondrial respiration, is particularly susceptible to oxidative stress. Neurons, the fundamental units of the brain, are rich in polyunsaturated fatty acids, making them prone to interactions with ROS. This interaction can initiate peroxidation of lipid and lead to apoptosis. Additionally, the decreased level of glutathione in neurons contributes significantly to oxidative stress-related damage [13].

Metal Ion Hypothesis

Imbalances in metal ions play a role in the expansion and progression of various diseases, together with neurodegenerative disorders and cancer. Ionosphere and metal chelators are recognized as key regulators of transition metal balance, with several of these compounds undergoing clinical trials. However, the drugs that bind to metals are not the only options for addressing transition metal homeostasis [14]. Current research shows that there are disruptions in the equilibrium of redox-active transition metals, especially copper, iron, and other trace elements. Elevated levels of these metals have been detected in the individual's brain with AD. Additionally, in other neurodegenerative conditions, metals such as copper, aluminium, and zinc are implicated [15].

Cholinergic Hypothesis

In cases of mild to moderate AD, there is a notable drop in cholinergic receptor binding within certain brain regions, which is associated with neuropsychiatric symptoms. In cognitively healthy older adults, this decline may result in slower cognitive processing. *In vivo* measurement of cholinergic receptor binding may indicate broader neurological changes related to aging and AD, positioning it as a potential target for molecular therapies. Clinical decline is associated with a significant loss of cholinergic neurons in the medial forebrain nuclei, resulting in disrupted acetylcholine-mediated neurotransmission. For more than two decades, treatments designed to sustain acetylcholine levels, including cholinesterase inhibitors and donepezil, have played a vital role in the management of AD symptoms [16, 17].

EPIDEMIOLOGY

Incidence and Prevalence

Alzheimer's Disease International [2018] reported that around 50 million individuals globally were affected by dementia, with projections signifying that this figure could triple by 2050 [18], predominantly in low- and middle-income nations. Current estimates indicate that the prevalence of dementia in Europe is expected to double by 2050 [19]. Accumulating research indicates a decrease in the prevalence of dementia in high-income countries; however, the signal supporting a decay in prevalence is less compelling [20, 21].

Mortality

Mortality rates remain relatively stable despite a decline in incidence, which may be attributed to the prolonged duration of the disease; however, existing lessons on mortality do not substantiate this idea. One of the studies conducted in the United States, which assessed survival following a dementia diagnosis among nearly 60,000 participants, indicated survival periods ranging from 3 to 4 years [22]. In a cohort from a European memory clinic, the median survival duration after a diagnosis of AD dementia was found to be around six years, with a range of 6.0 to 6.5 years [23]. This observation is consistent with results from a multicentre study that evaluated not only the dementia phase but also the earlier prodromal and preclinical stages of the condition [24]. For an individual aged 70, the estimated durations are approximately 10 years for the preclinical stage, 4 years for the prodromal stage, and 6 years for the dementia stage, culminating in a total of 20 years. An initial effort to estimate prevalence using a biological definition, as opposed to a clinical one, indicated that by the age of 85, biologically defined Alzheimer's disease is three times more common than the clinically diagnosed form [25].

Risk factors

Non-Modifiable Factors: Age, genetic predisposition, and family history

The primary risk factors associated with AD include progressive age, particularly in individuals over the age of 65, although this age threshold is not so strictly defined [26]. Furthermore, research indicates that females are at a greater risk of developing Alzheimer's disease compared to males, particularly after reaching the age of 80 [27]. Notably, women tend to exhibit a higher tau protein load, even when their amyloid β levels are comparable to those of men.

A total of six comprehensive genome-wide association studies have been carried out to identify novel genetic variants associated with Alzheimer's disease. The most recent study involved approximately 150,000 individuals diagnosed with AD, along with matched age controls, and over 3,00,000 participants with a proxy phenotype for AD defined as having a parental history of the condition compared to controls without such a history. This research has been directed towards the identification of more than 40 risk alleles linked with AD. Notably, while the communal APOE ε4 risk allele is linked to an assessed 3 to 4 times higher risk of developing AD across various studies, other identified risk alleles contribute significantly less to the overall risk [28].

Modifiable Factors: Cardiovascular disease, diabetes, hypertension, obesity, smoking, and low physical or cognitive activity

Various neurological conditions, including multiple sclerosis, schizophrenia, depression, and autism, are shaped by gender-related developmental and physiological variations. Regarding Alzheimer's disease, although it impacts both genders, women account for almost two-thirds of all diagnosed instances. Research indicates that women not only have an increased likelihood of developing Alzheimer's as they age but also generally experience a more rapid transition from mild cognitive impairment to advanced clinical dementia in comparison to men [29].

Cardiovascular risk features and unhealthy routine choices have been linked to an elevated risk of dementia. According to the Lancet Commission on Dementia Prevention, around 35% of the global risk for any form of dementia can be attributed to 12 modifiable risk factors. These discoveries underscore the importance of preventive measures targeting those factors, even though a significant percentage of dementia cases may not be preventable through lifestyle changes alone [30, 31]. However, current evidence indicates that vascular risk factors do not seem to increase the likelihood of AD pathology, as indicated by cerebrospinal fluid biomarkers and PET imaging. This implies that although lifestyle and vascular risk factors may affect the risk of dementia, they do not do so *via* the AD mechanism [32].

Also, one of the studies conducted on animals has demonstrated that insulin deficiency can trigger the activation of β as well as γ secretase, while also diminishing the concentration of amyloid-beta, leading to its accretion within brain tissue contributes to the acquired factors of increased hazard of developing AD [33].

Various potential modifiable risk factors and protective factors of Alzheimer's disease are shown in Fig. (**3**).

GLOBAL IMPACT OF ALZHEIMER'S DISEASE

Economic Burden

The GBD 2021 research on AD incorporated both prevalence and, for the first time, incidence data, drawing from 81 data points across 60 different locations. The estimates for incidence were enhanced by utilizing cohort studies that analysed mortality rates among individuals with and without dementia. To mitigate bias, claims data from the United States were omitted due to their potential for skewing results. The study employed matched datasets and sophisticated models such as CODEm and DisMod-MR 2.1 to maintain consistency and address data deficiencies, leveraging covariates and geographic trends to generate dependable global estimates [34].

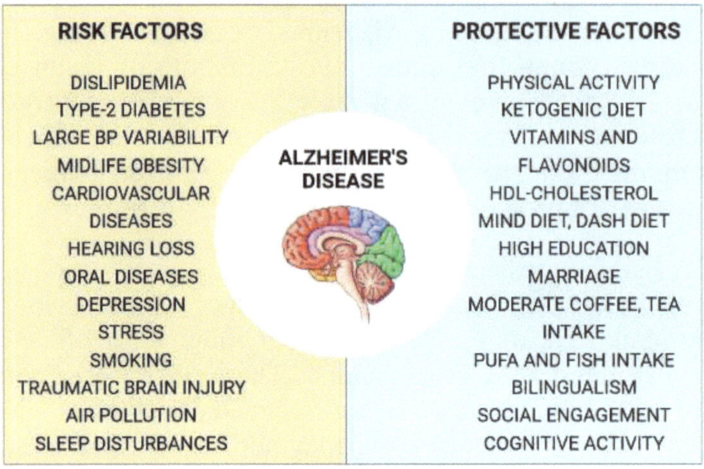

Fig. (3). Potential modifiable risk factors and protective factors of Alzheimer's disease.

Societal and Caregiver Burden

The increasing pervasiveness of AD is leading to an upsurge in associated care costs [35]. The profound effects on both caregivers and the public health system are significant. In terms of societal expenses, the AD ranks as the third most expensive disease globally, following heart disease and cancer [36]. The estimated total economic burden of AD in 2020 reached USD 305 billion [37]. This figure encompasses indirect as well as direct costs. Direct costs comprise various therapeutic expenses, such as medications, consultations with physicians, and hospital stay expenses, nursing home care, as well as non-medical care expenses like private social care and adult day care services. Indirect costs represent the estimated value of possessions lost due to the disease, together with premature mortality, decreased productivity of patients and caregivers. Direct medical expenses are distributed between families and public health systems; however, the responsibility for indirect costs rests entirely with families, leading to a significant economic burden on them. Intangible costs include the pain and suffering endured by patients and their families, along with the deterioration in the quality of life for both patients and caregivers. The existence of these conditions notably escalates the costs that are associated with the care of patients suffering from AD, as many individuals with AD also suffer from at least one additional health issue [38]. Consequently, there is a pressing need to grow effective therapeutic strategies for AD. Currently, various therapies, including both active as well as passive inoculation approaches targeting amyloid-β, are undergoing clinical trials [39]. There are seven clinical stages of Alzheimer's disease, as per the Global Deterioration Scale, as shown in Table **2**.

Table 2. Seven clinical stages of Alzheimer's disease as per the Global Deterioration Scale.

STAGE	CHARACTERISTICS	REFERENCES
Stage-1	Individuals exhibit usual cognitive function; however, underlying pathological changes are occurring in the brain.	[40]
Stage-2	The prodromic phase is characterized by a slight loss of memory, which is often vague from typical absentmindedness.	[40]
Stage-3	Transition to mild cognitive impairment, where the entities may experience disorientation or struggle to find the right words.	[40]
Stage-4	Moderate dementia is evident, marked by significant short-term memory loss, leading persons to forget aspects of their personal history.	[40]
Stage-5	Cognitive deterioration progresses, necessitating assistance in daily activities. Individuals experience confusion and forget numerous personal details.	[40]
Stage-6	Severe dementia manifests, necessitating continuous care under supervision. Patients may not identify their family members and friends and may exhibit variations in personality.	[40]
Stage-7	Entities approach the end of life, displaying motor symptoms, communication difficulties, incontinence, and requiring help with feeding.	[40]

VARIOUS GLOBAL INITIATIVES AIM TO ADDRESS THE CHALLENGES POSED BY ALZHEIMER'S DISEASE

World Health Organization [WHO]

The vision of the GAPD is to establish a biosphere where the prevention of dementia allows individuals with dementia and their caregivers to thrive and obtain the necessary care and provision to realize their potential with poise, respect, independence, and equality [41]. This comprehensive public health initiative builds upon earlier WHO policy documents focused on dementia, such as "Dementia: A Public Health Priority" [42], and the foundational 2015 Ministerial conference on global action against dementia, which was joined by representatives from nearly ninety countries worldwide. The formulation of this initiative was informed by contributions from Member States, UN agencies, and various non-state actors. A significant influence on the creation of the Global Action Plan on Dementia was the collaborative feedback provided by Alzheimer's Disease International, Alzheimer Europe, Dementia Alliance International, and the Global Alzheimer's and Dementia Action Alliance in response to the draft plan [43]. The Strategy delineates 7 critical action areas, each accompanied by specific goals for countries to attain between 2017 and 2025. The prioritized actions concerning dementia include:

1. **Action 1:** Recognizing dementia as a public health priority,
2. **Action 2:** Promoting awareness and creating friendly environments,
3. **Action 3:** Implementing risk reduction strategies,
4. **Action 4:** Enhancing diagnosis, treatment, care, and support,
5. **Action 5:** Providing support for caregivers of individuals with dementia,
6. **Action 6:** Developing information systems,
7. **Action 7:** Fostering research and innovation.

Concerning Actions 1 and 5, the established targets indicate that by 2025, 75% of nations will have either developed or revised national dementia policies, whether as standalone documents or integrated into broader health strategies. Additionally, these countries are expected to offer support and training programs for caregivers and families of individuals affected by dementia. For Action 2, the targets stipulate that by 2025, every country will have implemented at least one effective public awareness campaign, and half of all countries will have initiated at least one dementia friendly initiative. Concerning the action of implementing risk reduction strategies, the targets concerning modifiable risk factors are numerous, with a primary goal of achieving a 25% decrease in premature deaths caused by dementia, which is anticipated by the year 2025.The objectives for Actions 4 and 6 specify that by 2025, at least 50% of countries should have diagnosed half of the estimated population living with dementia, and that 50% of countries will have gathered essential dementia-related data biannually. Finally, the objective of Action 7 is to increase global research initiatives on dementia twofold by the year 2025 [44].

Alzheimer's Disease International [ADI]

The primary goals of the ADI organization include prioritizing dementia as a worldwide health issue, bolstering and auxiliary its member associations, and increasing worldwide awareness of the condition. To reach these objectives, various initiatives are implemented, such as an annual international conference and the training program at Alzheimer's University. Since 1996, ADI has maintained official relations with the World Health Organization and has been associated with the United Nations since 2012 [45].

To improve worldwide public policy regarding dementia, ADI has published several rumours, including the World Alzheimer Report 2009, which addresses the global prevalence of dementia; the report of 2010, which discusses the financial impact of dementia; and the report of 2011, which highlights the advantages of early diagnosis and intervention. In partnership with the World Health Organization, ADI played a role in the report named Dementia: A Public Health Priority, which was launched in Geneva in April 2012. This report reaffirms earlier findings and offers a comprehensive summary of the disease. It advocates for each country to formulate a national Alzheimer's and dementia strategy and provides a structured outline for action, which encompasses several key stages. The initial phase of the framework emphasizes advocacy and raising awareness. Advocacy efforts are directed at governments at all levels to promote policies that enhance dementia care and services. Meanwhile, awareness initiatives aim to educate the general public, families, and healthcare professionals to foster a better understanding of dementia and to shift attitudes and practices.

The formulation and execution of dementia guidelines and plans should involve collaboration crossways various governmental divisions, addressing all therapeutic, social, legal, and financial dimensions of the disease. These plans ought to be developed and implemented in partnership with academic institutions, non-governmental organizations, professional bodies, and governmental agencies. The third phase of the agenda focuses on the enhancement of health and social systems. These systems must be adequately prepared to deliver the comprehensive care and services required by individuals with dementia and the caregivers of their families. This preparation involves building capacity and providing education for healthcare professionals, as well as investing in health information systems.

Lastly, the framework should be complemented by research and evaluation efforts. Each nation ought to establish a research agenda that prioritizes global cooperation and public-private partnerships to advance fundamental scientific research and facilitate the discovery of more effective treatments. This agenda will encompass multiple disciplines. Additionally, countries must track the progression of the dementia epidemic to observe any changes in prevalence and incidence, which may reflect the effectiveness or shortcomings of implemented measures [45].

Precision Medicine and Biomarker Research

The pressing necessity for the identification, analysis, and measurement of biomarkers defined as quantifiable substances or cognitive impairments in individuals predisposed to AD is essential for various reasons. First, these biomarkers can facilitate more accurate diagnoses of AD, particularly in the early 'prodromal' phase, thereby enabling the implementation of defensive measures or besieged therapeutic policies at the initial signs of the disease. Additionally, they can improve the efficient and dependable integration of multi-modal biomarkers for AD, which is essential for streamlining, supporting, and strengthening diagnostic and therapeutic decision-making processes. Notably, since 1983, peer-reviewed literature has produced nearly fifty-three thousand original research articles and reviews that discuss the intersection of 'Alzheimer's disease' and 'biomarkers' [46 - 48]. This body of research encompasses both established and emerging biomarkers for AD. It includes genetic variations identified through genome-wide association studies, mutations, and changes like methylation and post-transcriptional modifications. It also investigates neurotoxic and disease-related metabolic by-products that accumulate in the AD brain, such as various forms of tau aggregates and amyloid-beta species, and plaques. AD biomarkers span a wide range of substances—including proteins, lipids, proteolipids, inflammatory cytokines, chemokines, carbohydrates, microRNAs, and messenger RNAs alongside a growing array of neuroimaging and neuro-radiological tools [49, 50].

These biomarkers are invaluable for detecting dementia throughout its progression, as their presence and levels correlate dynamically with cognitive decline, enabling clinicians to assess responses to therapeutic interventions against the backdrop of the aging AD patient.

A variety of analytical molecular-genetic techniques, progress in geriatric psychiatry and clinical evaluations, enhancements in neuro-radiological labelling methods and neuroimaging technologies, along with integrated diagnostic and predictive approaches, have resulted in significant improvements in methodology. These advancements have enabled the identification of detailed pathophysiological profiles linked to complex, multifactorial neurodegenerative disorders. In particular, these developments: [i] are now well within the reach of contemporary clinical, medical, and diagnostic neurology; [ii] are producing substantial biomarker data for individual AD patients, large groups of AD cases, and age- and gender-matched controls; and [iii] are laying a data-driven groundwork for the shift towards a 'precision medicine' model in the prevention, diagnosis, prognosis, and treatment of AD [51 - 55]. Additionally, less common clinical manifestations of AD are gaining recognition, contributing to the growing body of AD biomarker data [56 - 58]. A fundamental aspect of precision medicine is the utilization of biomarker-driven medical data. Therefore, progress in the collection, integration, and analysis of clinical information, coupled with advancements in bioinformatics and the coordination of clinical, laboratory, molecular-genetic, neuroimaging, geriatric, and psychological data, will significantly improve diagnostic precision for individuals at risk of AD. Given the intricate and diverse characteristics of AD, a wide range of treatment approaches and data-informed, precision-oriented pharmacological strategies will be particularly advantageous and [ii] the therapeutic designs based on biomarkers will significantly enhance the current healthcare landscape, leading to more effective treatments and the advancement of disease-modifying medications for patients with Alzheimer's disease, regardless of the stage of their condition [59].

CURRENT CHALLENGES AND FUTURE DIRECTIONS

Research on Alzheimer's Disease [AD] is advancing rapidly, yet it faces several challenges and potential future directions as identified in recent investigations:

Current Challenges

a. Early Diagnosis: Detecting Alzheimer's during its preclinical phase is challenging, as many symptoms can be misinterpreted as typical signs of aging.
b. Data Limitations: The lack of adequate datasets for training sophisticated diagnostic models, such as deep learning algorithms, impedes progress in the field.
c. Complex Pathophysiology: The intricate molecular mechanisms and pathways involved in Alzheimer's are not completely understood, making it difficult to create targeted therapies [60].
d. Treatment Gaps: Although there have been advancements, there remains a shortage of effective treatments for persons in the later phases of Alzheimer's.
e. Biomarker Integration: There are significant challenges in combining various biomarkers to enhance diagnostic precision [61].

Future Directions

a. Artificial Intelligence: Utilizing AI, particularly deep learning models, to analyse neuroimaging data and enhance early diagnostic capabilities.
b. Biomarker Research: Innovating new biomarkers to identify Alzheimer's at its earliest stages.
c. Preventative Strategies: Emphasizing non-pharmacological methods like cognitive training, physical exercise, and behavioural alterations [61].
d. Personalized Medicine: Tailoring therapies according to the unique genetic and molecular traits of each individual [62].
e. Global Collaboration: Promoting interdisciplinary and international research initiatives to tackle the complex nature of AD [62].

CONCLUSION

Alzheimer's disease [AD] has emerged as a global health concern of profound complexity and magnitude, warranting a multi-dimensional approach to understanding and addressing its impact. From its historical discovery by Dr. Alois Alzheimer in the early 20th century to the progressive advancements in research, the study of Alzheimer's has illuminated the interplay of biology, genetics, and environmental factors underlying dementia. This journey has paved the way for a deeper comprehension of the pathophysiology of the disease, although effective cures remain elusive. Epidemiologically, AD poses a growing challenge, especially with the aging of global populations. It is now recognized as a foremost cause of infirmity and dependency among grown-up adults, impacting millions worldwide. In exploring the pathophysiology, AD is characterized by hallmark changes in the brain with amyloid beta plaque and neurodegeneration. These pathological features provide serious insights into the mechanisms of disease progression and probable therapeutic goals. Yet, the complex nature of Alzheimer's calls for continued exploration to unravel the complexities of its onset and progression.

The global impact of Alzheimer's is both unescapable and multifaceted. The economic costs, caregiver burden, and societal challenges associated with the disease demand a coordinated international response. Public health campaigns, community support, as well as global collaboration are vigorous in reducing stigma and ensuring impartial access to care for affected populations. Present challenges in Alzheimer's research and care highlight significant gaps in understanding and resources. The lack of an ultimate cure or defensive treatment, coupled with the challenges of early detection, emphasizes the need for intensified research and innovation. Equally, addressing discrepancies in access to care and support services remains a demanding priority. Looking forward, the future of Alzheimer's research and care holds great promise. Advances in technology, such as artificial intelligence and biomarkers, are dignified to revolutionize early diagnosis and personalized treatment. The growing prominence of interdisciplinary collaboration, international partnerships, and increased funding will likely accelerate progress in both therapeutic development and comprehensive care models.

REFERENCES

[1] Masters CL, Bateman R, Blennow K, Rowe CC, Sperling RA, Cummings JL. Alzheimer's disease. Nat Rev Dis Primers 2015; 1(1): 15056. [http://dx.doi.org/10.1038/nrdp.2015.56] [PMID: 27188934]

[2] Duncan GW, Khoo TK, Coleman SY, *et al.* The incidence of Parkinson's disease in the North-East of England. Age Ageing 2014; 43(2): 257-63.

[http://dx.doi.org/10.1093/ageing/aft091] [PMID: 23880142]

[3] Ryan NS, Rossor MN, Fox NC. Alzheimer's disease in the 100 years since Alzheimer's death. Brain 2015; 138(12): 3816-21.
[http://dx.doi.org/10.1093/brain/awv316] [PMID: 26541346]

[4] Jellinger KA. Alzheimer 100 – highlights in the history of Alzheimer research. J Neural Transm (Vienna) 2006; 113(11): 1603-23.
[http://dx.doi.org/10.1007/s00702-006-0578-3] [PMID: 17039299]

[5] Agarwal M, Alam MR, Haider MK, Malik MZ, Kim DK. Alzheimer's disease: An overview of major hypotheses and therapeutic options in nanotechnology. Nanomaterials (Basel) 2020; 11(1): 59.
[http://dx.doi.org/10.3390/nano11010059] [PMID: 33383712]

[6] Dubois B, Hampel H, Feldman HH, *et al.* Preclinical Alzheimer's disease: Definition, natural history, and diagnostic criteria. Alzheimers Dement 2016; 12(3): 292-323.
[http://dx.doi.org/10.1016/j.jalz.2016.02.002] [PMID: 27012484]

[7] Albert MS, DeKosky ST, Dickson D, *et al.* The diagnosis of mild cognitive impairment due to Alzheimer's disease: Recommendations from the National Institute on Aging□Alzheimer's Association workgroups on diagnostic guidelines for Alzheimer's disease. Alzheimers Dement 2011; 7(3): 270-9.
[http://dx.doi.org/10.1016/j.jalz.2011.03.008] [PMID: 21514249]

[8] Kivipelto M, Mangialasche F, Ngandu T. Lifestyle interventions to prevent cognitive impairment, dementia and Alzheimer disease. Nat Rev Neurol 2018; 14(11): 653-66.
[http://dx.doi.org/10.1038/s41582-018-0070-3] [PMID: 30291317]

[9] Dafsari FS, Jessen F. Depression—an underrecognized target for prevention of dementia in Alzheimer's disease. Transl Psychiatry 2020; 10(1): 160.
[http://dx.doi.org/10.1038/s41398-020-0839-1] [PMID: 32433512]

[10] Kumar A, Singh A, Ekavali . A review on Alzheimer's disease pathophysiology and its management: an update. Pharmacol Rep 2015; 67(2): 195-203.
[http://dx.doi.org/10.1016/j.pharep.2014.09.004] [PMID: 25712639]

[11] Mishra S, Palanivelu K. The effect of curcumin (turmeric) on Alzheimer's disease: An overview. Ann Indian Acad Neurol 2008; 11(1): 13-9.
[http://dx.doi.org/10.4103/0972-2327.40220] [PMID: 19966973]

[12] Liu Z, Zhang A, Sun H, Han Y, Kong L, Wang X. Two decades of new drug discovery and development for Alzheimer's disease. RSC Advances 2017; 7(10): 6046-58.
[http://dx.doi.org/10.1039/C6RA26737H]

[13] Mohamed T, Shakeri A, Rao PPN. Amyloid cascade in Alzheimer's disease: Recent advances in medicinal chemistry. Eur J Med Chem 2016; 113: 258-72.
[http://dx.doi.org/10.1016/j.ejmech.2016.02.049] [PMID: 26945113]

[14] Weekley CM, He C. Developing drugs targeting transition metal homeostasis. Curr Opin Chem Biol 2017; 37: 26-32.
[http://dx.doi.org/10.1016/j.cbpa.2016.12.011] [PMID: 28040658]

[15] Prakash A, Dhaliwal GK, Kumar P, Majeed ABA. Brain biometals and Alzheimer's disease – boon or bane? Int J Neurosci 2017; 127(2): 99-108.
[http://dx.doi.org/10.3109/00207454.2016.1174118] [PMID: 27044501]

[16] Sultzer DL, Marder SR. Older brains are different: brain–behavior studies and their CLinical UtiLity. Am J Geriatr Psychiatry 2017; 25(1): 11-2.
[http://dx.doi.org/10.1016/j.jagp.2016.10.002] [PMID: 28231870]

[17] Chase TN, Farlow MR, Clarence-Smith K. Donepezil plus solifenacin [CPC-201] treatment for Alzheimer's disease. Neurotherapeutics 2017; 14(2): 405-16.
[http://dx.doi.org/10.1007/s13311-016-0511-x] [PMID: 28138837]

[18] Patterson C. World Alzheimer Report 2018: The State of the Art of Dementia Research [Internet]. London: Alzheimer's Disease International (ADI); 2018.

[19] Kumari R, Sharma A. A Review on New Drug Development for the Treatment of Neurodegenerative Disease. Asian J Res Pharmaceut Sci 2024; 14(3): 304-12.

[20] Wu YT, Beiser AS, Breteler MMB, *et al.* The changing prevalence and incidence of dementia over time — current evidence. Nat Rev Neurol 2017; 13(6): 327-39.
[http://dx.doi.org/10.1038/nrneurol.2017.63] [PMID: 28497805]

[21] Prince M, Ali GC, Guerchet M, Prina AM, Albanese E, Wu YT. Recent global trends in the prevalence and incidence of dementia, and survival with dementia. Alzheimers Res Ther 2016; 8(1): 23.
[http://dx.doi.org/10.1186/s13195-016-0188-8] [PMID: 27473681]

[22] Mayeda ER, Glymour MM, Quesenberry CP, Johnson JK, Pérez-Stable EJ, Whitmer RA. Survival after dementia diagnosis in five racial/ethnic groups. Alzheimers Dement 2017; 13(7): 761-9.
[http://dx.doi.org/10.1016/j.jalz.2016.12.008] [PMID: 28174069]

[23] Rhodius-Meester HFM, Tijms BM, Lemstra AW, *et al.* Survival in memory clinic cohort is short, even in young-onset dementia. J Neurol Neurosurg Psychiatry 2019; 90(6): 726-8.
[http://dx.doi.org/10.1136/jnnp-2018-318820] [PMID: 30045942]

[24] Vermunt L, Sikkes SAM, van den Hout A, *et al.* Duration of preclinical, prodromal, and dementia stages of Alzheimer's disease in relation to age, sex, and *APOE* genotype. Alzheimers Dement 2019; 15(7): 888-98.
[http://dx.doi.org/10.1016/j.jalz.2019.04.001] [PMID: 31164314]

[25] Jack CR Jr, Therneau TM, Weigand SD, *et al.* Prevalence of biologically vs clinically defined Alzheimer spectrum entities using the National Institute on Aging–Alzheimer's Association research framework. JAMA Neurol 2019; 76(10): 1174-83.
[http://dx.doi.org/10.1001/jamaneurol.2019.1971] [PMID: 31305929]

[26] Buckley RF, Mormino EC, Rabin JS, *et al.* Sex differences in the association of global amyloid and regional tau deposition measured by positron emission tomography in clinically normal older adults. JAMA Neurol 2019; 76(5): 542-51.
[http://dx.doi.org/10.1001/jamaneurol.2018.4693] [PMID: 30715078]

[27] van der Lee SJ, Wolters FJ, Ikram MK, *et al.* The effect of APOE and other common genetic variants on the onset of Alzheimer's disease and dementia: a community-based cohort study. Lancet Neurol 2018; 17(5): 434-44.
[http://dx.doi.org/10.1016/S1474-4422(18)30053-X] [PMID: 29555425]

[28] Jansen IE, Savage JE, Watanabe K, *et al.* Genome-wide meta-analysis identifies new loci and functional pathways influencing Alzheimer's disease risk. Nat Genet 2019; 51(3): 404-13.
[http://dx.doi.org/10.1038/s41588-018-0311-9] [PMID: 30617256]

[29] Pike CJ. Sex and the development of Alzheimer's disease. J Neurosci Res 2017; 95(1-2): 671-80.
[http://dx.doi.org/10.1002/jnr.23827] [PMID: 27870425]

[30] Babapour Mofrad R, Tijms BM, Scheltens P, *et al.* Sex differences in CSF biomarkers vary by Alzheimer disease stage and *APOE* ε4 genotype. Neurology 2020; 95(17): e2378-88.
[http://dx.doi.org/10.1212/WNL.0000000000010629] [PMID: 32788242]

[31] Livingston G, Huntley J, Sommerlad A, *et al.* Dementia prevention, intervention, and care: 2020 report of the Lancet Commission. Lancet 2020; 396(10248): 413-46.
[http://dx.doi.org/10.1016/S0140-6736(20)30367-6] [PMID: 32738937]

[32] Gustavsson AM, van Westen D, Stomrud E, Engström G, Nägga K, Hansson O. Midlife atherosclerosis and development of Alzheimer or vascular dementia. Ann Neurol 2020; 87(1): 52-62.
[http://dx.doi.org/10.1002/ana.25645] [PMID: 31721283]

[33] Silva MVF, Loures CMG, Alves LCV, de Souza LC, Borges KBG, Carvalho MG. Alzheimer's disease: risk factors and potentially protective measures. J Biomed Sci 2019; 26(1): 33. [http://dx.doi.org/10.1186/s12929-019-0524-y] [PMID: 31072403]

[34] GBD 2015 Eastern Mediterranean Region Adolescent Health Collaborators. Adolescent health in the Eastern Mediterranean Region: findings from the global burden of disease 2015 study. Int J Public Health 2018; 63(1): 79–96.

[35] Marasco RA. Current and evolving treatment strategies for the Alzheimer disease continuum. Am J Manag Care 2020; 26(8) (Suppl.): S167-76. [PMID: 32840330]

[36] Shi J, Sabbagh MN, Vellas B. Alzheimer's disease beyond amyloid: strategies for future therapeutic interventions. BMJ 2020; 371: m3684. [http://dx.doi.org/10.1136/bmj.m3684] [PMID: 33036984]

[37] Olivari BS, Jeffers EM, Tang KW, McGuire LC. Improving brain health for populations disproportionately affected by Alzheimer's disease and related dementias. Clin Gerontol 2023; 46(2): 128-32. [http://dx.doi.org/10.1080/07317115.2022.2043977] [PMID: 35235500]

[38] Maciejewska K, Czarnecka K, Szymański P. A review of the mechanisms underlying selected comorbidities in Alzheimer's disease. Pharmacol Rep 2021; 73(6): 1565-81. [http://dx.doi.org/10.1007/s43440-021-00293-5] [PMID: 34121170]

[39] Song C, Shi J, Zhang P, *et al.* Immunotherapy for Alzheimer's disease: targeting β-amyloid and beyond. Transl Neurodegener 2022; 11(1): 18. [http://dx.doi.org/10.1186/s40035-022-00292-3] [PMID: 35300725]

[40] Rostagno AA. Pathogenesis of Alzheimer's disease. Int J Mol Sci 2022; 24(1): 107. [http://dx.doi.org/10.3390/ijms24010107] [PMID: 36613544]

[41] Cahill S. WHO's global action plan on the public health response to dementia: some challenges and opportunities. Aging Ment Health 2020; 24(2): 197-9. [http://dx.doi.org/10.1080/13607863.2018.1544213] [PMID: 30600688]

[42] World Health Organization, Alzheimer's Disease International. Dementia: A Public Health Priority [Internet]. Geneva: World Health Organization; 2012.

[43] Rees G. World Health Organization endorses global action plan on rising incidence of dementia. Nurs Older People. 2017 Jun 30; 29(6): 7. [PMID: 28664801]

[44] Wimo A, Prince MJ. World Alzheimer Report 2010: The Global Economic Impact of Dementia [Internet]. London: Alzheimer's Disease International (ADI); 2010 Sep 21. Available from: https://www.alzint.org/u/WorldAlzheimerReport2010.pdf

[45] Wortmann M. Dementia: a global health priority - highlights from an ADI and World Health Organization report. Alzheimers Res Ther 2012; 4(5): 40. [http://dx.doi.org/10.1186/alzrt143] [PMID: 22995353]

[46] Arvanitakis Z, Shah RC, Bennett DA. Diagnosis and management of dementia. JAMA 2019; 322(16): 1589-99. [http://dx.doi.org/10.1001/jama.2019.4782] [PMID: 31638686]

[47] Patnode CD, Perdue LA, Rossom RC, *et al.* Screening for Cognitive Impairment in Older Adults: An Evidence Update for the U.S. Preventive Services Task Force [Internet]. Rockville (MD): Agency for Healthcare Research and Quality (US); 2020 Feb. (Evidence Synthesis, No. 189.) Available from: https://www.ncbi.nlm.nih.gov/books/NBK554654/

[48] Lukiw WJ, Vergallo A, Lista S, Hampel H, Zhao Y. Biomarkers for Alzheimer's disease [AD] and the application of precision medicine. J Pers Med 2020; 10(3): 138.

[http://dx.doi.org/10.3390/jpm10030138] [PMID: 32967128]

[49] Blennow K, Hampel H, Weiner M, Zetterberg H. Cerebrospinal fluid and plasma biomarkers in Alzheimer disease. Nat Rev Neurol 2010; 6(3): 131-44. [http://dx.doi.org/10.1038/nrneurol.2010.4] [PMID: 20157306]

[50] Hampel H, Frank R, Broich K, *et al.* Biomarkers for Alzheimer's disease: academic, industry and regulatory perspectives. Nat Rev Drug Discov 2010; 9(7): 560-74. [http://dx.doi.org/10.1038/nrd3115] [PMID: 20592748]

[51] Hampel H, O'Bryant SE, Molinuevo JL, *et al.* Blood-based biomarkers for Alzheimer disease: mapping the road to the clinic. Nat Rev Neurol 2018; 14(11): 639-52. [http://dx.doi.org/10.1038/s41582-018-0079-7] [PMID: 30297701]

[52] Hampel H, Vergallo A, Afshar M, *et al.* Blood-based systems biology biomarkers for next-generation clinical trials in Alzheimer's disease. Dialogues Clin Neurosci 2019; 21(2): 177-91. [http://dx.doi.org/10.31887/DCNS.2019.21.2/hhampel] [PMID: 31636492]

[53] Lista S, Khachaturian ZS, Rujescu D, Garaci F, Dubois B, Hampel H. Application of Systems Theory in Longitudinal Studies on the Origin and Progression of Alzheimer's Disease. Methods Mol Biol 2016; 1303: 49–67. [http://dx.doi.org/10.1007/978-1-4939-2627-5_2]

[54] Hampel H, Vergallo A, Caraci F, *et al.* Future avenues for Alzheimer's disease detection and therapy: liquid biopsy, intracellular signaling modulation, systems pharmacology drug discovery. Neuropharmacology 2021; 185108081 [http://dx.doi.org/10.1016/j.neuropharm.2020.108081] [PMID: 32407924]

[55] Sims R, Hill M, Williams J. The multiplex model of the genetics of Alzheimer's disease. Nat Neurosci 2020; 23(3): 311-22. [http://dx.doi.org/10.1038/s41593-020-0599-5] [PMID: 32112059]

[56] Habes M, Grothe MJ, Tunc B, McMillan C, Wolk DA, Davatzikos C. Disentangling heterogeneity in Alzheimer's disease and related dementias using data-driven methods. Biol Psychiatry 2020; 88(1): 70-82. [http://dx.doi.org/10.1016/j.biopsych.2020.01.016] [PMID: 32201044]

[57] DeTure MA, Dickson DW. The neuropathological diagnosis of Alzheimer's disease. Mol Neurodegener 2019; 14(1): 32. [http://dx.doi.org/10.1186/s13024-019-0333-5] [PMID: 31375134]

[58] Veitch DP, Weiner MW, Aisen PS, *et al.* Understanding disease progression and improving Alzheimer's disease clinical trials: Recent highlights from the Alzheimer's Disease Neuroimaging Initiative. Alzheimers Dement 2019; 15(1): 106-52. [http://dx.doi.org/10.1016/j.jalz.2018.08.005] [PMID: 30321505]

[59] Jellinger KA. Neuropathological assessment of the Alzheimer spectrum. J Neural Transm (Vienna) 2020; 127(9): 1229-56. [http://dx.doi.org/10.1007/s00702-020-02232-9] [PMID: 32740684]

[60] Liu H, Fei Z, Zhou W. Deep Learning Models in Early Diagnosis of Alzheimer's Disease: A Systematic Review of Current Applications and Challenges. In: Life System Modeling and Simulation. LSMS 2024. Communications in Computer and Information Science, vol 2187. Singapore: Springer; 2024. p. 289–303.

[61] Abdul Manap AS, Almadodi R, Sultana S, *et al.* Alzheimer's disease: a review on the current trends of the effective diagnosis and therapeutics. Front Aging Neurosci 2024; 16: 1429211. [http://dx.doi.org/10.3389/fnagi.2024.1429211] [PMID: 39185459]

[62] Liu GP, Wang S. Editorial: New challenges and future perspectives in Alzheimer's disease and related dementias. Front Aging Neurosci 2023; 15: 1345560. [http://dx.doi.org/10.3389/fnagi.2023.1345560] [PMID: 38152603]

CHAPTER 2

Pathophysiology and Molecular Mechanisms of Alzheimer's Disease

Chandragiri Siva Sai[1], **Manish Mathur**[2] and **Neha Mathur**[3,*]

[1] *Department of Pathology, The University of Oklahoma Health Sciences Center, Oklahoma, OK 73104, USA*

[2] *Department of Academic Affairs(E), Amity University, Lucknow Uttar Pradesh 226010, India*

[3] *Amity Institute of Pharmacy, Amity University, Lucknow, Uttar Pradesh 226010, India*

Abstract: Alzheimer's Disease (AD), the leading cause of dementia worldwide, remains without an effective cure, largely due to an incomplete understanding of its underlying mechanisms. Recent research highlights the central roles of amyloid β (Aβ) and tau proteins, alongside glial cell dysfunction, in the pathogenesis of AD. This chapter reviews advances in Aβ- and tau-related mechanisms, focusing on neuronal and glial receptors mediating Aβ toxicity and the role of glial cells in neurodegeneration. Early neurodegeneration in AD is marked by neuronal loss and synaptic impairment. Despite significant progress in uncovering AD's molecular mechanisms, the disease's complex pathogenesis and limited diagnostic and therapeutic tools have hindered the development of effective treatments. Comprehensive disease modelling is imperative to clarify AD's mechanisms and facilitate the creation of targeted therapies. Mitochondrial dysfunction, oxidative stress, and chronic neuroinflammation emerge as critical mediators of neuronal damage, emphasizing the multifactorial nature of AD. This chapter reviews the interplay between inherited risk, aging, lifestyle factors, and environmental exposures in influencing AD pathogenesis. Additionally, it will explore how these factors contribute to the broader molecular and cellular pathways involved in disease progression. Microglial activation and astrocytic dysfunction significantly influence the progression of neuronal injury and synaptic loss, underscoring their importance in AD pathogenesis. Emerging evidence underscores the importance of understanding glial contributions to neuronal dysfunction, offering new perspectives on disease progression and potential intervention points. The transition from early-stage molecular alterations to widespread neuronal dysfunction and clinical decline underscores the importance of early pathophysiological insights. Recent advancements in understanding AD's complex pathophysiology provide promising avenues for enhanced diagnostics, assessing risks more accurately, and novel prevention strategies. These discoveries bring hope for improved management and therapeutic options for the millions affected by this debilitating condition.

[*] **Corresponding author Neha Mathur:** Amity Institute of Pharmacy, Amity University, Lucknow, Uttar Pradesh 226010, India; E-mails: nmathur1@amity.edu; neha07.mathur@gmail.com

Prashant Kumar Dhakad, Raghav Mishra, Shweta Sharma, Sudhanshu Mishra & Nishant Gaur (Eds.)

Keywords: Alzheimer's disease, AD, Molecular mechanisms, Pathophysiology alzheimer's.

INTRODUCTION

Disease Prevalence rate in India and other countries

Alzheimer's Disease (AD) is a disorder of the brain. It accounts for about 75% of all dementia cases and is the most prevalent type. Associated dementia is amongst the chief causes of debilitation in the elderly (>65 years of age); however, early onset forms of AD are also known, and these comprise ~2-5% of all Alzheimer's disease cases. It is progressive and goes undetected in its prodromal phases since there are no obvious symptoms. However, as the disease worsens and symptoms start to show, the level of neuronal damage becomes irreversible. Neuronal cells necessary for performing even basic body functions, like walking and swallowing, eventually deteriorate, leaving the patient immobile.

AD is a complicated illness that is caused by several pathogenic factors. Researchers have thus far been unable to arrive at a definite conclusion regarding what initiates the chain of events that culminates in the atrophy of the brain. The accumulation of Amyloid beta (Aβ) peptide and tau protein into highly stable aggregates termed beta-amyloid plaques and tau tangles has long been held as the central feature of this disease pathogenesis, but has been challenged in recent times. Aggregated Aβ and Tau get deposited in the neurons of the brain, extracellularly (in case of Aβ) and intracellularly (in case of tau). This is suggested to contribute to cellular and inflammatory stress, from which cells are unable to recover, and consequently, they die. Furthermore, the brain's ability to metabolize glucose also gets impaired in this disease. The disease remains incurable as of now and is a cause of anguish for millions around the globe. The multifactorial nature of this disease is the reason behind failure in contriving equate therapeutic and diagnostic measures against it. It forms a major burden to the worldwide healthcare system, and it is estimated that in the coming decade, the AD burden will vastly impact developing countries like India, which are currently lacking adequate awareness and infrastructural support to combat the rising disease cases of AD.

Alzheimer's Disease is the leading cause of dementia

In 1906, after reporting a case of dementia, Dr. Alois Alzheimer believed it to be a novel form of illness. Dr. Emil Kraepelin later gave it the name Alzheimer's disease. After a century, this disease has emerged as the most prevalent form of dementia that afflicts the elderly and poses a significant health burden. Nevertheless, there are currently no disease-modifying treatments that stop, prevent, or even delay the disease's progression, and the pathophysiology of the condition is still unknown.

Dr. Alzheimer initially noticed that the brain of the first patient with both intracellular neurofibrillary tangles and extracellular plaques, as were seen in AD. Senile plaques, the sole distinct clinical hallmark of AD, were named after the fibrous β-amyloid peptide (Ab) that was shown to be the plaques' constituent

Probable Causes of Alzheimer's Disease

A three-part process for AD induction is put forth:

1. Increased blood ceramide.
2. Decreased blood folic acid.
3. Decreased blood lactic acid

Age-related alterations in these systems are caused by decreasing muscle mass, increased visceral fat, and nutritional changes. This procedure also explains why many people do not develop AD. It is brought on by an AD case of events that damage neurons and the blood-brain barrier. The blood–brain barrier keeps dangerous chemicals out of the brain while allowing necessary molecules to remain inside. Lactic acid, which is produced during exercise, is a brain nutrient. Endothelial cells and pericytes are harmed by lactic acid AD, which is comparable to brain damage and blood-brain barrier impairment. The blood–brain barrier is harmed in AD by adequate folate consumption and oxidative stress brought on by endothelial nitric oxide synthase and transient receptor potential cation channel activation. The final processes that lead to AD neuronal death could be modifications in microRNA levels. IL-1β, Lactic acid, ceramide, folate, Tumor necrosis factor α, kynurenine metabolites, nicotinamide, and microRNA are all implicated in a novel mode of Alzheimer's disease induction.

CORE PATHOLOGICAL FEATURES

Amyloid-beta (Aβ) plaques and their role

AD is characterized by the aberrant deposition of amyloid-beta (Aβ) protein in the brain, a process tightly linked to neuronal injury and cognitive decline [1]. Aβ, initially identified in cerebrovascular amyloid deposits, later emerged as the primary element of amyloid plaques in AD and Down syndrome [2, 3]. Aβ is generated through the enzymatic cleavage of APP via two pathways: the amyloidogenic pathway, which leads to Aβ plaques, and the non-amyloidogenic pathway, which does not [4]. In the amyloidogenic pathway, β-secretase (BACE1) and γ-secretase produce neurotoxic Aβ42, while the non-amyloidogenic pathway, involving α-secretase, generates neuroprotective fragments [5 - 8].

Monomeric Aβ aggregates into soluble oligomers and insoluble fibrils that form plaques, disrupting synaptic function and causing neurotoxicity. Cryo-electron microscopy differentiates Aβ42 filaments in sporadic and familial AD [9]. Aβ pathology spreads prion-like, where misfolded Aβ seeds amplify aggregation, accelerating disease progression [10]. Experimental models confirm that Aβ-containing extracts from AD brains induce amyloid pathology [11 - 13]. Though rare and under investigation, cases involving cadaver-derived pituitary hormone treatments also suggest human-to-human Aβ transmission [14].

Aβ accumulates extracellularly in neuritic plaques, where it binds to various neuronal receptors and triggers a cascade of neurotoxic signaling pathways. These interactions contribute to synaptic dysfunction, cellular damage, and ultimately, disease progression. A detailed understanding of Aβ-receptor dynamics is essential for elucidating AD pathophysiology and identifying therapeutic targets. As shown in Table **1**, several receptors implicated in AD are influenced by Aβ, each playing distinct roles in mediating its toxic effects [1].

Table 1. Receptors influenced by Aβ in Alzheimer's disease.

Receptor	Role in AD	Effect of Aβ Interaction	References
p75 Neurotrophin Receptor (P75NTR)	Activated by Aβ, triggering signaling pathways associated with neuronal death.	Contributes to neurodegeneration and progressive neuronal loss.	[26 - 28]
Low-Density Lipoprotein Receptor-Related Protein (LRP)	Involved in Aβ clearance, its dysfunction leads to accumulation.	Impairment results in excessive Aβ deposition and plaque formation.	[29, 30]
Cellular Prion Protein (PrPc)	Interacts with Aβ, leading to potential neurotoxicity.	May contribute to Aβ aggregation and toxicity-related cellular events.	[31, 32]
Metabotropic Glutamate Receptor 5 (mGluR5)	Regulates synaptic transmission and function.	Aβ interaction disrupts synaptic communication, leading to cognitive impairment.	[31, 33, 34]
α7 Nicotinic Acetylcholine Receptor (α7nAChR)	Modulates cholinergic signaling and neurotransmission.	Aβ binding impairs signaling, contributing to memory deficits.	[35]
N-Methyl-D-Aspartate Receptor (NMDAR)	Vital to cognitive function and synaptic plasticity.	Aβ interaction leads to synaptic dysfunction and cognitive decline.	[36, 37]

(Table 1) cont.....

Receptor	Role in AD	Effect of Aβ Interaction	References
β-Adrenergic Receptor (β-AR)	Plays a role in neuronal signaling.	Aβ interaction triggers neurotoxic pathways, leading to cellular damage.	[38, 39]
Erythropoietin-Producing Hepatoma Receptor (EphR)	Supports synaptic function and neuronal connectivity.	Aβ disrupts receptor function, impairing neuronal communication.	[40 - 42]
Paired Immunoglobulin-Like Receptor B (PirB)	Regulates synaptic plasticity and neuronal function.	Aβ interaction leads to synaptic impairment and altered connectivity.	[42]

Several cofactors, including metal ions, APOE, α-synuclein, and β2-microglobulin (B2M), contribute to Aβ aggregation and toxicity [15 - 18]. Elevated B2M levels are linked to increased Aβ aggregation and cognitive decline. Targeting peripheral B2M with antibodies may reduce amyloid pathology and cognitive deficits in AD models, presenting a potential therapy that bypasses blood-brain barrier limitations [19, 20].

The amyloid cascade hypothesis posits that Aβ accumulation triggers AD, leading to tau pathology, neuronal loss, and cognitive decline [21]. Genetic evidence links Down syndrome (extra APP gene) and familial AD mutations in the APP and presenilin genes to increased Aβ production [22]. Monoclonal antibodies like lecanemab and donanemab show promise in slowing early AD progression [23 - 25]. However, cognitive decline correlates more with tau pathology than Aβ burden [26], and regional brain hypometabolism appears independent of Aβ levels [27]. This suggests that while Aβ aggregation initiates AD, tau pathology and neuroinflammation drive neurodegeneration [1]. Further research is needed to explore their interplay for effective treatments.

Tau Tangles and their Contribution

The microtubule-associated protein tau, which is mostly present in neurones, is essential for maintaining microtubule stability and controlling intracellular transport [43]. Tau is hyperphosphorylated under pathological circumstances, which results in the development of paired helical filaments (PHF) and neurofibrillary tangles (NFTs), two important characteristics of AD [44]. These tau aggregates disrupt neuronal function, impair axonal transport, and contribute to neurodegeneration [45]. While Aβ accumulation is an early event in AD, tau pathology is more closely associated with disease progression and cognitive decline [46].

In contrast to the more diffuse Aβ pathology, NFTs follow a predictable pattern of spread, first appearing in the entorhinal cortex and hippocampus before extending to the neocortex, as described in Braak staging [47]. This progression correlates with symptom severity, from mild memory loss in early stages to widespread cortical involvement in advanced AD [48]. Tau pathology is also implicated in other tauopathies, including Progressive Supranuclear Palsy (PSP) and corticobasal degeneration (CBD), further highlighting its significance in neurodegenerative disorders [49].

Post-translational modifications, particularly hyperphosphorylation—though acetylation and truncation also contribute—impair tau's normal function and promote its aggregation, resulting in toxicity [50]. Phosphorylated tau (P-tau) disrupts mitochondrial function, synaptic activity, and neuronal integrity, further exacerbating AD pathology [51]. Additionally, evidence suggests that Aβ and tau interact synergistically, with Aβ triggering tau pathology, which in turn amplifies neurodegeneration [52]. Therapeutic strategies targeting tau include kinase inhibitors, immunotherapies, and aggregation blockers aimed at reducing tau pathology and preserving neuronal function [53]. Tau-directed medicines show promise as prospective treatments for AD and kindred tauopathies because of the substantial link between tau buildup and cognitive deterioration [54]. Further research into tau's role in disease progression and its interactions with Aβ is essential to developing effective therapeutic interventions.

Neuroinflammation Hypothesis

A well-balanced interaction between neurons, glial cells, and the immune system is essential for maintaining cognitive and behavioural function. Although the Central Nervous System (CNS) was once considered immune-privileged, recent research has demonstrated active bidirectional communication between the CNS and the peripheral immune system [55]. Neuroinflammation refers to a sustained immune response within the CNS, triggered by endogenous stressors such as ischemia and cellular injury, as well as exogenous factors like infections, trauma, and environmental toxins [56]. This response involves activation of astrocytes and microglia, disruption of the endothelial barrier, and infiltration of peripheral immune cells. Fig. (**1**) illustrates this pathological cascade, showing the transition from healthy neuronal architecture supported by tau-stabilized microtubules and anti-inflammatory microglia to the Alzheimer's disease state, marked by tau aggregation, microtubule disintegration, and activated microglia that release inflammatory mediators and drive neurodegeneration.

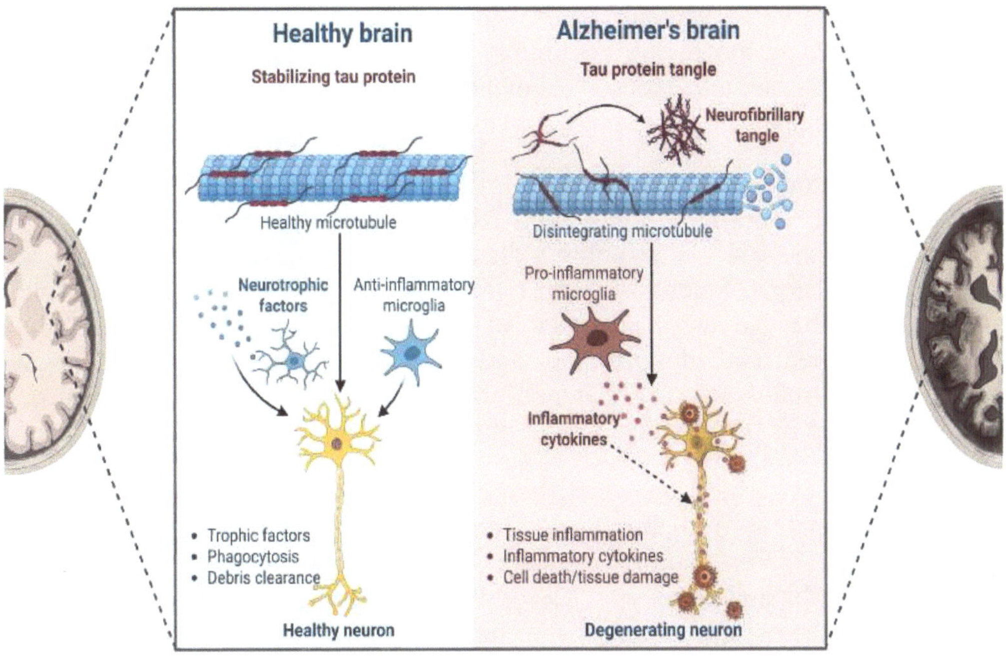

Fig. (1). Tau pathology and microglial activation in Alzheimer's disease.

This comparative diagram highlights the differences between a healthy brain and a brain affected by Alzheimer's. In the healthy brain, tau stabilizes microtubules and supports neurotrophic functions, with anti-inflammatory microglia maintaining tissue homeostasis. In contrast, AD is marked by tau hyperphosphorylation and neurofibrillary tangle formation, leading to microtubule disintegration. This pathological shift activates pro-inflammatory microglia, promoting the release of inflammatory cytokines and contributing to neuronal degeneration.

These processes culminate in the release of pro-inflammatory cytokines (*e.g.,* IL-1β, IL-6, IL-18), chemokines (*e.g.,* CCL1, CCL5, CXCL1), and Reactive Oxygen Species (ROS), all of which contribute to neuronal damage through mitochondrial dysfunction, oxidative stress, synaptic loss, and apoptosis [57].

The role of neuroinflammation in Alzheimer's Disease (AD) has been recognized for decades, with persistent inflammation strongly associated with disease progression [58, 59]. Notably, microglial activation occurs years before the onset of clinical symptoms and is a significant contributor to the accumulation of tau and Aβ, further exacerbating neurodegeneration [60]. Aβ plaques are known to activate the NLRP3 inflammasome in microglia, which accelerates Aβ aggregation and promotes tau pathology [61 - 63]. Additionally, chronically reactive microglia mediate excessive synaptic pruning, a process that significantly contributes to cognitive decline.

Neuroinflammatory mechanisms are implicated in a range of Central Nervous System (CNS) disorders beyond Alzheimer's Disease (AD), including Parkinson's disease [64], Huntington's disease [64], Multiple Sclerosis (MS), and Amyotrophic Lateral Sclerosis (ALS) [65]. A common feature among these conditions is the elevated presence of pro-inflammatory cytokines, which contribute to neuronal dysfunction and cell death. Neuropathological and neuroimaging studies suggest that neuroinflammation often precedes neuronal loss, highlighting its potential role as an initiating factor in disease pathogenesis.

In AD, specific cytokines and reactive species have been shown to directly alter intracellular signaling, mitochondrial function, and structural integrity in neurons and glial cells. Table **2** summarizes key inflammatory mediators released by activated microglia and astrocytes, along with their associated neurotoxic effects. Understanding the complex relationship between these mediators and the cellular damage they induce may reveal critical therapeutic targets to mitigate neuroinflammation and slow neurodegeneration [66].

Table 2. Inflammatory mediators released by microglia and astrocytes in neuroinflammation.

Pro-Inflammatory mediators	Effects
Chemokines	Dysfunction, apoptosis and necrosis of neurons, microglia and astrocytes
IL-1β, IL-6, IL-12, INF, TNF	Microglia and astrocytes activation, dysfunction, apoptosis and necrosis of microglia, neurons, and astrocytes
NO, ROS, O2	Oxidative stress in cells: dysfunction, apoptosis and necrosis of neurons, astrocytes, and microglia

Key Mechanisms of Neuronal Damage

Mitochondrial dysfunction

Mitochondrial integrity is essential for maintaining cellular vitality; dysfunction in these organelles disrupts energy metabolism, exacerbates oxidative stress, dysregulates calcium signaling, and activates apoptosis processes implicated in neurodegenerative diseases like Alzheimer's [67, 68]. Reduced glucose metabolism and changed mitochondrial enzyme activity are seen in AD neuroimaging, which contributes to neuronal dysfunction and neurodegeneration

[69, 70]. AD pathogenesis is made worse by deficiencies in energy generation, oxidative stress, calcium imbalance, and altered mitochondrial dynamics [68, 70].

Aβ directly impairs mitochondrial function by inhibiting key enzymes, such as electron transport chain complex IV, pyruvate dehydrogenase (PDH), and α-ketoglutarate dehydrogenase (αKGDH), leading to reduced ATP synthesis and increased reactive oxygen species (ROS) production [71]. Additionally, Aβ interacts with mitochondrial Aβ-binding alcohol dehydrogenase (ABAD), disrupting NAD binding and enhancing oxidative stress [72, 73]. The accumulation of ROS damages mitochondrial DNA, proteins, and lipids, triggering apoptosis and neuronal death [68, 74].

Calcium homeostasis is crucial for mitochondrial stability, and its disruption plays a key role in AD [75]. Aβ promotes cytosolic calcium influx through membrane receptors and channels [76], impairs mitochondrial calcium buffering, and enhances endoplasmic reticulum calcium release [77]. This overload activates cyclophilin D (CypD), leading to mitochondrial permeability transition pore (mPTP) formation, which further disrupts ATP production and promotes oxidative stress [78]. Tau pathology also contributes to mitochondrial calcium dysregulation and impairs axonal transport, affecting energy distribution within neurons [79, 80].

Mitochondrial fission, fusion, and mitophagy (removal of damaged mitochondria) are also altered in AD [81]. Dysregulation of fission/fusion proteins such as Drp1, Opa1, Fis1, and MFN1/2 promotes excessive mitochondrial fragmentation, leading to bioenergetic deficits and mitochondrial DNA damage [82 - 84]. PINK1/parkin-regulated mitophagy, a key pathway for mitochondrial quality control, is impaired, resulting in the accumulation of defective mitochondria [85]. Aβ, APP-CTFs, tau, and APOE4 contribute to mitophagy inhibition, exacerbating AD progression [86 - 89]. Addressing mitochondrial dysfunction represents a promising avenue for therapeutic intervention in AD [68, 87].

Oxidative stress

Oxidative stress has emerged as a pivotal contributor to Alzheimer's Disease (AD), closely tied to mitochondrial dysfunction. Reactive Oxygen Species (ROS) are spontaneously produced by mitochondria as a consequence of cellular respiration, but excessive ROS accumulation results in oxidative damage that disrupts neuronal function and accelerates neurodegeneration [87]. Studies have shown that ROS-induced oxidative stress promotes tau hyperphosphorylation and aggregation. For instance, mice treated with mitochondria-targeted antioxidants like mito-TEMPO exhibited reduced tau pathology, highlighting the critical role of mitochondrial ROS in AD pathogenesis [88].

In addition to promoting tau pathology, oxidative stress in AD is linked to impaired glucose metabolism. Neurons rely heavily on glucose for energy, and mitochondrial dysfunction compromises this supply. Disruption of insulin signaling pathways—particularly the IRS/PI3K/Akt cascade—induces metabolic stress [89]. Under insulin resistance, hyperactivated FOXO transcription factors upregulate heme oxygenase-1 (HMOX1), impairing the electron transport chain (ETC) due to the loss of heme, a vital ETC component [90]. Consequently, mitochondrial ROS levels rise, leading to oxidative DNA damage, apoptosis, and further neuronal injury [91].

Age-related mitochondrial DNA damage compounds this effect, especially in the absence of robust repair mechanisms or histone protection [92]. Oxidative stress also disrupts nuclear-cytoplasmic transport by altering nuclear pore proteins such as Nup93, further impairing neuronal communication [93]. Moreover, oxidative phosphorylation is compromised, leading to ATP depletion and a shift toward anaerobic metabolism. Elevated expressions of glycolytic enzymes like hexokinase and lactate dehydrogenase in AD cortical regions support this compensatory shift [94].

Ultimately, oxidative stress from mitochondrial dysfunction results in decreased ATP synthesis, calcium imbalance, and activation of the mitochondrial permeability transition pore (mPTP) [95, 96]. This cascade leads to synaptic disruption and neuronal degeneration [97, 98]. Additionally, abnormal calcium influx and impaired mitochondrial calcium buffering exacerbate ROS production and trigger apoptotic pathways [99, 100]. These findings underscore the need to develop therapeutic strategies that directly target oxidative damage and restore mitochondrial health in AD [101 - 109].

Glial Cell-Mediated Neuroinflammation

Microglia and Astrocytes: Key Regulators of Neuroinflammation

Glial cells, primarily microglia and astrocytes, serve as the central nervous system's primary immune responders. These cells orchestrate the inflammatory response and are deeply involved in the pathogenic mechanisms underlying Alzheimer's Disease (AD) [110]. Depending on the disease stage and regional context, glial cells can assume both neuroprotective and neurotoxic roles [111]. Their diverse functional states contribute to the complexity of AD pathogenesis and may explain the limited success of anti-inflammatory therapies in clinical trials [112]. A deeper understanding of glial cell function in AD is crucial for the development of effective therapeutic interventions [113].

Microglia: The Immune Sentinels of the Brain

Microglia, constituting approximately 10% of central nervous system cells, play a critical role in immune surveillance, homeostasis, and neurodevelopment [114]. These cells detect environmental changes through a specialized cluster of genes termed the "sensome" and respond to pathological stimuli by migrating to injury sites, releasing cytokines, and clearing debris [115]. In AD, microglia undergo morphological and functional alterations, shifting from a homeostatic state to a reactive phenotype [116]. This transformation is characterized by changes in cell shape, surface receptor expression, and proliferation, ultimately contributing to neuroinflammation [117].

Mediators of Microglial Activation in AD

Several molecular pathways mediate microglial activation in AD. The purinergic receptor P2X7 (P2X7R) is upregulated in microglia surrounding amyloid-beta (Aβ) plaques [118]. Activation of P2X7R by ATP, released in response to Aβ accumulation, triggers the release of inflammatory cytokines, exacerbating neuroinflammation [119]. Studies using P2X7R inhibitors have shown a reduction in microglial-mediated inflammation and improved phagocytic capacity, suggesting P2X7R as a potential therapeutic target [120].

The Receptor for Advanced Glycation End products (RAGE), which promotes oxidative stress and mitochondrial dysfunction and makes it easier for Aβ to pass the Blood-Brain Barrier (BBB), is another important component of microglial activation [121]. RAGE-mediated Aβ transport correlates with increased inflammatory cytokine production, including TNF-α and IL-6, further amplifying neuroinflammation [122]. Targeting RAGE has been proposed as a strategy to mitigate microglial-driven inflammation and Aβ accumulation [123].

Inflammasome Activation and Microglial Inflammation

Microglial activation in AD is also associated with the NLRP3 inflammasome, a multiprotein complex that triggers the release of IL-1β and IL-18, key proinflammatory cytokines [124]. Aβ deposition activates NLRP3, contributing to tau hyperphosphorylation and neurodegeneration [125]. Studies have demonstrated that inhibition of NLRP3 inflammasome activity reduces neuroinflammation, ameliorates cognitive deficits, and protects against tau pathology, highlighting its potential as a therapeutic target [126].

The immune receptor TREM2 further modulates microglial function by facilitating Aβ phagocytosis [127]. Loss-of-function mutations in TREM2 impair microglial clustering around plaques, leading to unchecked neuroinflammation

[128]. Interestingly, TREM2 function appears to shift during disease progression, reducing inflammatory cytokine secretion in early AD but exacerbating neuroinflammation in later stages [129]. Therapeutic modulation of TREM2 expression may offer a means of restoring beneficial microglial responses [130].

Astrocytes: Regulators of Brain Homeostasis and Inflammation

Astrocytes play a pivotal role in maintaining neuronal function, BBB integrity, and neurotransmitter homeostasis [131]. These cells also participate in Aβ clearance via receptor-mediated uptake and lysosomal degradation [132]. However, glial fibrillary acidic protein (GFAP) expression is elevated, and inflammatory cytokines, including TNF-α, IL-1β, and IL-6, are released when astrocytes undergo reactive remodelling in AD [133]. This reactive condition causes neurodegeneration and interferes with regular astrocytic processes [134].

Astrocytes and Aβ Clearance

In a healthy brain, astrocytes assist in clearing Aβ through transport proteins such as LRP1 and RAGE [135]. Additionally, they secrete ApoE, which promotes Aβ degradation via LRP1-mediated pathways [136]. However, in AD, reactive astrocytes exhibit reduced Aβ clearance efficiency, leading to increased Aβ accumulation and plaque formation [137].

Neurovascular Dysfunction in AD

Astrocytes are integral to neurovascular coupling—a process that ensures Cerebral Blood Flow (CBF) is matched to neuronal activity [138]. This function is mediated by astrocytic Ca^{2}-dependent signaling and the release of vasoactive mediators [139]. In AD, amyloid accumulation disrupts astrocytic Ca^{2+} dynamics, leading to impaired neurovascular responses and chronic hypoperfusion [140]. This vascular dysfunction exacerbates Aβ accumulation and neuronal damage, contributing to disease progression [141].

Heterogeneity of Glial Activation in AD

Microglia and astrocytes exhibit a spectrum of activation states, transitioning between neuroprotective and neurotoxic phenotypes based on disease progression [142]. While early glial activation may serve a protective role, prolonged activation results in chronic inflammation and neuronal damage [143]. This dynamic regulation underscores the need for targeted therapeutic strategies that modulate glial function to restore homeostasis while preventing excessive inflammation [144].

Importance of Understanding Glial Contributions

Glial cells, particularly microglia and astrocytes, play a dual role in AD pathogenesis. Their activation contributes to both protective and detrimental outcomes, depending on the disease stage and molecular context [145]. There are encouraging opportunities for therapeutic intervention when the processes behind glial-mediated neuroinflammation are understood. Targeting key pathways, including RAGE, P2X7R, NLRP3, and TREM2, may provide effective strategies to mitigate neuroinflammation and slow disease progression in AD [146].

Key Determinants of AD: The Interplay of Genetic, Environmental, and Lifestyle Factors

Genetic predisposition accounts for approximately 70% of the risk of developing (AD). However, numerous modifiable environmental and lifestyle factors—including obesity, dyslipidemia, diabetes, hypertension, and cerebrovascular diseases substantially influence disease onset and progression. These elements, along with aging, psychological stress, and metabolic dysfunction, interact in complex ways to accelerate neurodegeneration. As summarized in Fig. (**2**), these diverse risk factors act through overlapping mechanisms such as chronic inflammation, oxidative stress, impaired cerebral perfusion, and metabolic dysregulation.

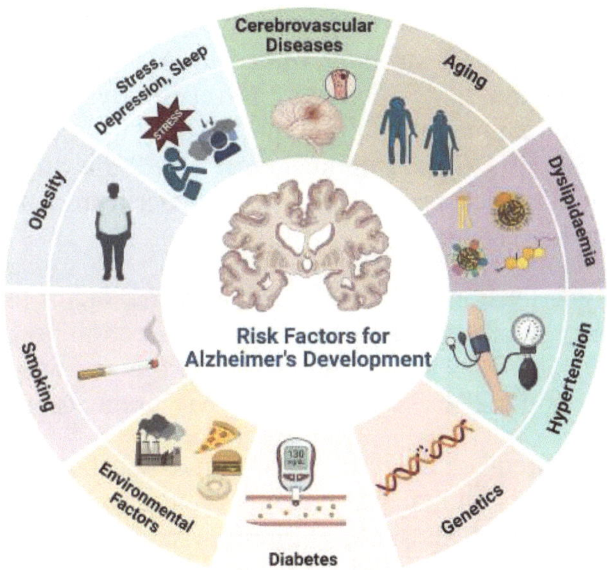

Fig. (2). Risk factors for the development of Alzheimer's disease.

Aging: Aging is the most significant risk factor for AD, with the disease predominantly affecting individuals over 65 [147]. Brain atrophy, synaptic loss, and the buildup of amyloid plaques (SP) and neurofibrillary tangles (NFT) are among the complex interplay of physiological changes linked to ageing [148]. The pathophysiology of AD is also influenced by age-related glucose hypometabolism, dysregulation of cholesterol, mitochondrial dysfunction, depression, and cognitive impairment [149, 150].

Cerebrovascular Diseases: Cerebrovascular abnormalities such as ischemic infarcts, vasculopathy, and hemorrhagic strokes are commonly observed in individuals with AD [151, 152]. Research by Liu et al. revealed that 6–47% of dementia patients also exhibit cerebrovascular disease [153]. The "double stroke" hypothesis suggests that vascular impairment ("first stroke") compromises the Blood-Brain Barrier (BBB), leading to reduced cerebral blood flow (oligemia), which in turn promotes neuroinflammation and amyloid accumulation ("second stroke"), exacerbating neuronal degeneration [154, 155].

Obesity: The association between obesity and AD remains debated. While some studies link metabolic changes in obesity to neuronal death and neurodegeneration [156, 157], others suggest no direct relationship between obesity and AD risk. However, midlife obesity has been significantly correlated with an increased likelihood of developing dementia, likely due to its impact on metabolic and vascular pathways [158, 159].

Diabetes: Diabetes has been strongly linked to AD due to its role in insulin resistance, hyperglycemia, and inflammation. Insulin dysfunction contributes to amyloid accumulation and tau pathology, particularly in individuals carrying the APOE ε4 allele [160, 161]. Studies indicate that type 2 diabetes increases the risk of dementia in middle-aged populations, with hyperinsulinemia promoting Aβ deposition and cognitive decline [162 - 164].

Hypertension: Hypertension is a well-established risk factor for AD, as chronic high blood pressure leads to vascular damage, cerebral ischemia, and BBB dysfunction [165, 166]. These alterations promote amyloid deposition, inflammation, and neuronal loss, accelerating AD progression [167].

Dyslipidemia: Elevated cholesterol levels have been implicated in AD pathogenesis. Studies reveal that AD patients often have cholesterol levels 10% higher than healthy individuals [168, 169]. Hypercholesterolemia contributes to BBB disruption, increased amyloid deposition, and neuroinflammation [170, 171].

Depression: Depression has been associated with a heightened risk of cognitive decline and AD development [172 - 174]. Studies suggest that hippocampal shrinkage, altered protein metabolism, and increased cortisol levels in depressed individuals contribute to neurodegeneration [175, 176].

Stress: Because prolonged activation of the hypothalamic-pituitary-adrenal (HPA) axis causes oxidative stress, inflammation, and dysregulation of glucocorticoids, chronic stress has been recognised as a risk factor for AD [177 - 180]. Persistent HPA axis activation contributes to hippocampal neuronal damage, supporting the glucocorticoid cascade hypothesis of AD [181 - 184].

Sleep Disorders: Sleep disturbances have been identified as both a precursor and a consequence of AD [185]. Studies indicate that individuals with chronic insomnia or sleep apnea have a heightened risk of developing dementia, particularly AD [186].

Smoking: Smoking increases oxidative stress, promotes inflammation, and contributes to cerebrovascular damage, thereby heightening AD risk [187, 188].

Genetics: Genetic factors play a crucial role in AD, with hereditary factors accounting for up to 70% of cases. Early-onset AD is associated with mutations in APP, PSEN1, and PSEN2, while late-onset AD is linked to the APOE ε4 allele, which increases disease risk up to 12-fold in homozygous individuals [189 - 191]. Genetic mutations affect amyloid processing, leading to early Aβ accumulation and neurodegeneration [192 - 195].

Environmental Risk Factors: Environmental factors such as air pollution, diet, heavy metals, and chronic infections may contribute to AD by promoting oxidative stress and inflammation [196, 197]. Infections, including Chlamydia pneumoniae and Herpes Simplex Virus (HSV-1), have been linked to AD pathology through mechanisms involving Aβ accumulation and neuroinflammation [198, 199]. Heavy metals such as aluminum, lead, and cadmium have been shown to promote protein misfolding and amyloid aggregation, further increasing AD susceptibility [200 - 202]. Additionally, air pollution has been associated with increased amyloid deposition and cognitive decline [203, 204].

This schematic representation summarizes key modifiable and non-modifiable risk factors implicated in the development and progression of Alzheimer's Disease (AD). Non-modifiable factors include aging and genetic predisposition, which are strongly associated with increased disease susceptibility. Modifiable risk factors encompass metabolic and vascular conditions such as diabetes, hypertension, dyslipidemia, and cerebrovascular disease, as well as lifestyle and environmental influences, including smoking, obesity, poor diet, pollution, and psychosocial stressors (e.g., depression, sleep disturbances). These factors contribute to AD pathophysiology through interconnected mechanisms such as chronic neuroinflammation, oxidative stress, impaired cerebral perfusion, and neurovascular dysfunction. Understanding and targeting these risk factors offers potential strategies for early intervention and prevention of AD.

Challenges in AD Research and Treatment

Complex Pathogenesis

AD is an advancing neurological condition that has a major social impact and high epidemiological relevance. By 2050, dementia instances are predicted to have doubled or tripled, with the disease accounting for the bulk of these cases. Like other low-income regions, Latin America faces challenges beyond the rising incidence of dementia cases, and also the underfunding of epidemiological research and health professional training, which perpetuates long-standing barriers related to resources, culture, and stigmas. Nonetheless, there is increasing recognition that biomarkers may be crucial. Experts in the field concur that using laboratory testing of cerebrospinal fluid (CSF) and blood-based biomarkers to identify the pathogenic process or molecular imaging techniques can infer the etiology of the underlying disease.

When compared to biomarkers utilized in other medical specialties, the present generation of AD biomarkers cannot be regarded as reliable. However, the accuracy of identifying brain pathology has significantly improved due to biomarker studies. Current dementia recommendations highlight the importance of rigorous and systematic clinical and cognitive research, as well as the use of biomarker approaches. On the other hand, the creation of medications to treat this disease has not yet proved successful. There is no solid proof that any of the licensed medications will slow the progression of the disease; all are symptomatic [205].

Lately, practically all clinical trials for anti-amyloid drugs have failed. The idea behind this research line was that individuals with clinical symptoms of AD should benefit in terms of their cognitive abilities and functional capacity if Ab peptide deposits were removed from brain tissue. As is now evident, the majority of clinical studies have tested possible disease-modifying treatment drugs too late in the pathophysiological course of AD. Even early interventions and other goals should be the focus of the upcoming success tactics.

Limited Diagnostic and Therapeutic Tools

Scientific evidence suggests that the initiating molecular events in the pathogenic cascade of AD are alterations in the synthesis or degeneration of amyloid-β peptide (Aβ) to the creation of aggregated structures, even though the exact mechanism of the disease is yet unknown. The hallmarks of Aβ misfolding and aggregation, as well as how these toxic and supramolecular aggregates cause cell damage and death pathways.

A small percentage of all AD cases are caused by dominant genetic mutations in the three genes PSEN1 and PSEN2, which code for the amyloid precursor protein. These mutations are usually linked to early-onset forms of the disease, where clinical symptoms start to show before the age of 65 [206]. Late-onset AD, which manifests later in life, affects most patients. Although the disease is not hereditary in these situations and does not show a single genetic cause, the most important genetic risk factor is the E4 allele in the ApoE (apolipoprotein E) gene, which is found in about 16% of the population.

A higher risk for the disease is generally associated with lifestyle choices like poor eating habits and a lack of physical activity, as well as metabolic and environmental risk factors like cerebrovascular disease, diabetes, stress and AD injury. The intricate molecular pathways underlying the pathophysiology of AD remain unclear, even though the disease has been known for more than a century. Oxidative stress, vascular abnormalities, mitochondrial harmful changes, decreased brain glucose utilization, and neuroinflammation are among the various multifactorial pathways impacted by the disease and are currently thought to be significant contributors to its onset and/or progression.

Need for Early Diagnosis and Intervention

A multitude of disorders can cause dementia, with AD being the most common. Since it is frequently challenging to pinpoint the exact dementia subtype, it is challenging to calculate an accurate estimate of the prevalence of Alzheimer's dementia. Nonetheless, it is widely acknowledged that approximately two-thirds of dementia diagnoses are caused by Alzheimer's disease [207].

A growing percentage of society will likely be impacted, according to studies that have estimated the frequency of dementia in the future using mathematical models that account for variables including rising life expectancy, changes in mortality, and the incidence of cardiovascular disease. According to UK projections, the number of dementia sufferers in England and Wales is expected to rise by 57% between 2016 and 2040, reaching over 1.2 million by that time. This growth is primarily due to longer life expectancies. According to US projections, there will be 15 million Americans living with Alzheimer's disease or mild cognitive impairment by 2060, up from 6.08 million in 2017. However, new research indicates that dementia incidence rates may be falling in some nations, such as the UK. Evidence of a decline in the age-specific incidence of dementia has been discovered in a two-decade AD study of people 65 and older from three different regions of England. These findings are promising because they demonstrate that Alzheimer's and other types of dementia are not an inevitable byproduct of ageing. They also imply that, for a subset of the population, dementia may be prevented or delayed by reducing exposure to risk factors like diabetes, smoking, high blood pressure, and obesity [207]. The focus has shifted to detect people considerably earlier in the disease process as a result of the understanding that immediate action is needed to lessen the burden of AD, a disorder with rising expenses and few available treatment choices. There is a pressing need to increase diagnosis rates so that those who are most at risk can be identified early and steps taken to slow or stop further progression, even though there is currently no cure or prevention for dementia. Not everyone with mild cognitive impairment will develop dementia.

CONCLUSION

Alzheimer's disease (AD) is a complicated neurological illness with a complex etiology. Although age continues to be the predominant risk factor, its onset and progression are probably influenced by a mix of lifestyle decisions, environmental variables, and genetic predispositions. The amyloid cascade theory and tau pathology are key components of our current knowledge of the complex mechanisms underlying AD. The main pathogenic characteristic is the buildup of amyloid-beta (Aβ) plaques and neurofibrillary tangles (NFTs), which are made of hyperphosphorylated tau. Synaptic dysfunction, neuronal loss, and eventually cognitive decline are caused by these mechanisms that impair neural function. Future research directions are crucial for developing effective treatments and preventive strategies. However, a deeper understanding of the molecular mechanisms underlying AD is essential to unravel its complexities and pave the way for therapeutic interventions that can slow down, prevent, or even reverse this devastating disease.

Future Directions and Hope

Particular attention is paid to recent developments in the creation of efficient molecular tools and methods for both diagnostic and treatment in AD, particularly those that are still in the early phases of research (clinical trials). Many facets of the molecular mechanisms underlying the pathophysiology of multifactorial AD, as well as the advancements made thus far and potential roadblocks that have impeded the creation of efficient medications and early diagnostic tools, are up for debate. The integration of many disease pathways for the successful creation of an early diagnosis and efficient medications for the treatment of AD is one of the challenges and opportunities for the future.

In addition to amyloid tangles and indicators, research is focusing on inflammation, mitochondrial dysfunction, and synaptic abnormalities as important therapeutic targets. Monoclonal antibodies (such as those used to treat AD) are one treatment for the amyloid pathway. Antisense oligonucleotides and small molecules are used in small molecule techniques to lower tau hyperphosphorylation and its levels. To balance the immune response, future therapies for neurodegenerative illnesses will try to modify cytokine signaling and microglial activation, utilizing possible NLRP3 inhibitors, for instance. Neural progenitor cells (NPCs) and biological mesenchymal stem cells (MSCs) are being considered for stem cell therapy and neurodegenerative disease treatment. Drug transport via the blood-brain barrier has been enhanced by nanotechnology, leading to better targeting and decreased toxicity. The genetic foundation of AD will soon be targeted by gene editing using CRISPR-Cas9 gene therapy. In AD, biomarkers help to identify and track treatment response, treatment options, management techniques, and possibly even patient compliance.

REFERENCES

[1] Sperling RA, Jack CR Jr, Aisen PS. Testing the right target and right drug at the right stage. Sci Transl Med 2011; 3(111): 111cm33.
[http://dx.doi.org/10.1126/scitranslmed.3002609]

[2] Glenner GG, Wong CW. Alzheimer's disease: Initial report of the purification and characterization of a novel cerebrovascular amyloid protein. Biochem Biophys Res Commun 1984; 120(3): 885-90.
[http://dx.doi.org/10.1016/S0006-291X(84)80190-4]

[3] Masters CL, Simms G, Weinman NA, Multhaup G, McDonald BL, Beyreuther K. Amyloid plaque core protein in Alzheimer disease and Down syndrome. Proc Natl Acad Sci USA 1985; 82(12): 4245-4249.
[http://dx.doi.org/10.1073/pnas.82.12.4245]

[4] Paula VJR, Guimarães FM, Diniz BS, Forlenza OV. Neurobiological pathways to Alzheimer's disease: Amyloid-beta, TAU protein or both? Dement Neuropsychol 2009; 3(3): 188-94.
[http://dx.doi.org/10.1590/S1980-57642009DN30300003]

[5] Roher AE, Lowenson JD, Clarke S, *et al.* beta-Amyloid-(1-42) is a major component of cerebrovascular amyloid deposits: implications for the pathology of Alzheimer disease. Proc Natl Acad Sci USA 1993; 90(22): 10836-40.
[http://dx.doi.org/10.1073/pnas.90.22.10836]

[6] Ring S, Weyer SW, Kilian SB, *et al.* The Secreted β-Amyloid Precursor Protein Ectodomain APPsα Is Sufficient to Rescue the Anatomical, Behavioral, and Electrophysiological Abnormalities of APP-Deficient Mice. J Neurosci 2007; 27(29): 7817-26.
[http://dx.doi.org/10.1523/JNEUROSCI.1026-07.2007]

[7] Yang Y, Arseni D, Zhang W, *et al.* Cryo-EM structures of amyloid-β 42 filaments from human brains. Science 2022; 375(6577): 167-72.

[http://dx.doi.org/10.1126/science.abm7285]

[8] Walker LC, Jucker M. Neurodegenerative diseases: expanding the prion concept. Annu Rev Neurosci 2015; 38(1): 87-103.
[http://dx.doi.org/10.1146/annurev-neuro-071714-033828]

[9] Eisele YS, Bolmont T, Heikenwalder M, *et al.* Induction of cerebral beta-amyloidosis by Abeta inoculation. Proc Natl Acad Sci USA 2009; 106(31): 12926-31.
[http://dx.doi.org/10.1073/pnas.0903200106]

[10] Bussian TJ, Aziz A, Meyer CF, Swenson BL, van Deursen JM, Baker DJ. Clearance of senescent glial cells prevents tau-dependent pathology and cognitive decline. Nature 2018; 562(7728): 578-82.
[http://dx.doi.org/10.1038/s41586-018-0543-y]

[11] Purro SA, Farrow MA, Linehan J, *et al.* Transmission of amyloid-β protein pathology from cadaveric pituitary growth hormone. Nature 2018; 564(7736): 415-9.
[http://dx.doi.org/10.1038/s41586-018-0790-y]

[12] Azargoonjahromi A. The duality of amyloid-β: its role in normal and Alzheimer's disease states. Mol Brain 2024; 17(1): 44.
[http://dx.doi.org/10.1186/s13041-024-01118-1]

[13] Abelein A. Metal Binding of Alzheimer's Amyloid-β and Its Effect on Peptide Self-Assembly. Acc Chem Res 2023; 56(19): 2653-63.
[http://dx.doi.org/10.1021/acs.accounts.3c00370]

[14] Chia S, Flagmeier P, Habchi J, Lattanzi V, Linse S, Dobson CM, *et al.* α-synuclein influences Aβ42 aggregates. Proc Natl Acad Sci USA 2017; 114(30): 8005-10.
[http://dx.doi.org/10.1073/pnas.1700239114]

[15] Garai K, Verghese PB, Baban B, Holtzman DM, Frieden C. Apolipoprotein E affects amyloid aggregation. Biochemistry 2014; 53(40): 6323-31.
[http://dx.doi.org/10.1021/bi5008172]

[16] Zhao Y, Zheng Q, Hong Y, *et al.* β2-Microglobulin coaggregates with Aβ and contributes to amyloid pathology and cognitive deficits in Alzheimer's disease model mice. Nat Neurosci 2023; 26(7): 1170-84.
[http://dx.doi.org/10.1038/s41593-023-01352-1]

[17] Gao Y, Hong Y, Huang L, Zheng S, Zhang H, Wang S, *et al.* β2-microglobulin impairs synaptic function. Cell 2023; 186(5): 1026-38.e20.
[http://dx.doi.org/10.1016/j.cell.2023.01.021]

[18] Hardy JA, Higgins GA. Alzheimer's disease: The amyloid cascade hypothesis. Science 1992; 256(5054): 184-5.
[http://dx.doi.org/10.1126/science.1566067]

[19] Tcw J, Goate AM. Genetics of β-Amyloid Precursor Protein in Alzheimer's Disease. Cold Spring Harb Perspect Med 2017; 7(6)a024539
[http://dx.doi.org/10.1101/cshperspect.a024539]

[20] Budd Haeberlein S, Aisen PS, Barkhof F, *et al.* Two Randomized Phase 3 Studies of Aducanumab in Early Alzheimer's Disease. J Prev Alzheimers Dis 2022; 9(2): 197-210.
[http://dx.doi.org/10.14283/jpad.2022.30]

[21] Sims JR, Zimmer JA, Evans CD, *et al.* Donanemab in Early Symptomatic Alzheimer Disease. JAMA 2023; 330(6): 512-27.
[http://dx.doi.org/10.1001/jama.2023.13239]

[22] van Dyck CH, Swanson CJ, Aisen P, *et al.* Lecanemab in Early Alzheimer's Disease. N Engl J Med 2023; 388(1): 9-21.
[http://dx.doi.org/10.1056/NEJMoa2212948]

[23] Nelson MR, Liu P, Agrawal A, *et al.* APOE-R136S mutation protects against Tau pathology. Nat Neurosci 2023; 26(12): 2104-21.
[http://dx.doi.org/10.1038/s41593-023-01480-8]

[24] Altmann A, Ng B, Landau SM, Jagust WJ, Greicius MD. Brain hypometabolism is unrelated to plaque burden. Brain 2015; 138(Pt 12): 3734-46.
[http://dx.doi.org/10.1093/brain/awv278]

[25] Long JM, Holtzman DM. Alzheimer Disease: An Update on Pathobiology and Treatment Strategies. Cell 2019; 179(2): 312-39.
[http://dx.doi.org/10.1016/j.cell.2019.09.001]

[26] Naveilhan P, Neveu I, Baudet C, Funakoshi H, Wion D, Brachet P, *et al.* Vitamin D regulates p75NTR expression. Brain Res Mol Brain Res 1996; 41(1–2): 259-68.
[http://dx.doi.org/10.1016/0169-328X(96)00103-9]

[27] Hashimoto Y, Kaneko Y, Tsukamoto E, Frankowski H, Kouyama K, Kita Y, *et al.* Amyloid-β induces death via p75NTR. J Neurochem 2004; 90(3): 549-58.
[http://dx.doi.org/10.1111/j.1471-4159.2004.02513.x]

[28] Perini G, Della-Bianca V, Politi V, Della Valle G, Dal-Pra I, Rossi F, *et al.* Role of p75NTR in Aβ neurotoxicity. J Exp Med 2002; 195(7): 907-18.
[http://dx.doi.org/10.1084/jem.20011797]

[29] Etique N, Verzeaux L, Dedieu S, Emonard H. LRP-1: a checkpoint for ECM proteolysis. BioMed Res Int 2013; 20: 13152163.
[http://dx.doi.org/10.1155/2013/152163]

[30] Wijnberg MJ, Quax PH, Nieuwenbroek NM, Verheijen JH. Migration of human smooth muscle cells is plasminogen-mediated. Thromb Haemost 1997; 78(2): 880-6.
[http://dx.doi.org/10.1055/s-0038-1657646]

[31] Selkoe DJ. Hypothesis: Alzheimer's disease is caused by Aβ accumulation. Ann N Y Acad Sci 2000; 924: 17-25.
[http://dx.doi.org/10.1111/j.1749-6632.2000.tb05554.x]

[32] Linden R, Martins VR, Prado MAM, Cammarota M, Izquierdo I, Brentani RR. Physiology of prion protein. Physiol Rev 2008; 88(2): 673-728.
[http://dx.doi.org/10.1152/physrev.00007.2007]

[33] Ribeiro FM, Paquet M, Cregan SP, Ferguson SS. Group I mGluR signaling in neurological disease. CNS Neurol Disord Drug Targets 2010; 9(5): 574-95.
[http://dx.doi.org/10.2174/187152710793361612]

[34] Bordi F, Ugolini A. Group I mGluRs: implications for brain diseases. Prog Neurobiol 1999; 59(1): 55-79.
[http://dx.doi.org/10.1016/S0301-0082(98)00095-1]

[35] Wang HY, Stucky A, Liu J, Shen C, Trocme-Thibierge C, Morain P. Novel therapeutic agent S 24795 normalizes receptor function. J Neurosci 2009; 29(35): 10961-73.
[http://dx.doi.org/10.1523/JNEUROSCI.6088-08.2009]

[36] Olsen KM, Sheng M. NMDA receptors and BAX are essential for Aβ impairment of LTP. Sci Rep 2012; 2(1): 225.
[http://dx.doi.org/10.1038/srep00225]

[37] Zhang H, Jiang X, Ma L, *et al.* Role of Aβ in synaptic dysfunction. Front Cell Dev Biol 2022; 10: 964075.
[http://dx.doi.org/10.3389/fcell.2022.964075]

[38] Wang D, Govindaiah G, Liu R, De Arcangelis V, Cox CL, Xiang YK. Aβ binds β2-adrenergic receptor and induces hyperactivity. FASEB J 2010; 24(9): 3511-21.

[http://dx.doi.org/10.1096/fj.10-156661]

[39] Kandouz M. Dying to communicate: apoptotic functions of Eph/Ephrin proteins. Apoptosis 2018; 23(5-6): 265-89.
[http://dx.doi.org/10.1007/s10495-018-1458-7]

[40] Sinclair P, Baranova A, Kabbani N. Mitochondrial disruption by Aβ42. Cells 2021; 10(9): 2380.
[http://dx.doi.org/10.3390/cells10092380]

[41] Zhong S, Pei D, Shi L, Cui Y, Hong Z. Ephrin-B2 inhibits Aβ-induced apoptosis. Neurosci Lett 2019; 704: 50-6.
[http://dx.doi.org/10.1016/j.neulet.2019.03.028]

[42] Kawaguchi Y, Matsubayashi J, Kawakami Y, Nishida R, Kurihara Y, Takei K. LOTUS blocks Aβ binding to PirB. Mol Med 2022; 28(1): 154.
[http://dx.doi.org/10.1186/s10020-022-00581-7]

[43] Vershinin M, Carter BC, Razafsky DS, King SJ, Gross SP. Multiple-motor based transport and its regulation by Tau. Proc Natl Acad Sci USA 2007; 104(1): 87-92.
[http://dx.doi.org/10.1073/pnas.0607919104]

[44] Hanseeuw BJ, Betensky RA, Jacobs HI, *et al.* Amyloid and tau associated with cognition. JAMA Neurol 2019; 76(8): 915-24.
[http://dx.doi.org/10.1001/jamaneurol.2019.1424]

[45] Wang Y, Mandelkow E. Tau in physiology and pathology. Nat Rev Neurosci 2016; 17(1): 22-35.
[http://dx.doi.org/10.1038/nrn.2015.1]

[46] Giannakopoulos P, Herrmann FR, Bussière T, *et al.* Tangles and neuron numbers predict cognition. Neurology 2003; 60(9): 1495-500.
[http://dx.doi.org/10.1212/01.WNL.0000063311.58879.01]

[47] Braak H, Braak E. Neuropathological stageing of Alzheimer-related changes. Acta Neuropathol 1991; 82(4): 239-59.
[http://dx.doi.org/10.1007/BF00308809]

[48] Nelson PT, Alafuzoff I, Bigio EH, *et al.* Correlation of AD neuropathology with cognition. J Neuropathol Exp Neurol 2012; 71(5): 362-81.
[http://dx.doi.org/10.1097/NEN.0b013e31825018f7]

[49] Crary JF, Trojanowski JQ, Schneider JA, *et al.* Primary age-related tauopathy (PART): a common pathology associated with human aging. Acta Neuropathol 2014; 128(6): 755-66.
[http://dx.doi.org/10.1007/s00401-014-1349-0]

[50] Morris M, Maeda S, Vossel K, Mucke L. The many faces of tau. Neuron 2011; 70(3): 410-26.
[http://dx.doi.org/10.1016/j.neuron.2011.04.009]

[51] Wesseling H, Mair W, Kumar M, *et al.* Tau PTM profiles identify patient heterogeneity and stages of Alzheimer's disease. Cell 2020; 183(6): 1699-1713.e13.
[http://dx.doi.org/10.1016/j.cell.2020.10.029]

[52] Goedert M, Eisenberg DS, Crowther RA. Propagation of tau aggregates and neurodegeneration. Annu Rev Neurosci 2017; 40(1): 189-210.
[http://dx.doi.org/10.1146/annurev-neuro-072116-031153]

[53] Congdon EE, Sigurdsson EM. Tau-targeting therapies for Alzheimer disease. Nat Rev Neurol 2018; 14(7): 399-415.
[http://dx.doi.org/10.1038/s41582-018-0013-z]

[54] Spillantini MG, Goedert M. Tau protein pathology in neurodegenerative diseases. Trends Neurosci 1998; 21(10): 428-33.
[http://dx.doi.org/10.1016/S0166-2236(98)01337-X]

[55] Sandrone S, Kipnis J, Grice EA. The gut–brain axis and the immune system: A coordinated

communication network. Trends Neurosci 2019; 42(10): 728-36.

[56] Fernández JA, Rojo L, Kuljis RO, Maccioni RB. The damage signals hypothesis of Alzheimer's disease pathogenesis. J Alzheimers Dis 2008; 14(3): 329-33. [http://dx.doi.org/10.3233/JAD-2008-14307]

[57] DiSabato DJ, Quan N, Godbout JP. Neuroinflammation: the devil is in the details. J Neurochem 2016; 139(S2): 136-53. [http://dx.doi.org/10.1111/jnc.13607]

[58] Eikelenboom P, Veerhuis R, Van Gool WA. Neuroinflammation in Alzheimer's disease: An immunopathological view. Trends Immunol 1994; 15(11): 495-9.

[59] Rogers J, Lue LF, Walker DG. Microglia and inflammatory mechanisms in the clearance of amyloid beta peptide. Glia 1996; 16(3): 282-6.

[60] Robbins JR. Microglial priming in Alzheimer's disease: A critical role in the initiation of neuroinflammation. Front Aging Neurosci 2021; 13: 644450.

[61] McGeer PL, Rogers J. Neuroinflammation and Alzheimer's disease: The role of activated microglia and cytokines. Neurology 1990; 40(5): 755-61.

[62] Wang Y, Mandelkow E. Tau in physiology and pathology. Nat Rev Neurosci 2015; 17(1): 5-21.

[63] Leng F, Edison P. Neuroinflammation and microglial activation in Alzheimer disease: where do we go from here? Nat Rev Neurol 2021; 17(3): 157-72. [http://dx.doi.org/10.1038/s41582-020-00435-y]

[64] Ferrari CC, Tarelli R. Parkinson's disease and systemic inflammation. Parkinsonism Relat Disord 2006; 12 (Suppl. 2): S110-3.

[65] Hsiao HY. Microglial activation and damage in Huntington's disease models. Front Aging Neurosci 2013; 5: 12.

[66] Kassa RM. Neuroinflammation in amyotrophic lateral sclerosis and multiple sclerosis. Brain Res Brain Res Rev 2009; 60(1): 234-44.

[67] Frank-Cannon TC, Alto LT, McAlpine FE, Tansey MG. Does neuroinflammation fan the flame in neurodegenerative diseases? Mol Neurodegener 2009; 4(1): 47. [http://dx.doi.org/10.1186/1750-1326-4-47]

[68] Wang W, Zhao F, Ma X, Perry G, Zhu X. Mitochondria dysfunction in the pathogenesis of Alzheimer's disease: recent advances. Mol Neurodegener 2020; 15(1): 30. [http://dx.doi.org/10.1186/s13024-020-00376-6]

[69] Gowda P, Reddy PH, Kumar S. Deregulated mitochondrial microRNAs in Alzheimer's disease: Focus on synapse and mitochondria. Ageing Res Rev 2022; 73101529 [http://dx.doi.org/10.1016/j.arr.2021.101529]

[70] Kapogiannis D, Mattson MP. Disrupted energy metabolism and neuronal circuit dysfunction in cognitive impairment and Alzheimer's disease. Lancet Neurol 2011; 10(2): 187-98. [http://dx.doi.org/10.1016/S1474-4422(10)70277-5]

[71] Godoy JA, Rios JA, Zolezzi JM, Braidy N, Inestrosa NC. Signaling pathway cross talk in Alzheimer's disease. Cell Commun Signal 2014; 12(1): 23. [http://dx.doi.org/10.1186/1478-811X-12-23]

[72] Eckert A, Schmitt K, Götz J. Mitochondrial dysfunction - the beginning of the end in Alzheimer's disease? Separate and synergistic modes of tau and amyloid-β toxicity. Alzheimers Res Ther 2011; 3(2): 15. [http://dx.doi.org/10.1186/alzrt74]

[73] Lustbader JW, Cirilli M, Lin C, *et al.* ABAD directly links Abeta to mitochondrial toxicity in Alzheimer's disease. Science 2004; 304(5669): 448-52.

[http://dx.doi.org/10.1126/science.1091230]

[74] Ye Z, Liu Y, Jin X, *et al.* Aβ-binding with alcohol dehydrogenase drives Alzheimer's disease pathogenesis: A review. Int J Biol Macromol 2024; 264130580 [http://dx.doi.org/10.1016/j.ijbiomac.2024.130580]

[75] Misrani A, Tabassum S, Yang L. Mitochondrial dysfunction and oxidative stress in Alzheimer's disease. Front Aging Neurosci 2021; 13617588 [http://dx.doi.org/10.3389/fnagi.2021.617588]

[76] De Nicolo B, Cataldi-Stagetti E, Diquigiovanni C, Bonora E. Calcium and Reactive Oxygen Species Signaling Interplays in Cardiac Physiology and Pathologies. Antioxidants 2023; 12(2): 353. [http://dx.doi.org/10.3390/antiox12020353]

[77] Bezprozvanny I, Mattson MP. Neuronal calcium mishandling and the pathogenesis of Alzheimer's disease. Trends Neurosci 2008; 31(9): 454-63. [http://dx.doi.org/10.1016/j.tins.2008.06.005]

[78] Ansari MA, Scheff SW, *et al.* Oxidative stress in the progression of Alzheimer disease in the frontal cortex. J Neuropathol Exp Neurol 2010; 69(2): 155-67. [http://dx.doi.org/10.1097/NEN.0b013e3181cb5af4]

[79] Calvo-Rodriguez M, Bacskai BJ. High mitochondrial calcium levels precede neuronal death in AD. Cell Stress 2020; 4(7): 187-90. [http://dx.doi.org/10.15698/cst2020.07.226]

[80] Dowdy SF. Overcoming cellular barriers for RNA therapeutics. Nat Biotechnol 2017; 35(3): 222-229. [http://dx.doi.org/10.1038/nbt.3802]

[81] Atamna H, Killilea DW, Killilea AN, Ames BN. Heme deficiency may be a factor in the mitochondrial and neuronal decay of aging. Proc Natl Acad Sci USA 2002; 99(23): 14807-12. [http://dx.doi.org/10.1073/pnas.192585799]

[82] Calvo-Rodriguez M, Bacskai BJ. Mitochondria and Calcium in AD. Trends Neurosci 2021; 44(2): 136-51. [http://dx.doi.org/10.1016/j.tins.2020.10.004]

[83] Millecamps S, Julien JP. Axonal transport deficits and neurodegenerative diseases. Nat Rev Neurosci 2013; 14(3): 161-76. [http://dx.doi.org/10.1038/nrn3380]

[84] Chen W, Zhao H, Li Y. Mitochondrial dynamics in health and disease: mechanisms and potential targets. Signal Transduct Target Ther 2023; 8(1): 333. [http://dx.doi.org/10.1038/s41392-023-01547-9]

[85] Cho D-H, Nakamura T, Fang J, Cieplak P, Godzik A, Gu Z, Lipton SA. S-nitrosylation of Drp1 mediates Aβ-related mitochondrial fission. Science 2009; 324(5923): 102-5. [http://dx.doi.org/10.1126/science.1171091]

[86] Bossy B, Petrilli A, Klinglmayr E, *et al.* S-Nitrosylation of DRP1 not specific to AD. J Alzheimers Dis 2010; 20 (Suppl. 2): S513-26. [http://dx.doi.org/10.3233/JAD-2010-100552]

[87] Resende R, Ferreiro E, Pereira C, Resende de Oliveira C. Neurotoxic effect of Aβ1-42 species: ER calcium release involvement. Neuroscience 2008; 155(3): 725-37. [http://dx.doi.org/10.1016/j.neuroscience.2008.06.036]

[88] Eckert A, Nisbet R, Grimm A, Götz J. March separate, strike together—role of phosphorylated tau in mitochondrial dysfunction in Alzheimer's disease. Biochim Biophys Acta 2014; 1842(8): 1258-66. [http://dx.doi.org/10.1016/j.bbadis.2013.08.013]

[89] Calvo-Rodriguez, M, Hou SS., Snyder AC. *et al.* Increased mitochondrial calcium and neuronal death in AD mouse model. Nat Commun 2020; 11: 2146.

[http://dx.doi.org/10.1038/s41467-020-16074-2]

[90] Du H, Yan SS. Unlocking the Door to Neuronal Woes in AD: Aβ and mPTP. Pharmaceuticals 2010; 3(6): 1936-48.
[http://dx.doi.org/10.3390/ph3061936]

[91] Wang X, Su B, Lee HG,, *et al.* Impaired balance of mitochondrial fission and fusion in Alzheimer's disease. J Neurosci 2009; 29(28): 9090-103.
[http://dx.doi.org/10.1523/JNEUROSCI.1357-09.2009]

[92] Joshi AU, Saw NL, Shamloo M, Mochly-Rosen D. Drp1/Fis1 interaction mediates mitochondrial dysfunction, bioenergetic failure and cognitive decline in Alzheimer's disease. Oncotarget 2017; 9(5): 6128-43.
[http://dx.doi.org/10.18632/oncotarget.23640]

[93] Chen H, Vermulst M, Wang YE, *et al.* Mitochondrial fusion is required for mtDNA stability in skeletal muscle and tolerance of mtDNA mutations. Cell 2010; 141(2): 280-289.
[http://dx.doi.org/10.1016/j.cell.2010.02.026]

[94] Martin-Maestro P, Gargini R, Perry G, Avila J, García-Escudero V, *et al.* Parkin-mediated mitophagy is altered in Alzheimer's disease. J Alzheimers Dis 2016; 56(1): 109-20.
[http://dx.doi.org/10.3233/JAD-160461]

[95] Bossy EF, Hou Y, Palikaras K, *et al.* Mitophagy inhibits amyloid-β and tau pathology and reverses cognitive deficits in models of Alzheimer's disease. Nat Neurosci 2019; 22(3): 401-12.
[http://dx.doi.org/10.1038/s41593-018-0332-9]

[96] Deane R, Du Yan S, Submamaryan RK, *et al.* RAGE mediates amyloid-β peptide transport across the blood-brain barrier and accumulation in brain. Nat Med 2003; 9(7): 907-13.
[http://dx.doi.org/10.1038/nm890]

[97] Fang F, Lue LF, Yan S, *et al.* RAGE-dependent signaling in microglia contributes to neuroinflammation and Aβaccumulation in AD models. Glia 2010; 58(9): 1047-57.

[98] Heneka MT, Kummer MP, Stutz A, *et al.* NLRP3 is activated in Alzheimer's disease and contributes to pathology in APP/PS1 mice. Nature 2013; 493(7434): 674-8.
[http://dx.doi.org/10.1038/nature11729]

[99] Brookes PS, Yoon Y, Robotham JL, Anders MW, Sheu SS. Calcium, ATP, and ROS: a mitochondrial love-hate triangle. Am J Physiol Cell Physiol 2004; 287(4): C817-33.
[http://dx.doi.org/10.1152/ajpcell.00139.2004]

[100] Ulrich JD, Finn MB, Wang Y, *et al.* Altered microglial response to Aβ plaques in APPPS1-21 mice lacking TREM2. J Neurosci 2014; 34(7): 2141-53.

[101] Meilandt WJ. TREM2-dependent microglial functions in AD pathogenesis. Cell Rep 2020; 33(3)108398.

[102] Sofroniew MV, Vinters HV. Astrocytes: biology and pathology. Acta Neuropathol 2010; 119(1): 7-35.
[http://dx.doi.org/10.1007/s00401-009-0619-8]

[103] Zamanian JL, Xu L, Foo LC, *et al.* Genomic analysis of reactive astrogliosis. J Neurosci 2012; 32(18): 6391-410.
[http://dx.doi.org/10.1523/JNEUROSCI.6221-11.2012]

[104] Wyss-Coray T, Loike JD, Brionne TC, *et al.* Adult mouse astrocytes degrade amyloid-β in vitro and in situ. Nat Med 2003; 9(4): 453-7.
[http://dx.doi.org/10.1038/nm838]

[105] Koistinaho M, Lin S, Wu X, *et al.* Apolipoprotein E promotes astrocyte colocalization and degradation of deposited amyloid-β peptides. Nat Med 2004; 10(7): 719-26.
[http://dx.doi.org/10.1038/nm1058]

[106] Iadecola C. The neurovascular unit coming of age: a journey through neurovascular coupling in health

and disease. Neuron 2017; 96(1): 17-42.
[http://dx.doi.org/10.1016/j.neuron.2017.07.030]

[107] Takano T, Tian G-F, Peng W, *et al.* Astrocyte-mediated control of cerebral blood flow. Nat Neurosci 2006; 9(2): 260-7.
[http://dx.doi.org/10.1038/nn1623]

[108] Chandragiri SS, Nyul-Toth A, Negri S, *et al.* Functional ultrasound imaging reveals microvascular rarefaction, decreased cerebral blood flow, and impaired neurovascular coupling in a mouse model of paclitaxel-induced chemobrain. Geroscience 2025; 47(3): 4813-25.
[http://dx.doi.org/10.1007/s11357-025-01624-7]

[109] Nyúl-Tóth Á, Negri S, Sanford M, *et al.* Novel intravital approaches to quantify deep vascular structure and perfusion in the aging mouse brain using ultrasound localization microscopy (ULM). J Cereb Blood Flow Metab 2024; 44(11): 1378-96.
[http://dx.doi.org/10.1177/0271678X241260526]

[110] Salter MW, Stevens B. Microglia emerge as central players in brain disease. Nat Med 2017; 23(9): 1018-27.
[http://dx.doi.org/10.1038/nm.4397]

[111] Hickman S, Izzy S, Sen P, Morsett L, El Khoury J. Microglia in neurodegeneration. Nat Neurosci 2018; 21(10): 1359-69.
[http://dx.doi.org/10.1038/s41593-018-0242-x]

[112] Heneka MT, Golenbock DT, Latz E. Innate immunity in Alzheimer's disease. Nat Immunol 2015; 16(3): 229-36.
[http://dx.doi.org/10.1038/ni.3102]

[113] Hansen DV, Hanson JE, Sheng M. Microglia in Alzheimer's disease. J Cell Biol 2018; 217(2): 459-72.
[http://dx.doi.org/10.1083/jcb.201709069]

[114] Ransohoff RM, Cardona AE. The myeloid cells of the central nervous system parenchyma. Nature 2010; 468(7321): 253-62.
[http://dx.doi.org/10.1038/nature09615]

[115] Hickman SE, Kingery ND, Ohsumi TK, *et al.* The microglial sensome revealed by direct RNA sequencing. Nat Neurosci 2013; 16(12): 1896-905.
[http://dx.doi.org/10.1038/nn.3554]

[116] Keren-Shaul H, Spinrad A, Weiner A, *et al.* A unique microglia type associated with restricting development of Alzheimer's disease. Cell 2017; 169(7): 1276-1290.e17.
[http://dx.doi.org/10.1016/j.cell.2017.05.018]

[117] Streit WJ, Sammons NW, Kuhns AJ, Sparks DL. Dystrophic microglia in the aging human brain. Glia 2004; 45(2): 208-12.
[http://dx.doi.org/10.1002/glia.10319]

[118] Martin E, El Khoury J. The P2X7 receptor in Alzheimer's disease. Brain Pathol 2019; 29(1): 16-25.

[119] Ribeiro DE, Roncalho AL, Glaser T, Ulrich H, Wegener G, Joca SRL. Purinergic receptors in neurodegenerative diseases: a review from basic to clinical applications. Pharmacol Res 2019; 147: 104373.

[120] Parvathenani LK, Tertyshnikova S, Greco CR, Roberts SB, Robertson B, Posmantur R. P2X7 mediates superoxide production in primary microglia and is up-regulated in a transgenic mouse model of Alzheimer's disease. J Biol Chem 2003; 278(15): 13309-17.
[http://dx.doi.org/10.1074/jbc.M209478200]

[121] Bernardi P, Rasola A, Forte M, Lippe G. The mitochondrial permeability transition pore: channel formation by F-ATP synthase, integration in signal transduction, and role in pathophysiology. Physiol Rev 2015; 95(4): 1111-55.
[http://dx.doi.org/10.1152/physrev.00001.2015]

[122] Fang F, Lue LF, Yan S, *et al.* RAGE-dependent signaling in microglia contributes to neuroinflammation, Aβ accumulation, and impaired learning/memory in a mouse model of Alzheimer's disease. FASEB J 2010; 24(4): 1043-55. [http://dx.doi.org/10.1096/fj.09-139634]

[123] Bierhaus A, Humpert PM, Morcos M, *et al.* Understanding RAGE, the receptor for advanced glycation end products. J Mol Med (Berl) 2005; 83(11): 876-86. [http://dx.doi.org/10.1007/s00109-005-0688-7]

[124] Spires-Jones TL, Hyman BT. The intersection of amyloid beta and tau at synapses in Alzheimer's disease. Neuron 2014; 82(4): 756-71. [http://dx.doi.org/10.1016/j.neuron.2014.05.004]

[125] Ising C, Venegas C, Zhang S, *et al.* NLRP3 inflammasome activation drives tau pathology. Nature 2019; 575(7784): 669-73. [http://dx.doi.org/10.1038/s41586-019-1769-z]

[126] Daniels MJD, Rivers-Auty J, Schilling T, *et al.* Fenamate NSAIDs inhibit the NLRP3 inflammasome and protect against Alzheimer's disease in rodent models. Nat Commun 2016; 7(1): 12504. [http://dx.doi.org/10.1038/ncomms12504]

[127] Bateman RJ, Xiong C, Benzinger TL, *et al.* Clinical and biomarker changes in dominantly inherited Alzheimer's disease. N Engl J Med 2012; 367(9): 795-804. [http://dx.doi.org/10.1056/NEJMoa1202753]

[128] Wang Y, Cella M, Mallinson K, *et al.* TREM2 lipid sensing sustains the microglial response in an Alzheimer's disease model. Cell 2015; 160(6): 1061-71. [http://dx.doi.org/10.1016/j.cell.2015.01.049]

[129] Meilandt WJ, Ngu H, Gogineni A, Lalehzadeh G, Lee SH, Srinivasan K. Trem2-dependent microglial surveillance in Alzheimer's disease models. Cell Rep 2020; 33(3)108398.

[130] Parhizkar S, Arzberger T, Brendel M, *et al.* Loss of TREM2 function increases amyloid seeding but reduces plaque-associated ApoE. Nat Neurosci 2019; 22(2): 191-204. [http://dx.doi.org/10.1038/s41593-018-0296-9]

[131] Butterfield DA, Halliwell B. Oxidative stress, dysfunctional glucose metabolism and Alzheimer disease. Nat Rev Neurosci 2019; 20(3): 148-60. [http://dx.doi.org/10.1038/s41583-019-0132-6]

[132] Onyango IG, Bennett JP, Stokin GB. Mitochondrially-targeted therapeutic strategies for Alzheimer's disease. Curr Alzheimer Res 2021; 18(10): 753-71. [http://dx.doi.org/10.2174/1567205018666211208125855]

[133] Area-Gomez E, Schon EA. Mitochondrial genetics and Alzheimer's disease. Biochim Biophys Acta 2017; 1863(5): 1429-36. [http://dx.doi.org/10.1016/j.bbadis.2016.12.012]

[134] Reichenbach N, Delekate A, Plescher M, *et al.* Inhibition of Stat3-mediated astrogliosis ameliorates pathology in an Alzheimer's disease model. EMBO Mol Med 2019; 11(2)e9665 [http://dx.doi.org/10.15252/emmm.201809665]

[135] Swerdlow RH. Mitochondria and mitochondrial cascades in Alzheimer's disease. J Alzheimers Dis 2018; 62(3): 1403-16. [http://dx.doi.org/10.3233/JAD-170585]

[136] Liu CC, Zhao N, Fu Y, *et al.* ApoE4 accelerates early seeding of amyloid pathology. Neuron 2017; 96(5): 1024-1032.e3. [http://dx.doi.org/10.1016/j.neuron.2017.11.013]

[137] Osborn LM, Kamphuis W, Wadman WJ, Hol EM. Astrogliosis: An integral player in the pathogenesis of Alzheimer's disease. Prog Neurobiol 2016; 144: 121-41.

[http://dx.doi.org/10.1016/j.pneurobio.2016.01.001]

[138] Picone P, Nuzzo D, Caruana L, Scafidi V, Di Carlo M. Mitochondrial dysfunction: different routes to Alzheimer's disease therapy. Oxid Med Cell Longev 2014; 2014: 780179. [http://dx.doi.org/10.1155/2014/780179]

[139] Wang X, Wang W, Li L, Perry G, Lee HG, Zhu X. Oxidative stress and mitochondrial dysfunction in Alzheimer's disease. Biochim Biophys Acta 2014; 1842(8): 1240-7. [http://dx.doi.org/10.1016/j.bbadis.2013.10.015]

[140] Tarantini S, Yabluchanskiy A, Csipo T, Fulop GA, Kiss T, Balasubramanian P, *et al.* Neurovascular uncoupling in aging and Alzheimer's disease: contribution of astrocyte dysfunction and endothelial impairment to cognitive decline. Geroscience 2021; 43(1): 107-38.

[141] Nortley R, Korte N, Izquierdo P, *et al.* Amyloid β oligomers constrict human capillaries in Alzheimer's disease via signaling to pericytes. Science 2019; 365(6450)eaav9518. [http://dx.doi.org/10.1126/science.aav9518]

[142] Escartin C, Galea E, Lakatos A, *et al.* Reactive astrocyte nomenclature, definitions, and future directions. Nat Neurosci 2021; 24(3): 312-25. [http://dx.doi.org/10.1038/s41593-020-00783-4]

[143] Liddelow SA, Guttenplan KA, Clarke LE, *et al.* Neurotoxic reactive astrocytes are induced by activated microglia. Nature 2017; 541(7638): 481-7. [http://dx.doi.org/10.1038/nature21029]

[144] Olmos-Alonso A, Schetters STT, Sri S, et al. Pharmacological targeting of CSF1R inhibits microglial proliferation and prevents the progression of Alzheimer's-like pathology. Brain 2016; 139(3): 891-907. [http://dx.doi.org/10.1093/brain/awv379]

[145] Kaur D, Sharma V, Deshmukh R. Role of microglia and astrocytes in Alzheimer's disease. Int J Neurosci 2019; 129(6): 681-96.

[146] Villemagne VL, Doré V, Burnham SC, Masters CL, Rowe C. Imaging tau and amyloid-β proteinopathies in Alzheimer disease and other conditions. Nat Rev Neurol 2018; 14(4): 225-36. [http://dx.doi.org/10.1038/nrneurol.2018.9]

[147] Gatz M, Reynolds CA, Fratiglioni L, *et al.* Role of genes and environments for explaining Alzheimer's disease. Arch Gen Psychiatry 2006; 63(2): 168-74. [http://dx.doi.org/10.1001/archpsyc.63.2.168]

[148] Livingston G, Huntley J, Sommerlad A, *et al.* Dementia prevention, intervention, and care: 2020 report of the Lancet Commission. Lancet 2020; 396(10248): 413-46. [http://dx.doi.org/10.1016/S0140-6736(20)30367-6]

[149] Norton S, Matthews FE, Barnes DE, Yaffe K, Brayne C. Potential for primary prevention of Alzheimer's disease: an analysis of population-based data. Lancet Neurol 2014; 13(8): 788-94. [http://dx.doi.org/10.1016/S1474-4422(14)70136-X]

[150] Khan SH, Zaidi SK, Hasan M, Ashraf GM. Inflammation and Alzheimer's disease: Mechanistic pathways and therapeutic approaches. Neuropsychiatr Dis Treat 2020; 16: 681-703.

[151] Hou Y, Dan X, Babbar M, *et al.* Ageing as a risk factor for neurodegenerative disease. Nat Rev Neurol 2019; 15(10): 565-81. [http://dx.doi.org/10.1038/s41582-019-0244-7]

[152] Jack CR Jr, Bennett DA, Blennow K, *et al.* NIA-AA Research Framework: Toward a biological definition of Alzheimer's disease. Alzheimers Dement 2018; 14(4): 535-62. [http://dx.doi.org/10.1016/j.jalz.2018.02.018]

[153] Kalaria RN, Akinyemi R, Ihara M. Stroke injury, cognitive impairment and vascular dementia. Biochim Biophys Acta Mol Basis Dis 2016; 1862(5): 915-25.

[http://dx.doi.org/10.1016/j.bbadis.2016.01.015]

[154] Iadecola C. The pathobiology of vascular dementia. Neuron 2013; 80(4): 844-66. [http://dx.doi.org/10.1016/j.neuron.2013.10.008]

[155] Gorelick PB, Scuteri A, Black SE, *et al.* Vascular contributions to cognitive impairment and dementia: A statement for healthcare professionals. Stroke 2011; 42(9): 2672-713. [http://dx.doi.org/10.1161/STR.0b013e3182299496]

[156] Fitzpatrick AL, Kuller LH, Lopez OL, *et al.* Midlife and late-life obesity and the risk of dementia. Arch Neurol 2009; 66(3): 336-42. [http://dx.doi.org/10.1001/archneurol.2008.582]

[157] Profenno LA, Porsteinsson AP, Faraone SV. Meta-analysis of Alzheimer's disease risk with obesity, diabetes, and related disorders. Neurobiol Aging 2010; 31(10): 1516-30.

[158] Pedditzi E, Peters R, Beckett N. The risk of overweight/obesity in midlife and late life for the development of dementia: A systematic review and meta-analysis of cohort studies. Int J Obes 2016; 40(1): 1402-9.

[159] Biessels GJ, Despa F. Cognitive decline and dementia in diabetes mellitus: mechanisms and clinical implications. Nat Rev Endocrinol 2018; 14(10): 591-604. [http://dx.doi.org/10.1038/s41574-018-0048-7]

[160] De Felice FG, Ferreira ST. Inflammation, defective insulin signaling, and neuronal dysfunction in Alzheimer's disease. Alzheimers Dement 2014; 10(1) (Suppl.): S26-32.

[161] Arnold SE, Arvanitakis Z, Macauley-Rambach SL, *et al.* Brain insulin resistance in type 2 diabetes and Alzheimer disease: concepts and conundrums. Nat Rev Neurol 2018; 14(3): 168-81. [http://dx.doi.org/10.1038/nrneurol.2017.185]

[162] Launer LJ, Ross GW, Petrovitch H, Nelson J, Masaki K, Foley D, *et al.* Midlife blood pressure and dementia: the Honolulu–Asia aging study. JAMA 2000; 284(23): 3103-7.

[163] Skoog I, Gustafson D. Hypertension and related factors in the etiology of Alzheimer's disease. Ann N Y Acad Sci 2002; 977(1): 29-36. [http://dx.doi.org/10.1111/j.1749-6632.2002.tb04796.x]

[164] Qiu C, Winblad B, Fratiglioni L. The age-dependent relation of blood pressure to cognitive function and dementia. Lancet Neurol 2005; 4(8): 487-99. [http://dx.doi.org/10.1016/S1474-4422(05)70141-1]

[165] Puglielli L, Tanzi RE, Kovacs DM. Alzheimer's disease: the cholesterol connection. Nat Neurosci 2003; 6(4): 345-51. [http://dx.doi.org/10.1038/nn0403-345]

[166] Cramer PE, Cirrito JR, Wesson DW, *et al.* ApoE-directed therapeutics rapidly clear β-amyloid and reverse deficits in AD mouse models. Science 2012; 335(6075): 1503-6. [http://dx.doi.org/10.1126/science.1217697]

[167] Reitz C. Dyslipidemia and the risk of Alzheimer's disease. Curr Atheroscler Rep 2013; 15(3): 307. [http://dx.doi.org/10.1007/s11883-012-0307-3]

[168] Ownby RL, Crocco E, Acevedo A, John V, Loewenstein D. Depression and risk for Alzheimer disease: systematic review, meta-analysis, and metaregression analysis. Arch Gen Psychiatry 2006; 63(5): 530-8. [http://dx.doi.org/10.1001/archpsyc.63.5.530]

[169] Zunszain PA, Anacker C, Cattaneo A, Carvalho LA, Pariante CM. Glucocorticoids, cytokines and brain abnormalities in depression. Prog Neuropsychopharmacol Biol Psychiatry 2011; 35(3): 722-9. [http://dx.doi.org/10.1016/j.pnpbp.2010.04.011]

[170] McEwen BS. Physiology and neurobiology of stress and adaptation: central role of the brain. Physiol Rev 2007; 87(3): 873-904.

[http://dx.doi.org/10.1152/physrev.00041.2006]

[171] Lupien SJ, McEwen BS, Gunnar MR, Heim C. Effects of stress throughout the lifespan on the brain, behaviour and cognition. Nat Rev Neurosci 2009; 10(6): 434-45. [http://dx.doi.org/10.1038/nrn2639]

[172] Ju YES, Lucey BP, Holtzman DM. Sleep and Alzheimer disease pathology—a bidirectional relationship. Nat Rev Neurol 2014; 10(2): 115-9. [http://dx.doi.org/10.1038/nrneurol.2013.269]

[173] Xie L, Kang H, Xu Q, *et al.* Sleep drives metabolite clearance from the adult brain. Science 2013; 342(6156): 373-7. [http://dx.doi.org/10.1126/science.1241224]

[174] Cataldo JK, Prochaska JJ, Glantz SA. Cigarette smoking is a risk factor for Alzheimer's disease: An analysis controlling for tobacco industry affiliation. J Alzheimers Dis 2010; 19(2): 465-80. [http://dx.doi.org/10.3233/JAD-2010-1240]

[175] Durazzo TC, Meyerhoff DJ, Nixon SJ. Chronic cigarette smoking: implications for neurocognition and brain neurobiology. Int J Environ Res Public Health 2010; 7(10): 3760-91. [http://dx.doi.org/10.3390/ijerph7103760]

[176] Campion D, Dumanchin C, Hannequin D, *et al.* Early-onset autosomal dominant Alzheimer disease: prevalence, genetic heterogeneity, and mutation spectrum. Am J Hum Genet 1999; 65(3): 664-70. [http://dx.doi.org/10.1086/302553]

[177] Strittmatter WJ, Saunders AM, Schmechel D, *et al.* Apolipoprotein E: high-avidity binding to β-amyloid and increased frequency of type 4 allele in late-onset familial Alzheimer disease. Proc Natl Acad Sci USA 1993; 90(5): 1977-81. [http://dx.doi.org/10.1073/pnas.90.5.1977]

[178] Corder EH, Saunders AM, Strittmatter WJ, *et al.* Gene dose of apolipoprotein E type 4 allele and the risk of Alzheimer's disease in late onset families. Science 1993; 261(5123): 921-3. [http://dx.doi.org/10.1126/science.8346443]

[179] Goate A, Chartier-Harlin MC, Mullan M, *et al.* Segregation of a missense mutation in the amyloid precursor protein gene with familial Alzheimer's disease. Nature 1991; 349(6311): 704-6. [http://dx.doi.org/10.1038/349704a0]

[180] Calderón-Garcidueñas L, Reed W, Maronpot RR, *et al.* Brain inflammation and Alzheimer's-like pathology in individuals exposed to severe air pollution. Toxicol Pathol 2004; 32(6): 650-8. [http://dx.doi.org/10.1080/01926230490520232]

[181] Itzhaki RF, Lin WR, Shang D, Wilcock GK, Faragher B, Jamieson GA. Herpes simplex virus type 1 in brain and risk of Alzheimer's disease. Lancet 1997; 349(9047): 241-4. [http://dx.doi.org/10.1016/S0140-6736(96)10149-5]

[182] Balin BJ, Gérard HC, Arking EJ, *et al.* Identification and localization of Chlamydia pneumoniae in the Alzheimer's brain. Med Microbiol Immunol (Berl) 1998; 187(1): 23-42. [http://dx.doi.org/10.1007/s004300050071]

[183] Exley C. Aluminium should now be considered a primary etiological factor in Alzheimer's disease. J Alzheimers Dis Rep 2017; 1(1): 23-5. [http://dx.doi.org/10.3233/ADR-170010]

[184] Liu Y, Chen K, Li F, Gu Z, Liu Q, He Q, *et al.* Lead exposure and its association with dementia risk: a systematic review and meta-analysis. Sci Total Environ 2020; 707135333

[185] Chen P, Miah MR, Aschner M. Metals and neurodegeneration. F1000Res 2016 ;5: F1000 Faculty Rev-366.

[186] Wu Y, Zhang Y, Wang L, Wang Y, Xu D. Exposure to air pollution and cognitive decline: A systematic review and meta-analysis of prospective cohort studies. Environ Int 2021; 147: 106357.

[187] Maher BA, Ahmed IAM, Karloukovski V, *et al.* Magnetite pollution nanoparticles in the human brain. Proc Natl Acad Sci USA 2016; 113(39): 10797-801.
[http://dx.doi.org/10.1073/pnas.1605941113]

[188] Block ML, Calderón-Garcidueñas L. Air pollution: mechanisms of neuroinflammation and CNS disease. Trends Neurosci 2009; 32(9): 506-16.
[http://dx.doi.org/10.1016/j.tins.2009.05.009]

[189] Riediker M, Devlin RB, Griggs TR, *et al.* Cardiovascular effects in patrol officers are associated with fine particulate matter from brake wear and engine emissions. Part Fibre Toxicol 2004; 1(1): 2.
[http://dx.doi.org/10.1186/1743-8977-1-2]

[190] Wilker EH, Preis SR, Beiser AS, *et al.* Long-term exposure to fine particulate matter, residential proximity to major roads and measures of brain structure. Stroke 2015; 46(5): 1161-6.
[http://dx.doi.org/10.1161/STROKEAHA.114.008348]

[191] Power MC, Adar SD, Yanosky JD, Weuve J, Kelly J, Hajat A, *et al.* Exposure to air pollution and cognitive function, cognitive decline, and dementia: a systematic review. Environ Health Perspect 2016; 124(3): 416-22.

[192] Peters R, Ee N, Peters J, Booth A, Mudway I, Anstey KJ. Air pollution and dementia: A systematic review. J Alzheimers Dis 2019; 70(s1): S145-63.
[http://dx.doi.org/10.3233/JAD-180631]

[193] Cacciottolo M, Wang X, Driscoll I, *et al.* Particulate air pollutants, APOE alleles and their contributions to cognitive impairment in older women and to amyloidogenesis in experimental models. Transl Psychiatry 2017; 7(1)e1022
[http://dx.doi.org/10.1038/tp.2016.280]

[194] Costa LG, Cole TB, Dao K, Chang YC, Garrick JM. Developmental impact of air pollution on brain function. Neurochem Int 2019; 131: 104580.
[http://dx.doi.org/10.1016/j.neuint.2019.104580]

[195] Ranft U, Schikowski T, Sugiri D, Krutmann J, Krämer U. Long-term exposure to traffic-related particulate matter impairs cognitive function in the elderly. Environ Res 2009; 109(8): 1004-11.
[http://dx.doi.org/10.1016/j.envres.2009.08.003]

[196] Power MC, Weisskopf MG, Alexeeff SE, Coull BA, Spiro A III, Schwartz J. Traffic-related air pollution and cognitive function in a cohort of older men. Environ Health Perspect 2011; 119(5): 682-7.
[http://dx.doi.org/10.1289/ehp.1002767]

[197] Cacciottolo M, Morgan TE, Saffari A, Shirmohammadi F, Forman HJ, Sioutas C, *et al.* Traffic-related air pollutants alter brain structure and function in mice. PLoS One 2014; 9(7)e101431

[198] Calderón-Garcidueñas L, Mora-Tiscareño A, Ontiveros E, *et al.* Air pollution, cognitive deficits and brain abnormalities: A pilot study with children and dogs. Brain Cogn 2008; 68(2): 117-27.
[http://dx.doi.org/10.1016/j.bandc.2008.04.008]

[199] Block ML, Elder A, Auten RL, *et al.* The outdoor air pollution and brain health workshop. Neurotoxicology 2012; 33(5): 972-84.
[http://dx.doi.org/10.1016/j.neuro.2012.08.014]

[200] Zhao Z, Zhang Y, Liu H, Yan H, Wang Y, Liu D, *et al.* Cadmium induces Alzheimer-like pathology in rat hippocampus. Biol Trace Elem Res 2017; 177(2): 295-302.

[201] Walton JR. Aluminum involvement in the progression of Alzheimer's disease. J Alzheimers Dis 2010; 20(2): 403-13.

[202] Liu CM, Ma JQ, Sun YZ. Protective role of puerarin on lead-induced alteration of synaptic plasticity in rat hippocampus. Toxicol Appl Pharmacol 2012; 263(2): 274-82.

[203] Tanzi RE. The genetics of Alzheimer disease. Cold Spring Harb Perspect Med 2012; 2(10): a006296.

[http://dx.doi.org/10.1101/cshperspect.a006296]

[204] Calderón-Garcidueñas L, González-Maciel A, Mukherjee PS, *et al.* Combustion- and friction-derived magnetic air pollution nanoparticles in human hearts. Environ Res 2019; 176108567 [http://dx.doi.org/10.1016/j.envres.2019.108567]

[205] Fagan AM, Mintun MA, Mach RH, *et al.* Inverse relation between in vivo amyloid imaging load and cerebrospinal fluid $A\beta_{42}$ in humans. Ann Neurol 2006; 59(3): 512-9. [http://dx.doi.org/10.1002/ana.20730]

[206] Hampel H, Frank R, Broich K, *et al.* Biomarkers for Alzheimer's disease: Alzimers Disease, industry and regulatory perspectives. Natl Rev 2010; 9: 560-74.

[207] World Health Organization. Dementia, Geneva: World Health Organization; 2017.

CHAPTER 3

Role of Biomarker in Treating Alzheimer's Disease

Piyush Anand[1], **Deepak Kumar**[1], **Supriya Gupta**[2], **Juhi Tiwari**[1] and **Shashi Kant Singh**[1,*]

[1] *Faculty of Pharmaceutical Sciences, Mahayogi Gorakhnath University, Gorakhpur, Uttar Pradesh 273007, India*

[2] *Faculty of Nursing & Paramedical, Mahayogi Gorakhnath University, Gorakhpur, Uttar Pradesh 273007, India*

Abstract: Alzheimer's Disease (AD) is a multifactorial neurodegenerative disease associated with progressive decline in cognitive functions. Recent advances have been made in diagnostic techniques for AD by using Biomarkers. Biomarkers are not mere indicators of their pathophysiology but are the key to optimizing and making clinical decisions and therapeutic strategies. Biomarkers have thus become crucial elements in the diagnosis, monitoring, and treatment of AD. Early detection of AD is also essential for interventions, and biomarkers in AD, such as the development of senile plaques through aggregation of amyloid-beta and neurofibrillary tangle formation through tau proteins, and in recent years, new trends have been made towards the detection of Ocular biomarkers are considered important contributors in the diagnosis of AD. Ocular biomarker is an important biomarker because the nervous system of the eyes is considered an extension of the CNS. So, biomarkers help in distinguishing the cause of dementia in cognitive disorders like AD and thereby provide a target for therapeutic strategies. As the disease progresses, levels of biomarkers can be monitored for clinical manifestation, thus informing healthcare providers about disease trajectories in patients and changing the therapeutic approach correspondingly. "For instance, *via* examination of the cerebrospinal fluid, neuroimaging, or related techniques, valuable information on the pathology of AD may be obtained, which helps in tracking neurocognitive decline. Biomarkers also play a critical role in the assessment of treatment efficacy". Objective measures, reduction in amyloid burden as measured by PET scans, can be an indication of good response to amyloid-targeting therapies and direct further treatment. Further studies on new biomarkers, including neuroinflammatory markers and genetic variants, will provide further refinement of personalized treatment approaches, thereby helping improve the outcomes of the patients. To enhance the appropriate early diagnosis and proper monitoring plus targeted treatment interventions, these biomarkers should be integrated into the clinical care of AD. It is now clear from other various studies that this might change the way AD is treated and improve the standard of care.

[*] **Corresponding author Shashi Kant Singh:** Faculty of Pharmaceutical Sciences, Mahayogi Gorakhnath University, Gorakhpur, Uttar Pradesh 273007, India; E-mails: principal.fps@mgug.ac.in

Prashant Kumar Dhakad, Raghav Mishra, Shweta Sharma, Sudhanshu Mishra & Nishant Gaur (Eds.)

Keywords: Alzheimer's disease, Amyloid-beta, Biomarkers, CNS, CSF, Dementia, Genetic variants, Neuroimaging, Ocular Biomarkers, Tau proteins.

INTRODUCTION

Alzheimer's Disease (AD) remains a public health challenge, with a current prevalence of 10% in the 65 to 74-year-old population and up to 40% post age 85. At the molecular level, AD is characterized by extracellular aggregation of beta-amyloid plaques; intracellular aggregation of hyperphosphorylated tau protein, resulting in the formation of neurofibrillary tangles; synaptic dysfunction; and neurodegeneration, which impairs cognitive function. It has been more than a century since Alzheimer's disease was discovered, and we have a sophisticated understanding of disease pathology and progression. However, due to the lack of a complete understanding of the etiology, development, and toxic pathways, the clinical toolkit for diagnosing and treating patients with AD today is still very limited [1, 2]. Currently, there is no cure for this disease. Disease-modifying treatments could theoretically be established before the onset of symptoms or at an early stage, but this may necessitate the use of risk-linked biomarkers for therapeutic trials and interventions. The discovery of these risk-linked biomarkers is in part due to a deeper understanding of the disease and the subsequent identification and development of potential AD therapies [3, 4].

The discovery of biomarkers of pathophysiological alterations, often referred to as 'amyloid toxicity' or 'neurodegeneration', has modified the pathological framework of AD, which was initially based on the clinical, imaging, and neuropathological evidence of Aβ deposition and NFT buildup in animal models and individuals. These insights allow a much earlier and more accurate detection and risk estimation of AD dementia—the main clinical target being tests for Aβ and tau under an increasing global challenge resulting from the growing underestimation and underdiagnosis of the disease [5]. This review describes the roles and the required characteristics of risk-linked and diagnostic AD biomarkers in relation to the different predefined stages of the disease, detailing several alternative methods through which these biomarkers can be detected in various biological matrices; it refers to the complementary use that these biomarkers should have for correctly diagnosing an individual in routine clinical practice and during and after clinical trials, and presents an algorithm based on a new staging system of AD that underlines three preclinical and three dementia-specific stages in these cases [6, 7].

A Brief on Alzheimer's Disease

Alzheimer's Disease (AD), which affects up to two-thirds of dementia cases, is characterized by the appearance of two classes of brain lesions in histopathological examination, namely senile plaques and neurofibrillary tangles. Plaques are primarily compounds of insoluble beta-amyloid, whereas the neurofibrillary tangles are composed of an abnormal form of a microtubule-associated protein. The current breakthrough descriptions of AD pathophysiology have centered on changes secondary to the sarcoplasmic-endoplasmic reticulum [8]. The most important of these descriptions is secondary impairment of intraneuronal calcium homeostasis; this has served as the foundation of a firm biochemical AD model. The development makes possible advances beyond post-mortem secular AD practice and understanding pre-mortem AD biology. In combination with other data suggesting a critical role of the endoplasmic reticulum, calcium biology, tau biology, inflammation, oxidation, or glucose and lipid utilization in the predisposition to cognitive disorders, the development forms the pillars of a credible hypothesis regarding the early pathogenesis of cognitive disorders [9, 10].

Current attempts to treat AD are controversial. They rely on a "monolithic" model where dementia is the endpoint of pathology and all individuals may benefit from slowing the progression of pathology regardless of etiology. Finally, the "monolithic" model does not require population risk stratification. While this is a convenient public health model, it contradicts our concepts of geriatric medicine and personalized AD treatment. The recent description of sarcoplasmic-endoplasmic reticulum calcium homeostasis impairment as a signature of AD biology and as an early event associated with the biological pathogenesis of cognitive disorders may represent a substantial advance in cognitive disorder care. In combination with other data suggesting critical roles for the endoplasmic reticulum, tau biology, inflammation, glucose, and lipid metabolism in the predisposition to cognitive disorders, the endoplasmic reticulum hypothesis may serve as a major regulatory tool, guiding the development of pertinent biomarkers toward early identification of preclinical AD biology and targeted early AD treatment [11, 12].

Biomarkers as an Efficient Tool in AD Management

Introduction and Aim: Accurate and efficient tools for diagnosing and treating Alzheimer's disease (AD) are essential for controlling the economic and medical burden of AD and for extending a patient's quality of life. The review article aims to create an in-depth and comprehensive resource for clinicians, researchers, and policymakers to understand the role of cerebrospinal fluid (CSF), plasma, and neuroimaging biomarkers in the management of AD. Such an understanding will help to implement the current and future biomarkers efficiently and successfully at the individual, local, and population levels. Methods: The author performed systematic literature searches on databases using relevant Medical Subject Headings (MeSH) and related search terms. The thematic search focused on the role of biomarkers, mainly concentrated on specific CSF and plasma markers and neuroimaging modalities, which include amyloid and tau pathology tracking, neural structure, and glucose metabolism in relation to Alzheimer's disease, either independently or through comparative studies with a broader range of related themes including diagnostic and prognostic tools, treatment decisions, efficiency, cost-effectiveness, acceptance, and implementation. Fig. (**1**) basically depicts the role of biomarkers in physiology, diagnosis, and management of AD. This figure was prepared by using ChemDraw [13, 14].

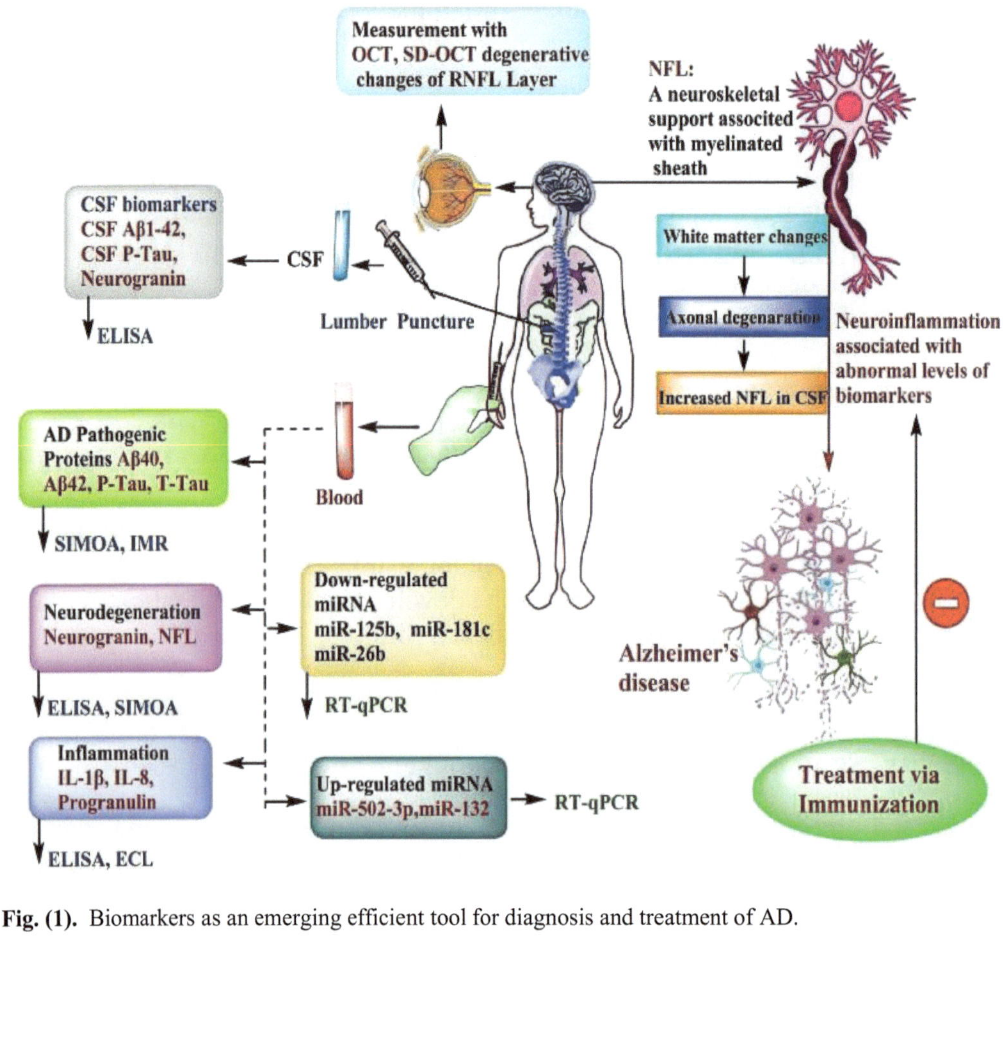

Fig. (1). Biomarkers as an emerging efficient tool for diagnosis and treatment of AD.

CLASSIFICATION OF BIOMARKERS IN AD

Biomarkers have been defined as "a characteristic that can be objectively measured and evaluated as an indicator of normal biological processes, pathogenic processes, or pharmacologic responses to a therapeutic intervention." This definition is broad, as it also includes the practical concept of "surrogate biomarkers" (disease activity and risk, pharmacologic response; defining clinical disease and staging). In Alzheimer's disease (AD), biomarkers represent objective means toward a more accurate diagnosis: since both the clinical and neuropathologic criteria for the diagnosis of AD are relatively nonspecific, and the presentation of AD varies substantially, several research efforts have looked to imaging and biochemical biomarkers to increase the ability to diagnose AD accurately and early in its course to identify preclinical AD for the enrollment of very early AD patients in clinical trials [15, 16].

Biomarkers can be classified in different ways. When researchers refer to "biomarkers" of AD or any other condition, they are often referring to biomarkers of AD pathogenesis, that is, to indicators of the primary phenomena that lead to the disease or condition. These can have several forms: the genes whose abnormal regulation causes the disease, the products of these genes, examples of products of genes able to interfere with processes involved in the initiation or progression of AD, such as the Aβ peptide, or to induce neuronal death, noise-oriented genes, ceramide, dysregulation of ceramide kinase, *etc*. These markers are the primary targets for the development of disease-modifying drugs, which are directed against these early phenomena in the hope that blocking them will block the progression of the disease and neuronal death, in order to modify the course of AD [17, 18].

TYPES OF "BIOMARKERS"

Biomarkers are needed to define the Alzheimer's disease process in the brain by use of molecular, structural, and/or functional characteristics that can be assessed either *In vivo* or post-mortem. Our concept of what the term "biomarker" includes has been changing in recent years. The traditional use of the term as a characteristic that can be objectively measured as an indicator of normal biological, pathogenic processes, or pharmacologic responses to a therapeutic intervention is still important, but advances in usable technologies for characterizing the molecular content of blood, cerebrospinal fluid, and urine, and in brain imaging make it possible to establish characteristics at multiple levels. The term has thus been expanded in its broadest sense and is applied to fluid and imaging characteristics that likely provide a measure of the extent of brain tissue injury. The use of biomarkers in the progression of Alzheimer's disease is described in a conceptual framework with proposed stages for Alzheimer's disease research and clinical development [19, 20].

The term "biomarker" typically refers to a characteristic that is objectively measured and evaluated as an indicator of normal biological processes, pathogenic processes, or pharmacologic responses to a therapeutic intervention. Biomarkers are "traits" that represent the change in biological systems and express a change in health status. Biomarkers provide the ability to identify characteristics of the progression of disease that can be used to diagnose the early onset of disease and to monitor the progression of disease throughout life. They also provide tools to better understand how the progression of disease occurs and why. The development and validation of biomarkers at the molecular, structural, and functional levels, including cerebrospinal fluid and/or peripheral blood for research related to dementia, are addressed. Therefore, in this paper, we address biomarkers from the perspective of those that provide the ability to monitor disease progression. Table **1** is describing about different biomarkers, their method of diagnosis, the specificity used in the diagnosis and treatment of AD [21, 22].

Table 1. Various biomarkers used in the diagnosis and treatment of AD.

Sr. No.	Biomarker	Method of Diagnosis	Specificity	Reference
1.	Amyloid Beta	Measures the Amyloid beta protein in CSF through lumbar puncture by using a PET scan to visualize brain plaques.	80-90%	[23, 24]
2.	Tau Protein	A lumbar puncture can "evaluate the amount of phosphorylated tau (p-tau) in cerebrospinal fluid (CSF), and more recent blood tests (such as p-tau217) can identify aberrant tau aggregates associated with Alzheimer's disease".	48% to 88%	[25]
3.	S100B	An early-stage Alzheimer's disease biomarker may be found in cerebrospinal fluid (CSF) with high S100B levels.	17-20%	[26]
4.	Advanced glycosylation end products	To diagnose Alzheimer's disease using advanced glycosylation end products (AGEs), researchers typically measure the level of "AGEs in a patient's blood serum or cerebrospinal fluid (CSF) using techniques like ELISA (enzyme-linked immunosorbent assay) or immunochemistry", as elevated levels of AGEs are associated with the disease and can be detected in brain tissue	50-60%	[27, 28]

(Table 1) cont.....

Sr. No.	Biomarker	Method of Diagnosis	Specificity	Reference
5.	Oxidized Low-Density Lipoprotein	Measuring oxLDL may be a "factor in determining the risk of Alzheimer's disease, and elevated oxLDL levels may be linked to an increased risk of the disease".	17-20%	[29]
6.	TNF-alpha	While a direct "TNF-alpha test" isn't considered a conclusive diagnostic technique for Alzheimer's disease, measuring levels of Tumor Necrosis Factor-alpha (TNF-alpha) in blood or cerebrospinal fluid (CSF) by an enzyme-linked immunosorbent assay (ELISA) can be used as a supportive signal	50-60%	[30]
7.	Neurofilament Light Chain (NfL)	The amount of NfL protein in the blood is measured by a neurofilament light chain (NfL) test, which may be a useful biomarker for Alzheimer's disease diagnosis and progression tracking.	70-90%	[31]
8.	C Reactive Protein	A higher risk of Alzheimer's disease is "increasingly associated with evidence of inflammation in the body, which elevated CRP levels can be utilized to detect".	Less than 50%	[32, 33]
9.	Interleukin -6	An increase in level shows the presence of diseases.	23-35%	[34]
10.	Brain-derived Neurotrophic Factor	An elevated risk for Alzheimer's disease may be indicated by low blood levels of BDNF, which may also serve as a potential marker for tracking the illness's course.	33%	[35]

BIOLOGICAL MARKERS

Mutations in certain genes are associated with a higher risk of developing Alzheimer's disease, especially those that encode the amyloid protein precursor, apolipoprotein E, and the presenilins. Thus, the presence of these mutations can also be considered a biological marker of this pathology [36]. The scientific advances in translational administrative processes have made it possible to use a series of substances as imaging markers or biological materials to indicate the probability of the onset of this scourge. These include established biomarkers such as the detection of certain proteins in cerebrospinal fluid or the use of specific molecules that bind to functional and structural elements of the disease's protein signature as a reference for imaging, advancing in different time spans and with different levels of invasiveness of each measure [37, 38].

Recent advances in the understanding of the pathophysiological processes that support the appearance of clinical alterations of Alzheimer's disease suggest that the success of treatments aimed at overcoming these processes will be greater the earlier initiation occurs. For this reason, it is reasonable to think that reliable methods that allow us to recognize biological markers of the preclinical stage or its immediate antecedent will have a marked positive impact on the evaluation and subsequent treatment of this increasingly frequent disorder [39]. This article reviews the role that imaging and biological markers play in the current algorithm used for diagnosing Alzheimer's disease. Data on the composition of the main Alzheimer's disease syndromes are reviewed, with special emphasis on the influence of the temporal sequence of the appearance of different clinical, biological, and imaging variables on these profiles. The unavailability of these markers is discussed, and the proposed beneficiary and demand population is identified, both if these markers were available in the clinical setting or strictly in the research environment. The final section evaluates the utility of studying genetic and biological markers in the diagnosis and management of Alzheimer's disease, as well as their possible future applications [40, 41].

IMAGING "BIOMARKERS"

The role of imaging findings in the investigation of Alzheimer's disease began about four decades ago, with the introduction of computed tomography as a clinical examination method for the head. Magnetic resonance imaging of the brain improved our diagnostic potential, not only by signal modifications on neuroimaging, but also with the use of diffusion criteria for various brain regions and the cerebrospinal fluid in the differential diagnosis [41]. The development of radioisotope imaging using single-photon emission or positron emission tomography allowed us to recognize the lack of molecular imaging. The use of cerebral perfusion modifications, starting with SPECT using tracer agents and moving on to fluorodeoxyglucose while using PET, provides not only a means of differential diagnosis but also support for the precision of the neuropsychological assessment. Finally, thanks to the advances in genetic, molecular biology, and molecular imaging fields, it has been proven that the focal amyloid deposits can be identified *In vivo*, which allowed for the creation of classification criteria for Alzheimer's disease [42, 43].

GENETIC BIOMARKERS

The identification of a combination of genetic, imaging, and cerebrospinal fluid (CSF) markers that sharply measure the conversion from prodromal Alzheimer's Disease (AD) to AD dementia, even before CSF Abeta reaches the abnormal level, in the general population is important for the selection of subjects for secondary prevention trials and to optimize the effect of AD-modifying drugs [44]. The changes of CSF Ab1-40 and Ab1-42 are the most sensitive biochemical markers of AD pathology, as the Abeta accumulation is the earliest biomarker event during the pathogenesis of sporadic AD. Since cerebrospinal fluid (CSF)

Abeta 1-42 is considered to reflect the cognitive alterations and conversion from MCI to AD, by contrast. Several studies reported a very useful model, based on A1-42, for the recognition of subjects at increased risk for developing MCI or dementia due to probable AD [45, 46].

Genetic biomarkers of AD have received great attention as AD susceptibility genes, such as apolipoprotein E (APOE), ABCA7, SORL1, and PICALM, are currently being intensively studied in European populations [47]. Some nonsynonymous mutations in the TREM2 gene are reported to be associated with a significantly increased risk of AD. Such genetic biomarkers will become extremely valuable for the generation of risk assessment schemes and strategies for the diagnosis and detection of people at risk of developing AD, even in the absence of any significant cognitive decline [48]. In the future, studies with a larger sample size and greater representativeness of Japanese populations are needed to comprehensively screen the roles of major and minor genetic variants in these neuroinflammatory pathways that ultimately affect AD initiation [49, 50].

EMERGING BIOMARKERS IN ALZHEIMER'S RESEARCH

The search for biomarkers of Alzheimer's Disease (AD) is an important area of research in both diagnostic and therapeutic terms. Remarkable advances have been made in the field, especially in the determination of the levels of Aβ and tau protein. In addition, neuropsychological tests, cerebrospinal fluid, and neuroimaging are commonly used to help make a diagnosis. These diagnostic approaches are highly specialized and difficult to carry out, and the tools depend on the country's infrastructure and technology. Finding biomarkers that are easily accessible and can perform this type of classification or any kind of diagnostic workup is of utmost importance and would improve the way that treatments are provided to AD patients. This review covers the state of the art in recent biomarkers that reflect signs of inflammation, lipidosis, or energy metabolism. Many new and accessible biomarkers have been proposed or discovered for the diagnosis of dementia or the cognitive decline that predicts AD [51].

While these new biomarkers may be useful, not all of the methods proposed are reliable and consistently yield the same results for each patient. Nevertheless, AD biology has mainly considered proteins, and no study to date has reported the usefulness of these markers to follow a therapeutic progression or treatment. These biomarkers require validation by clinical trials, which can be very arduous [52]. These problems make it difficult to reproduce any research findings and to validate the results obtained. All biomarkers not validated by these parameters should not be considered biomarkers or diagnostic tools to validate using a clinical trial. Currently proposed diagnostic tools are great aids, not an irrefutable method for diagnosing or treating Aβ. AD will still be diagnosed early based on the symptoms described by patients and relatives. Symptoms, neuroimaging, and other parameters are highly relevant for its construction, complementary diagnosis, or medical intervention for stakeholders. Table **2** is describing about current updates on available therapeutic agents targeting different biomarkers for AD treatment [53, 54].

Table 2. Current updates on different biomarkers for AD treatment.

Sr. No.	Therapeutic Agents	Targeted Biomarker	Mechanism of Action	Reference
1.	Donepezil	Acetylcholinesterase (AChE), Tau Protein, Amyloid Beta	Inhibits the breakdown of acetylcholine in synapses, increasing cholinergic neurotransmission	[55]
2.	Rivastigmine	Tau Protein, Neuroinflammation, Amyloid Beta	Rivastigmine may also impact tau protein phosphorylation and aggregation, increase BDNF levels, and Reduce Oxidative stress.	[56]
3.	Galantamine	TNF-α, IL-6, Aβ, α7nAChR	It lowers the expression of proinflammatory cytokines	[57]
4.	Aducanumab	Phosphorylated tau (p-tau), Amyloid-beta (Aβ) plaques, Tau PET imaging	It "targets phosphorylated tau (p-tau) and amyloid-beta (Aβ) plaques in the brain and cerebrospinal fluid (CSF)".	[58]
5.	Lecanemab	Aβ	It targets the amyloid-beta (Aβ) protein in the brain	[59]
6.	Verubecestat	(Aβ)40, Aβ42	It decreases the levels of amyloid-beta (Aβ)40, Aβ42, and sAPPβ in the cerebrospinal fluid (CSF	[60]
7.	Memantine	Glutamate excitotoxicity and extrasynaptic NMDARs	Memantine inhibits the effects of glutamate, a substance that communicates between brain nerve cells.	[61]
8.	Resveratrol	Aβ40, Aβ42, tau, and phospho-tau 181	Increase the clearance of beta-amyloid	[62]
9.	Gantenerumab	Tau (t-tau), Aβ, NfL	It targets phosphorylated tau (p-tau) and amyloid-beta (Aβ) plaques in the brain and cerebrospinal fluid.	[63]
10.	NSAIDs	Amyloid-β (Aβ), cyclooxygenase (COX) enzymes, and neuroinflammation.	Reduces the Amyloid Beta Production, Inhibits Amyloid Beta Aggregation	[63]

BIOMARKER'S ROLE IN THE DIAGNOSIS OF AD

Diagnosis is the starting point of the journey of memory loss. With the help of excellent clinicians in the memory clinic, caregivers can obtain a reliable diagnosis and also valuable treatments and care if they are suspected of having Alzheimer's disease. Both the diagnosis and differential diagnosis of dementia and the clinical drugs approved for Alzheimer's disease were recalled. Patients and families affected by Alzheimer's should be mindful. No drug can alter the outcome of Alzheimer's disease; the expectation is that the signs of Alzheimer's be suppressed and that existing functions be maintained with drug therapy. If a test produces a measure of change in brain or cognitive function over time, it should be considered in the biomarker synergy model [64, 65].

Accumulation of objective biomarker data would be the foundation of a future personalized, family-centered, efficient intervention and monitoring for the prevention of Alzheimer's disease. In both research and clinical settings, the use of biomarkers as research tools should inevitably be centered on a more functional clinical model that correlates Alzheimer's disease with variations in brain and cognitive capability known to occur over time. What is it that we want to do about Alzheimer's today and over the next two decades most urgently? The benefit of symptom management is becoming obvious [66]. To more immediately identify and refine care best practices, we need to use biomarkers at this point to coordinate various bedside diagnostic observations into a comprehensive model of patients. To focus on our treatment objectives and the patients we treat, we need to use a biomarker synergy model. Ultimately, most of the research questions that remain are medical, genetic, and therefore quantitative. We must focus our efforts on questing for reliable functional biomarkers. Physicians, patients, and families should all come together to solve the mystery of Alzheimer's disease [67, 68].

Detection at early stages of Alzheimer's Disease

Detection at early stages of Alzheimer's disease is essential for future therapeutic interventions, as it would allow smaller and more precise target populations. Of paramount importance, it could correspond to the phase of the disease where currently available treatments may still have a chance to be effective in delaying the course of the disease. In the last eight years, research on Alzheimer's disease pathology and its dynamics throughout the natural course of the disease has identified some critical periods for therapeutic intervention. According to the two-stage model of Alzheimer's biomarker progression, the disease currently has two important levels of severity: the first level corresponds to biomarker positivity, usually when there is detectable Alzheimer's disease pathology as seen by the appearance of fibrillar Aβ amyloid in PET or a change in cerebrospinal fluid from below to above a cut-off level [69, 70].

This event may happen ten to fifteen years before the symptomatic phase of the disease. Moreover, good precision in cut points can be obtained not by dynamics of progression or disease spreading, but by measuring consecutive thresholds that correspond to an abnormality in the respective biological marker, such as a biophysical limit detected by PET when the concentration of fibrillar Aβ is above a certain level [71]. The second threshold is the level of significant cognitive decline to the level of mild cognitive impairment, which will be detected later by neuroimaging techniques or other cognitive assessments. The study demonstrates that, despite methodological constraints, important final conclusions can be obtained. It showed that good sensitivity and specificity are achieved by the quantification of CSF proteins with an AUC between certain values [72, 73].

Differentiating Alzheimer's from FTD and Other Dementias

Differential diagnosis is also important, as different types of dementia have different courses and treatment plans. Common symptoms of FTLD include changes in personality and behavior or difficulties with language. Both AD and FTLD share difficulties with memory and motor functions [74]. There are a couple of different methods to differentiate FTLD from AD. FTLD is usually diagnosed in patients in their late 40s and 50s as opposed to late-onset AD, which occurs in the 60s, 70s, and 80s. If the patient has lost interest in hobbies that he or she used to enjoy, has bad hygiene, a decrease in sympathy or empathy, a change in eating habits, and language problems, the patient may have FTLD [75, 76].

Methods to differentiate FTLD from AD by neuropsychological testing and different types of examinations to observe and test the patient's behavior, such as naming objects, changing the task, attentive behaviors, the response to commands, and changing the question for naming objects [77]. However, these methods are simple observations, and the results are sometimes not consistent among observers. Diagnosis. Try to diagnose if the patient may have AD, vascular dementia, or AD with cerebrovascular disease [78, 79].

Biomarkers Use in Clinical Guidelines

Nowadays, we have one set of general guidelines that are established rules and regulations that a buyer and seller agree on before a transaction, concerning the significance of the use of biomarker evidence in Alzheimer's disease. The working groups responsible for coordinating the next biomarker-related work recommend that cerebrospinal fluid tests for protein or amyloid imaging commence in limited cases and show evidence of amyloid deposition. These guidelines for biomarkers in Alzheimer's disease are prefixed by the statement that they were using only current test results. They review the role of Alzheimer's disease-related trends during biomarker progression until finally establishing a diagnosis, noting an increase in work focusing on preclinical Alzheimer's disease as well as the lack of established clinical guidelines for when and how to use biomarkers before the diagnosis of clinical Alzheimer's disease, especially in the primary care clinical setting [80]. A second set of guidelines was developed through a structured consensus process based on a grading system. This document forms a framework for implementing appropriate investigational drugs and biometric guidelines when new knowledge is gained while studying Alzheimer's disease. The guidelines seek to encourage not just a symmetrical review but also open questions and promote the legitimate use of the terminology preclinical Alzheimer's disease in a group of patients who are similar to those term clinical Alzheimer's disease. They describe patient selection, a diagnostic algorithm, and patient care. The aim of these guidelines is explicitly to enable a faster response to developments in funding, regulations, and science, and thus to cover a wider range of knowledge for preclinical disease stages [81, 82].

DISEASE PROGRESSION MONITORING AND MANAGEMENT

Diseases like diabetes and hypertension greatly increase the risk of developing Alzheimer's disease in the future, and the management of these diseases and risk factors can help to reduce the chances of getting Alzheimer's disease. Currently available treatments can relieve some symptoms and generally help slow the pace of cognitive decline. However, these drugs only help maintain cognitive function for a limited time, and they do not help stop the progression of the disease. Research to investigate the different phases of disease development and to identify the warning signs will help to find solutions to slow down the progression of Alzheimer's disease and reduce the risk of developing the disease in the future. New approaches to the development of drugs that are based on understanding the pathophysiological pathways of Alzheimer's disease exhibit potential for effective intervention. Biomarkers that can accurately detect Alzheimer's disease conditions have been identified and are being used in clinical trials [83].

Monitoring the progression of diseases by measuring indicators of various changes related to health is important for making decisions regarding the treatment approach, and some prognostic and diagnostic tests are utilized in the management process. In particular, in the case of chronic diseases, monitoring the progression of the disease's status at the earliest possible time is important for healthcare treatment. Conditioning these biomarkers on their highly predictive capabilities of predicting the possibility of developing diseases and related problems is important for developing personalized therapy programs before the development of a complex condition [84]. In terms of healthcare services and their impact, early prediction of developing and managing health services for people regarding their future conditions is important, as it can help the relevant authorities in healthcare services find resources and inform stakeholders and policymakers about their medical planning as required. Early information about the related disease progression is an important issue for empowering individuals, as it helps in ultimately controlling healthcare expenditures. With the development of scientific research, researchers often discover and use many different types of biomarkers, but the uptake of these biomarkers for use in clinical practice is a difficult and time-consuming process. In addition, legal, ethical, or economic challenges may limit the availability and use of these biomarkers [85, 86].

Biomarkers as cognitive impairment trackers

The main clinical need for Alzheimer's disease is a biomarker to follow cognitive impairment or to track disease progression. Biologically, mild cognitive impairment sits between normal older adults and Alzheimer's disease, so it is important to determine if different molecular differences can distinguish the various types of mild cognitive impairment. However, at present, there are no accurate biomarkers to detect it, but only clinical diagnosis through evaluation of cognitive characteristics [87, 88].

In individuals with prodromal Alzheimer's disease, it has been demonstrated that cerebrospinal fluid concentration of tau-based biomarkers, for example, total tau, phosphorylated tau, or Aβ42, is a good biomarker that can accurately predict future clinical diagnosis. Patients with a greater level of tau-based biomarkers also have a more severe level of atrophy, hypometabolism, and impairment. Furthermore, tau-based biomarkers are supposed not only to detect early changes but also to improve differential diagnosis. Many studies have demonstrated that increased levels or a decrease of Aβ42 within the cerebrospinal fluid in all patients with probable Alzheimer's disease can be used as a diagnostic criterion [89, 90].

However, there is less difference between patients with non-Alzheimer's disease and patients with Alzheimer's disease, with only some amyloid-β plaques in the brain. Moreover, a significantly decreased level of T-tau in the cerebrospinal fluid and a low level of Aβ42 have been reported in patients with early or prodromal Alzheimer's, which is not sufficient to be used as a diagnostic test in a clinical setting due to the stage of disease tissues and other confounding co-pathologies [91]. Currently, the diagnosis is essentially clinical and based on progressive memory loss and executive functions using a combination of imaging and molecular biomarkers. Furthermore, the CSF is a more concentrated version of the fluid surrounding the brain. In this study, we can show the utility of protein markers for progressive mild cognitive impairment and distinguishing mild cognitive impairment subtypes [92, 93].

Biomarkers as an indicator for neurological changes

Biomarkers have become a popular area in the study of neurodegenerative diseases, including Alzheimer's disease. Biomarkers are proxies for the pathophysiological changes associated with Alzheimer's disease. Some biomarkers can predict the disease at preclinical stages, where clinical symptoms have not yet developed. Alzheimer's disease is the most common type of dementia and is currently a huge burden for affected individuals, but also their families. The cause of Alzheimer's disease is a large number of senile plaques and neurofibrillary tangles. One of the limitations of the treatments for patients with Alzheimer's disease is that the medication is a symptomatic treatment. Therefore, early detection and intervention are important. Cerebrospinal fluid biomarkers include amyloid β, total tau protein, and phosphorylated tau protein, which are considered to be quite useful for the prediction of conversion from mild cognitive impairment to Alzheimer's disease. These cerebrospinal fluid biomarkers are the most reliable because of the high sample stability. However, it is invasive. Blood, which is easy to obtain, can detect a lot of biomarkers in plasma, serum, and whole blood, but some blood biomarkers are not stable. The confirmation of the stability of biomarkers is also required. The other problem with blood biomarkers is that the levels often do not show a marked change [94, 95].

The development of diagnostic markers for Alzheimer's disease detection is of interest to patients and their relatives. Through further research, more useful individuals would be detected as asymptomatic, and a great contribution would be required only from people who have become symptomatic. Then, various kinds of approaches, such as cell therapy and biogen, can be provided that are suitable for the individual's constitution. Due to these reasons, the understanding of the pathology, manipulation, and diagnosis of biomarkers in Alzheimer's disease is significant for the medical field [95].

Utilization in comparative clinical studies for AD and other forms of dementia

Comparative clinical studies are one way to comprehend the characteristics of the groups in order to find the best therapy for patients. The choice of a diagnostic test may depend on the condition of interest; for example, one biomarker could be a sensitive marker of the subtypes of Alzheimer's disease, whereas another could also mirror the effective outcome in any dementia. The creation of a ranking list of how different biomarkers perform compared to other applicable methods in distinguishing between these dementias was aimed at, and the best biomarkers for the comparative study of Alzheimer's disease and other dementias were listed. These biomarkers should enable us to comprehend these diseases and the heterogeneity of Alzheimer's disease better, and make decisions about the treatment of patients based on this data [96, 97].

It is important to know if an evaluated biomarker has totally consistent performance in different clinical studies before using it in clinical decision-making, especially if the clinical study has potential effects. However, in this case, we needed consistent performance across different study groups, including different patients and different clinical sites, different patient numbers, for example, a large or small sample, or data collection tools, also across different levels of disease severity. However, the results obtained by the biomarker are very promising, but only validated for one sex, age, or molecular subtype of disease; they should be judged with caution in clinical practice. Biomarker variability was another challenge for the considered diagnostic markers because most of the diagnostic markers had a substantial degree of variability and a non-negligible number of studies. Data sparsity was another serious problem in this field [98, 99].

Utilization in the evaluation of the Therapeutic Efficacy of available treatment

The situation in the pharmaceutical armory of dementia is quite difficult now. No medication was able to significantly reverse or modify the clinical course of dementia, including Alzheimer's disease. In advanced cases, despite treatment with high doses of procholinesterase inhibitors, the therapeutic effect is modest and the duration of the favorable action is not long. Treatment guidelines on mild cognitive impairment recommend initiation of cholinergic medication for subjects with amnestic mild cognitive impairment who are functionally dependent and have a Clinical Dementia Rating of 0.5. Unfortunately, cholinomimetics, just like antioxidants, anti-inflammatory remedies, and other potential treatments, are prescribed at the stage of moderate dementia, which is the time of significant reduction of brain volume and the number of neurons. It is the latest neurodegenerative indication of dementia that typically comes to the specialist. AD-related structural changes in the brain precede this indication by a few to several decades [79, 100].

AD starts in the fourth decade as a subclinical organic brain syndrome and is pathologically presented by amyloid deposits and neurofibrillary tangles. There are no methods for fighting these changes or biomarkers indicating these changes. CSF Ab1-42 is a marker of beta-amyloid, CSF Tau-protein, and PET markers of brain structural changes. The localization of neurofibers relates to the cholinergic pathogenesis of the disease and demonstrates the degree of dementia severity. Now, the most promising target for early treatment of AD using the biomarker concept is the monoclonal anti-amyloid antibodies. Their efficiency has been shown in models with transgenic mice in which these antibodies stopped the spread of amyloid deposits and brain atrophy when the formation of biological and histopathological markers was reduced. Diffusion tensor MRI showed that these antibodies curtailed the destructive white matter changes. *In vitro* tests proved the effectiveness of these antibodies not only in stopping but also in eliminating the amyloid deposits. Now, the efficiency of these new medicinal agents is being tested in different clinical phases. However, there are strong grounds to suppose that their performance will decrease rather than increase with the patient's age and the progression of degenerative changes that impede the absorption of these agents. Since the results of treatment of AD with other methods cannot be immediately noticed, it is more important to find the factors reliably indicating the therapeutic effect and the remission period after discontinuation of therapy. In this respect, the perspectives of the development of pharmaceutical agents whose effectiveness can be predicted and measured through biomarkers are promising [101, 102].

Four groups of pharmaceutical agents for anti-amyloid therapy may be conditionally singled out. The first group is the anti-amyloid antibodies. The second group consists of agents that favor the transformation of the amyloid precursor protein into less amyloid with the help of alpha-enzyme. The potential targets for combating NFT formation are the agents that normalize tau-protein synthesis and its complete inactivation. Now, the validity of the concept itself, biomarkers, and measurement methods indicating the treatment effect are being discussed. The fact that most of these potential pharmaceutical agents are also target-specific medicines allows one to suppose that these biological markers might be quite different. As a rule, these markers are divided into the "delayed" and the "early" ones. The "early" ones reflect the response of the main target to the applied treatment. The "delayed" markers demonstrate the completeness of neuroprotection and neurorestoration [103, 104]

Utilization in Clinical Trials through Biomarker-driven therapeutic approaches

While it is true that success in preclinical forms was not as good as expected in Alzheimer's disease prevention trials, it is also true that, before the arrival of fluid biomarkers, PET, and post-mortem amyloid imaging, these prevention trials provided benefits on the results of neuroscientific research. PET tracers showed a pattern of standard amyloid-beta deposition pathology in mild cognitive impairment and AD patients, being especially useful in relatively early phases of pre-dementia. They also changed not only current knowledge but also clinical use by detecting patients with high amyloid-beta levels as asymptomatic preclinical or at-risk phase AD continuum stages, warning about the safety of a pharmacological approach that could generate ARIA-E and dystrophic calcification in region-specific new microhemorrhages in patients at these early disease stages or cause the same damage. Biomarkers also add valuable information on tau deposition and illustrated patterns as they differ from non-AD tau pathology, highlighting their potential to target immunotherapy [105, 106].

Understanding that ARIA could be potentiated beyond acceptable levels with established therapies, targeted through biomarkers during clinical development, and validated in real clinical practice, combination studies of cholinesterase inhibitors cause not only ARIA-E but widespread allergy, making clinical entities registered for cholinesterase inhibitors subject to regulations. Aggressive elimination. These findings have been validated by the PET and post-mortem studies, confirming that it is indeed testing the underlying AD mechanism. Therefore, patients diagnosed in all countries with clinical abnormalities and similar neuropathological findings are considered affected by the same nosological entity and characterized as the AD [107, 108].

Utilization in the Case study of treatment using Biomarkers

For patient care, it is necessary not only to diagnose but also to manage how we understand diseases, the understanding of the disease itself, and the treatment and care in detail. Diagnosis is easily considered, as medical imaging is commonly used in various areas. Biomarkers make diagnosis by medical imaging useful in Alzheimer's disease at an early stage of the disease. In addition, symptomatic treatment will be more effective on the basis of the pathology than to diagnose as MCI or other disease pre-symptom. In addition to their studies, meta-analyses of gene expression profiles and protein levels of tissues like the brain, CSF, serum, *etc.* have been performed before and after treatment and administration of cytokines in an Alzheimer's disease model mouse. As a result, studies to know the action target or working mechanism of administered cytokines and studies to reveal shared pathways, which affect tissues and diseases or lead to treatment effects in Alzheimer's disease, were reported [109].

The examination of serially appearing signals that changed over time by administering cytokines into the cerebroventricle has been reported in treatment studies using biomarkers of AD models. There is a case report of the immediate effect of flickering light illuminating the eyes on the amount of Aβ in the eyes after several minutes, or after thirty minutes elapse, the difference in the amount of change of the protein is assessed by the thickness of the amyloid plaque in the brain after 5 and 10 days. Alternatively, the difference in the thickness of the amyloid plaque in the brain after 5 days is not a signal for future dementia, while after 10 days it has been observed in APP/PS1 [110]. Secondly, the difference in the amount of change of the protein assessed by the thickness of the amyloid plaque in the brain after 5 and 10 days has been observed in APPswe as well. Third, in the animals treated with the flicker light for 30 minutes, there was no significant difference in blood plasma between the control and the treatment, but in the sham animals, the serum Aβ1-42 level was significantly higher. Therefore, some reports reflect the reaction of the fusion gene polypeptide. These reports show that the biological markers increase, that the distribution of treatment and treatment effects are detected, but time courses and detection limits are not known. The mechanism of the immediate effect is reported by observing the regulation mediated through retinal cells, DR, SCN, and paraventricular nuclei of the hypothalamus [111, 112].

According to these studies that designate neuropeptide Y as a controlling neurotransmitter, Y1 receptors were selectively targeted. As a result, there is a decrease in Aβ release into the ISF, a reduction in Aβ levels in the ISF, and overexpression of AβPP is reduced in Tg, and a guideline has been shown [113, 114].

Comparative studies on the Therapeutic Response of available treatment

The focus of the present analysis is to identify the comparative studies on the therapeutic response of available treatments for chronic and progressive brain disorders like Alzheimer's disease, which is required to recognize the optimum treatment regimen with the least adverse effect. Many traditional plant extracts provide additive or synergistic activity on simultaneous administration with conventional drugs, especially when the treatments are aimed against different pathogenic events. However, in the case of drugs for psychiatric disorders, the need for a standardized treatment regimen determines acceptance among the patients and health care providers. Selective electrolyte insufficiencies are corrected using suitable cyanobacteria to exploit the rapidly increasing research interest in the field. Analysis of existing health-related databases showed a direct correlation between deficiencies of certain ions, like Zn, and neuropsychological disorders [115, 116].

There is a growing interest in determining the contributions of BACE1 inhibitors to Alzheimer's disease therapy. BACE1 inhibitors are being examined in clinical settings for the treatment of early-stage Alzheimer's Disease. Along with the study of BACE, it is important to consider BACE2, a close homolog of BACE1, which is described as a key regulator of AP processing and a potential drug target. In the present work, we have used the available data of adhesive work performed to analyze the affinity pattern of BACE1 inhibitors on its homolog BACE2. The results showed better rendering if compound similarity strategies are utilized to more effectively search for lead compounds. Thus, compound similarity-based strategies are potential tools for developing inhibitors that are selective for BACE in preference to BACE2 [117, 118].

LIMITATIONS ON IMPLEMENTATION OF BIOMARKER

The biomarkers that are being mainly used now in Alzheimer's disease fall into three categories: cerebrospinal fluid measurements taken by a spinal tap; PET scans that provide images of the brain; and tactile carbon-based biosensor technology that can measure certain proteins in a simple blood sample. Although the use of amyloid PET scans is now approved and reimbursement strategies are being established for certain patient subgroups, not all patients with dementia or mild cognitive impairments will qualify for such testing. As for CSF testing, while some methods are improving at estimating these measures, the need for spinal taps limits the number of individuals willing to undergo this type of test. Blood-based protein measures are even less developed and will require time to develop further and to validate in independent cohorts [119].

While the shift toward prodromal stages may signal a change in the role of biomarkers, and the aforementioned data make this transition somewhat palatable, several factors remain limitations, and we wish to discuss these with the hope of invigorating creative thinking in this space. First and foremost, our ability to discriminate biological signals related to the earliest disease stages in symptomatic or somewhat more mildly symptomatic individuals may be riddled with gray areas. Individuals in the cognitive concerns stage may indeed be on a trajectory of very early disease change, but they may also harbor additional biological challenges unrelated to Alzheimer's disease progression due to the very complex milieu of late life and the complicated comorbidities common in this age group. Indeed, until we can dissect out a signature of disease followed by several years of well-functioning cognitive and physical health, within which we can observe normal non-Alzheimer's aging and harmonized peripheral health, such concerns will persist. Second, as it becomes clear that prodromal and presymptomatic individuals may not have the type of amyloid-driven disruption of quality of life that initially triggered approval of amyloid PET, the general structure of how measurements that lack highly specific clinical indications fit in the health system may need rethinking and reengineering. We believe that biomarkers intended to identify and track stages on the very front end of disease manifestation will require a partnership approach to management in which joint decision-making with patients and clinicians shapes and defines what data use is appropriate and potentially beneficial. Third, we introduce our use of clinical implementation science in designing process improvements for integrating techniques intended to drive practice change and provide suggestions for areas where such efforts can be initiated [120, 121].

ISSUES RELATED TO STANDARDIZATION AND VALIDATION

Role of Biomarkers in Treating Alzheimer's Disease

To approach these issues, all signs suggest that it is convenient to carry out a profound classification of the different biomarkers that exist and that act from different aspects in relation to the disease. Their use, both as a diagnostic tool and as a marker to monitor the patient's clinical course, and hence the effectiveness of therapeutic interventions, is not exempt from medical questions associated with the marginal differences of these markers in relation to age, sex, genetic, and environmental influences that eventually condition the useful range of the parameters and the cut-off point to consider. From the possible biological markers that express the underlying neurodegeneration, the most used are the quantification of proteins in cerebrospinal fluid, with the determination of the total tau protein, the phosphorylated protein, and the amyloid peptide. With regard to protein markers, such as GFAP, S100β, lactate, and cytokines, their presence without amyloid changes made them lose weight for the diagnosis. The interest in the use of drugs that diminish both the tau protein and the amyloid beta peptide is not free of issues related to their pharmacological use [122, 123].

Cost-related concern and Accessibility

The cost and accessibility of a test are major determinants of how frequently that test is likely to be deployed. Despite that, much discussion of the utility of diagnostic tests for different conditions and diseases focuses only on the benefit aspect and not on the cost and accessibility. Given the considerable number of individuals with subjective cognitive impairment or mild cognitive impairment, there might be less interest in testing beyond clinical evaluation in settings where the cost is high. People who know that there is no effective treatment directed at the underlying cause of a prodromal condition might also be less inclined to use a less accessible test if it is expensive. These factors are pertinent both for primary and secondary care settings, including general practice. Who will be able to perform such tests or oversee them in each setting is likely to differ; in resource-poor settings, it may be necessary to upskill health care workers working in general practice to perform some of these tests, at least initially, and to provide them with appropriate oversight to ensure that the appropriate test is done and that the interpretation of the results is correct, while in wealthier settings, such testing may be embedded into general practice at intake or run by practice nurses or physician assistants. Addressing the cost of appropriate interpretation of such tests and their accessibility to the target populations is a third concern, recognizing that healthy aging is the primary goal of these populations [124, 125].

Ethical issues related to Implications of Biomarker Use

Realizing the promise of these new biomarkers will require widespread incorporation of molecular diagnosis and disease staging into clinical and research settings. The use of these biomarkers can play a critical role in the development and testing of novel therapeutic interventions; bringing them into the clinic for safety and efficacy is paramount. Ethical issues associated with biomarker use in clinical trials or treatment trials for Alzheimer's disease using diagnostic and prognostic brain biomarkers deserve prompt attention. This chapter suggests principles for addressing these issues and recommends strategies to facilitate the ethical use of biomarkers in these clinical trials. A hypothetical case is used to illustrate the issues at stake. The use of diagnostic brain biomarkers for early detection of Alzheimer's disease is becoming more common. Programs for the very early identification of likely end-of-life phenomena have encountered and will continue to face a number of difficult ethical issues, but these concerns are incompletely identified, and principles for their resolution are expressed to guide eventual judgment. Methods include brain autopsy studies that assess biomarker performance, annual informant questionnaires, and spectroscopy or magnetic resonance imaging. The optimal approach of using neurobehavioral biomarkers in clinical research in this rapidly advancing field is not exactly clear. Clearly, these biomarkers are invaluable tools for conducting research. Their use raises new ethical issues of potential significance. The potential effects of their use with respect to methods used for clinical research participant recruitment, involvement, and follow-up during early-phase clinical trials need to be considered. Brain biomarker evaluation also carries non-clinical relevance, such as use in qualifying individuals for a police officer or a power of attorney [125].

FUTURE ASPECTS IN THE ADVANCEMENT OF BIOMARKERS

Biomarker research will continue to expand the definition of Alzheimer's disease, potentially revising clinical diagnostic criteria. Incorporation of multiple biomarkers of diverse types that represent the system biology of Alzheimer's disease should lead to an understanding of the basis for and evolution of multiple cellular, molecular, and synaptic causes of the disease. Currently, the field is in early exploratory phases, but analytical platforms, biological statistical frameworks, application to remote and inexpensive testing, increased biological understanding of the risk function of individual preclinical forms, potential for easily accessible targets, and plans for patient clinical trials or for early prevention trials are all being evaluated or developed to take full advantage of this new powerhouse [126].

We believe that the incorporation of both well-vetted, high-performing model-to-disease-type biomarkers and surrogates of the broadest domain of endophenotypes of the specific molecular pathogeneses into clinical diagnostics and presymptomatic systems, thus changing the nature of research and healthcare delivery, should result in more specific and healthcare-directed clinical drug development. If the disease-modifying agent is given at a time when clinical progression and severity of its natural history have been modulated, then the endpoints of treatment may be potentially neutralized unless they correspond directly to the specific pathogenic process. However, with better marker enrichment in terms of identifying and stratifying eligible patients to potentially benefit from the experimental medication, the analysis benefits apply. The recognition that these drugs may reflect races against time to impact the spread cascade of neurodegenerative disease necessitates that efficient clinical trials enroll the highest percentage of preclinical subjects that can be achieved, as the involvement of the stages [127, 128].

Advancements in the Discovery of Biomarkers

A biochemical substance whose presence or level might indicate a disease process related to the presence of disease in the body is called a biomarker. The newest techniques have elaborated on the advancement in biomarker discovery. CSF biomarkers, derived from the CNS, capture information related to pathogenesis and the course of the underlying disease. In the case of Alzheimer's disease, they have been shown to aid in diagnosis, identify preclinical disease, identify different brain disorders, predict the course of the disorder, assess patient response to treatment, and discover new treatments. Many instances show interest in CSF biomarkers in Alzheimer's disease that support and explore different studies. This interest must always take into consideration the heterogeneity of Alzheimer's disease. Progress in understanding the more basic forms and pathways of the disease has led to advances in more selective biomarkers. A large number of proteomic strategies allow for the detection and recognition of low-abundant proteins and the characterization of post-translational modifications, which might be altered in neurological diseases. These studies also provide information about the relevance of the pathway-specific changes observed during neurological diseases [129, 130].

Use as a Potential for Personalized Medicine

There is data to suggest that the use of biomarkers may have a role not just in the development of new treatments. For example, accumulating pathological amyloid or tau burden in amyloid-positive patients over amyloid-negative patients might stratify other groups of patients who may benefit from treatment [131, 132]. This concept of 'precision' or 'personalized' medicine, where the choice of treatment is determined by the patient's unique biology, is appealing. Hypothetical live markers of indirect synaptic markers may help enrich an early Alzheimer's clinical trial population by identifying those subjects exhibiting potential demonstrable abnormal synaptic function and those who do not [133].

One of the challenges of precision medicine, and more generally the use of biomarkers, is that it remains to be demonstrated that the level, or more importantly, the rate of change in such biomarkers, correlates with future cognitive decline reliably and over a long period. Here, the evidence is either missing or there are differences of opinion on this. Nonetheless, as the field advances, the potential role for biomarkers, especially those based on new technology, should become more apparent. The standard of proof that will be expected will also likely change for pathophysiological hypothesis testing, needing histological confirmation with some measures in order to show that they actually predict future clinically relevant change [134].

Integrating Biomarkers into Clinical Practice

Increasing evidence indicates that preventive interventions should begin before significant cellular synthesis, suggesting a potentially important role for biomarkers. Once therapies work for Alzheimer's disease, the primary care physician will probably need simple, rapid in-office tests to screen patients. Any patient suspected of being ill could be sent for more sophisticated tests. In the meantime, a patient who does not have evidence of major brain damage could benefit from a "wait and see" approach, provided we have safe treatment options. Furthermore, substitutes for the biomarkers of Alzheimer's disease might be used to better select patients for research studies. Biomarkers might be used for monitoring Alzheimer's disease and intervention in everyday clinical practice. Primary care physicians do not, however, perform brain imaging in the office. If an in-office test were developed, performing such imaging might not be cost-effective. Additionally, the information gathered *via* such tests would not generally alter the clinician's treatment. It is unlikely that even good biomarkers will ever consistently predict who is going to develop Alzheimer's disease. Therefore, testing positron emission tomography, hippocampal measurement, and cerebrospinal fluid analyses are not fundamental or mobilized, as is presently the case. It would be reasonable, though, to consider how to implement such testing for guiding the management of therapeutic decisions for individuals suspected of having any cognitive impairment before the century is over. Ultimately, brain imaging might supplant pen-and-paper testing, as the question now is less about the science and more about the technology and how many individuals the healthcare system should be subjected to [135].

CONCLUSION

The successful application of biomarkers in Alzheimer's disease drug development promises to decrease development costs and time by a combination of more effective proof-of-mechanism studies, better target patient stratification, and simpler, validated clinical endpoints. By providing evidence that pathophysiological products are being altered, it is more likely that they reflect greater changes in clinical outcomes than more indirect measures, thereby raising the probability that results link to functional improvements. Furthermore, the value of using a battery of complementary markers to provide both a profile of the drug's activity and increased statistical power to detect alterations in the chosen set of specific domains is likely to be important. It is increasingly apparent that AD pharmacotherapy will require the use of combination treatments, and both the mechanism of drug action and the development of suitable biomarkers will impact this process. Consequently, the inclusion of pharmacogenomic measures within clinical trials to help explain the response of different AD patient subgroups or to exclude a poor-responder subgroup is of particular value. Such biomarker information will then directly drive the use of rational combinations to maximize treatment benefit. In conclusion, definite progress in the development and validation of AD biomarkers has occurred, such that it is likely that AD trials, with a more detailed understanding of patient population heterogeneity and rising placebo group event rates, can be more effectively managed in the future.

SUMMARY

This review of state-of-the-art and emerging fluid biomarkers in Alzheimer's disease has shown that these biomarkers offer clinical benefits in diagnosis, prognosis, tracking disease progression, and assessing treatment effects. In addition, these biomarkers are increasingly utilized in diagnostic criteria for the disease. Although most of these biomarkers are not used clinically, due to the increasing number of clinical trials, utilization, and need for accurate diagnostic criteria, they are likely to be added to future diagnostic criteria for the disease. To achieve this, however, several issues need to be addressed, including validation of the data among different test systems by different research groups; standardization of pre-analytic and analytic factors and cutoff levels for each biomarker; exploration of new biomarkers; and cost and insurance issues. In conclusion, fluid biomarkers for Alzheimer's disease are entering a new phase in which they will play a central role in drug development and as a companion diagnostic for new drugs.

The Future Aspects in Alzheimer's Treatment

The sensitivity and earliest appearance of neuropathological fibrillar tangle and plaque changes in Alzheimer's disease make it a suitable condition for trials of imaging therapies, as well as earlier diagnosis of the disease. In later stages, imaging trials could evaluate the influence of the compounds on the progression of Alzheimer's disease, its time course, and dose-related changes, thus providing useful information for future treatment trials. Realizing the growing cholinergic deficit in the patients' cognitive functions, in the protoneuaceuticals field, EEG biofeedback and electrical stimulating strategies supported by imaging the signal modifications can monitor the changes in acetylcholine levels related to the cholinergic system. These tools are also assessed both in longitudinal and short-term placebo-controlled clinical trials in terms of safety aspects, side effects, and behavioral and imaging modulations.

Future successful disease-modifying studies may need to incorporate a prolonged chronic administration of agents that, at an increased level, might produce toxic changes in the trials or even increase the vascular risk in Alzheimer's disease patients. Other strategies currently examined in human studies are based on the effectiveness of Aβ clearance. These gather in the field of active and passive immunization along with other immune modulators associated with amyloid-related imaging abnormalities and discontinuing the drugs administered, together with appropriate dosing schedules, concerning overexpression of aggressive pathologic activated microglial response. Still interesting is the other complementary approach obtained from the action of the endogenous Aβ-clearing molecule focused on Aβ-neuropeptide fragments and receptor for advanced glycation end product, such as Aβ Jab peptides. Another line of studies is targeting the Aβ precursor protein cleaving enzyme inhibitors. This approach, although promising, has not provided good results in treatment yet, being still in the preclinical or early research studies. Furthermore, other compounds targeting tau hyperphosphorylation, abnormal folding patterns, deposition, and spread in the brain have not produced significant results in the treatment or delay of the disease, showing negative results or leading to a significant worsening of cognitive functions.

AUTHORS' CONTRIBUTIONS

Conceptualization: All of the Authors contributed to the development of this book's chapter framework and concept. This collaborative effort ensured that a well-rounded perspective was preserved throughout the study.

LIST OF ABBREVIATIONS

AD - Alzheimer's disease

Aβ - Amyloid beta

NFT - Neurofibrillary tangles

CSF - Cerebrospinal fluid

PET - Positron Emission Tomography

MeSH - Medical Subject Headings

FTLD - Frontotemporal lobar degeneration

APOE - Apolipoprotein E

MCI - Mild Cognitive Impairment

NfL - Neurofilament Light Chain

AGEs - Advanced Glycosylation End products

ELISA - Enzyme-linked immunosorbent assay

oxLDL - Oxidized Low-density Lipoprotein

TNF-α - Tumor Necrosis Factor-alpha

BDNF - Brain-Derived Neurotrophic Factor

AChE - Acetylcholinesterase

IL-6 - Interleukin-6

p-tau - Phosphorylated tau

SCN - Suprachiasmatic nucleus

ISF - Interstitial fluid

ACKNOWLEDGEMENTS

The authors are thankful to Mahayogi Gorakhnath University, Gorakhpur, for providing the necessary facilities to execute this manuscript.

REFERENCES

[1] Schroeder KM, Afonso AS, Wang H, Grace S, Phipps A, Sims JR. Demographics, comorbidities, and comedications in newly diagnosed patients with Alzheimer's disease and related dementias: Findings from United States Medicare claims data. J Alzheimers Dis 2024; 102(2): 506-19. [http://dx.doi.org/10.3233/JAD-231488] [PMID: 39539059]

[2] Toppala S, Ekblad LL, Viitanen M, Rinne JO, Jula A. Impaired Early Insulin Response to Glucose Load Predicts Episodic Memory Decline: A 10-Year Population-Based Cohort Follow-Up of 45–74-Year-Old Men and Women. J Alzheimers Dis 2023; 92(1): 349-59. [http://dx.doi.org/10.3233/JAD-220894] [PMID: 36744339]

[3] Tahami Monfared AA, Tafazzoli A, Ye W, Chavan A, Deger KA, Zhang Q. A simulation model to evaluate the potential impact of disease-modifying treatments on burden of illness in Alzheimer's disease. Neurol Ther 2022; 11(4): 1609-23. [http://dx.doi.org/10.1007/s40120-022-00393-1] [PMID: 35960495]

[4] Lu YR, Wu TH, Jeng Y, *et al.* The impact of active community-based survey on dementia detection ratio in Taiwan: A cohort study with historical control. Front Public Health 2023; 101005252 [http://dx.doi.org/10.3389/fpubh.2022.1005252] [PMID: 36684977]

[5] Klyucherev TO, Olszewski P, Shalimova AA, *et al.* Advances in the development of new biomarkers for Alzheimer's disease. Transl Neurodegener 2022; 11(1): 25. [http://dx.doi.org/10.1186/s40035-022-00296-z] [PMID: 35449079]

[6] Boccardi M, Dodich A, Albanese E, *et al.* The strategic biomarker roadmap for the validation of Alzheimer's diagnostic biomarkers: methodological update. Eur J Nucl Med Mol Imaging 2021; 48(7): 2070-85. [http://dx.doi.org/10.1007/s00259-020-05120-2] [PMID: 33688996]

[7] Bai B, Vanderwall D, Li Y, *et al.* Proteomic landscape of Alzheimer's Disease: novel insights into pathogenesis and biomarker discovery. Mol Neurodegener 2021; 16(1): 55. [http://dx.doi.org/10.1186/s13024-021-00474-z] [PMID: 34384464]

[8] Gustavsson A, Norton N, Fast T, *et al.* Global estimates on the number of persons across the Alzheimer's disease continuum. Alzheimers Dement 2023; 19(2): 658-70. [http://dx.doi.org/10.1002/alz.12694] [PMID: 35652476]

[9] Sehar U, Kopel J, Reddy PH. Alzheimer's disease and its related dementias in US Native Americans: A major public health concern. Ageing Res Rev 2023; 90: 102027. [http://dx.doi.org/10.1016/j.arr.2023.102027] [PMID: 37544432]

[10] Mills ZB, Faull RLM, Kwakowsky A. Is hormone replacement therapy a risk factor or a therapeutic option for Alzheimer's disease? Int J Mol Sci 2023; 24(4): 3205. [http://dx.doi.org/10.3390/ijms24043205] [PMID: 36834617]

[11] Huang S, Li Y, Shen J, Liang W, Li C. Identification of a diagnostic model and molecular subtypes of major depressive disorder based on endoplasmic reticulum stress-related genes. Front Psychiatry 2023; 14: 1168516. [http://dx.doi.org/10.3389/fpsyt.2023.1168516] [PMID: 37649561]

[12] Böhme R, Schmidt AW, Hesselbarth N, *et al.* Induction of oxidative- and endoplasmic-reticulu--stress dependent apoptosis in pancreatic cancer cell lines by DDOST knockdown. Sci Rep 2024; 14(1): 20388.
[http://dx.doi.org/10.1038/s41598-024-68510-8] [PMID: 39223141]

[13] Schöll M, Verberk IMW, del Campo M, *et al.* Challenges in the practical implementation of blood biomarkers for Alzheimer's disease. Lancet Healthy Longev 2024; 5(10)100630.
[http://dx.doi.org/10.1016/j.lanhl.2024.07.013] [PMID: 39369727]

[14] Buccellato FR, D'Anca M, Fenoglio C, Scarpini E, Galimberti D. Role of oxidative damage in Alzheimer's disease and neurodegeneration: From pathogenic mechanisms to biomarker discovery. Antioxidants 2021; 10(9): 1353.
[http://dx.doi.org/10.3390/antiox10091353] [PMID: 34572985]

[15] Dubois B, Villain N, Frisoni GB, *et al.* Clinical diagnosis of Alzheimer's disease: recommendations of the International Working Group. Lancet Neurol 2021; 20(6): 484-96.
[http://dx.doi.org/10.1016/S1474-4422(21)00066-1] [PMID: 33933186]

[16] Diogo VS, Ferreira HA, Prata D. Early diagnosis of Alzheimer's disease using machine learning: a multi-diagnostic, generalizable approach. Alzheimers Res Ther 2022; 14(1): 107.
[http://dx.doi.org/10.1186/s13195-022-01047-y] [PMID: 35922851]

[17] Zetterberg H, Bendlin BB. Biomarkers for Alzheimer's disease—preparing for a new era of disease-modifying therapies. Mol Psychiatry 2021; 26(1): 296-308.
[http://dx.doi.org/10.1038/s41380-020-0721-9] [PMID: 32251378]

[18] Oosthoek M, Vermunt L, de Wilde A, *et al.* Utilization of fluid-based biomarkers as endpoints in disease-modifying clinical trials for Alzheimer's disease: a systematic review. Alzheimers Res Ther 2024; 16(1): 93.
[http://dx.doi.org/10.1186/s13195-024-01456-1] [PMID: 38678292]

[19] Hansson O, Edelmayer RM, Boxer AL, *et al.* The Alzheimer's Association appropriate use recommendations for blood biomarkers in Alzheimer's disease. Alzheimers Dement 2022; 18(12): 2669-86.
[http://dx.doi.org/10.1002/alz.12756] [PMID: 35908251]

[20] Moloney CM, Lowe VJ, Murray ME. Visualization of neurofibrillary tangle maturity in Alzheimer's disease: A clinicopathologic perspective for biomarker research. Alzheimers Dement 2021; 17(9): 1554-74.
[http://dx.doi.org/10.1002/alz.12321] [PMID: 33797838]

[21] Bodaghi A, Fattahi N, Ramazani A. Biomarkers: Promising and valuable tools towards diagnosis, prognosis and treatment of Covid-19 and other diseases. Heliyon 2023; 9(2)e13323
[http://dx.doi.org/10.1016/j.heliyon.2023.e13323] [PMID: 36744065]

[22] Abi-Dargham A, Moeller SJ, Ali F, *et al.* Candidate biomarkers in psychiatric disorders: state of the field. World Psychiatry 2023; 22(2): 236-62.
[http://dx.doi.org/10.1002/wps.21078] [PMID: 37159365]

[23] Lee M, Guo JP, Kennedy K, McGeer EG, McGeer PL. A method for diagnosing Alzheimer's disease based on salivary amyloid-β protein 42 levels. J Alzheimers Dis 2016; 55(3): 1175-82.
[http://dx.doi.org/10.3233/JAD-160748] [PMID: 27792013]

[24] Lewczuk P, Lelental N, Spitzer P, Maler JM, Kornhuber J. Amyloid-β 42/40 cerebrospinal fluid concentration ratio in the diagnostics of Alzheimer's disease: validation of two novel assays. J Alzheimers Dis 2014; 43(1): 183-91.
[http://dx.doi.org/10.3233/JAD-140771] [PMID: 25079805]

[25] Sadeghzadeh J, Shahabi P, Farhoudi M, Ebrahimi-Kalan A, Mobed A, Shahpasand K. Tau protein biosensors in the diagnosis of neurodegenerative diseases. Adv Pharm Bull 2023; 13(3): 502-11.
[http://dx.doi.org/10.34172/apb.2023.061] [PMID: 37646056]

[26] Christl J, Verhülsdonk S, Pessanha F, *et al.* Association of cerebrospinal fluid S100B protein with core biomarkers and cognitive deficits in prodromal and mild Alzheimer's disease. J Alzheimers Dis 2019; 72(4): 1119-27.
[http://dx.doi.org/10.3233/JAD-190550] [PMID: 31683478]

[27] Srikanth V, Maczurek A, Phan T, *et al.* Advanced glycation endproducts and their receptor RAGE in Alzheimer's disease. Neurobiol Aging 2011; 32(5): 763-77.
[http://dx.doi.org/10.1016/j.neurobiolaging.2009.04.016] [PMID: 19464758]

[28] Kothandan D, Singh DS, Yerrakula G, *et al.* Advanced glycation end products-induced Alzheimer's disease and its novel therapeutic approaches: A comprehensive review. Cureus 2024; 16(5)e61373.
[http://dx.doi.org/10.7759/cureus.61373] [PMID: 38947632]

[29] D'Introno A, Solfrizzi V, Colacicco AM, *et al.* Polymorphisms in the oxidized low-density lipoprotein receptor-1 gene and risk of Alzheimer's disease. J Gerontol A Biol Sci Med Sci 2005; 60(3): 280-4.
[http://dx.doi.org/10.1093/gerona/60.3.280] [PMID: 15860461]

[30] Decourt B, Lahiri DK, Sabbagh MN. Targeting tumor necrosis factor alpha for Alzheimer's disease. Curr Alzheimer Res 2017; 14(4): 412-25.
[http://dx.doi.org/10.2174/1567205013666160930110551] [PMID: 27697064]

[31] Shim Y. Clinical application of plasma neurofilament light chain in a memory clinic: a pilot study. Dement Neurocognitive Disord 2022; 21(2): 59-70.
[http://dx.doi.org/10.12779/dnd.2022.21.2.59] [PMID: 35585907]

[32] Tao Q, Alvin Ang TF, Akhter-Khan SC, *et al.* Impact of C-reactive protein on cognition and Alzheimer disease biomarkers in homozygous APOE ε4 carriers. Neurology 2021; 97(12): e1243-52.
[http://dx.doi.org/10.1212/WNL.0000000000012512] [PMID: 34266923]

[33] Noble JM, Manly JJ, Schupf N, Tang MX, Mayeux R, Luchsinger JA. Association of C-reactive protein with cognitive impairment. Arch Neurol 2010; 67(1): 87-92.
[http://dx.doi.org/10.1001/archneurol.2009.308] [PMID: 20065134]

[34] Bishnoi RJ, Palmer RF, Royall DR. Serum interleukin (IL)-15 as a biomarker of Alzheimer's disease. PLoS One 2015; 10(2)e0117282.
[http://dx.doi.org/10.1371/journal.pone.0117282] [PMID: 25710473]

[35] O'Bryant SE, Hobson V, Hall JR, *et al.* Brain-derived neurotrophic factor levels in Alzheimer's disease. J Alzheimers Dis 2009; 17(2): 337-41.
[http://dx.doi.org/10.3233/JAD-2009-1051] [PMID: 19363274]

[36] Prokopenko D, Morgan SL, Mullin K, *et al.* Whole-genome sequencing reveals new Alzheimer's disease–associated rare variants in loci related to synaptic function and neuronal development. Alzheimers Dement 2021; 17(9): 1509-27.
[http://dx.doi.org/10.1002/alz.12319] [PMID: 33797837]

[37] Soheili-Nezhad S, van der Linden RJ, Olde Rikkert M, Sprooten E, Poelmans G. Long genes are more frequently affected by somatic mutations and show reduced expression in Alzheimer's disease: Implications for disease etiology. Alzheimers Dement 2021; 17(3): 489-99.
[http://dx.doi.org/10.1002/alz.12211] [PMID: 33075204]

[38] Chen ZY, Zhang Y. Animal models of Alzheimer's disease: Applications, evaluation, and perspectives. Zool Res 2022; 43(6): 1026-40.
[http://dx.doi.org/10.24272/j.issn.2095-8137.2022.289] [PMID: 36317468]

[39] Veitch DP, Weiner MW, Aisen PS, *et al.* Using the Alzheimer's Disease Neuroimaging Initiative to improve early detection, diagnosis, and treatment of Alzheimer's disease. Alzheimers Dement 2022; 18(4): 824-57.
[http://dx.doi.org/10.1002/alz.12422] [PMID: 34581485]

[40] Vrahatis AG, Skolariki K, Krokidis MG, Lazaros K, Exarchos TP, Vlamos P. Revolutionizing the early detection of Alzheimer's disease through non-invasive biomarkers: the role of artificial

intelligence and deep learning. Sensors (Basel) 2023; 23(9): 4184. [http://dx.doi.org/10.3390/s23094184] [PMID: 37177386]

[41] Porsteinsson AP, Isaacson RS, Knox S, Sabbagh MN, Rubino I. Diagnosis of early Alzheimer's disease: clinical practice in 2021. J Prev Alzheimers Dis 2021; 8(3): 371-86. [http://dx.doi.org/10.14283/jpad.2021.23] [PMID: 34101796]

[42] Li J, Liu C, Ang TFA, Au R. BMI decline patterns and relation to dementia risk across four decades of follow-up in the Framingham Study. Alzheimers Dement 2023; 19(6): 2520-7. [http://dx.doi.org/10.1002/alz.12839] [PMID: 36519667]

[43] Zhang XX, Tian Y, Wang ZT, Ma YH, Tan L, Yu JT. The epidemiology of Alzheimer's disease modifiable risk factors and prevention. J Prev Alzheimers Dis 2021; 8(3): 313-21. [http://dx.doi.org/10.14283/jpad.2021.15] [PMID: 34101789]

[44] Schraen-Maschke S, Duhamel A, Vidal JS, *et al.* The free plasma amyloid $A\beta_{1-42}/A\beta_{1-40}$ ratio predicts conversion to dementia for subjects with mild cognitive impairment with performance equivalent to that of the total plasma $A\beta_{1-42}/A\beta_{1-40}$ ratio. The BALTAZAR study. Neurobiol Dis 2024; 193106459 [http://dx.doi.org/10.1016/j.nbd.2024.106459] [PMID: 38423192]

[45] Huszár Z, Engh MA, Pavlekovics M, *et al.* Risk of conversion to mild cognitive impairment or dementia among subjects with amyloid and tau pathology: a systematic review and meta-analysis. Alzheimers Res Ther 2024; 16(1): 81. [http://dx.doi.org/10.1186/s13195-024-01455-2] [PMID: 38610055]

[46] Fowler CJ, Stoops E, Rainey-Smith SR, *et al.* Plasma p-tau181/$A\beta_{1-42}$ ratio predicts Aβ-PET status and correlates with CSF-p-tau181/$A\beta_{1-42}$ and future cognitive decline. Alzheimers Dement (Amst) 2022; 14(1)e12375 [http://dx.doi.org/10.1002/dad2.12375] [PMID: 36447478]

[47] Viqas Shafi, Ifrah Siddiqui. Alzheimer's Disease Investigated *via* Gene-Environment Interactions, Biochemical Pathways, Cellular Processes, and Disease Phenotype Variability, 14 May 2024, PREPRINT (Version 1) Available at Research Square: https://doi.org/10.21203/rs.3.rs-4408464/v1

[48] Seto M, Weiner RL, Dumitrescu L, Hohman TJ. Protective genes and pathways in Alzheimer's disease: moving towards precision interventions. Mol Neurodegener 2021; 16(1): 29. [http://dx.doi.org/10.1186/s13024-021-00452-5] [PMID: 33926499]

[49] Bagyinszky E, An SSA. Haploinsufficiency and Alzheimer's Disease: The Possible Pathogenic and Protective Genetic Factors. Int J Mol Sci 2024; 25(22): 11959. [http://dx.doi.org/10.3390/ijms252211959] [PMID: 39596030]

[50] Vigneswaran J, Muthukumar SA, Shafras M, Pant G. An insight into Alzheimer's disease and its on-setting novel genes. Egypt J Neurol Psychiat Neurosurg 2021; 57(1): 160. [http://dx.doi.org/10.1186/s41983-021-00420-2]

[51] Zetterberg H, Blennow K. Moving fluid biomarkers for Alzheimer's disease from research tools to routine clinical diagnostics. Mol Neurodegener 2021; 16(1): 10. [http://dx.doi.org/10.1186/s13024-021-00430-x] [PMID: 33608044]

[52] Hong R, Xu B. Breast cancer: an up-to-date review and future perspectives. Cancer Commun (Lond) 2022; 42(10): 913-36. [http://dx.doi.org/10.1002/cac2.12358] [PMID: 36074908]

[53] Lakkireddy M, Gadiga SG, Malathi RD, *et al.* RETRACTED ARTICLE: Impact of daily high dose oral vitamin D therapy on the inflammatory markers in patients with COVID 19 disease. Sci Rep 2021; 11(1): 10641. [http://dx.doi.org/10.1038/s41598-021-90189-4] [PMID: 34017029]

[54] Taieb J, Svrcek M, Cohen R, Basile D, Tougeron D, Phelip JM. Deficient mismatch repair/microsatellite unstable colorectal cancer: Diagnosis, prognosis and treatment. Eur J Cancer 2022; 175: 136-57.

[http://dx.doi.org/10.1016/j.ejca.2022.07.020] [PMID: 36115290]

[55] Birks, J.S. and R.J. Harvey, Donepezil for dementia due to Alzheimer's disease. Cochrane Database of systematic reviews, 2018(6).
[http://dx.doi.org/10.1002/14651858.CD001190.pub3]

[56] Onor ML, Trevisiol M, Aguglia E. Rivastigmine in the treatment of Alzheimer's disease: an update. Clin Interv Aging 2007; 2(1): 17-32.
[http://dx.doi.org/10.2147/ciia.2007.2.1.17] [PMID: 18044073]

[57] Ajenikoko MK, Ajagbe AO, Onigbinde OA, Okesina AA, Tijani AA. Review of Alzheimer's disease drugs and their relationship with neuron-glia interaction. IBRO Neuroscience Reports 2023; 14: 64-76.
[http://dx.doi.org/10.1016/j.ibneur.2022.11.005] [PMID: 36593897]

[58] Tampi RR, Forester BP, Agronin M. Aducanumab: evidence from clinical trial data and controversies. Drugs Context 2021; 10: 1-9.
[http://dx.doi.org/10.7573/dic.2021-7-3] [PMID: 34650610]

[59] McDade E, Cummings JL, Dhadda S, *et al.* Lecanemab in patients with early Alzheimer's disease: detailed results on biomarker, cognitive, and clinical effects from the randomized and open-label extension of the phase 2 proof-of-concept study. Alzheimers Res Ther 2022; 14(1): 191.
[http://dx.doi.org/10.1186/s13195-022-01124-2] [PMID: 36544184]

[60] Kennedy ME, Stamford AW, Chen X, et al. The BACE1 inhibitor verubecestat (MK-8931) reduces CNS β-amyloid in animal models and in Alzheimer's disease patients. Sci Transl Med. 2016 Nov 2; 8(363): 363ra150.

[61] Molinuevo JL, Lladó A, Rami L. Memantine: Targeting glutamate excitotoxicity in Alzheimer's disease and other dementias. Am J Alzheimers Dis Other Demen 2005; 20(2): 77-85.
[http://dx.doi.org/10.1177/153331750502000206] [PMID: 15844753]

[62] Sawda C, Moussa C, Turner RS. Resveratrol for Alzheimer's disease. Ann N Y Acad Sci 2017; 1403(1): 142-9.
[http://dx.doi.org/10.1111/nyas.13431] [PMID: 28815614]

[63] Wagemann O, Liu H, Wang G, *et al.* Downstream biomarker effects of gantenerumab or solanezumab in dominantly inherited alzheimer disease: the DIAN-TU-001 randomized clinical trial. JAMA Neurol 2024; 81(6): 582-93.
[http://dx.doi.org/10.1001/jamaneurol.2024.0991] [PMID: 38683602]

[64] Assunção SS, Sperling RA, Ritchie C, *et al.* Meaningful benefits: a framework to assess disease-modifying therapies in preclinical and early Alzheimer's disease. Alzheimers Res Ther 2022; 14(1): 54.
[http://dx.doi.org/10.1186/s13195-022-00984-y] [PMID: 35440022]

[65] Olloquequi J, Ettcheto M, Cano A, *et al.* Impact of new drugs for therapeutic intervention in Alzheimer's disease. Frontiers in Bioscience-Landmark 2022; 27(5): 146.
[http://dx.doi.org/10.31083/j.fbl2705146] [PMID: 35638413]

[66] Jain N, Nagaich U, Pandey M, Chellappan DK, Dua K. Predictive genomic tools in disease stratification and targeted prevention: a recent update in personalized therapy advancements. EPMA J 2022; 13(4): 561-80.
[http://dx.doi.org/10.1007/s13167-022-00304-2] [PMID: 36505888]

[67] van den Brink W, Bloem R, Ananth A, *et al.* Digital resilience biomarkers for personalized health maintenance and disease prevention. Frontiers in Digital Health 2021; 2: 614670.
[http://dx.doi.org/10.3389/fdgth.2020.614670] [PMID: 34713076]

[68] Duan N, Cheung K, Kravitz RL. Personalized data science and personalized (N-of-1) trials: Promising paradigms for individualized health care. Harv Data Sci Rev. 2022; 4(SI3).

[69] Ogulur I, Pat Y, Ardicli O, *et al.* Advances and highlights in biomarkers of allergic diseases. Allergy 2021; 76(12): 3659-86.

[http://dx.doi.org/10.1111/all.15089] [PMID: 34519063]

[70] Verduci L, Tarcitano E, Strano S, Yarden Y, Blandino G. CircRNAs: role in human diseases and potential use as biomarkers. Cell Death Dis 2021; 12(5): 468.
[http://dx.doi.org/10.1038/s41419-021-03743-3] [PMID: 33976116]

[71] Zhang L, Liang X, Zhang Z, Luo H. Cerebrospinal fluid and blood biomarkers in the diagnostic assays of Alzheimer's disease. J Innov Opt Health Sci 2022; 15(1)2230001.
[http://dx.doi.org/10.1142/S1793545822300014]

[72] Krause K, Wulf M, Sommer P, *et al.* CSF diagnostics: a potentially valuable tool in neurodegenerative and inflammatory disorders involving motor neurons: a review. Diagnostics (Basel) 2021; 11(9): 1522.
[http://dx.doi.org/10.3390/diagnostics11091522] [PMID: 34573864]

[73] Halbgebauer S, Steinacker P, Riedel D, *et al.* Visinin-like protein 1 levels in blood and CSF as emerging markers for Alzheimer's and other neurodegenerative diseases. Alzheimers Res Ther 2022; 14(1): 175.
[http://dx.doi.org/10.1186/s13195-022-01122-4] [PMID: 36419075]

[74] Chouliaras L, O'Brien JT. The use of neuroimaging techniques in the early and differential diagnosis of dementia. Mol Psychiatry 2023; 28(10): 4084-97.
[http://dx.doi.org/10.1038/s41380-023-02215-8] [PMID: 37608222]

[75] Bousiges O, Blanc F. Biomarkers of dementia with Lewy bodies: differential diagnostic with Alzheimer's disease. Int J Mol Sci 2022; 23(12): 6371.
[http://dx.doi.org/10.3390/ijms23126371] [PMID: 35742814]

[76] Perovnik M, Vo A, Nguyen N, *et al.* Automated differential diagnosis of dementia syndromes using FDG PET and machine learning. Front Aging Neurosci 2022; 141005731
[http://dx.doi.org/10.3389/fnagi.2022.1005731] [PMID: 36408106]

[77] Benussi A, Cantoni V, Rivolta J, *et al.* Classification accuracy of blood-based and neurophysiological markers in the differential diagnosis of Alzheimer's disease and frontotemporal lobar degeneration. Alzheimers Res Ther 2022; 14(1): 155.
[http://dx.doi.org/10.1186/s13195-022-01094-5] [PMID: 36229847]

[78] Li P, Quan W, Wang Z, *et al.* Early-stage differentiation between Alzheimer's disease and frontotemporal lobe degeneration: Clinical, neuropsychology, and neuroimaging features. Front Aging Neurosci 2022; 14: 981451.
[http://dx.doi.org/10.3389/fnagi.2022.981451] [PMID: 36389060]

[79] Dickerson BC, Atri A, Clevenger C, *et al.* The Alzheimer's Association clinical practice guideline for the Diagnostic Evaluation, Testing, Counseling, and Disclosure of Suspected Alzheimer's Disease and Related Disorders (DETeCD-ADRD): Executive summary of recommendations for specialty care. Alzheimers Dement 2025; 21(1)e14337
[http://dx.doi.org/10.1002/alz.14337] [PMID: 39713957]

[80] Haque R, Watson CM, Liu J, *et al.* A protein panel in cerebrospinal fluid for diagnostic and predictive assessment of Alzheimer's disease. Sci Transl Med 2023; 15(712)eadg4122.
[http://dx.doi.org/10.1126/scitranslmed.adg4122] [PMID: 37672565]

[81] Bouwman FH, Frisoni GB, Johnson SC, *et al.* Clinical application of CSF biomarkers for Alzheimer's disease: From rationale to ratios. Alzheimers Dement (Amst) 2022; 14(1)e12314.
[http://dx.doi.org/10.1002/dad2.12314] [PMID: 35496374]

[82] Alawode DOT, Heslegrave AJ, Ashton NJ, *et al.* Transitioning from cerebrospinal fluid to blood tests to facilitate diagnosis and disease monitoring in Alzheimer's disease. J Intern Med 2021; 290(3): 583-601.
[http://dx.doi.org/10.1111/joim.13332] [PMID: 34021943]

[83] Liss JL, Seleri Assunção S, Cummings J, *et al.* Practical recommendations for timely, accurate diagnosis of symptomatic Alzheimer's disease (MCI and dementia) in primary care: a review and

synthesis. J Intern Med 2021; 290(2): 310-34.
[http://dx.doi.org/10.1111/joim.13244] [PMID: 33458891]

[84] Garcia-Pavia P, Bengel F, Brito D, *et al.* Expert consensus on the monitoring of transthyretin amyloid cardiomyopathy. Eur J Heart Fail 2021; 23(6): 895-905.
[http://dx.doi.org/10.1002/ejhf.2198] [PMID: 33915002]

[85] Schmidt-Erfurth U, Reiter GS, Riedl S, *et al.* AI-based monitoring of retinal fluid in disease activity and under therapy. Prog Retin Eye Res 2022; 86: 100972.
[http://dx.doi.org/10.1016/j.preteyeres.2021.100972] [PMID: 34166808]

[86] Maj M, van Os J, De Hert M, *et al.* The clinical characterization of the patient with primary psychosis aimed at personalization of management. World Psychiatry 2021; 20(1): 4-33.
[http://dx.doi.org/10.1002/wps.20809] [PMID: 33432763]

[87] Tahami Monfared AA, Byrnes MJ, White LA, Zhang Q. Alzheimer's disease: epidemiology and clinical progression. Neurol Ther 2022; 11(2): 553-69.
[http://dx.doi.org/10.1007/s40120-022-00338-8] [PMID: 35286590]

[88] Hampel H, Au R, Mattke S, *et al.* Designing the next-generation clinical care pathway for Alzheimer's disease. Nature Aging 2022; 2(8): 692-703.
[http://dx.doi.org/10.1038/s43587-022-00269-x] [PMID: 37118137]

[89] Mankhong S, Kim S, Lee S, *et al.* Development of Alzheimer's disease biomarkers: From CSF-to blood-based biomarkers. Biomedicines 2022; 10(4): 850.
[http://dx.doi.org/10.3390/biomedicines10040850] [PMID: 35453600]

[90] Halicki MJ, Hind K, Chazot PL. Blood-based biomarkers in the diagnosis of chronic traumatic encephalopathy: research to date and future directions. Int J Mol Sci 2023; 24(16): 12556.
[http://dx.doi.org/10.3390/ijms241612556] [PMID: 37628736]

[91] Corriveau-Lecavalier N, Machulda MM, Botha H, *et al.* Phenotypic subtypes of progressive dysexecutive syndrome due to Alzheimer's disease: a series of clinical cases. J Neurol 2022; 269(8): 4110-28.
[http://dx.doi.org/10.1007/s00415-022-11025-x] [PMID: 35211780]

[92] Zhang S, Shen L, Jiao B. Cognitive dysfunction in repeat expansion diseases: a review. Front Aging Neurosci 2022; 14841711
[http://dx.doi.org/10.3389/fnagi.2022.841711] [PMID: 35478698]

[93] Corbo I, Casagrande M. Higher-level executive functions in healthy elderly and mild cognitive impairment: a systematic review. J Clin Med 2022; 11(5): 1204.
[http://dx.doi.org/10.3390/jcm11051204] [PMID: 35268294]

[94] Long JM, Coble DW, Xiong C, *et al.* Preclinical Alzheimer's disease biomarkers accurately predict cognitive and neuropathological outcomes. Brain 2022; 145(12): 4506-18.
[http://dx.doi.org/10.1093/brain/awac250] [PMID: 35867858]

[95] Li TR, Yang Q, Hu X, Han Y. Biomarkers and tools for predicting Alzheimer's disease in the preclinical stage. Curr Neuropharmacol 2022; 20(4): 713-37.
[http://dx.doi.org/10.2174/1570159X19666210524153901] [PMID: 34030620]

[96] Andersen E, Casteigne B, Chapman WD, *et al.* Diagnostic biomarkers in Alzheimer's disease. Biomarkers in Neuropsychiatry 2021; 5: 100041.
[http://dx.doi.org/10.1016/j.bionps.2021.100041]

[97] Dubois B, von Arnim CAF, Burnie N, Bozeat S, Cummings J. Biomarkers in Alzheimer's disease: role in early and differential diagnosis and recognition of atypical variants. Alzheimers Res Ther 2023; 15(1): 175.
[http://dx.doi.org/10.1186/s13195-023-01314-6] [PMID: 37833762]

[98] Angioni D, Delrieu J, Hansson O, *et al.* Blood biomarkers from research use to clinical practice: what must be done? A report from the EU/US CTAD task force. J Prev Alzheimers Dis 2022; 9(4): 569-79.

[http://dx.doi.org/10.14283/jpad.2022.85] [PMID: 36281661]

[99] Zarkali A, Thomas GEC, Zetterberg H, Weil RS. Neuroimaging and fluid biomarkers in Parkinson's disease in an era of targeted interventions. Nat Commun 2024; 15(1): 5661. [http://dx.doi.org/10.1038/s41467-024-49949-9] [PMID: 38969680]

[100] Hong S, Baek SH, Lai MKP, Arumugam TV, Jo DG. Aging-associated sensory decline and Alzheimer's disease. Mol Neurodegener 2024; 19(1): 93. [http://dx.doi.org/10.1186/s13024-024-00776-y] [PMID: 39633396]

[101] Giannini G, Baiardi S, Dellavalle S, *et al. In vivo* assessment of Lewy body and beta-amyloid copathologies in idiopathic normal pressure hydrocephalus: prevalence and associations with clinical features and surgery outcome. Fluids Barriers CNS 2022; 19(1): 71. [http://dx.doi.org/10.1186/s12987-022-00368-2] [PMID: 36071460]

[102] Bougea A, Gourzis P. Biomarker-Based Precision Therapy for Alzheimer's Disease: Multidimensional Evidence Leading a New Breakthrough in Personalized Medicine. J Clin Med 2024; 13(16): 4661. [http://dx.doi.org/10.3390/jcm13164661] [PMID: 39200803]

[103] Rafii MS, Aisen PS. Detection and treatment of Alzheimer's disease in its preclinical stage Nature Aging 2023; 3: 520–531 [http://dx.doi.org/10.1038/s43587-023-00410-4]

[104] Cullen NC, Leuzy A, Janelidze S, *et al.* Plasma biomarkers of Alzheimer's disease improve prediction of cognitive decline in cognitively unimpaired elderly populations. Nat Commun 2021; 12(1): 3555. [http://dx.doi.org/10.1038/s41467-021-23746-0] [PMID: 34117234]

[105] McCarter SJ, Lesnick TG, Lowe V, *et al.* Cerebral amyloid angiopathy pathology and its association with amyloid-β PET signal. Neurology 2021; 97(18): e1799-808. [http://dx.doi.org/10.1212/WNL.0000000000012770] [PMID: 34504022]

[106] Wang M, Yan Z, Zhang H, *et al.* Parametric estimation of reference signal intensity in the quantification of amyloid-beta deposition: an ^{18}F-AV-45 study. Quant Imaging Med Surg 2021; 11(1): 249-63. [http://dx.doi.org/10.21037/qims-20-110] [PMID: 33392026]

[107] Villain N, Planche V, Levy R. High-clearance anti-amyloid immunotherapies in Alzheimer's disease. Part 1: Meta-analysis and review of efficacy and safety data, and medico-economical aspects. Rev Neurol (Paris) 2022; 178(10): 1011-30. [http://dx.doi.org/10.1016/j.neurol.2022.06.012] [PMID: 36184326]

[108] Ailioaie LM, Ailioaie C, Litscher G. Photobiomodulation in Alzheimer's disease—A complementary method to state-of-the-art pharmaceutical formulations and nanomedicine? Pharmaceutics 2023; 15(3): 916. [http://dx.doi.org/10.3390/pharmaceutics15030916] [PMID: 36986776]

[109] van Oostveen WM, de Lange ECM. Imaging techniques in Alzheimer's disease: a review of applications in early diagnosis and longitudinal monitoring. Int J Mol Sci 2021; 22(4): 2110. [http://dx.doi.org/10.3390/ijms22042110] [PMID: 33672696]

[110] Basak JM, Ferreiro A, Cohen LS, *et al.* Bacterial sepsis increases hippocampal fibrillar amyloid plaque load and neuroinflammation in a mouse model of Alzheimer's disease. Neurobiol Dis 2021; 152: 105292. [http://dx.doi.org/10.1016/j.nbd.2021.105292] [PMID: 33556539]

[111] Rossini PM, Miraglia F, Vecchio F. Early dementia diagnosis, MCI-to-dementia risk prediction, and the role of machine learning methods for feature extraction from integrated biomarkers, in particular for EEG signal analysis. Alzheimers Dement 2022; 18(12): 2699-706. [http://dx.doi.org/10.1002/alz.12645] [PMID: 35388959]

[112] van Dyck CH, Mecca AP, O'Dell RS, *et al.* A pilot study to evaluate the effect of CT1812 treatment on synaptic density and other biomarkers in Alzheimer's disease. Alzheimers Res Ther 2024; 16(1):

20.
[http://dx.doi.org/10.1186/s13195-024-01382-2] [PMID: 38273408]

[113] Yang Q, Wei X, Deng B, *et al.* Cerebral small vessel disease alters neurovascular unit regulation of microcirculation integrity involved in vascular cognitive impairment. Neurobiol Dis 2022; 170105750
[http://dx.doi.org/10.1016/j.nbd.2022.105750] [PMID: 35580816]

[114] Que M, Li Y, Wang X, Zhan G, Luo X, Zhou Z. Role of astrocytes in sleep deprivation: accomplices, resisters, or bystanders? Front Cell Neurosci 2023; 171188306
[http://dx.doi.org/10.3389/fncel.2023.1188306] [PMID: 37435045]

[115] Husain SA, Ahmad S, Abass S, Parveen R, Irfan M, Jan B. Synergy based extracts of medicinal plants: Future antimicrobials to combat multidrug resistance. Curr Pharm Biotechnol 2022; 23(13): 1527-40.
[http://dx.doi.org/10.2174/1389201023666220126115656] [PMID: 35081888]

[116] Parveen A, Zahiruddin S, Agarwal N, Akhtar Siddiqui M, Husain Ansari S, Ahmad S. Modulating effects of the synergistic combination of extracts of herbal drugs on cyclophosphamide-induced immunosuppressed mice. Saudi J Biol Sci 2021; 28(11): 6178-90.
[http://dx.doi.org/10.1016/j.sjbs.2021.06.076] [PMID: 34764748]

[117] Hrabinova M, Pejchal J, Kucera T, Jun D, Schmidt M, Soukup O. Is it the twilight of BACE1 inhibitors? Curr Neuropharmacol 2020; 19(1): 61-77.
[http://dx.doi.org/10.2174/1570159X18666200503023323] [PMID: 32359337]

[118] Xu Y, Jiang H, Zhu B, *et al.* Advances and applications of fluids biomarkers in diagnosis and therapeutic targets of Alzheimer's disease. CNS Neurosci Ther 2023; 29(8): 2060-73.
[http://dx.doi.org/10.1111/cns.14238] [PMID: 37144603]

[119] Smirnov DS, Ashton NJ, Blennow K, *et al.* Plasma biomarkers for Alzheimer's Disease in relation to neuropathology and cognitive change. Acta Neuropathol 2022; 143(4): 487-503.
[http://dx.doi.org/10.1007/s00401-022-02408-5] [PMID: 35195758]

[120] Llitjos JF, Carrol ED, Osuchowski MF, *et al.* Enhancing sepsis biomarker development: key considerations from public and private perspectives. Crit Care 2024; 28(1): 238.
[http://dx.doi.org/10.1186/s13054-024-05032-9] [PMID: 39003476]

[121] del Campo M, Zetterberg H, Gandy S, *et al.* New developments of biofluid-based biomarkers for routine diagnosis and disease trajectories in frontotemporal dementia. Alzheimers Dement 2022; 18(11): 2292-307.
[http://dx.doi.org/10.1002/alz.12643] [PMID: 35235699]

[122] Ashton NJ, Leuzy A, Karikari TK, *et al.* The validation status of blood biomarkers of amyloid and phospho-tau assessed with the 5-phase development framework for AD biomarkers. Eur J Nucl Med Mol Imaging 2021; 48(7): 2140-56.
[http://dx.doi.org/10.1007/s00259-021-05253-y] [PMID: 33677733]

[123] Wegmann S, Biernat J, Mandelkow E. A current view on Tau protein phosphorylation in Alzheimer's disease. Curr Opin Neurobiol 2021; 69: 131-8.
[http://dx.doi.org/10.1016/j.conb.2021.03.003] [PMID: 33892381]

[124] Nachega JB, Atteh R, Ihekweazu C, *et al.* Contact tracing and the COVID-19 response in Africa: best practices, key challenges, and lessons learned from Nigeria, Rwanda, South Africa, and Uganda. Am J Trop Med Hyg 2021; 104(4): 1179-87.
[http://dx.doi.org/10.4269/ajtmh.21-0033] [PMID: 33571138]

[125] Lane J, Greig C, Garnett A. Uncertain storage prospects create a conundrum for carbon capture and storage ambitions. Nat Clim Chang 2021; 11(11): 925-36.
[http://dx.doi.org/10.1038/s41558-021-01175-7]

[126] Erickson CM, Chin NA, Johnson SC, Gleason CE, Clark LR. Disclosure of preclinical Alzheimer's disease biomarker results in research and clinical settings: Why, how, and what we still need to know. Alzheimers Dement (Amst) 2021; 13(1)e12150.

[http://dx.doi.org/10.1002/dad2.12150] [PMID: 33665341]

[127] Ramesh M, Govindaraju T. Multipronged diagnostic and therapeutic strategies for Alzheimer's disease. Chem Sci (Camb) 2022; 13(46): 13657-89. [http://dx.doi.org/10.1039/D2SC03932J] [PMID: 36544728]

[128] Juganavar A, Joshi A, Shegekar T. Navigating early Alzheimer's diagnosis: a comprehensive review of diagnostic innovations. Cureus 2023; 15(9)e44937. [http://dx.doi.org/10.7759/cureus.44937] [PMID: 37818489]

[129] Dulewicz M, Kulczyńska-Przybik A, Mroczko P, Kornhuber J, Lewczuk P, Mroczko B. Biomarkers for the diagnosis of Alzheimer's disease in clinical practice: the role of CSF biomarkers during the evolution of diagnostic criteria. Int J Mol Sci 2022; 23(15): 8598. [http://dx.doi.org/10.3390/ijms23158598] [PMID: 35955728]

[130] Amft M, Ortner M, Eichenlaub U, *et al.* The cerebrospinal fluid biomarker ratio Aβ42/40 identifies amyloid positron emission tomography positivity better than Aβ42 alone in a heterogeneous memory clinic cohort. Alzheimers Res Ther 2022; 14(1): 60. [http://dx.doi.org/10.1186/s13195-022-01003-w] [PMID: 35473631]

[131] Abdelnour C, Agosta F, Bozzali M, *et al.* Perspectives and challenges in patient stratification in Alzheimer's disease. Alzheimers Res Ther 2022; 14(1): 112. [http://dx.doi.org/10.1186/s13195-022-01055-y] [PMID: 35964143]

[132] Vogel JW, Young AL, Oxtoby NP, *et al.* Four distinct trajectories of tau deposition identified in Alzheimer's disease. Nat Med 2021; 27(5): 871-81. [http://dx.doi.org/10.1038/s41591-021-01309-6] [PMID: 33927414]

[133] Ossenkoppele R, Pichet Binette A, Groot C, *et al.* Amyloid and tau PET-positive cognitively unimpaired individuals are at high risk for future cognitive decline. Nat Med 2022; 28(11): 2381-7. [http://dx.doi.org/10.1038/s41591-022-02049-x] [PMID: 36357681]

[134] Aschenbrenner AJ, Li Y, Henson RL, *et al.* Comparison of plasma and CSF biomarkers in predicting cognitive decline. Ann Clin Transl Neurol 2022; 9(11): 1739-51. [http://dx.doi.org/10.1002/acn3.51670] [PMID: 36183195]

[135] Assfaw AD, Schindler SE, Morris JC. Advances in blood biomarkers for Alzheimer disease (AD): A review. Kaohsiung J Med Sci 2024; 40(8): 692-8. [http://dx.doi.org/10.1002/kjm2.12870] [PMID: 38888066]

CHAPTER 4

Neuroimaging and Advanced Techniques in Alzheimer's Diagnosis

Snigdha Srivastava[1,3,*], **Rajan Kumar Kurmi**[2,3] and **Shiva Mishra**[4,5]

[1] *Galgotias College of Pharmacy, Greater Noida, Uttar Pradesh 201306, India*

[2] *Metro College of Health Sciences and Research, Greater Noida, Uttar Pradesh 201308, India*

[3] *Department of Pharmaceutical Chemistry, Jamia Hamdard (Hamdard University), New Delhi 110062, India*

[4] *Lloyd Institute of Management and Technology, Greater Noida, Uttar Pradesh 201306, India*

[5] *Lloyd School of Pharmacy, Greater Noida, Uttar Pradesh 201306, India*

Abstract: Globally, dementia is a major cause of death and disability. The early, precise, and differential diagnosis of dementia and Alzheimer's disease depends on neuroimaging or brain imaging. Because of structural and molecular imaging, which is also increasingly being employed in clinical practice for early and accurate diagnosis, the pathogenesis of neurodegenerative dementias has been better understood. Several neuroimaging techniques that are regarded as state-of-the-art diagnostic aids were highlighted. Most of the research that uses imaging modalities for early diagnosis is preventive studies. The accurate identification of AD biomarkers is made possible by AI-enhanced neuroimaging methods such as MRI, PET, and CT scans. With technologies that improve the precision of neuropsychological testing and examine speech and language patterns for early indicators of dementia, artificial intelligence has also enhanced cognitive and behavioural examinations. The MRI modalities discussed encompass structural MRI, functional MRI, diffusion tensor imaging, magnetic resonance spectroscopy, and arterial spin labelling. These advanced techniques enable the identification of presymptomatic diagnostic biomarkers in the brains of cognitively normal older adults and can also be utilized to track Alzheimer's disease progression following the emergence of clinical symptoms. Advanced MRI techniques, such as arterial spin labelling and Diffusion Tensor Imaging (DTI), are also being explored to better understand Alzheimer's disease. In PET imaging, ongoing efforts focus on developing high selectivity and affinity ligands. As no single imaging method is adequate for early diagnosis, this has led to the emergence of a new approach known as "multi-modal imaging."

[*] **Corresponding author Snigdha Srivastava:** Galgotias College of Pharmacy, Greater Noida, Uttar Pradesh 201306, India, Department of Pharmaceutical Chemistry, Jamia Hamdard (Hamdard University), New Delhi 110062, India; E-mail: snigdhasrivastava2247@gmail.com

Prashant Kumar Dhakad, Raghav Mishra, Shweta Sharma, Sudhanshu Mishra & Nishant Gaur (Eds.)

Keywords: AI-enhanced diagnosis, Diffusion tensor imaging, Magnetic resonance imaging, MRI modalities, Neuroimaging.

INTRODUCTION

The most common type of dementia, Alzheimer's Disease (AD), is defined by a progressive loss of cognitive abilities that limits a person's ability to carry out daily tasks by impacting memory, thinking, and social skills. The pathological basis of Alzheimer's Disease (AD) is the accumulation of neurofibrillary tangles and amyloid plaques, primarily composed of hyperphosphorylated tau proteins and amyloid-beta (Aβ) peptides, respectively. These pathological processes drive neurodegeneration and synaptic loss, ultimately leading to profound cognitive and behavioral impairments [1, 2].

As the global population ages, the prevalence of AD is expected to increase markedly. Current estimates suggest that approximately 47 million individuals are affected worldwide, with a new case diagnosed every 35 seconds. This demographic trend highlights the urgent need for more effective diagnostic tools and therapeutic strategies to improve patient outcomes [3].

Early diagnosis is critical in the management of AD. Evidence indicates that cognitive decline can precede the onset of overt clinical symptoms by several years. Detecting the disease in its early stages allows clinicians to initiate interventions that may prolong independence and enhance quality of life for both patients and their families [4, 5]. Initiation of symptomatic treatment soon after diagnosis may also delay progression [6 - 8]. Furthermore, early diagnosis reduces the likelihood of misdiagnosis and inappropriate treatment, which are more common in later stages. It also enables patient participation in clinical trials, providing access to emerging therapies with the potential to modify disease course. Thus, early detection is essential not only for optimizing individual care but also for addressing the broader public health challenges associated with AD [9].

NEUROIMAGING'S ROLE IN DIAGNOSING ALZHEIMER'S DISEASE

Neuroimaging provides essential information on the structural and functional alterations of the brain in Alzheimer's Disease (AD). Fig. (**1**) illustrates the major neuroimaging modalities—Computed Tomography (CT), Magnetic Resonance Imaging (MRI), functional MRI (fMRI), positron emission tomography (PET), and Single-Photon Emission Computed Tomography (SPECT)—each of which has distinct applications in the clinical evaluation of AD. The figure also summarizes the characteristic symptoms and brain regions typically assessed by these techniques (Table **1**).

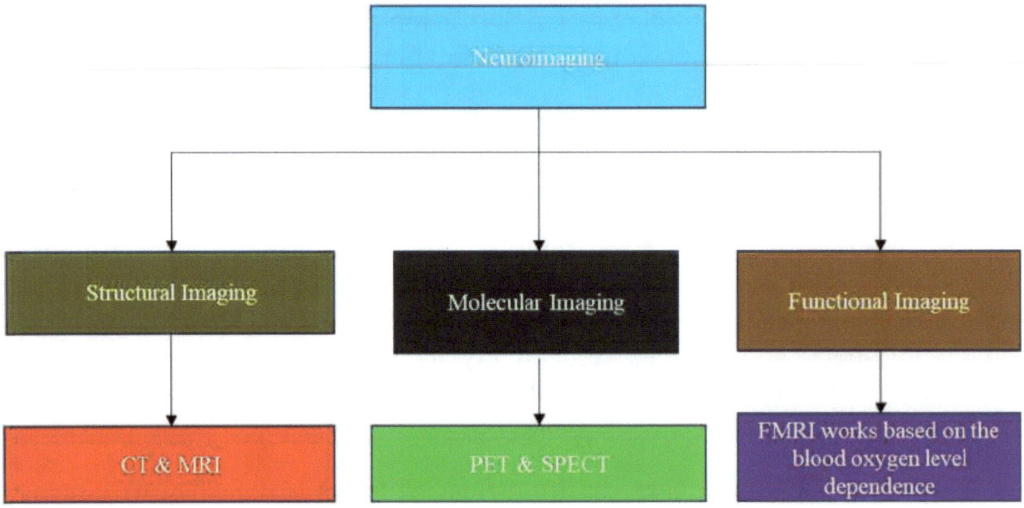

Fig. (1). Classification of neuroimaging techniques. The chart is prepared using MS PowerPoint.

Table 1. Types of neuroimaging, classifications, symptoms, and areas involved.

Neuroimaging Type	Classification	Types	Common Symptoms	Areas Involved
FMRI	Non-Invasive	Functional Imaging	Cognitive impairments, mood disorders	Prefrontal Cortex, Temporal Lobe
MRI	Non-Invasive	Functional	Memory Issues, Cognitive Deficits	Prefrontal Cortex, Temporal Lobe
CT	Non-Invasive	Structural Imaging	Confusion, Strokes, Trauma, Tumors, Dizziness	Various Brain Sites
PET	Non-Invasive	Metabolic Imaging	Alzheimer's, cancer	Hippocampus region
SPECT	Non-Invasive	Perfusion Imaging	Neurodegenerative Disorders	Various Brain Regions
DTI	Non-Invasive	Structural (White Matter)	Chronic Pain Reduction in Cognitive Ability	Anterior-Posterior Brain Regions
DWI	Non-Invasive	Diffusive Restriction Imaging	Fractional Anisotropy, Neuronal Loss, Hippocampal Atrophy	Corpus Callosum, Posterior Cingulate Cortex
NFTS	Non-Invasive	NFT Distribution and Tau Pathology Visualization	Memory and Language Impairment, Visuospatial Dysfunction, Tremors, and Hallucinations	Entorhinal Cortex, Parahippocampal Gyrus, Primary Sensor and Motor Cortex

Structural Imaging

Magnetic Resonance Imaging (MRI) is particularly valuable for assessing the hippocampus, a region critical for memory and cognitive function. Studies have shown that MRI can reliably detect hippocampal atrophy, a hallmark feature of Alzheimer's Disease (AD) pathology. According to the National Institute on Aging–Alzheimer's Association (NIA-AA) guidelines, MRI plays an important role in distinguishing AD from other neurodegenerative disorders, especially when combined with complementary diagnostic modalities [10, 11]. Advanced approaches such as Diffusion Tensor Imaging (DTI) further enhance diagnostic accuracy by revealing early microstructural changes in white matter tracts, often detectable before the onset of clinical symptoms [12]. Fig. (**2**) compares neuroimaging findings in the brain of a patient with AD to those of a healthy individual.

Functional Imaging

Advances in functional neuroimaging, particularly functional MRI (fMRI), have enabled the investigation of altered brain connectivity in Alzheimer's Disease (AD). Resting-state fMRI (rs-fMRI) is especially useful for assessing spontaneous brain activity and has been employed to detect functional connectivity changes within the Default Mode Network (DMN), a network critical for memory and self-referential processing [13]. Notably, alterations in low-frequency oscillations of the BOLD signal measured by rs-fMRI have been associated with cognitive decline in AD, suggesting that this technique may serve as a potential biomarker for early diagnosis [14].

Molecular Imaging

AT the molecular level, PET imaging is essential because it makes it possible to see the buildups of tau protein and amyloid plaques, two pathological indicators. Before clinical dementia, the significance of PET in early diagnosis is shown by studies that confirm the existence of amyloid plaques up to 10–30 years. Better understanding and improved patient care of the course of the disease are made possible by the use of different PET tracers, such as those that target tau and amyloid beta deposits. SPECT imaging provides an additional layer of functional analysis; it is frequently utilized to evaluate cerebral blood flow and metabolic activity in AD patients [15, 16].

Emerging Techniques

Machine learning algorithms and Dynamic contrast-enhanced MRI (DCE-MRI) applied to fMRI data are two recent developments in neuroimaging techniques that are opening up new possibilities for classification and AD diagnosis. For example, machine learning classifiers built on fMRI data have demonstrated encouraging outcomes in distinguishing between normal aging and Alzheimer's disease. DCE-MRI also sheds light on blood-brain barrier integrity, which is weakened in AD, demonstrating the wide range of applications for contemporary neuroimaging methods [17, 18].

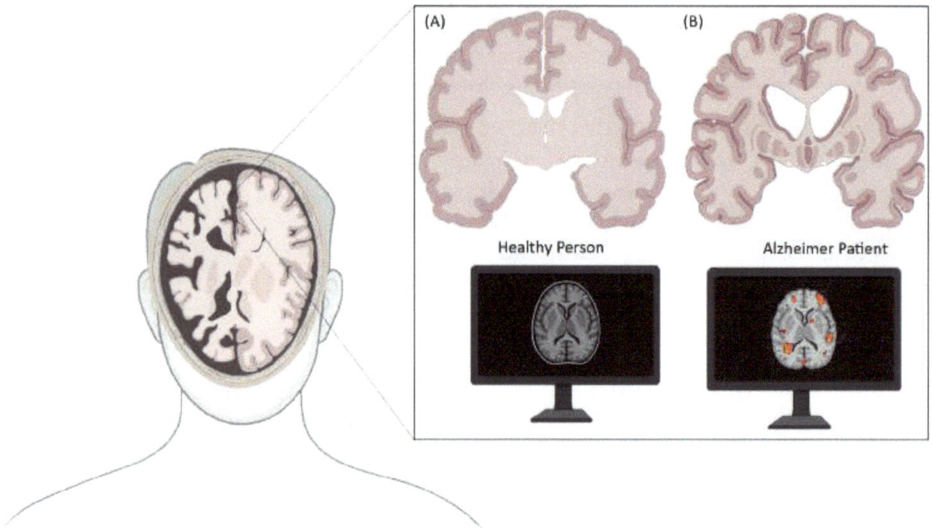

Fig. (2). Neuroimaging of a human brain (**A**) Healthy person's brain, and (**B**) Alzheimer's patient's brain. "Figure created with BioRender.com."

Computed Tomography (CT)

Computed Tomography (CT) uses X-rays to generate high-resolution cross-sectional images of the brain. Its rapid image acquisition and ability to detect acute intracranial conditions make it particularly valuable in clinical practice. CT is often employed in initial evaluations to promptly identify hemorrhages, tumors, or strokes, providing critical information for timely medical decision-making. Studies have demonstrated that CT effectively determines the location and extent of pathological changes in the brain, supporting the assessment of various neurological disorders, including structural abnormalities and neurological emergencies [19, 20].

CT also holds historical significance as a standard brain imaging modality, especially in pediatric patients, where minimizing radiation exposure is crucial. While MRI may offer greater sensitivity in certain contexts, CT remains preferred in acute situations due to its speed and reliability, making it indispensable in emergency care. Additionally, CT has been utilized in psychiatric research to differentiate between organic and non-organic origins of psychopathology, including in individuals experiencing first-episode psychosis [21, 22].

Magnetic Resonance Imaging (MRI)

Magnetic Resonance Imaging (MRI) is a non-invasive technique that provides high-resolution images of the brain, revealing both macrostructural and microstructural changes in neuronal architecture. Unlike CT, MRI does not use ionizing radiation, allowing for repeated imaging without associated risks. By employing radio waves and strong magnetic fields, MRI generates detailed images of delicate brain tissues [23, 24].

MRI encompasses several modalities, including structural MRI (sMRI), diffusion tensor imaging (DTI), and functional MRI (fMRI). Structural MRI is primarily used to evaluate brain morphology, while fMRI analyzes brain activity by detecting changes in the blood oxygen level-dependent (BOLD) signal. In conditions such as Alzheimer's disease, where structural atrophy and functional connectivity alterations coexist, combining these modalities enables a comprehensive assessment of brain networks and has expanded research opportunities [25, 26].

MRI has also been instrumental in identifying biomarkers associated with cognitive dysfunction, advancing our understanding of both neurodegenerative and neurodevelopmental disorders, such as neurological manifestations in systemic lupus erythematosus [27, 28]. Its unmatched ability to visualize brain structure and function has established MRI as the "gold standard" in numerous research and clinical settings, supporting the development of diagnostic criteria and therapeutic strategies [29 - 32].

Positron Emission Tomography (PET) and Single Photon Emission Computed Tomography (SPECT)

PET and SPECT have significantly advanced our understanding of neurological disorders by providing functional imaging of the brain. SPECT detects gamma rays emitted from radiolabeled compounds to evaluate cerebral blood flow and brain activity, whereas PET employs positron-emitting radiotracers to visualize metabolic processes. These modalities are essential in both clinical practice and research, aiding in the diagnosis and management of conditions such as epilepsy and Alzheimer's disease [29 - 31]. Unlike structural imaging techniques such as CT or MRI, PET and SPECT offer real-time insights into molecular and neurochemical processes, allowing analysis of disease mechanisms at the biochemical level. For instance, radiotracers that selectively bind to neuroreceptors can provide detailed information on the neurochemistry of disorders such as depression and schizophrenia, where serotonin dysregulation is implicated [33 - 35].

Recent advances in hybrid imaging, including PET/CT and SPECT/CT, enhance diagnostic accuracy by combining anatomical and functional information. This multimodal approach improves sensitivity and specificity, enabling the detection of subtle changes that may be missed with a single modality. In epilepsy, interictal PET and ictal SPECT have demonstrated high effectiveness in localizing seizure foci, and patients undergoing surgery guided by these imaging results often experience significant improvements in seizure control [33, 34]. Statistical parametric mapping further enhances data interpretation, allowing more detailed analysis of regional brain activity [36, 37].

In neurodegenerative disorders, PET and SPECT are invaluable for early, non-invasive diagnosis. Specific radioligands can visualize beta-amyloid plaques in Alzheimer's disease, while dopaminergic tracers assess vesicular monoamine transporter function in Parkinson's disease, aiding early detection and therapeutic planning [36]. These imaging methods also provide insights into the Blood-Brain Barrier (BBB), particularly in epilepsy, where they can reveal pathophysiological changes during seizures and guide novel therapeutic strategies [38, 39].

Ongoing innovations in molecular probes and radiopharmaceuticals continue to expand the utility of PET and SPECT, clarifying neurotransmitter signaling pathways and advancing understanding of psychiatric and neurodegenerative conditions [40]. Integration with machine learning and artificial intelligence offers further potential, improving image reconstruction, noise reduction, artifact correction, and processing speed. Such approaches could facilitate more personalized and efficient patient care, enabling rapid decision-making in clinical and emergency settings [40 - 43].

The convergence of advanced imaging technologies, molecular probes, and AI presents an unprecedented opportunity to redefine neuroimaging standards, paving the way for personalized medicine where imaging informs individualized diagnostic and therapeutic strategies.

FMRI

Over the past few decades, functional imaging has advanced significantly, largely due to improvements in Magnetic Resonance Imaging (MRI) technology. These developments have led to the creation of numerous neuroimaging techniques, each with specific methodologies and clinical or research applications, fulfilling the demand for accurate, non-invasive diagnostic tools.

Functional MRI (fMRI), one of the most widely used functional imaging modalities, provides insights into brain activity by detecting changes in blood flow. This technique relies on the Blood-Oxygen-Level-Dependent (BOLD) effect, which highlights regions of the brain that are active during specific tasks or at rest. Fig. (**3**) illustrates the pathway of magnetic field utilization in the fMRI detection process. fMRI is particularly valuable for identifying biomarkers in psychiatric disorders such as schizophrenia and for understanding cognitive processes, as it reveals disrupted connectivity within well-defined brain networks. Recent advancements have further enhanced the sensitivity of neuroimaging data analysis, enabling integration of fMRI with machine learning techniques to improve diagnostic accuracy and classification of neurological and psychiatric conditions.

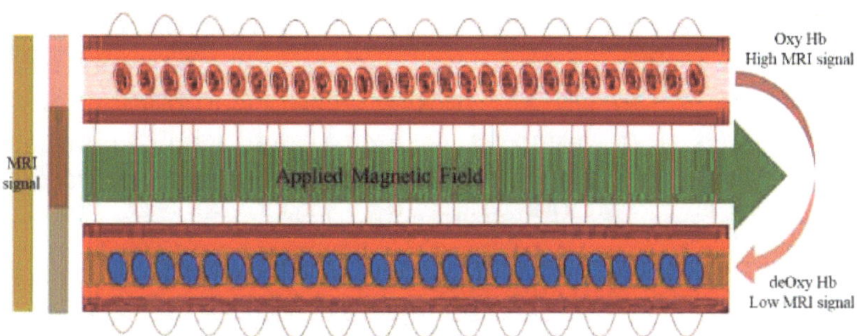

Fig. (3). Image showing the magnetic signal pathway used in FMRI detection. The image is prepared using MS PowerPoint.

Diffusion Weighted Imaging

Diffusion-Weighted Imaging (DWI), which measures the diffusion of water molecules within tissue, provides critical information about tissue integrity and pathology. DWI and its derivative, Diffusion Tensor Imaging (DTI), are highly effective in evaluating white matter integrity and detecting abnormalities associated with neurodegenerative disorders and traumatic brain injury. Combining DWI with functional MRI (fMRI) further enhances understanding of the brain's functional connectivity.

Dynamic contrast-enhanced MRI (DCE-MRI) is another important modality, particularly in oncology, as it assesses tissue perfusion by analyzing contrast agent uptake over time. DCE-MRI is essential for tumor characterization, treatment planning, and outcome evaluation in malignancies such as cervical and prostate cancer. Multiparametric MRI (MP-MRI), which integrates multiple imaging approaches, has demonstrated improved diagnostic accuracy and sensitivity across a range of clinical scenarios.

Emerging functional imaging techniques, including Arterial Spin Labeling (ASL) and hyperpolarized gas MRI, allow non-invasive assessment of cerebral blood flow and lung function, respectively, providing direct evaluation of physiological processes. Furthermore, hybrid imaging modalities, such as PET/MRI, enable comprehensive assessments by combining anatomical, functional, and molecular information. These multimodal approaches enhance the understanding of neurological and psychiatric disorders, facilitate accurate diagnosis, and support more effective treatment planning. The field of functional neuroimaging continues to evolve rapidly, with ongoing advancements prioritizing precision, efficiency, and patient safety.

Neurofibrillary Tangles (NFTs) and Their Role in Neuroimaging Techniques

In neuroimaging, these can be crucial for monitoring the origin of medical imaging procedures and preserving data integrity, especially for diseases where neurofibrillary tangles (NFTs) are present. The incorporation of NFTs into healthcare processes can improve patient data management, encourage transparency in research, and enable the safe interchange of neuroimaging data, given the growing complexity of neuroimaging techniques.

Applications of NFTs in Neuroimaging

Non-fungible tokens (NFTs) are emerging as practical tools for monitoring and validating medical imaging data. In neuroimaging, they have been proposed to manage the lifecycle of medical devices, such as PET scanners used in Alzheimer's disease research, and to document the steps involved in image processing. By securely encapsulating metadata from each imaging session, NFTs enhance traceability and reduce the risk of data manipulation, thereby improving the reliability of neuroimaging results.

NFTs can also serve as digital twins of medical equipment, recording characteristics from manufacturing through current use. This functionality allows researchers to authenticate devices and ensure that imaging data acquired from them is trustworthy. In addition, NFT-based frameworks can support the entire research process—from data collection to analysis and reporting—by guaranteeing secure access to validated data for all stakeholders.

Beyond device and data management, NFTs are gaining traction in the interpretation of neuroimaging results. For example, NFT-linked radiotracers in PET imaging have been applied to tauopathies, selectively binding to neurofibrillary tangles and revealing pathogenic pathways in Alzheimer's disease. Tracking these radiotracers with NFT technology enables researchers to monitor their performance over time and more accurately correlate imaging findings with clinical outcomes [40 - 43].

Neurofeedback Training (NFT) and Its Role in Neuroimaging

Neurofeedback training (NFT) represents a novel application of neuroimaging aimed at enhancing cognitive abilities and treating neurological deficits. This non-invasive approach provides real-time feedback on physiological signals, most commonly electroencephalogram (EEG) activity, enabling individuals to learn how to modulate their own brain function. By creating a feedback loop in which brainwave patterns are adjusted in response to external cues, NFT promotes neuroplasticity and supports improvements in cognitive performance [44 - 47]. Recent studies have demonstrated its effectiveness in various contexts, particularly in the rehabilitation of cognitive impairments following brain injury.

Advanced neuroimaging techniques are often integrated into NFT protocols to visualize neural changes associated with training. Alterations in Default Mode Network (DMN) activity, for example, have been correlated with improvements in cognitive control [48]. This integration not only deepens understanding of the physiological mechanisms underlying cognitive rehabilitation but also facilitates the development of personalized treatment strategies.

The benefits of NFT extend beyond clinical rehabilitation. In sports science, for instance, NFT has been applied to enhance performance by training athletes to achieve greater focus and self-regulation, as reflected in specific brainwave modifications [49, 50]. Overall, NFT complements traditional neuroimaging by adding an interactive training dimension that holds promise for both therapeutic and performance-enhancement applications.

Standardizing Neuroimaging with Blockchain and NFTs

By leveraging blockchain technology, Non-Fungible Tokens (NFTs) can provide a tamper-proof and transparent ledger for documenting every stage of neuroimaging data management, from initial acquisition to final analysis. This capability holds particular promise for facilitating multi-institutional collaborations, streamlining cooperative research efforts, and ultimately improving the reproducibility of neuroimaging findings [51, 52]. In addition, the integration of NFTs with blockchain systems enables secure sharing of neuroimaging datasets while maintaining strict patient confidentiality. By tokenizing sensitive data and enforcing controlled access protocols, researchers can balance the need for data availability with the increasing demands of privacy regulations and ethical research practices [53, 54].

Despite these advantages, several challenges must be addressed before NFTs can be widely adopted in neuroimaging. Key concerns include the environmental impact of blockchain technologies, the lack of unified regulatory frameworks governing data ownership, and interoperability issues across different blockchain platforms. Moreover, the absence of universally accepted standards for NFT use in medical imaging limits scalability and consistency.

Looking forward, NFTs may also serve as a foundation for integrating Artificial Intelligence (AI) into neuroimaging workflows. AI, particularly deep learning, has already shown transformative potential in image analysis by enhancing accuracy and efficiency in classification and segmentation tasks. When paired with NFTs, AI-driven pipelines could produce more reproducible, auditable, and generalizable outcomes across diverse datasets [55 - 57]. Together, these innovations highlight the potential of NFTs not only as tools for data security and governance but also as enablers of more robust and transparent neuroimaging research.

BIOMARKERS IN ALZHEIMER'S DISEASE

Cerebrospinal Fluid (CSF) Biomarkers

Cerebrospinal fluid (CSF) is a clear, colorless fluid that surrounds the brain and spinal cord, providing mechanical protection, facilitating nutrient transport, and supporting the clearance of metabolic waste. It is primarily produced by the choroid plexus within the brain's ventricles, circulates through the ventricular system and subarachnoid space, and is ultimately reabsorbed into the bloodstream via the arachnoid granulations. Because CSF directly reflects biochemical processes within the central nervous system and maintains a close connection with peripheral circulation, it serves as a valuable medium for detecting metabolic and pathological changes associated with neurological disorders. Consequently, CSF analysis has become a central approach in biomarker research. A range of methodologies—such as proteomics, metabolomics, and immunoassays—are employed to identify and categorize biomarkers into diagnostic, prognostic, and therapeutic-response indicators. These approaches provide critical insights into disease mechanisms, support early and accurate diagnosis, and help monitor treatment efficacy (Table **2**).

Role of CSF Biomarkers in Neurological Diseases

CSF biomarkers are essential for neurological illness diagnosis, prognosis, and follow-up.

• **Neurodegenerative diseases:** Biomarkers such as tau, α-synuclein, and Aβ42 help diagnose Parkinson's disease, Alzheimer's disease, and other neurodegenerative disorders.

• **Neuroinflammatory disorders:** Cytokines and oligoclonal bands (OCBs) aid in the diagnosis of multiple sclerosis and other autoimmune diseases that impact the central nervous system. Proteins like Glial Fibrillary Acidic Protein (GFAP) and neurofilament light chain (NfL) are markers of astrocytic and neuronal damage in traumatic brain injury (TBI).
• **Infections of the central nervous system:** Pathogen-specific proteins and nucleic acids in CSF can identify bacterial, fungal, and viral meningitis or encephalitis.

Characteristics of an Ideal CSF Biomarker

• The following characteristics are appropriate for a CSF biomarker:

• **High specificity:** It needs to be able to discern between various neurological disorders with accuracy.
• **High sensitivity:** It ought to identify the illness before any outward signs appear.
• **Reproducibility:** Findings need to hold true for various patient groups and labs.
• **Clinical utility:** The biomarker should help with prognostication, illness diagnosis, or therapy response monitoring.

Table 2. Classification of type and category of biomarker.

Type of Biomarker	Category of Biomarker	Method to Analyze	Outcomes in Alzheimer's Disease
Hippocampal Atrophy	Neurodegeneration	MRI	Present
CSF & Plasma NfL		Blood & CSF Test	Elevated
PET Hypometabolism		PET Imaging	Reduced Glucose Metabolism
CSF AB42	Amyloid	Lumbar Puncture	Decreased
CSF AB42/AB40		Lumbar Puncture	Decreased
Amyloid-PET		PET Imaging	Increased Signals
CSF p-Tau181,p-Tau217	Tau	Lumbar Puncture	Elevated
CSF t-Tau		Lumbar Puncture	Elevated
Tau-PET		PET Imaging	Increased Binding in the Effected Region

Amyloid-β (Aβ) Peptides

One of the primary diagnostic criteria for Alzheimer's Disease (AD) is the presence of amyloid-β (Aβ) plaques in the brain, which is the traditional way that AD pathology is defined. One crucial step in the diagnostic procedure is still measuring the amount of Aβ peptides, namely Aβ1-42, in cerebrospinal fluid (CSF). Research indicates that a disrupted amyloid metabolism frequently seen in AD patients is indicated by reduced CSF Aβ1-42 levels, which correspond with greater amyloid deposition in the brain. Furthermore, the cutoff values for Aβ1-42 levels might change based on the stage of the illness or the presence of tau pathologies; advanced AD frequently exhibits noticeably lower levels, which validate the existence of neurodegeneration [58, 59]. Aβ peptides represent the underlying neurodegenerative mechanisms that cause cognitive impairment in addition to acting as important biomarkers [60, 61].

Tau Proteins

Tau proteins, particularly hyperphosphorylated tau, are important indicators in the diagnosis of Alzheimer's disease in addition to Aβ peptides. Increased levels of phosphorylated tau (p-tau) and total tau (t-tau) in CSF are linked to neurofibrillary tangles, a defining characteristic of AD pathogenesis. Numerous studies indicate that p-tau levels—particularly p-tau181—can significantly predict the start of cognitive impairment and the course of Alzheimer's disease, exceeding tests of total tau [62, 63]. Additionally, P-tau is a marker for tau deposition and can reveal the disease's pathological stage, offering crucial information for early identification and patient classification in clinical settings [64, 65].

BLOOD-BASED BIOMARKERS

Neurofilament Light Chain (NfL)

NfL, or neurofilament light chain, has become a potential blood-based Alzheimer's disease biomarker. Neuronal damage and the severity of the disease are reliably indicated by elevated NfL levels in plasma, which correspond with neurodegeneration and cognitive impairment. NfL levels may be able to differentiate Alzheimer's disease from other neurodegenerative diseases, according to research, providing a non-invasive testing method that improves diagnostic specificity in clinical settings [66, 67]. The increasing amount of data demonstrates NfL's capacity for diagnosis, tracking the course of the illness, and evaluating the effectiveness of treatment measures [68 - 70].

Emerging Biomarkers

New biomarkers, including phosphorylated tau (p-tau181 and p-tau217), have been developed and validated in recent blood-based biomarker research, and their diagnostic efficacy has been equivalent to that of conventional CSF biomarkers. The evaluation of people with Mild Cognitive Impairment (MCI) who are at risk of developing AD may be streamlined by these biomarkers, which also aid in early identification. Furthermore, current research investigates new cytokine profiles as possible biomarkers that might improve diagnostic precision and provide insight into the inflammatory mechanisms underlying Alzheimer's disease pathogenesis. All things considered, using blood-based biomarkers in a clinical setting makes them more accessible and useful, encouraging early intervention and tailored medicine approaches for Alzheimer's patients [71, 72].

Integration of Neuroimaging and Biomarkers in Clinical Research

A promising path toward the development of diagnostic and treatment approaches in neurology and psychiatry is the combination of biomarkers and neuroimaging techniques. Researchers improve the specificity and accuracy of their evaluations of a range of neurological disorders by fusing biological markers with imaging modalities.

Techniques and Approaches

A wide range of neuroimaging techniques, including Positron Emission Tomography (PET) and Magnetic Resonance Imaging (MRI), have become central tools for investigating brain structure and function. Advances in MRI, for instance, now enable the detection of subtle tissue alterations associated with cerebrovascular disease and neurodegeneration. Parallel progress in biomarker research, particularly those derived from blood and cerebrospinal fluid (CSF), has deepened understanding of the pathophysiological mechanisms underlying conditions such as Alzheimer's disease and vascular cognitive impairment. Current research trends emphasize the integration of imaging and biomarker data within multimodal frameworks, which enhances both diagnostic accuracy and disease monitoring. For example, the TRACK-FA multicenter study applies harmonized imaging protocols across different scanners to validate biomarkers, thereby generating robust datasets with improved reliability. Similarly, incorporating biomarkers into clinical assessments has strengthened the evaluation of Huntington's disease by linking neuroimaging findings with functional and cognitive measures, providing a more comprehensive picture of disease progression [73, 74].

Clinical Trials and Case Studies

To improve patient stratification and treatment efficacy, clinical trials are increasingly using this combined approach. Recent studies—such as those examining the use of T2-FLAIR and amyloid PET imaging—show how whole-brain white matter patterns correlate with clinical characteristics when assessing cognitive impairment in older adults. Additionally, several cohort studies are showing that neuroimaging biomarkers may accurately forecast clinical outcomes. For example, in the setting of stroke recovery, neuroimaging data improves the prediction accuracy of conventional clinical assessments. New research suggests that neuroimaging and biomarkers may help guide individualized treatment plans, especially for conditions like major depressive disorder, where finding cross-modal biomarkers may help explain how each patient responds to treatment. In addition to promoting a better comprehension of disease mechanisms, this integration improves the planning of interventional studies meant to create novel treatments [75, 76].

Challenges and Limitations

The integration of neuroimaging and biomarkers holds great promise, but there are several obstacles that prevent its widespread use. One of the main ones is the variation in imaging techniques and protocols, which can impact study comparability and reproducibility. Furthermore, because biomarkers are dynamic, it might be challenging to develop uniform categorization standards, especially when monitoring the course of a disease. Furthermore, there are also ethical and practical issues with collecting longitudinal data and obtaining patient permission in multicenter research. For example, research on complicated diseases like chronic traumatic encephalopathy emphasizes the necessity of organized procedures to properly evaluate the interactions between neuroimaging and biomarker data from various populations [77, 78].

FUTURE DIRECTIONS AND INNOVATIONS IN NEUROIMAGING TECHNIQUES

Advancements in Technology

Over the past few decades, neuroimaging has advanced significantly, moving from crude methods to complex modalities that improve our comprehension of the structure and function of the brain. Several important topics, including the development of new imaging methods, the shift to personalized care, and the use of artificial intelligence and machine learning, have the potential to significantly advance neuroimaging in the future [77].

Artificial Intelligence and Machine Learning

Artificial Intelligence (AI) and Machine Learning (ML) have started to change the field of neuroimaging. Because of their ability to analyze large amounts of data, AI and ML algorithms can find intricate patterns that may be invisible to human observers. These technologies are especially useful for automating image segmentation, classification, and analysis, which improves the accuracy and efficiency of neuroimaging studies. Significant progress has been achieved in the use of deep learning methods, which improve picture interpretation by employing multi-layer neural networks. For example, Convolutional Neural Networks (CNNs) can improve diagnosis accuracy by analyzing neuroimaging data to find biomarkers linked to neurological illnesses. Additionally, based on baseline neuroimaging data, predictive modeling made possible by machine learning algorithms may estimate the course of a disease and the results of treatment. We believe that as these technologies develop, AI will become more smoothly integrated into neuroimaging operations, enabling real-time analysis and speedy diagnosis. Furthermore, our knowledge of brain health and illness will be greatly expanded by AI's capacity to help identify novel biomarkers, opening the door to creative treatment strategies [79].

Novel Imaging Techniques

Exploring new imaging modalities that provide previously unheard-of insights into the complexity of the brain is also crucial to the future of neuroimaging. Utilizing magnetic fields higher than the standard 3 Tesla, ultra-high-field Magnetic Resonance Imaging (MRI) is one potential field. To better comprehend a variety of neuropsychiatric disorders, researchers can see smaller anatomical features and changes in brain activity thanks to ultra-high-field MRI's improved spatial resolution and sensitivity. In addition to these methods, a more thorough understanding of brain function and disease is made possible by the integration of multimodal imaging, which combines structural (like MRI), functional (like fMRI), and molecular imaging (like PET). These approaches facilitate the examination of complex interactions between brain structure, function, and chemical processes, and are increasingly important in both research and clinical settings. Functional near-infrared spectroscopy (fNIRS) is another innovative technology that can measure cerebral oxygenation and hemodynamics in real-time, making it suitable for a variety of applications, including children's brain development studies or monitoring brain function during rehabilitation [80].

Personalized Medicine

Neuroimaging has evolved beyond its traditional diagnostic role to become a cornerstone of personalized medicine. Capturing individual variability in brain structure and function enables the design of treatment strategies tailored to each patient. Such personalization improves therapeutic efficacy, reduces adverse effects, and supports more precise clinical decision-making. In research, neuroimaging enhances patient stratification in clinical trials, increasing the likelihood of successful outcomes and accelerating drug development. For example, imaging biomarkers can identify patients most likely to benefit from specific pharmacological interventions, thereby facilitating targeted therapeutic approaches. Integrating genetic, environmental, and lifestyle factors into neuroimaging studies further strengthens this precision framework. As knowledge of brain mechanisms deepens, evaluating how these variables influence brain health and disease will advance the field and drive the development of more effective, individualized interventions [81].

CONCLUDING REMARKS

The establishment of several neuroimaging modalities and analytical approaches holds great promise for improving Alzheimer's disease early diagnosis, enabling prompt therapies and enhancing management approaches in general. Combining several neuroimaging methods helps us gain a thorough picture of Alzheimer's disease. Structural abnormalities can be found with CT and MRI, while fMRI and PET allow for the investigation of brain function and molecular alterations, respectively. The fundamental methods of neuroimaging are CT and MRI, each of which has special advantages that complement the others. CT scans provide for quick evaluation, especially in emergencies. The combination of PET and SPECT in neuroimaging not only improves the diagnostic landscape for neurological disorders but also offers a dynamic platform for continued research into the underlying mechanisms of brain function and disease, even though MRI offers thorough insights into brain structure and function, supporting both clinical and research applications. Because they improve data verification, facilitate neurofeedback training, and enhance medical imaging and device monitoring, nonfibrillar tangles have great potential in the field of neuroimaging. Researchers and clinicians may enhance the validity of their findings, promote teamwork in research, and develop individualized treatment plans by fusing this cutting-edge technology with neuroimaging procedures. An important advancement in our knowledge and treatment of Alzheimer's disease is the combination of blood-based diagnostics with cerebrospinal fluid biomarkers. Blood-based biomarkers like NfL and p-tau are prospective substitutes for more widely available, non-invasive testing techniques, whereas CSF biomarkers like tau and Aβ proteins are proven instruments essential for conclusive diagnosis and staging. Addressing these issues will be essential to successfully integrating research findings into clinical practice. Neuroimaging and biomarkers together have incredible prospects for improving clinical neurology and psychiatry. To fully realize the promise of this integrated strategy, ongoing efforts to standardize procedures and guarantee the validity of biomarkers will enhance the upcoming studies.

ACKNOWLEDGEMENTS

All authors were involved in the conception and design of the study. Material preparation, data collection, and analysis were carried out by Snigdha Srivastava and Rajan Kumar Kurmi. Shiva Mishra has reviewed and cited the manuscript. All authors reviewed and approved the final manuscript.

REFERENCES

[1] Thomas A. Comparative study of CNN Models for detection of Alzheimer's disease 2023. https://www.irjmets.com/uploadedfiles/paper//issue_10_october_2023/45247/final/fin_irjmets1697250911.pdf

[2] Gavini V, Kethar J, Appavu R. Norman High School L. Neuropathological Mechanisms of Alzheimer's Disease: Complications and Potential Treatments. J Student Res 2023; 12(4) https://www.jsr.org/hs/index.php/path/article/view/5645

[3] Huang CC, Isidoro C. Raman Spectrometric Detection Methods for Early and Non-Invasive Diagnosis of Alzheimer's Disease. J Alzheimers Dis 2017; 57(4): 1145-56. [http://dx.doi.org/10.3233/JAD-161238] [PMID: 28304304]

[4] Gamal A, Elattar M, Selim S. Automatic Early Diagnosis of Alzheimer's Disease Using 3D Deep Ensemble Approach. IEEE Access 2022; 10: 115974-87. [http://dx.doi.org/10.1109/ACCESS.2022.3218621]

[5] Porsteinsson AP, Isaacson RS, Knox S, Sabbagh MN, Rubino I. Diagnosis of Early Alzheimer's Disease: Clinical Practice in 2021. J Prev Alzheimers Dis 2021; 8(3): 371-86. [http://dx.doi.org/10.14283/jpad.2021.23] [PMID: 34101796]

[6] Gayathri P, Geetha N, Sridhar M, Kuchipudi R, Suresh Babu K, Maguluri LP, *et al.* Deep Learning Augmented with SMOTE for Timely Alzheimer's Disease Detection in MRI Images 2024. [http://dx.doi.org/10.14569/IJACSA.2024.0150253]

[7] Subburaj N, Rani P. A meta-analysis method for identifying potential blood based protein markers associated with Alzheimer's disease. In: Proceedings of the 1st International Conference on Cognitive and Intelligent Computing (ICCAP 2021). Ghent: EAI; 2021.

[8] Reiss AB, Raphael D de L, Chin NA, Sinha V, Reiss AB, Raphael D de L, et al. The physician's Alzheimer's disease management guide: Early detection and diagnosis of cognitive impairment, Alzheimer's disease and related dementia. AIMS Public Health 2022 4:661 [Internet]. 2022 [cited 2025 Mar 15]; 9(4): 661–89. Available from: http://www.aimspress.com/article/doi/10.3934/publichealth.2022047

[9] Albon E, Tsourapas A, Frew E, *et al.* Structural neuroimaging in psychosis: a systematic review and economic evaluation. Health Technol Assess 2008; 12(18) iii–iv, ix–163 [http://dx.doi.org/10.3310/HTA12180]

[10] Besson F, Egret S. Cognitive and Brain Profiles Associated with Current Neuroimaging Biomarkers of Preclinical Alzheimer' s Disease. Journal of Neuroscience [Internet]. 2015 Jan 1 [cited 2025 Mar 15]; Available from: https://www.academia.edu/20115746/Cognitive_and_Brain_Profiles_Associated_with_Current_Neuroimaging_Biomarkers_of_Preclinical_Alzheimers_Disease

[11] Villamizar J, Uribe L, Cerquera A, Prada E, Prada D, Alvarez M. Fractal analysis of neuroimaging: comparison between control patients and patients with the presence of Alzheimer's disease. J Phys

Conf Ser 2022; 2159(1): 012011.
[http://dx.doi.org/10.1088/1742-6596/2159/1/012011]

[12] Korthauer LE, Zhan L, Ajilore O, Leow A, Driscoll I. Disrupted topology of the resting state structural connectome in middle-aged APOE ε4 carriers. Neuroimage 2018; 178: 295-305.
[http://dx.doi.org/10.1016/j.neuroimage.2018.05.052] [PMID: 29803958]

[13] Mazaheri A, Segaert K, Olichney J, *et al.* EEG oscillations during word processing predict MCI conversion to Alzheimer's disease. Neuroimage Clin 2018; 17: 188-97.
[http://dx.doi.org/10.1016/j.nicl.2017.10.009] [PMID: 29159036]

[14] Kazemifar S, Manning KY, Rajakumar N, *et al.* Spontaneous low frequency BOLD signal variations from resting-state fMRI are decreased in Alzheimer disease. PLoS One 2017; 12(6): e0178529.
[http://dx.doi.org/10.1371/journal.pone.0178529] [PMID: 28582450]

[15] Teunissen CE, Verberk IMW, Thijssen EH, *et al.* Blood-based biomarkers for Alzheimer's disease: towards clinical implementation. Lancet Neurol 2022; 21(1): 66-77.
[http://dx.doi.org/10.1016/S1474-4422(21)00361-6] [PMID: 34838239]

[16] Thientunyakit T, Shiratori S, Ishii K, Gelovani JG. Molecular PET Imaging in Alzheimer's Disease 2022; 42: 301–317.
[http://dx.doi.org/10.1007/s40846-022-00717-4]

[17] Li Y, Yao H, Lin P, *et al.* Frequency-dependent altered functional connections of default mode network in Alzheimer's disease. Front Aging Neurosci 2017; 9(AUG): 259.
[http://dx.doi.org/10.3389/fnagi.2017.00259] [PMID: 28824420]

[18] Noh JH, Kim JH, Yang HD. Classification of Alzheimer's Progression Using fMRI Data. Sensors 2023; 23(14): 6330.
[http://dx.doi.org/10.3390/s23146330]

[19] Coentre R, Silva-dos-Santos A, Talina MC. Retrospective study on structural neuroimaging in first-episode psychosis. PeerJ 2016; 4(26): e2069.
[http://dx.doi.org/10.7717/peerj.2069] [PMID: 27257547]

[20] Mandal T, Mishra SR, Singh V. Comprehensive advances in the synthesis, fluorescence mechanism and multifunctional applications of red-emitting carbon nanomaterials. Nanoscale Adv 2023; 5(21): 5717-65.
[http://dx.doi.org/10.1039/D3NA00447C] [PMID: 37881704]

[21] Benadiba M, Luurtsema G, Wichert-Ana L, Buchpigel CA, Busatto Filho G. New molecular targets for PET and SPECT imaging in neurodegenerative diseases. Rev Bras Psiquiatr 2012; 34(Suppl 2): 125–148.
[http://dx.doi.org/10.1016/j.rbp.2012.07.002]

[22] Oosterom N, Nijman J, Gunkel J, *et al.* Neuro-imaging findings in infants with congenital cytomegalovirus infection: relation to trimester of infection. Neonatology 2015; 107(4): 289-96.
[http://dx.doi.org/10.1159/000375439] [PMID: 25790782]

[23] Wu CL, Lin TJ, Chiou GL, *et al.* A Systematic Review of MRI Neuroimaging for Education Research. Front Psychol 2021; 12: 617599.
[http://dx.doi.org/10.3389/fpsyg.2021.617599] [PMID: 34093308]

[24] Shao X, Tan LH, He L. Physical activity and exercise alter cognitive abilities, and brain structure and activity in obese children. Front Neurosci 2022; 16: 1019129.
[http://dx.doi.org/10.3389/fnins.2022.1019129] [PMID: 36340766]

[25] Mandal PK, Goel A, Bush AI, *et al.* Hippocampal glutathione depletion with enhanced iron level in patients with mild cognitive impairment and Alzheimer's disease compared with healthy elderly participants. Brain Commun 2022; 4(5): fcac215.
[http://dx.doi.org/10.1093/braincomms/fcac215] [PMID: 36072647]

[26] Cummings J, Blennow K, Johnson K, *et al.* Anti-Tau Trials for Alzheimer's Disease: A Report from

the EU/US/CTAD Task Force. J Prev Alzheimers Dis 2019; 6(3): 157-163.
[http://dx.doi.org/10.14283/jpad.2019.14]

[27] Shimizu S, Hirose D, Hatanaka H, *et al.* Role of neuroimaging as a biomarker for neurodegenerative diseases. Front Neurol 2018; 9(APR): 265.www.frontiersin.org
[http://dx.doi.org/10.3389/fneur.2018.00265] [PMID: 29720959]

[28] Odusami M, Maskeliūnas R, Damaševičius R. Optimized Convolutional Fusion for Multimodal Neuroimaging in Alzheimer's Disease Diagnosis: Enhancing Data Integration and Feature Extraction. J Pers Med 2023; 13(10): 1496.
[http://dx.doi.org/10.3390/jpm13101496] [PMID: 37888107]

[29] Dark HE, Duggan MR, Walker KA. Plasma biomarkers for Alzheimer's and related dementias: A review and outlook for clinical neuropsychology. Arch Clin Neuropsychol 2024; 39(3): 313-324.
[http://dx.doi.org/10.1093/arclin/acae019]

[30] Magro-Checa C, Steup-Beekman GM, Huizinga TW, van Buchem MA, Ronen I. Laboratory and Neuroimaging Biomarkers in Neuropsychiatric Systemic Lupus Erythematosus: Where Do We Stand, Where To Go? Front Med (Lausanne) 2018; 5: 340.
[http://dx.doi.org/10.3389/fmed.2018.00340] [PMID: 30564579]

[31] Donald KA, Eastman E, Howells FM, *et al.* Neuroimaging effects of prenatal alcohol exposure on the developing human brain: a magnetic resonance imaging review. Acta Neuropsychiatr 2015; 27(5): 251-269.
[http://dx.doi.org/10.1017/NEU.2015.12]

[32] Valdés Cabrera D, El Tal T, Mohamed I, *et al.* Effects of systemic lupus erythematosus on the brain: a systematic review of structural MRI findings and their relationships with cognitive dysfunction. Lupus Sci Med 2024; 11(2): e001214.
[http://dx.doi.org/10.1136/lupus-2024-001214] [PMID: 39153821]

[33] Zhu L, Ploessl K, Kung HF. PET/SPECT imaging agents for neurodegenerative diseases. Chem Soc Rev 2014; 43(19): 6683-91.
[http://dx.doi.org/10.1039/C3CS60430F] [PMID: 24676152]

[34] Pimlott SL, Sutherland A. Molecular tracers for the PET and SPECT imaging of disease. Chem Soc Rev 2011; 40(1): 149-62.
[http://dx.doi.org/10.1039/B922628C] [PMID: 20818455]

[35] Phelps ME. Positron emission tomography provides molecular imaging of biological processes. Proc Natl Acad Sci USA 2000; 97(16): 9226-33.
[http://dx.doi.org/10.1073/pnas.97.16.9226] [PMID: 10922074]

[36] Dyrager C, Vieira RP, Nyström S, Nilsson KPR, Storr T. Synthesis and evaluation of benzothiazole-triazole and benzothiadiazole-triazole scaffolds as potential molecular probes for amyloid-β aggregation. New J Chem 2017; 41(4): 1566-73.
[http://dx.doi.org/10.1039/C6NJ01703G]

[37] Lin SH, Lee LT, Yang YK. Serotonin and mental disorders: a concise review on molecular neuroimaging evidence. Clin Psychopharmacol Neurosci 2014; 12(3): 196-202.
[http://dx.doi.org/10.9758/cpn.2014.12.3.196] [PMID: 25598822]

[38] Kaiboriboon K, Lowe VJ, Chantarujikapong SI, Hogan RE. The usefulness of subtraction ictal SPECT coregistered to MRI in single- and dual-headed SPECT cameras in partial epilepsy. Epilepsia 2002; 43(4): 408-14.
[http://dx.doi.org/10.1046/j.1528-1157.2002.21201.x] [PMID: 11952772]

[39] Lee K, Lee J, Kim G, *et al.* Rough-Surface-Enabled Capacitive Pressure Sensors with 3D Touch Capability. Small 2017; 13(43): 1700368.
[http://dx.doi.org/10.1002/smll.201700368] [PMID: 28524361]

[40] Breuer H, Meier M, Schneefeld S, *et al.* Multimodality imaging of blood–brain barrier impairment

during epileptogenesis. J Cereb Blood Flow Metab 2017; 37(6): 2049-61.
[http://dx.doi.org/10.1177/0271678X16659672] [PMID: 27435624]

[41] Dougherty G. Medical Image Processing 2011.https://link.springer.com/10.1007/978-1-4419-9779-1
[http://dx.doi.org/10.1007/978-1-4419-9779-1]

[42] Arabi H, AkhavanAllaf A, Sanaat A, Shiri I, Zaidi H. The promise of artificial intelligence and deep learning in PET and SPECT imaging. Phys Med 2021; 83: 122-37.
[http://dx.doi.org/10.1016/j.ejmp.2021.03.008] [PMID: 33765602]

[43] Frändin K, Grönstedt H, Helbostad JL, *et al.* Long-term effects of individually tailored physical training and activity on physical function, well-being and cognition in scandinavian nursing home residents: A randomized controlled trial. Gerontology 2016; 62(6): 571-80.
[http://dx.doi.org/10.1159/000443611]

[44] Ishibashi K, Tago T, Kameyama M, Toyohara J, Ishii K. 18F-THK5351 Uptake May Not Estimate Neurofibrillary Tangles in In Vivo Images. Clin Nucl Med 2024; 49(8): 754-6.
[http://dx.doi.org/10.1097/RLU.0000000000005316] [PMID: 38967508]

[45] Ng KP, Therriault J, Kang MS, *et al.* Rasagiline, a monoamine oxidase B inhibitor, reduces in vivo [^{18}F]THK5351 uptake in progressive supranuclear palsy. Neuroimage Clin 2019; 24: 102091.
[http://dx.doi.org/10.1016/j.nicl.2019.102091] [PMID: 31795034]

[46] Moloney CM, Lowe VJ, Murray ME. Visualization of neurofibrillary tangle maturity in Alzheimer's disease: A clinicopathologic perspective for biomarker research. Alzheimers Dement 2021; 17(9): 1554-74.https://pubmed.ncbi.nlm.nih.gov/33797838/
[http://dx.doi.org/10.1002/alz.12321] [PMID: 33797838]

[47] Teo ZL, Ting DSW. Non-fungible tokens for the management of health data. Nature Medicine 2023; 29: 287–288.
[http://dx.doi.org/10.1038/s41591-022-02125-2]

[48] Ali JI, Viczko J, Smart CM. Efficacy of Neurofeedback Interventions for Cognitive Rehabilitation Following Brain Injury: Systematic Review and Recommendations for Future Research. J Int Neuropsychol Soc 2020; 26(1): 31-46.
[http://dx.doi.org/10.1017/S1355617719001061] [PMID: 31983375]

[49] Martínez G, Vernooij RW, Fuentes Padilla P, *et al.* 18F PET with flutemetamol for the early diagnosis of Alzheimer's disease dementia and other dementias in people with mild cognitive impairment (MCI). Cochrane Database Syst Rev 2017; 11(11): CD012884.
[http://dx.doi.org/10.1002/14651858.CD012884]

[50] Gong A, Gu F, Nan W, Qu Y, Jiang C, Fu Y. A Review of Neurofeedback Training for Improving Sport Performance From the Perspective of User Experience. Front Neurosci 2021; 15: 638369.
[http://dx.doi.org/10.3389/fnins.2021.638369] [PMID: 34127921]

[51] Su KH, Hsueh JJ, Chen T, Shaw FZ. Validation of eyes-closed resting alpha amplitude predicting neurofeedback learning of upregulation alpha activity. Sci Rep 2021; 11(1): 19615.
[http://dx.doi.org/10.1038/s41598-021-99235-7] [PMID: 34608244]

[52] Gebreab SA, Salah K, Jayaraman R, Zemerly J. Trusted Traceability and Certification of Refurbished Medical Devices Using Dynamic Composable NFTs. IEEE Access 2023; 11: 30373-89.
[http://dx.doi.org/10.1109/ACCESS.2023.3261555]

[53] Gong A, Nan W, Yin E, Jiang C, Fu Y. Efficacy, Trainability, and Neuroplasticity of SMR vs. Alpha Rhythm Shooting Performance Neurofeedback Training. Front Hum Neurosci 2020; 14: 94.https://pubmed.ncbi.nlm.nih.gov/32265676/
[http://dx.doi.org/10.3389/fnhum.2020.00094] [PMID: 32265676]

[54] Nunes T, da Cunha PR, de Abreu JM, Duarte J, Corte-Real A. Non-Fungible Tokens (NFTs) in Healthcare: A Systematic Review. Int J Environ Res Public Health 2024; 21(8): 965.
[http://dx.doi.org/10.3390/ijerph21080965] [PMID: 39200576]

[55] Signaevsky M, Prastawa M, Farrell K, *et al.* Artificial intelligence in neuropathology: deep learning-based assessment of tauopathy Lab Invest 2019; 99(7): 1019-1029.
[http://dx.doi.org/10.1038/s41374-019-0202-4]

[56] Razi Q, Devrani A, Abhyankar H, Chalapathi GSS, Hassija V, Guizani M. Non-Fungible Tokens (NFTs)—Survey of Current Applications, Evolution, and Future Directions. IEEE Open J Commun Soc 2024; 5: 2765-91.
[http://dx.doi.org/10.1109/OJCOMS.2023.3343926]

[57] Gambril J, Boyd C, Egbaria J. Application of Nonfungible Tokens to Health Care. Comment on "Blockchain Technology Projects to Provide Telemedical Services: Systematic Review". J Med Internet Res 2022; 24(5): e34276.
[http://dx.doi.org/10.2196/34276] [PMID: 35635749]

[58] Abukuri DN. Novel Biomarkers for Alzheimer's Disease: Plasma Neurofilament Light and Cerebrospinal Fluid. Int J Alzheimers Dis. 2024; 2024: 6668159.
[https://dx.doi.org/10.1155/2024/6668159]

[59] Li M, Qin Y, Zhang H, *et al.* A small molecule based on dithieno[3,2-f:2′,3′-h]phthalimide enabling 19.22% efficiency by depressing non-radiative energy loss in ternary organic solar cells. J Energy Chem 2025; 104: 194-201.
[http://dx.doi.org/10.1016/j.jechem.2024.12.035]

[60] Grothe MJ, Moscoso A, Ashton NJ, *et al.* Associations of Fully Automated CSF and Novel Plasma Biomarkers With Alzheimer Disease Neuropathology at Autopsy. Neurology 2021; 97(12): e1229-42.
[http://dx.doi.org/10.1212/WNL.0000000000012513] [PMID: 34266917]

[61] Petersen RC. Alzheimer's disease: progress in prediction. Lancet Neurol 2010; 9(1): 4-5.https://pubmed.ncbi.nlm.nih.gov/20083022/
[http://dx.doi.org/10.1016/S1474-4422(09)70330-8] [PMID: 20083022]

[62] Karikari TK, Pascoal TA, Ashton NJ, *et al.* Blood phosphorylated tau 181 as a biomarker for Alzheimer's disease: a diagnostic performance and prediction modelling study using data from four prospective cohorts. Lancet Neurol 2020; 19(5): 422-33.
[http://dx.doi.org/10.1016/S1474-4422(20)30071-5] [PMID: 32333900]

[63] Zetterberg H, Burnham SC. Blood-based molecular biomarkers for Alzheimer's disease. Mol Brain 2019; 12(1): 26.
[http://dx.doi.org/10.1186/s13041-019-0448-1] [PMID: 30922367]

[64] Milà-Alomà M, Suárez-Calvet M, Molinuevo JL. Latest advances in cerebrospinal fluid and blood biomarkers of Alzheimer's disease. Ther Adv Neurol Disord 2019; 12: 1756286419888819.
[http://dx.doi.org/10.1177/1756286419888819] [PMID: 31897088]

[65] Stevenson-Hoare J, Heslegrave A, Leonenko G, *et al.* Plasma biomarkers and genetics in the diagnosis and prediction of Alzheimer's disease. Brain 2023; 146(2): 690-9.
[http://dx.doi.org/10.1093/brain/awac128] [PMID: 35383826]

[66] Abukuri DN. Novel Biomarkers for Alzheimer's Disease: Plasma Neurofilament Light and Cerebrospinal Fluid. Int J Alzheimers Dis. 2024;2024.

[67] Bateman RJ, Blennow K, Doody R, *et al.* Plasma Biomarkers of AD Emerging as Essential Tools for Drug Development: An EU/US CTAD Task Force Report. J Prev Alzheimers Dis 2019; 6(3): 169-73.
[http://dx.doi.org/10.14283/jpad.2019.21] [PMID: 31062827]

[68] Brickman AM, Manly JJ, Honig LS, *et al.* Plasma p-tau181, p-tau217, and other blood-based Alzheimer's disease biomarkers in a multi-ethnic, community study. Alzheimers Dement 2021; 17(8): 1353-64.
[http://dx.doi.org/10.1002/alz.12301] [PMID: 33580742]

[69] Brickman AM, Manly JJ, Honig LS, *et al.* Correlation of plasma and neuroimaging biomarkers in Alzheimer's disease. Ann Clin Transl Neurol 2022; 9(5): 756-61.

https://pubmed.ncbi.nlm.nih.gov/35306760/
[http://dx.doi.org/10.1002/acn3.51529] [PMID: 35306760]

[70] Melzer TR, Keenan RJ, Leeper GJ, *et al.* Test-retest reliability and sample size estimates after MRI scanner relocation. Neuroimage 2020; 211: 116608.
[http://dx.doi.org/10.1016/j.neuroimage.2020.116608]

[71] Ng KP, Pascoal TA, Mathotaarachchi S, *et al.* Monoamine oxidase B inhibitor, selegiline, reduces 18F-THK5351 uptake in the human brain. Alzheimers Res Ther 2017; 9: 25.
[http://dx.doi.org/10.1186/s13195-017-0253-y]

[72] Schöll M, Vrillon A, Ikeuchi T, *et al.* Cutting through the noise: A narrative review of Alzheimer's disease plasma biomarkers for routine clinical use. J Prev Alzheimers Dis 2025; 12(4): 100056.
[http://dx.doi.org/10.1016/j.tjpad.2024.100056] [PMID: 39814656]

[73] Lee H, Ugay D, Hong S, Kim Y. Alzheimer's Disease Diagnosis Using Misfolding Proteins in Blood. Dement Neurocogn Disord. 2020 Mar; 19(1): 1-18.
[http://dx.doi.org/10.12779/dnd.2020.19.1.1]

[74] Shukla S, Tiwari S, Bhattacharya R, Ojha S, Mishra S, Gupta SK. Nanotechnology-Based Approaches for Nose-to-Brain Drug Delivery in Neurodegenerative Diseases Lett Drug Des Discov 2024; (11): 1913-21.

[75] Krance SH, Cogo-Moreira H, Rabin JS, Black SE, Swardfager W. Reciprocal predictive relationships between amyloid and tau biomarkers in Alzheimer's disease progression: An empirical model. J Neurosci 2019; 39(37): 7428-37.
[http://dx.doi.org/10.1523/JNEUROSCI.1056-19.2019]

[76] Alawode DOT, Heslegrave AJ, Ashton NJ, *et al.* Transitioning from cerebrospinal fluid to blood tests to facilitate diagnosis and disease monitoring in Alzheimer's disease. J Intern Med 2021; 290(3): 583-601.
[http://dx.doi.org/10.1111/joim.13332] [PMID: 34021943]

[77] A Virata MC, Catahay JA, Lippi G, Henry BM. Neurofilament light chain: a biomarker at the crossroads of clarity and confusion for gene-directed therapies. Neurodegener Dis Manag 2024; 14(6): 227-239. [https://pubmed.ncbi.nlm.nih.gov/39545606/]

[78] Khalil M, Teunissen CE, Lehmann S, *et al.* Neurofilaments as biomarkers in neurological disorders — towards clinical application. Nature Reviews Neurology 2024; 20(5): 269–87. Available from: https://www.nature.com/articles/s41582-024-00955-x

[79] Davatzikos C. Machine learning in neuroimaging: Progress and challenges. Neuroimage 2019; 197: 652-6.
[http://dx.doi.org/10.1016/j.neuroimage.2018.10.003] [PMID: 30296563]

[80] Hu X, Yan CG. Current status and future challenges in brain imaging: Insights from leading experts. Medicine Plus 2024; 1(4): 100049.
[http://dx.doi.org/10.1016/j.medp.2024.100049]

[81] Yen C, Lin CL, Chiang MC. Exploring the Frontiers of Neuroimaging: A Review of Recent Advances in Understanding Brain Functioning and Disorders. Life (Basel) 2023; 13(7): 1472.https://pmc.ncbi.nlm.nih.gov/articles/PMC10381462/
[http://dx.doi.org/10.3390/life13071472] [PMID: 37511847]

CHAPTER 5

Digital Tools and Artificial Intelligence in Alzheimer's diagnosis

Shubh Deep Yadav[1], **Shubhanshu Goel**[1], **Ayushi Tyagi**[1,*] and **Khushboo Bansal**[2]

[1] *I.T.S College of Pharmacy, Ghaziabad, Uttar Pradesh 201206, India*

[2] *Sunder Deep Pharmacy College, Ghaziabad, Uttar Pradesh 201002, India*

Abstract: The integration of digital tools and Artificial Intelligence (AI) has brought significant advancements to the diagnosis of various diseases, particularly neurodegenerative disorders like Alzheimer's disease. AI, particularly Machine Learning (ML), enables the analysis of vast and complex datasets, such as neuroimaging, electronic health records, cognitive assessments, and biomarkers, which are crucial for early detection and accurate diagnosis. These technologies offer the potential to identify subtle changes in patients' conditions over time, improving the precision of diagnosis and facilitating the development of personalized treatment plans. Digital tools, including wearable devices, mobile applications, and fitness trackers, allow for continuous and passive monitoring of patient data, providing real-time insights into physiological parameters like gait, heart rate, and brain activity. These tools can detect early signs of cognitive decline and other abnormalities, aiding clinicians in making more informed decisions about disease progression. AI has demonstrated its potential in automating image analysis, identifying biomarkers, and predicting disease progression, which can expedite the diagnostic process and enhance the overall healthcare experience. Additionally, AI-driven models can integrate data from multiple sources, enabling comprehensive assessments and more accurate predictions about the onset and development of diseases like AD. However, there are several challenges associated with the widespread adoption of AI in clinical settings, such as concerns about data quality, model interpretability, ethical considerations, and cost. Overcoming these barriers is essential to ensuring that AI technologies become accessible, reliable, and effective in healthcare. This chapter explores the transformative potential of AI and digital tools in diagnostics, discussing their applications, benefits, challenges, and the future directions for improving patient care and outcomes.

Keywords: Alzheimer's disease, Artificial Intelligence, Biomarkers, Digital tools, Early diagnosis, Machine Learning, Neuroimaging.

[*] **Corresponding author Ayushi Tyagi:** I.T.S College of Pharmacy, Ghaziabad, Uttar Pradesh 201206, India; E-mail: arushivasisth18031993@gmail.com

Prashant Kumar Dhakad, Raghav Mishra, Shweta Sharma, Sudhanshu Mishra & Nishant Gaur (Eds.)

INTRODUCTION

Between 70 and 85% of dementia cases are caused by Alzheimer's Disease (AD), the most common neurological illness linked to dementia. Around 50 million individuals worldwide suffer from dementia at the moment, and as life expectancy rises, that figure is predicted to increase to 139 million by 2050, posing serious socioeconomic and medical issues. Cognitive deficits and gradual impairment of memory are the main characteristics of AD, including difficulties in language, spatial orientation, executive function, and behaviour, ultimately leading to a gradual loss of personal autonomy [1]. Approximately 95% of AD cases occur in individuals aged 60–65 and older, while 5% have an early onset. Key pathological features used to diagnose AD include amyloid angiopathy, tau protein-containing neurofibrillary tangles, and β-amyloid plaques. To distinguish AD from prion diseases, these plaques must stain negative for prion antibodies but positive for β-amyloid. The number of plaques and tangles in the brain must exceed that of age-matched individuals without dementia. In some cases, TDP-43 protein accumulation and Lewy bodies (alpha-synuclein aggregates) in the amygdala are also observed [2]. The incidence of AD increases with age, with women being at higher risk than men, as indicated by a European meta-analysis. AD is listed as a cause or contributing factor in many recorded deaths of affected patients. The disease progresses through stages, beginning with the absence of harmful Aβ species, advancing to normal cognition with Aβ deposition, and eventually leading to neurodegeneration [3]. Key pathogenic features of AD include amyloid-beta (Aβ) plaque buildup, neurofibrillary tangles (NFTs) formed by hyperphosphorylated tau proteins, synaptic dysfunction, and neuronal death. The Amyloid Precursor Protein (APP) cleaves abnormally, forming Aβ plaques that impair neuronal communication and trigger neuroinflammation. Tau proteins cause microtubule instability, leading to neuronal death. Cognitive decline is further exacerbated by oxidative stress and persistent neuroinflammation, contributing to neuronal damage. The cholinergic theory underscores how the depletion of acetylcholine and the death of cholinergic neurons impair memory and learning. Early synaptic loss, which is closely linked to cognitive symptoms, is a hallmark of AD [4, 5].

Current therapies for Alzheimer's disease focus on symptom management rather than changing the severity of the condition. Memantine and cholinesterase inhibitors are two medications that can improve memory and alertness, but do not increase life expectancy or decrease illness development. Lifestyle improvements, such as diet and exercise, are advised as first-line therapies because they have the potential to lower Alzheimer's risk and delay cognitive decline. Although limited therapeutic trials have yielded encouraging outcomes, larger research is needed to validate the pathological hallmarks of Alzheimer's disease (Fig. **1**), such as Aβ and phosphorylated tau. Although there is evidence that early detection of AD pathology can enhance treatment results, large-scale clinical studies are still required [6].

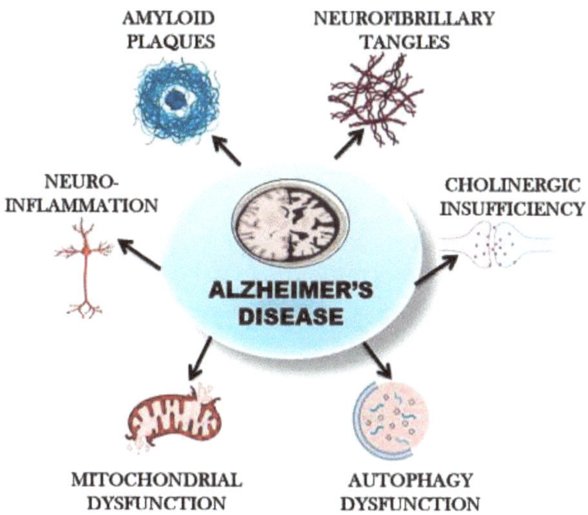

Fig. (1). Pathophysiology of alzheimer's disease [5] (Designed by Canva).

With growing healthcare expenditures, healthcare practitioners have been motivated to discover efficient methods for early detection and treatment of this degenerative neurological condition. Early brain abnormalities, including β-amyloid build-up and cell alterations, can develop during the preclinical period, even without clinical symptoms. Biological markers detected in cerebrospinal fluid (CSF) are used to identify whether a patient's condition has advanced to this stage. AD dementia is defined in several ways depending on the presence and certainty of clinical symptoms. It is described as a significant cognitive impairment that interferes with daily activities. Biomarkers such as reduced glucose absorption, lowering β-amyloid peptide levels, rising tau protein levels in CSF, and temporal lobe atrophy on magnetic resonance imaging can help identify Alzheimer's dementia more precisely [7].

TRADITIONAL *VS.* DIGITAL BIOMARKERS: ADVANCING ALZHEIMER'S DIAGNOSIS

Current diagnostic methods for Alzheimer's disease (AD) primarily rely on neurocognitive assessments, brain imaging techniques, and CSF analysis. In patients with AD, brain pathology often reveals amyloid plaque deposition, neurofibrillary tangles, and notable synapse loss. Diagnostic guidelines emphasize the use of CSF markers such as amyloid-β1-42 (Aβ42) and total tau protein. While Aβ42 and p-tau in CSF have been instrumental in the research, these diagnostic techniques are not only costly but also relatively invasive. Additionally, there are concerns regarding the clinical utility of these CSF biomarkers, as the sensitivity of CSF Aβ42 ranges between 0.70 and 0.82, and specificity varies from 0.45 to 0.90. Compounding this issue is the fact that AD is often diagnosed at a late stage when significant brain damage has already occurred. If the disease could be detected earlier, before major brain damage, patients would likely experience greater benefits from treatment. Hence, the identification of biomarkers that can detect AD in its early stages or at onset is crucial for improving diagnostics and allowing for timely intervention [8].

The FDA believes that the optimal biomarker should include specificity, sensitivity, predictability, robustness, simplicity, accuracy, and cost, as well as be easily quantifiable in common biological samples such as serum or urine. A biomarker for Alzheimer's disease is regarded as excellent when its sensitivity and specificity reach 80%. The National Institute on Aging and Alzheimer's Association have categorized AD diagnostic biomarkers into three categories: β-amyloid deposits (A), hyperphosphorylated tau clumps (T), and neurodegeneration or neuronal damage (N) [9].

Traditional Biomarkers

i. Positron Emission Tomography (PET) marker

PET imaging with various ligands plays an important role in establishing the pathophysiology behind the normal and atypical AD symptoms, supporting doctors in separating AD from other neurodegenerative diseases [10].

- FDG-PET acts as a topographic biomarker, helping to discriminate between typical and atypical Alzheimer's Disease (AD) by finding particular patterns of regional hypometabolism that correspond to clinical symptoms in various AD forms. In Alzheimer's disease patients, reduced metabolic activity is common on both sides of the parietal and medial temporal areas. Furthermore, FDG-PET can detect distinct hypometabolic patterns in various dementias, such as dementia with Lewy bodies and frontotemporal dementia (FTD), which aids in differential diagnosis. Although FDG-PET gives greater diagnostic information than MRI, its usage remains less common [10].
- Amyloid PET is mostly utilized in research to show fibrillar or insoluble Aβ plaques in the brain. However, it does not detect other Aβ peptides. It is the most accurate biomarker for diagnosing amyloid plaques, showing remarkable

accuracy in imaging-to-autopsy comparisons with a sensitivity of 92% and specificity of 100%. PET, unlike other imaging modalities, shows consistent amyloid distribution in both typical and atypical instances of Alzheimer's disease [10].

- Tau PET, used mainly in research, is the only autopsy-validated imaging tool that directly visualizes neurofibrillary tangles. Fluid-based indicators haven't proved comparable precision, making Tau PET a key marker for neurofibrillary tangles [10].

ii. **MRI Marker**

Structural MRI is one of the most often utilized neuroimaging methods for evaluating underlying Alzheimer's disease. Neurodegeneration can be evaluated with MRI by analysing the cortical grey matter. Minor alterations in the grey matter can be identified using a conventional T1-weighted MRI scan, which provides great contrast between the grey and white regions. Grey area volume, cortical thickness, and volumetric analysis of the cortex may now be objectively assessed owing to better spatial resolution on MRI scans. These evaluations were formerly accomplished visually using qualitative methods, but the introduction of automated tools to quantify volume and cortical thickness has made assessments more precise [9].

iii. **CSF Biomarker**

The diagnostic criteria for Alzheimer's Disease (AD) include both cerebrospinal fluid (CSF) biomarkers obtained through lumbar puncture and positron emission tomography (PET) imaging, which assesses amyloid accumulation in cortical brain tissue and different brain regions. Studies have shown that ratios of CSF biomarkers (Aβ1-42/1-40, p-tau/Aβ1-42, and t-tau/Aβ1-42) are more effective than individual values in detecting abnormalities. These abnormalities in core CSF and PET biomarkers are linked to the key AD pathologies of amyloid plaque buildup and neurofibrillary tangles. A reduction in CSF Aβ1-42 or increased uptake of specific PET tracers signals a higher amyloid burden, while elevated CSF tau levels indicate neuronal damage and degeneration, with increased tau PET pointing to tangle formation. These biomarker alterations, which begin long before clinical symptoms appear, may not happen simultaneously. For instance, amyloid deposition identified via PET provides distinct insights compared to amyloid processing captured by CSF biomarkers. CSF changes often precede PET changes; for example, elevated p-tau in the CSF may occur before PET-tracer positivity as AD progresses, contributing to cognitive decline [11].

iv. **Blood-based Biomarkers (BBBMs)**

Brain-derived biomarkers are frequently found in low concentrations in the blood due to the blood-brain barrier's restrictive nature, which hinders the free passage of molecules between the Central Nervous System (CNS) and the circulation. Some biomarkers linked with Alzheimer's Disease (AD) pathology are also generated in non-cerebral tissues, which may distort blood tests. Furthermore, the presence of heterophilic antibodies in blood—endogenous antibodies that might interfere with immunoassays used to detect these biomarkers—may result in incorrect findings, either falsely high or low. This problem is less severe in cerebrospinal fluid (CSF), where antibody levels are usually lower. Furthermore, the biomarker of interest in blood may be prone to proteolytic destruction by plasma proteases, reducing its detectability [12].

- Plasma Aβ
- Plasma tau
- Plasma neurofilament light
- Protein biomarker panels
- miRNA biomarker panels

Digital tools used in Alzheimer's diagnosis

Compared to traditional tools for Alzheimer's diagnosis, digital tools have the potential for more accurate and earlier detection by enabling continuous monitoring of cognitive function through data collected from wearable devices and digital cognitive assessments, allowing for a more nuanced picture of cognitive decline beyond a single clinical visit, while older tools typically rely on in-person cognitive tests and brain imaging, which may only capture a snapshot of a patient [13, 17].

Digital Cognitive Assessments

Digital cognitive assessments, which use digital technology to provide objective, standardized, and scalable methods for evaluating cognitive function, are frequently utilized in clinical settings, research studies, and even at-home monitoring. They are used to detect and monitor cognitive impairment, including Alzheimer's disease [13, 14, 20].

Types of Digital Cognitive Assessments

- **Computerized Cognitive Tests**
    - These tests are administered on computers, tablets, or smartphones and assess memory, attention, executive function, and processing speed.
    - Examples: Cogstate, CANTAB (Cambridge Neuropsychological Test Automated Battery), CNS Vital Signs [19, 21]
        - Mobile and Web-Based Apps
    - Apps provide quick and interactive cognitive testing, often incorporating gamification.
    - Examples: CognICA (by Cognetivity), Brain Check, Neurable [19, 21]
        - Digital Biomarkers & Passive Monitoring
    - Some tools use AI and machine learning to analyze speech patterns, eye movement, typing speed, and other digital interactions as indicators of cognitive decline.
    - Examples: Mindstrong Health, Neurotrack [23, 24].
        - Virtual Reality (VR) and Augmented Reality (AR) Assessments
    - VR and AR environments provide immersive cognitive testing that mimics real-world challenges.
    - Example: Altoida [23, 24].
        - Wearable & Sensor-Based Assessments
    - Smartwatches and fitness trackers collect movement and activity data to detect cognitive decline.
    - Example: Apple Watch cognitive tests, Fitbit cognitive health research [22, 24].

Advantages of Digital Cognitive Assessments

1. Early Detection: More sensitive than traditional paper-based tests in detecting early signs of Alzheimer's [22, 24].
2. Remote Monitoring: Allows for frequent testing without requiring hospital visits [23, 24].
3. Data Collection & AI Analysis: Enables continuous tracking and predictive analysis using AI and machine learning [26].
4. Personalization: Can adapt difficulty levels based on individual performance [26].

Limitations

1. Access and Digital Literacy: Not everyone (especially older adults) may be comfortable using digital tools [23, 27].
2. Validation and Accuracy: Some tools may lack extensive clinical validation compared to traditional neuropsychological tests [23, 27].

3. Data Privacy Concerns: Handling sensitive cognitive health data requires strict cybersecurity measures [27].
   i. Wearable sensors:

Devices like smartwatches or activity trackers that monitor movement patterns, sleep quality, and other physiological data. Wearable sensors are emerging as powerful tools in the initial finding and observation of Alzheimer's Disease (AD). These devices track physiological, cognitive, and behavioural changes that indicate neurodegeneration. Below are key details on wearable sensors used in Alzheimer's diagnosis [33]:

Types of Wearable Sensors for Alzheimer's Diagnosis

Wearable sensors can be categorized based on the physiological parameters they measure:

a. **Motion and Gait Sensors**

Purpose: Detects abnormalities in movement, gait, and balance—early indicators of cognitive decline.

Examples:

- **Accelerometers and Gyroscopes** (worn on the wrist, ankle, or integrated into shoes)
- **Inertial Measurement Units (IMU)** (tracks balance and walking patterns)
- **Smart Insoles** (pressure sensors detect walking irregularities) [29, 33]

b. **Heart Rate and Bioelectrical Sensors**

Purpose: Measures heart rate variability (HRV) and electrodermal activity, which correlate with cognitive decline.

Examples:

- **Electrocardiography (ECG) Patches** (monitors HRV)
- **Electrodermal Activity Sensors** (tracks stress response changes) [25, 30]

c. **Sleep Monitoring Sensors**

Purpose: Detects sleep disturbances, a common early sign of AD.

Examples:

- **Smartwatches (Fitbit, Apple Watch, Oura Ring)** (track sleep cycles)
- **Wearable EEG Headbands (Dreem, Muse)** (detect abnormal brain activity during sleep) [25, 31].

d. **Cognitive and Neurological Sensors**

Purpose: Measures brain activity changes related to cognitive decline.

Examples:

- **EEG Wearables (Electroencephalography headbands)** (detect abnormal neural oscillations)
- **Functional Near-Infrared Spectroscopy (fNIRS) Devices** (measures cerebral blood flow) [32, 33]

e. **Smart Textiles and Clothing Sensors**

Purpose: Integrates sensors into clothing to passively track physiological and behavioural changes.

Examples:

- **E-textiles with biosensors** (detect movement, temperature, and heart rate)
- **Smart Glasses** (eye tracking to measure cognitive processing speed) [28, 29]

Advantages of Wearable Sensors in Alzheimer's Diagnosis

1. **Early Detection** – Identifies signs before significant cognitive decline.
2. **Continuous Monitoring** – Tracks changes over time outside clinical settings.
3. **Non-invasive & Comfortable** – No need for complex hospital visits.
4. **Cost-effective** – Reduces dependency on expensive neuroimaging (MRI/PET scans) [28, 29]

Challenges and Limitations

1. **Data Accuracy and Noise** – Movement artifacts may affect readings.
2. **User Compliance** – Patients may forget or resist wearing devices.
3. **Privacy Concerns** – Data security and ethical concerns over continuous monitoring.
4. **Standardization Issues** – Different devices may not follow uniform protocols [28, 29].

Examples of Wearable Devices Used in Alzheimer's Research

1. **Dreem EEG Headband** – Measures brain waves during sleep.
2. **Empatica E4 Wristband** – Tracks electrodermal activity and heart rate.
3. **ActiGraph Motion Sensors** – Analyze gait and activity levels.
4. **Apple Watch & Fitbit** – Used for sleep and heart rate monitoring in AD studies [29, 33].

i. Virtual reality environments:

Virtual Reality (VR) is increasingly being used as a diagnostic and cognitive assessment tool for Alzheimer's Disease (AD). By simulating real-world environments, VR can detect early cognitive impairments, evaluate memory loss, and monitor disease progression more effectively than traditional pen-and-paper tests [30, 33].

How is VR used in Alzheimer's Diagnosis?

VR creates immersive environments where patients engage in cognitive, spatial, and motor tasks. These tasks help clinicians assess memory, attention, decision-making, and spatial navigation abilities, which are among the first to decline in AD [18].

a. **Cognitive Assessments** – Tests for memory, executive function, and problem-solving skills.
b. **Spatial Navigation Tests** – Evaluates the brain's ability to process space and orientation.
c. **Daily Life Simulations** – Measures how patients perform routine activities (e.g., shopping, cooking).
d. **Motor & Gait Analysis** – Detects subtle movement impairments linked to early AD [18].

Types of VR Environments Used in Alzheimer's Diagnosis

a. Virtual Navigation Tasks

Purpose: Tests spatial memory, which declines early in AD.

How It Works:

- Patients navigate virtual mazes, cities, or homes to test their ability to recall paths and landmarks.
- Examples: Morris Water Maze VR, Virtual Supermarket Test (VST), VR Wayfinding Tasks [21].
    a. Virtual Daily Activities Simulation

Purpose: Evaluate a patient's ability to perform everyday tasks.

How It Works:

- Patients interact with virtual objects and complete tasks like grocery shopping, cooking, or handling money.
- This assesses decision-making, memory recall, and executive function.
- Example: VR Kitchen Task (cooking simulation to assess planning and memory) [30].
    a. VR Cognitive Testing Games

Purpose: Tests memory, attention, and problem-solving skills in a fun, engaging way.

How It Works:

- Patients complete puzzle-solving, object recall, and reaction time tasks.
- Example: "Sea Hero Quest VR" (a spatial navigation game designed for Alzheimer's research) [30].
    a. VR-based Dual Tasking for Gait & Balance Analysis

Purpose: Assesses motor function along with cognitive load.

How It Works:

- Patients walk on a treadmill while performing cognitive tasks in VR.
- Monitors gait disturbances, balance, and reaction times, which are linked to cognitive decline.
- Example: VR Dual-Task Walking Tests (used in clinical research) [31].

Advantages of VR in Alzheimer's Diagnosis

1. **Early and Sensitive Detection** – Identifies cognitive decline earlier than traditional tests.

2. **Real-Life Scenario Testing** – Mimics natural environments for a more accurate assessment.
3. **Engaging and Non-Invasive** – More enjoyable and less stressful than standard cognitive tests.
4. **Remote and Continuous Monitoring** – VR can be used for home-based assessments [29].

Challenges & Limitations

1. **Cost and Accessibility** – VR headsets and development can be expensive.
2. **Patient Comfort and Motion Sickness** – Some elderly patients experience dizziness.
3. **Technological Adaptation** – Some older adults may struggle with VR interaction.
4. **Standardization Issues** – Different VR tests may yield inconsistent results [29].

Examples of VR Systems Used in Alzheimer's Research

1. **Sea Hero Quest VR"** – Tests spatial navigation abilities.
2. **Virtual Supermarket Test (VST)"** – Assesses memory and executive function.
3. **Smart Aging VR"** – Simulates daily life tasks for cognitive assessment.
4. **Alzheimer's Virtual Reality Maze"** – Evaluates spatial memory and navigation [29].

4. AI-powered analysis:

AI-powered analysis is revolutionizing Alzheimer's Disease (AD) diagnosis by improving early detection, accuracy, and efficiency. Here are the key areas where AI is making an impact [21]:

a. Neuroimaging Analysis

AI processes brain scans (MRI, CT, PET) to detect subtle changes associated with AD.

- **Deep Learning and CNNs**: Convolutional neural networks (CNNs) analyze MRI scans to detect hippocampal atrophy (a key AD biomarker).
- **PET scan Analysis**: AI detects abnormal amyloid-beta and tau protein deposits, which are hallmarks of AD.

- **Automated Segmentation**: AI tools segment brain structures to compare volume changes over time [31].
    a. Biomarker Detection from Body Fluids

AI models analyse biomarkers in blood and cerebrospinal fluid (CSF) for early diagnosis.

- **Machine Learning (ML) Models** predict AD by detecting protein abnormalities (e.g., amyloid-beta, tau, neurofilament light chain).
- AI-driven liquid biopsy tests can identify AD risk from blood samples [31].
    a. Cognitive and Behavioural Analysis

AI assesses speech, facial expressions, and handwriting to detect early cognitive decline.

- **Natural Language Processing (NLP)**: AI analyzes speech patterns, pauses, and sentence structure for AD markers.
- **Wearable and IoT Sensors**: Track movement patterns, daily activities, and sleep disturbances to detect early AD symptoms [30].
    a. Predictive and Personalized Diagnosis
- **Deep learning models** analyse multi-modal data (brain scans, genetic data, medical history) to predict AD progression.
- AI **personalizes treatment** by identifying individual risk factors and potential drug responses [30].
    a. AI-Assisted Drug Discovery and Clinical Trials
- AI **accelerates drug discovery** by predicting how molecules interact with AD-related proteins.
- AI-based **clinical trial matching** finds eligible AD patients based on genetic and medical data [30].

Popular AI Models and Tools in Alzheimer's Diagnosis

- **Deep AD** (Deep learning-based AD detection from MRI scans)
- **Brain Age** (AI-based brain age estimation for detecting neurodegeneration)
- **Google's DeepMind & IBM Watson** (AI-driven biomarker analysis)
- **Speech-to-text AI models** for linguistic analysis [32].

Future Prospects

- **Real-time home monitoring** using AI-powered smart devices.
- **Integration with wearables** (e.g., smartwatches detecting cognitive decline).

- **Federated learning** to train AI models on patient data while ensuring privacy [32].

Traditional tools for Alzheimer's diagnosis:

- **Neuropsychological tests:**

Standardized cognitive assessments administered by a clinician, such as the Mini-Mental State Examination (MMSE) [33].

- **Brain imaging techniques:**

MRI and PET scans to visualize brain structure and function, which can indicate potential abnormalities associated with Alzheimer's [32].

Important considerations:

- **Validation and standardization:** In clinical contexts, digital technologies must undergo thorough validation to guarantee accuracy and dependability [29].
- **Data privacy:** Protecting patient data collected through digital tools is crucial.
- **Patient acceptance and usability:** Design considerations are important to ensure patients can easily use and engage with digital tools [29].

Key differences between digital tools and old tools for Alzheimer's diagnosis:

- **Data collection:**

Digital tools can collect data passively and continuously through sensors in wearables, smartphones, and other devices, providing insights into daily activities, sleep patterns, movement, and cognitive performance, whereas traditional tools rely on structured assessments conducted in a clinical setting [14, 15].

- **Early detection:**

Digital tools can potentially identify subtle cognitive changes earlier than traditional methods by monitoring data over time, allowing for earlier intervention and treatment [14, 16].

- **Accessibility:**

Digital tools can be administered remotely, providing greater accessibility to assessments, especially aimed at those with immobility or breathing in rural regions [15, 17].

- **Objective measurement:**

Digital tools can objectively quantify cognitive abilities through metrics like reaction time and accuracy on tasks, reducing reliance on subjective patient reports [17, 19]. Table **1** includes the overview of sensors.

Table 1. Overview of sensors, their corresponding domains, and associated metrics.

Sensors	Metrics	Sense-Domain	Reference
Camera	Multiple face features	Behavior-facial expression	[35]
	Limitation of impulse in reaction to stimulus	Oculomotor-pupillary response	[35]
Microphone	Voice power spectrum and tremor	Speech and language- voice feature	[35]
	Breaks, linguistic and contextual characteristics, terminologies, and more.	Speech and language- cognition	[36]
Accel/Gyro	Gait metrics, distance, steps, symmetry, etc.	Movement	[35]
Barometer	Gait/ Climb information	Movement	[35]
Touchscreen	Swipe pattern efficiency	Movement-fine motor	[35]
	Frequency of keypad touching and texting		[37]
Geopositioned	Driving habits and the effectiveness of navigation	Cognitive feature: spatial retention and response time	[35]
Ballistocardiography	Sleep patterns, sleep phases, and efficiency	Sleep	[35]
PPG	cardiac rhythm, its variation	ANS function-Systemic circulation	[35]

A significant proportion of Alzheimer's Disease (AD) patients experience both pyramidal and extrapyramidal motor deficiencies initially in the illness, often occurring years before cognitive symptoms become evident. Gait characteristics such as speed, stride length, and symmetry tend to decrease, while gait speed variability increases. These changes can be assessed either actively through walking tests or passively with stationary or portable monitoring devices. Research by Buracchio et al. identified a critical point about 12 years prior to the diagnosis of Mild Cognitive Impairment (MCI), at which gait speed declines more sharply, shifting from a yearly decrease of 0.004 to 0.023 m/s. Step length variability is associated with sensory impairments, while variations in stance time are tied to central nervous system dysfunctions. Additionally, motor skills, particularly the accuracy and speed of finger tapping, serve as early indicators of Alzheimer's Disease (AD). The finger tapping test, which requires pressing a button as quickly as possible, shows a significant decline in speed approximately 2.7 years before the clinical symptoms emerge. Although the average finger tapping rate decreases by 0.03 taps per year with age, this rate accelerates to 0.15 taps per year after the inflection point. As AD advances, language difficulties, such as grammatical errors and altered speech patterns, become more evident. By examining speech metrics like segment durations, vocal response times, and conversational disruptions, researchers have been able to distinguish healthy individuals from those affected by AD. Virtual assistants or cell phone interactions offer useful voice data for these kinds of studies. Furthermore, eye movements and pupillary reflexes have long been investigated in neurological studies and provide information on dopamine activity in the brain, cholinergic pathways, cortical disease, and the medial temporal lobe memory system. Cardiac rhythm variation and heart rate may be passively and continuously monitored thanks to the increasing popularity of fitness trackers and smartwatches with photoplethysmography (PPG) technology. While the accuracy of these devices can vary when compared to ECG measurements, certain devices and conditions offer HRV data precise enough to assist in detection [34].

AI IN ALZHEIMER'S DIAGNOSIS: TRANSFORMING EARLY DETECTION AND TREATMENT

Alzheimer's Disease (AD) is a progressive neurological ailment that affects millions of people worldwide, causing memory impairment, cognitive decline, and, eventually, the inability to perform everyday tasks. Early and correct diagnosis is essential for successful illness care, slowing progression and increasing patient quality of life. However, traditional diagnostic methods, such as neuroimaging and cognitive testing, frequently fail to detect AD in its prodromal or early phases. Recently, Artificial Intelligence (AI) has emerged as a disruptive tool in healthcare, presenting unique and promising solutions for the early and precise identification of Alzheimer's disease. This study investigates the role of AI in advancing Alzheimer's disease diagnostics, reviews the current AI-driven methodologies in use, discusses the benefits and challenges of AI integration, and highlights future prospects for AI use in the diagnosis and management of neurodegenerative disorders [31].

AI, particularly Machine Learning (ML), specializes in managing complex, high-dimensional systems that are beyond human data processing capabilities. ML has been widely used in Computer-Aided Diagnosis (CAD) for a variety of disorders, including Alzheimer's disease, by combining data from biological markers, brain imaging, neuropsychological testing, electronic medical records, and executive function assessment approaches. Diffusion Tensor Imaging (DTI), Positron Emission Tomography (PET), FDG-PET, and Magnetic Resonance Imaging (MRI) may all offer precise information on the architecture and physiology of the brain. These modalities assist in identifying essential diagnostic markers, including atrophy, amyloid deposition, and microstructural damage, in order to diagnose Alzheimer's disease [38, 39]. Additionally, neuroimaging can be used to differentiate between pathologic processes that are not connected to AD but may possibly contribute to cognitive loss. Numerous studies have demonstrated that advanced diagnostic and prognostic tools may incorporate biomarkers linked to neurodegeneration, the primary pathology of AD, or biomarker combinations. The polymorphism in the apolipoprotein E (APOE) gene, which also has additional predictive value, is the most significant genetic risk factor for the sporadic form of AD. The first CAD tools for AD were developed by analyzing brain imaging data using AI algorithms. Although several machine learning models are developed using neuroimaging data to distinguish AD from normal controls, even when these models attain high accuracy, they still have problems, such as the high cost of neuroimaging and its limited availability in non-specialized locations (Fig. **2**). Finding fluid marker panels as possible screening tools has therefore been the subject of several investigations [40].

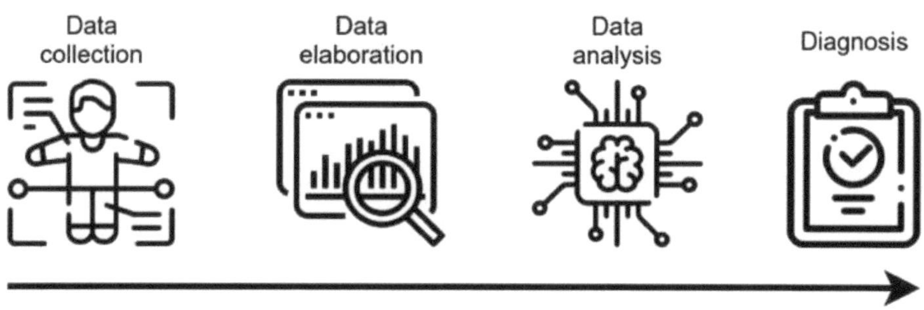

Fig. (2). Diagram representation of the functioning of CAD tools. After data collection, the data are processed and prepared for analysis using AI-based techniques [40] (Designed by Canva).

THE IMPORTANCE OF AI IN ALZHEIMER'S DIAGNOSIS

Traditional Alzheimer's diagnosis is based on cognitive evaluations, medical history, and neuroimaging methods such as Magnetic Resonance Imaging (MRI),

Positron Emission Tomography (PET), and cerebrospinal fluid (CSF) biomarkers. These procedures are frequently time-consuming, costly, and require the use of highly qualified medical personnel. Furthermore, by the time apparent symptoms show, considerable brain damage has already taken place. AI-driven technologies seek to diagnose the disease at a far earlier stage, allowing for quicker intervention and improved management [31].

AI Techniques in Alzheimer's Diagnosis

AI encompasses a variety of technologies, including Machine Learning (ML), Deep Learning (DL), and Natural Language Processing (NLP), all of which are critical for Alzheimer's disease detection. The succeeding sections cover several AI algorithms used to identify the illness [29].

1. Machine Learning and Deep Learning in Neuroimaging Analysis

One of the most promising applications of AI in Alzheimer's diagnosis is brain scan analysis. Machine learning algorithms can evaluate MRI and PET data to identify small changes in brain structure and function that may be early signs of Alzheimer's disease. Deep learning, a kind of machine learning, uses artificial neural networks to examine large datasets of neuroimaging scans and identify trends associated with the disease [29]. CNNs, or convolutional neural networks, have shown great promise in image processing. CNN-based models can accurately discriminate between patients with Alzheimer's disease, those with mild cognitive impairment (MCI), and healthy individuals. These models assist radiologists by reducing human error, increasing diagnosis accuracy, and providing automated reviews [33].

2. AI in Biomarker Analysis

Biomarkers such as tau tangles and beta-amyloid plaques in blood and cerebrospinal fluid (CSF) are important indicators of Alzheimer's disease. AI programs can examine complex biomarker data and spot abnormalities that traditional diagnostic methods might overlook. AI-powered algorithms have the potential to combine multi-omics data (genomic, proteomic, and metabolomic) to provide a comprehensive assessment of an individual's risk of Alzheimer's disease, paving the way for customized treatment regimens [33].

3. Natural Language Processing (NLP) for Cognitive Assessment

Language impairment is one of the first signs of Alzheimer's disease. Natural Language Processing (NLP), a branch of artificial intelligence that examines human language processing, has been used to analyse speech and writing patterns in an attempt to detect cognitive decline. AI algorithms trained on large datasets of written texts and recorded encounters may identify subtle changes in grammar, semantics, and coherence that indicate cognitive decline. Studies have demonstrated that NLP algorithms are capable of distinguishing between individuals with early-stage Alzheimer's disease and healthy individuals based on linguistic features. Digital health apps that include these AI approaches can prevent many hospital visits and enable remote patient monitoring [27].

Benefits of AI in Alzheimer's Diagnosis

The integration of AI in Alzheimer's diagnosis offers numerous advantages that improve early detection, accuracy, and patient outcomes.

1. **Early Detection and Diagnosis**: Compared to traditional approaches, AI models can identify Alzheimer's earlier, enabling prompt intervention and improved disease management [20].
2. **Higher Accuracy and Reliability**: AI-powered solutions lower diagnostic mistakes by accurately analysing large volumes of data [20].
3. **Personalized Treatment Plans**: AI systems can modify treatment plans in response to a patient's genetic, biomarker, and lifestyle information, enhancing therapy results [16].
4. **Cost Reduction**: AI-driven diagnosis lowers the cost of Alzheimer's care by eliminating the need for costly imaging tests and expert consultations [29].
5. **Remote Monitoring and Accessibility**: AI-powered digital technologies (Fig. **3**) make it possible to continuously monitor patients, which eases the strain on medical institutions and permits remote evaluations, particularly in rural locations [25].

CHALLENGES AND LIMITATIONS

The diagnosis, prognosis, and treatment of AD have all demonstrated significant promise with AI; however, some challenges and limitations need to be addressed in its practical implementation to ensure its reliability and effectiveness. The complexity of AD, data-related issues, ethical concerns, and high computing costs are the primary constraints and challenges associated with AI in AD research.

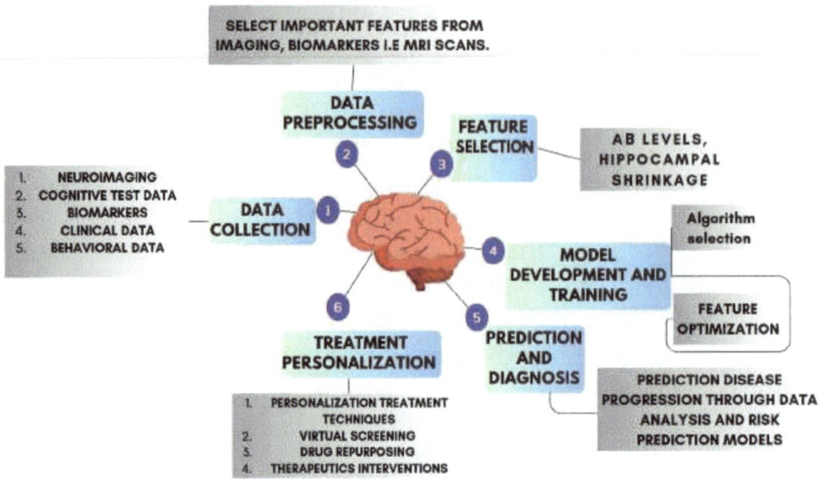

Fig. (3). How AI and ML are used in the treatment and diagnosis of Alzheimer's disease [29 - 33] (Designed by Canva).

1. Complexity and Heterogeneity of Alzheimer's Disease

AD is a very complex disease with many different molecular markers, progression patterns, and clinical symptoms. Models of artificial intelligence must account for genetic predispositions, differences in the rates of cognitive decline, and the presence of concurrent diseases such as vascular dementia and Lewy body dementia. Because of this heterogeneity, it is difficult to create a single AI model that can accurately predict when and how each patient's sickness will begin. It is challenging for AI to provide accurate and generally applicable predictions since disease presentations differ greatly [41].

2. Data Quality and Availability

For training and validation, AI models need big datasets, but problems with data availability and quality present major obstacles. The accuracy of models is impacted by the presence of partial or missing patient information in many clinical datasets. Inconsistencies are also produced by different institutions' approaches to data gathering, which makes it challenging to integrate and compare findings. The scope of AI training is therefore limited, and its efficacy in practical applications is diminished, as privacy rules frequently restrict access to important datasets [42].

3. Explainability and Trust in AI Decisions

One major drawback is the difficulty of describing AI models, particularly deep learning algorithms. As "black boxes," these models frequently make it difficult for medical experts to understand the decision-making process. Medical professionals can be reluctant to trust AI recommendations if there is no openness. Regulatory bodies also need interpretability before approving AI-based diagnostic systems for clinical use. Ethical concerns can arise when AI renders medical decisions without offering clear justifications, thereby limiting its application [43].

4. Overfitting and Model Generalization Issues

On training datasets, many AI models perform quite well, but often struggle to generalize to new, untested data. An issue known as overfitting occurs when models become overly dependent on certain biomarkers or imaging features from a dataset. Consequently, in different contexts or demographics, the AI system may struggle to produce accurate projections. This limitation makes AI models less reliable in practical settings and less applicable to therapy [44].

5. Need for Multimodal Data Integration

Connecting several data sources, including neuroimaging, cognitive test results, genetic markers, and lifestyle data, is required to construct successful AI models for AD research. Combining these many data sources, however, increases computational complexity and may result in conflicting conclusions. The inaccessibility and expense of particular data sets, such as PET scans, limit their application in non-AI models. Furthermore, contradictory data from diverse sources may make it difficult for AI to reach meaningful conclusions, lowering the overall accuracy of the model [45].

6. Ethical and Privacy Concerns

AI applications in healthcare present ethical problems about patient privacy, informed consent, and algorithmic discrimination. To secure huge datasets including sensitive patient information, strict data privacy standards such as GDPR and HIPAA must be adhered to. Another concern is bias in AI models, which can lead to unequal diagnosis and treatment recommendations by reflecting pre-existing disparities in training data. Concerns about openness and informed consent are also highlighted by the likelihood that patients are not completely aware of how AI is used in their medical choices [46].

7. High Costs and Infrastructure Requirements

Artificial Intelligence (AI) model construction, training, and long-term maintenance require large computing resources and financial commitments. Numerous AI-based frameworks depend on pricy neuroimaging techniques, including Positron Emission Tomography (PET), which are not always available in different healthcare systems. Furthermore, deep learning methods need cloud-based computing infrastructures and high-performance Graphics Processing Units (GPUs), which may be scarce or nonexistent in hospitals and research facilities with restricted funding. The necessity of constant algorithmic upgrades and retraining with changing datasets adds to operating expenses, making it financially difficult to deploy AI on a wide scale in clinical practice [47].

8. Limitations in Predicting MCI-to-AD Conversion

Predicting the development from Moderate Cognitive Impairment (MCI) to clinically evident AD is one of the most exciting uses of Artificial Intelligence (AI) in AD research. However, a number of restrictions limit AI's precision and dependability in this field. There are currently no well-recognized biomarkers that can reliably forecast the course of the illness, and some people with MCI do not develop AD, which makes categorization models more difficult. Furthermore, the robustness and generalizability of AI-based prediction models are further compromised by the heterogeneity in training data introduced by variations in neuropsychological assessment protocols and cognitive testing methodology among research [48].

9. Limited Clinical Validation and Real-World Implementation

Although research on artificial intelligence-based diagnostic techniques for Alzheimer's disease has shown promising findings, only a small number of these tools have been effectively used in clinical practice. A significant issue is that many AI models are trained on relatively limited or homogenous datasets and frequently lack large-scale, multicenter validation, lowering their dependability and therapeutic use. Furthermore, the regulatory licensing procedure for AI-driven medical technology is inherently time-consuming and complex, which contributes to clinical adoption delays. Certain healthcare professionals may be resistant to AI-based technologies, which may not integrate neatly with existing diagnostic procedures and traditional treatment paradigms. Without thorough validation, regulatory backing, and broad clinical adoption, AI remains primarily a research-oriented technology rather than a commonly applied therapeutic solution [49].

10. Challenges in AI-Driven Drug Discovery for AD

For the treatment of Alzheimer's Disease (AD), Artificial Intelligence (AI) has been thoroughly investigated as a technique for finding new therapeutic targets and repurposing already-approved pharmaceutical drugs. But there are still a lot of obstacles to overcome before AI-generated predictions can be turned into treatments that work in clinical settings. During clinical trials, the high attrition rate of AI-predicted treatment candidates is a result of the complex and diverse nature of AD pathophysiology. Further impeding the translational process is the need for thorough laboratory and preclinical validation of medicinal targets suggested by AI. Even while AI has the potential to speed up the medication research and development process, it hasn't yet produced game-changing therapeutic advances for AD [50].

11. Data Sharing

Data sharing has numerous possible dangers, which must be properly considered. However, increased access to research data can improve the repeatability of findings and help to build more rigorous approaches for AD identification. Open data sharing is frequently supported in the scientific community; yet, researchers have an ethical need to keep promises made to study participants. This involves protecting personal information and, in certain situations, limiting data access to specific requests. Importantly, it should not be assumed that all participants are against data sharing. While many people value secrecy, others may be encouraged to participate if they know their data will be widely publicized to enhance research, hence promoting scientific advancement [51].

12. Discrimination, Stigmatization, and the role of technology

When examining early indications of AD, stigmatization and discrimination are key issues even before clinical symptoms appear. Misconceptions and falsehoods about Alzheimer's disease frequently dehumanize those affected, promoting stigma in both personal and professional settings. Although several attempts have been implemented to lessen the worry and fear associated with Alzheimer's disease and related dementias, some of these treatments may paradoxically worsen stigmatization. This, in turn, raises the psychological load on people with dementia and their families while limiting their access to critical social support networks. To reduce discrimination, dementia should not be viewed simply as a danger to healthcare and social institutions. Discrimination against people in the pre-symptomatic or early stages of Alzheimer's disease can take many forms, including barriers to employment, difficulties obtaining legal recognition, denial of health insurance, restrictions on driving privileges, and exclusion from medical procedures such as organ donation. As predictive health technology progresses, these issues are expected to become more intense. The introduction of such technologies needs a thorough examination of their ethical implications, particularly in terms of possible misuse and dual-use problems. Historically, the term "dual use" applied to technology with both military and civilian purposes. However, the phrase has now expanded to include innovations that were initially intended for helpful reasons but may be repurposed or used in unethical or unexpected ways. Addressing these ethical concerns is therefore crucial in the development and deployment of new methods for early diagnosis and therapy of Alzheimer's disease [52].

13. Distress

Participating in cognitive testing and learning the research outcome can cause distress not only in the research subjects but also in other people involved in the process, such as families, clinicians, and researchers.

- **Cognitive testing:** Due to cognitive impairments and speech production issues, language examinations can be extremely challenging for people with Alzheimer's Disease (AD). Participants frequently see these tests or recordings as scenarios that highlight their deficiencies, which can cause emotions of humiliation or the conviction that the activities have no real significance, even though researchers may see them largely as objective measurements of cognitive performance. Frustration and confusion are often evoked when verbal fluency challenges include difficulties recalling relevant words. Furthermore, being asked to do tasks that were previously completed easily can cause intense emotional reactions, which can further impair cognitive function and have an adverse effect on test scores [53, 54].
- **Learning the research outcome:** Without proper treatment, being aware of the early warning symptoms of a growing neurological condition can be mentally stressful and may lead to low self-esteem or, in extreme situations, suicidal thoughts. The availability of therapeutic therapies has a significant impact on the psychological reaction to an early diagnosis. Early detection of illness symptoms may enhance clinical results and lessen adverse emotional reactions when therapeutic options are available. Early detection of Alzheimer's Disease (AD) symptoms can be beneficial even in the absence of effective treatments since it helps people better manage cognitive decline and make accurate choices about their future. On the other hand, those who are looking for a reason for their memory problems may find a negative test result upsetting. Additionally, the possibility of false-positive and false-negative results is still a serious worry, as it increases the worry and

uncertainty related to sharing early diagnostic results [55].

- **Distress in others involved:** Researchers, medical professionals, and study participants' families may experience psychological distress when looking into early signs of Alzheimer's Disease (AD). Families may feel more at ease if they understand the underlying reasons for symptoms they have seen, but they may also become more concerned about their own health or their loved ones, especially if there is a genetic component, which might strain family ties. The inherent unpredictability of early biomarker data and the ethical difficulties of sharing research results, especially choices about the degree of information disclosure, may provide difficulties for clinicians. In the same manner, researchers who work with emotionally sensitive data like voice samples that show cognitive decline may experience emotional strain and ought to have the choice to step away from the project at any time [56].

CONCLUSION

Although AI holds immense promise for improving Alzheimer's disease diagnosis, prognosis, and therapy, its current applications face a variety of challenges and limitations. To increase AI's reliability and clinical application, difficulties such as poor data quality, inexplicability, overfitting, ethical concerns, and high costs must be addressed. Future research should focus on making AI models more understandable, ensuring extensive validation methods, and improving the models' cost and usefulness. AI must solve these challenges in order to be an effective weapon in the fight against AD. This will enable earlier diagnosis, more personalized treatment approaches, and, eventually, the development of viable therapies.

REFERENCES

[1] Andrade-Guerrero J, Santiago-Balmaseda A, Jeronimo-Aguilar P, *et al.* Alzheimer's disease: an updated overview of its genetics. Int J Mol Sci 2023; 24(4): 3754. [http://dx.doi.org/10.3390/ijms24043754] [PMID: 36835161]

[2] Bird TD. Alzheimer disease overview 2018.

[3] Tahami Monfared AA, Byrnes MJ, White LA, Zhang Q. Alzheimer's disease: epidemiology and clinical progression. Neurol Ther 2022; 11(2): 553-69. [http://dx.doi.org/10.1007/s40120-022-00338-8] [PMID: 35286590]

[4] Henstridge CM, Hyman BT, Spires-Jones TL. Beyond the neuron–cellular interactions early in Alzheimer disease pathogenesis. Nat Rev Neurosci 2019; 20(2): 94-108. [http://dx.doi.org/10.1038/s41583-018-0113-1] [PMID: 30643230]

[5] Heneka MT, Carson MJ, Khoury JE, *et al.* Neuroinflammation in Alzheimer's disease. Lancet Neurol 2015; 14(4): 388-405. [http://dx.doi.org/10.1016/S1474-4422(15)70016-5] [PMID: 25792098]

[6] Weller J, Budson A. Current understanding of Alzheimer's disease diagnosis and treatment. F1000 Res 2018; 7: 1161.

[http://dx.doi.org/10.12688/f1000research.14506.1] [PMID: 30135715]

[7] McKhann GM, Knopman DS, Chertkow H, *et al.* The diagnosis of dementia due to Alzheimer's disease: Recommendations from the National Institute on Aging-Alzheimer's Association workgroups on diagnostic guidelines for Alzheimer's disease. Alzheimers Dement 2011; 7(3): 263-9. [http://dx.doi.org/10.1016/j.jalz.2011.03.005] [PMID: 21514250]

[8] Winchester LM, Harshfield EL, Shi L, *et al.* Artificial intelligence for biomarker discovery in Alzheimer's disease and dementia. Alzheimers Dement 2023; 19(12): 5860-5871. [http://dx.doi.org/10.1002/alz.13390]

[9] Khoury R, Ghossoub E. Diagnostic biomarkers of Alzheimer's disease: A state-of-the-art review. Biomark Neuropsychiatry 2019; 1100005 [http://dx.doi.org/10.1016/j.bionps.2019.100005]

[10] Dubois B, von Arnim CAF, Burnie N, Bozeat S, Cummings J. Biomarkers in Alzheimer's disease: role in early and differential diagnosis and recognition of atypical variants. Alzheimers Res Ther 2023; 15(1): 175. [http://dx.doi.org/10.1186/s13195-023-01314-6] [PMID: 37833762]

[11] Bouwman FH, Frisoni GB, Johnson SC, *et al.* Clinical application of CSF biomarkers for Alzheimer's disease: From rationale to ratios. Alzheimers Dement (Amst) 2022; 14(1)e12314 [http://dx.doi.org/10.1002/dad2.12314] [PMID: 35496374]

[12] Zetterberg H, Burnham SC. Blood-based molecular biomarkers for Alzheimer's disease. Mol Brain 2019; 12(1): 26. [http://dx.doi.org/10.1186/s13041-019-0448-1] [PMID: 30922367]

[13] van Gils AM, Rhodius-Meester HFM, Handgraaf D, *et al.* Use of a digital tool to support the diagnostic process in memory clinics–a usability study. Alzheimers Res Ther 2024; 16(1): 75. [http://dx.doi.org/10.1186/s13195-024-01433-8] [PMID: 38589933]

[14] Lott SA, Streel E, Bachman SL, *et al.* Digital Health Technologies for Alzheimer's Disease and Related Dementias: Initial Results from a Landscape Analysis and Community Collaborative Effort. J Prev Alzheimers Dis 2024; 11(5): 1480-9. [http://dx.doi.org/10.14283/jpad.2024.103] [PMID: 39350395]

[15] van Maurik IS, Broulikova HM, Mank A, *et al.* A more precise diagnosis by means of amyloid PET contributes to delayed institutionalization, lower mortality, and reduced care costs in a tertiary memory clinic setting. Alzheimers Dement 2023; 19(5): 2006-13. [http://dx.doi.org/10.1002/alz.12846] [PMID: 36419238]

[16] Nascimento IJ, Weerasekara I, Sharifan A, *et al.* Use of digital health technologies for dementia care: bibliometric analysis and report. JMIR Ment Health 2025; 12: e64445. [https://mental.jmir.org/2025/1/e64445]

[17] Balne S, Elumalai A. Machine learning and deep learning algorithms used to diagnosis of Alzheimer's: Review. Mater Today Proc 2021; 47: 5151-6. [http://dx.doi.org/10.1016/j.matpr.2021.05.499]

[18] Zhang W, Li Y, Ren W, Liu B. Artificial intelligence technology in Alzheimer's disease research. Intractable Rare Dis Res 2023; 12(4): 208-12. [http://dx.doi.org/10.5582/irdr.2023.01091] [PMID: 38024585]

[19] Cheung CY, Ran AR, Wang S, *et al.* A deep learning model for detection of Alzheimer's disease based on retinal photographs: a retrospective, multicentre case-control study. Lancet Digit Health 2022; 4(11): e806-15. [http://dx.doi.org/10.1016/S2589-7500(22)00169-8] [PMID: 36192349]

[20] Björngrim S, van den Hurk W, Betancort M, Machado A, Lindau M. Comparing traditional and digitized cognitive tests used in standard clinical evaluation–a study of the digital application minnemera. Front Psychol 2019; 10: 2327.

[http://dx.doi.org/10.3389/fpsyg.2019.02327] [PMID: 31681117]

[21] Seixas AA, Rajabli F, Pericak-Vance MA, Jean-Louis G, Harms RL, Tarnanas I. Associations of digital neuro-signatures with molecular and neuroimaging measures of brain resilience: The altoida large cohort study. Front Psychiatry 2022; 13899080 [http://dx.doi.org/10.3389/fpsyt.2022.899080] [PMID: 36061297]

[22] Staffaroni AM, Tsoy E, Taylor J, Boxer AL, Possin KL. Digital Cognitive Assessments for Dementia: Digital assessments may enhance the efficiency of evaluations in neurology and other clinics. Pract Neurol (Fort Wash Pa) 2020; 2020: 24-45. [PMID: 33927583]

[23] Kaye J, Aisen P, Amariglio R, *et al.* Using digital tools to advance Alzheimer's drug trials during a pandemic: the EU/US CTAD task force. J Prev Alzheimers Dis 2021; 8(4): 513-9. [http://dx.doi.org/10.14283/jpad.2021.36] [PMID: 34585227]

[24] Papp KV, Samaroo AH, Chou HCL, *et al.* Repeated memory-based assessments: Implications for clinical trials and practice. Alzheimers Dement 2020; 16(S6)e038143 [http://dx.doi.org/10.1002/alz.038143]

[25] Öhman F, Hassenstab J, Berron D, Schöll M, Papp KV. Current advances in digital cognitive assessment for preclinical Alzheimer's disease. Alzheimers Dement (Amst) 2021; 13(1)e12217 [http://dx.doi.org/10.1002/dad2.12217] [PMID: 34295959]

[26] Mc Carthy M, Schueler P. Can digital technology advance the development of treatments for Alzheimer's disease? J Prev Alzheimers Dis 2019; 6(4): 217-20. [http://dx.doi.org/10.14283/jpad.2019.32] [PMID: 31686090]

[27] Kale M, Wankhede N, Pawar R, *et al.* AI-driven innovations in Alzheimer's disease: Integrating early diagnosis, personalized treatment, and prognostic modelling. Ageing Res Rev 2024; 101102497 [http://dx.doi.org/10.1016/j.arr.2024.102497] [PMID: 39293530]

[28] Gao S, Lima D. A review of the application of deep learning in the detection of Alzheimer's disease. Int J Cognit Comput Engineer 2022; 3: 1-8. [http://dx.doi.org/10.1016/j.ijcce.2021.12.002]

[29] AbdulAzeem Y, Bahgat WM, Badawy M. A CNN based framework for classification of Alzheimer's disease. Neural Comput Appl 2021; 33(16): 10415-28. [http://dx.doi.org/10.1007/s00521-021-05799-w]

[30] Arbabyazd L, Shen K, Wang Z, Hofmann-Apitius M, Ritter P, McIntosh AR, Battaglia D, Jirsa V. Virtual connectomic datasets in alzheimer's disease and aging using whole-brain network dynamics modelling. eneuro. 2021 Jul 1;8(4). [http://dx.doi.org/10.1523/ENEURO.0475-20.2021]

[31] Atri A. The Alzheimer's disease clinical spectrum: diagnosis and management. Med Clin North Am 2019; 103(2): 263-93. [http://dx.doi.org/10.1016/j.mcna.2018.10.009] [PMID: 30704681]

[32] Basheera S, Ram MSS. Deep learning based Alzheimer's disease early diagnosis using T2w segmented gray matter MRI. Int J Imaging Syst Technol 2021; 31(3): 1692-710. [http://dx.doi.org/10.1002/ima.22553]

[33] Tanveer M, Richhariya B, Khan RU, *et al.* Machine learning techniques for the diagnosis of Alzheimer's disease: A review. ACM Trans Multimed Comput Commun Appl 2020; 16(1s): 1-35. [TOMM]. [http://dx.doi.org/10.1145/3344998]

[34] Angelucci F, Ai AR, Piendel L, Cerman J, Hort J. Integrating AI in fighting advancing Alzheimer: diagnosis, prevention, treatment, monitoring, mechanisms, and clinical trials. Curr Opin Struct Biol 2024; 87102857 [http://dx.doi.org/10.1016/j.sbi.2024.102857] [PMID: 38838385]

[35] Kourtis LC, Regele OB, Wright JM, Jones GB. Digital biomarkers for Alzheimer's disease: the mobile/wearable devices opportunity. NPJ Digit Med 2019; 2(1): 9.
[http://dx.doi.org/10.1038/s41746-019-0084-2] [PMID: 31119198]

[36] Pistono A, Jucla M, Barbeau EJ, *et al.* Pauses during autobiographical discourse reflect episodic memory processes in early Alzheimer's disease. J Alzheimers Dis 2016; 50(3): 687-98.
[http://dx.doi.org/10.3233/JAD-150408] [PMID: 26757034]

[37] Stringer G, Couth S, Brown LJE, *et al.* Can you detect early dementia from an email? A proof of principle study of daily computer use to detect cognitive and functional decline. Int J Geriatr Psychiatry 2018; 33(7): 867-74.
[http://dx.doi.org/10.1002/gps.4863] [PMID: 29424087]

[38] Spalletta G, Piras F, Piras F, *et al.* Neuroanatomical correlates of awareness of illness in patients with amnestic mild cognitive impairment who will or will not convert to Alzheimer's disease. Cortex 2014; 61: 183-95.
[http://dx.doi.org/10.1016/j.cortex.2014.10.010] [PMID: 25481475]

[39] Giulietti G, Torso M, Serra L, *et al.* Whole brain white matter histogram analysis of diffusion tensor imaging data detects microstructural damage in mild cognitive impairment and alzheimer's disease patients. J Magn Reson Imaging. 2018 Sep; 48(3): 755–764.

[40] Chang CH, Lin CH, Lane HY. Machine learning and novel biomarkers for the diagnosis of Alzheimer's disease. Int J Mol Sci 2021; 22(5): 2761.
[http://dx.doi.org/10.3390/ijms22052761] [PMID: 33803217]

[41] El-Sappagh S, Alonso JM, Islam SMR, Sultan AM, Kwak KS. A multilayer multimodal detection and prediction model based on explainable artificial intelligence for Alzheimer's disease. Sci Rep 2021; 11(1): 2660.
[http://dx.doi.org/10.1038/s41598-021-82098-3] [PMID: 33514817]

[42] Fabrizio C, Termine A, Caltagirone C, Sancesario G. Artificial Intelligence for Alzheimer's Disease: Promise or Challenge? Diagnostics (Basel) 2021; 11(8): 1473.
[http://dx.doi.org/10.3390/diagnostics11081473] [PMID: 34441407]

[43] Maudsley S, Devanarayan V, Martin B, Geerts H. Intelligent and effective informatic deconvolution of "Big Data" and its future impact on the quantitative nature of neurodegenerative disease therapy. Alzheimers Dement 2018; 14(7): 961-75.
[http://dx.doi.org/10.1016/j.jalz.2018.01.014] [PMID: 29551332]

[44] Valliani AAA, Ranti D, Oermann EK. Deep Learning and Neurology: A Systematic Review. Neurol Ther 2019; 8(2): 351-65.
[http://dx.doi.org/10.1007/s40120-019-00153-8] [PMID: 31435868]

[45] Singh A, Sengupta S, Lakshminarayanan V. Explainable Deep Learning Models in Medical Image Analysis. J Imaging 2020; 6(6): 52.
[http://dx.doi.org/10.3390/jimaging6060052] [PMID: 34460598]

[46] Feng X, Yang J, Lipton ZC, Small SA, Provenzano FA. Deep Learning on MRI Affirms the Prominence of the Hippocampal Formation in Alzheimer's Disease Classification. bioRxiv 2018.456277

[47] Lundberg S, Lee SI. A Unified Approach to Interpreting Model Predictions. In: Guyon I, Luxburg UV, Bengio S, et al., eds. Advances in Neural Information Processing Systems 30 (NIPS 2017). Red Hook (NY): Curran Associates, Inc.; 2017. p. 4765–4774.
[http://dx.doi.org/10.48550/arXiv.1705.07874]

[48] Selvaraju RR, Cogswell M, Das A, Vedantam R, Parikh D, Batra D. Grad-CAM: Visual Explanations from Deep Networks via Gradient-Based Localization. Proceedings of the 2017 IEEE International Conference on Computer Vision (ICCV) Venice, Italy. 2017; pp. 2017; 618-26.
[http://dx.doi.org/10.1109/ICCV.2017.74]

[49] Grapov D, Fahrmann J, Wanichthanarak K, Khoomrung S. Rise of Deep Learning for Genomic, Proteomic, and Metabolomic Data Integration in Precision Medicine. OMICS 2018; 22(10): 630-6. [http://dx.doi.org/10.1089/omi.2018.0097] [PMID: 30124358]

[50] Rajkomar A, Dean J, Kohane I. Machine Learning in Medicine. N Engl J Med 2019; 380(14): 1347-1358. [http://dx.doi.org/10.1056/NEJMra1814259] [PMID: 30943338]

[51] Meyer MN. Practical tips for ethical data sharing. Adv Methods Pract Psychol Sci 2018; 1(1): 131-44. [http://dx.doi.org/10.1177/2515245917747656]

[52] Gauthier S, Leuzy A, Racine E, Rosa-Neto P. Diagnosis and management of Alzheimer's disease: Past, present and future ethical issues. Prog Neurobiol 2013; 110: 102-13. [http://dx.doi.org/10.1016/j.pneurobio.2013.01.003] [PMID: 23578568]

[53] Chien YW, Hong SY, Cheah WT, Yao LH, Chang YL, Fu LC. An automatic assessment system for Alzheimer's disease based on speech using feature sequence generator and recurrent neural network. Sci Rep 2019; 9(1): 19597. [http://dx.doi.org/10.1038/s41598-019-56020-x] [PMID: 31862920]

[54] Lopez-de-Ipina K, Martinez-de-Lizarduy U, Calvo PM, *et al.* Advances on automatic speech analysis for early detection of Alzheimer disease: a non-linear multi-task approach. Curr Alzheimer Res 2018; 15(2): 139-48. [http://dx.doi.org/10.2174/1567205014666171120143800] [PMID: 29165084]

[55] Illes J, Rosen A, Greicius M, Racine E. Prospects for Prediction. Ann N Y Acad Sci 2007; 1097(1): 278-95. [http://dx.doi.org/10.1196/annals.1379.030] [PMID: 17413029]

[56] Smith AP, Beattie BL. Disclosing a diagnosis of Alzheimer's disease: patient and family experiences. Can J Neurol Sci 2001; 28(S1) (Suppl. 1): S67-71. [http://dx.doi.org/10.1017/S0317167100001220] [PMID: 11237313]

CHAPTER 6

Natural Products and Phytochemicals in Alzheimer's Management

Sakshi Tyagi[1,*], **Anubhav Tyagi**[2], **Baby Ilma**[1] and **Nikhila Shekhar**[3]

[1] *School of Pharmacy, Sharda University, Greater Noida, Uttar Pradesh 201310 India*

[2] *Department of Management Studies, Indian Institute of Technology, New Delhi, Delhi 110016, India*

[3] *Neuropharmacology Research Laboratory, School of Pharmaceutical Sciences, Delhi Pharmaceutical Sciences and Research University, New Delhi, Delhi 110016, India*

Abstract: Alzheimer's Disease (AD) is a progressive neurodegenerative condition that primarily impairs memory, cognition, and behavioural functions. Since no definitive cure currently exists, increasing attention has shifted toward lifestyle interventions and nutraceutical supplementation as supportive strategies for prevention and symptom management. Nutraceuticals such as polyphenols, omega-3 fatty acids, antioxidants, and vitamins have shown promise in alleviating oxidative stress, neuronal inflammation, and age-related neurodegeneration processes strongly implicated in AD pathology. Compounds like curcumin, resveratrol, and docosahexaenoic acid (DHA) have demonstrated neuroprotective properties in preclinical and clinical studies, suggesting their potential in moderating cognitive decline and supporting brain health. Lifestyle-based approaches, including regular physical activity, cognitive exercises, and dietary modifications, further enhance protective outcomes. Exercise promotes neuroplasticity, optimises cognitive performance, and reduces dementia risk, while social engagement and mental stimulation strengthen resilience to cognitive impairment. Nutritional patterns such as the Mediterranean diet, rich in plant-based foods and healthy fats, are associated with a reduced risk of developing AD, highlighting the significant role of diet in maintaining neural function. Although ongoing research continues to explore optimal doses, duration, and mechanisms of these approaches, integrating nutraceuticals with lifestyle modifications offers a practical and holistic strategy to delay or minimise AD progression and improve quality of life in at-risk individuals.

Keywords: Alzheimer's disease, Antioxidants, Cognition, Exercise, Mediterranean diet, Nutraceuticals.

* **Corresponding author Sakshi Tyagi:** School of Pharmacy, Sharda University, Greater Noida, Uttar Pradesh 201310 India; E-mail: Sakshi.tyagi@sharda.ac.in

Prashant Kumar Dhakad, Raghav Mishra, Shweta Sharma, Sudhanshu Mishra & Nishant Gaur (Eds.)

INTRODUCTION

Neurodegenerative disorders (NDD) are a group of conditions characterised by selective dysfunction and progressive loss of neurons, neuroglial cells, and neural pathways in the spinal cord and Central Nervous System (CNS) [1]. There are several types of NDDS based on their key clinical features, such as dementia, Parkinsonism [2], and Huntington's disease [3, 4]. Alzheimer's Disease (AD), the most prominent category of dementia in older people, is characterised by a progressive loss of memory and other cognitive functions [5, 6]. Cognitive impairment is one of the most common clinical manifestations of this disease [7, 8]. Amyloid beta (Aβ) and cholinergic hypotheses are the two leading theories for AD. Other risk factors include vascular illnesses, infections, ageing, head trauma, genetic disorders, and variables in the environment [8, 9]. Recent studies have shown that the repression of hyperphosphorylation of tau protein, accumulation of Aβ plaque, inflammatory responses, and unfolded protein reactions, as well as the reduction of oxidative injury and biological targets such as beta secretase (BACE-1), cholinesterase, and monoamine oxidase (MAO), have a significant impact on the genesis of AD [10].

Approximately 5 million new instances of AD are discovered each year, making it the most well-known and prevalent form of dementia, comprising 60 to 70% of all cases. Globally, there are already more than 50 million dementia patients, and by 2050, that Figure is expected to double, with AD accounting for the majority of cases [11, 12]. Compared to other nations worldwide, AD-related fatalities are more common in Libya, Lebanon, and Finland.

Many pathogenic mechanisms interact intricately to form the pathophysiology of AD. Extracellular accumulation of Aβ, anomalies in tau protein, oxidative destruction of neurons, intracellular neurofibrillary tangles (NFTs), neuronal inflammation, mutations in genes, dysregulation of neurotransmitters, and other molecular and cellular mechanisms are significant variables in AD, as represented in Fig. (**1**). The primary cause of AD is an extracellular aggregation of Aβ plaques, which causes neurotoxicity and finally leads to neuronal cell death [13 - 15].

The extracellular aggregation of Aβ significantly influences the pathophysiology of AD. The breakdown of Amyloid-β Protein Precursor (APP) yields Aβ peptides, which can subsequently unite to form insoluble plaques. These plaques cause inflammation and interfere with regular neural activity. Intermediate forms of Aβ accumulation, known as Aβ oligomers, are very detrimental to neurons and are implicated in neurotoxic cascades and synaptic malfunction. Anomalies of the tau protein contribute to the formation of intracellular NFTs. Tau (τ) is a microtubule-linked protein that helps preserve the shape and movement of neurons. Due to hyperphosphorylation, tau in AD detaches from microtubules and aggregates form NFTs. These tangles impede axonal transit, damage neurons, and cause cognitive impairment.

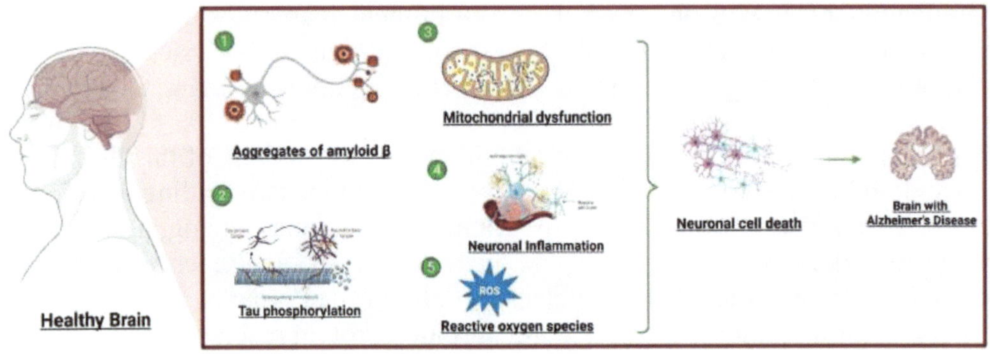

Fig. (1). Diagrammatic representative view of the pathophysiology of alzheimer's disease (Created by BioRender.com).

One important aspect of the pathophysiology of AD is oxidative stress. Overproduction of Reactive Oxygen Species (ROS) compromises antioxidant defences, ultimately resulting in oxidative neuronal injury. ROS are produced by oxidative stress caused by Aβ, inflammation, and mitochondrial impairment. Lipid oxidation, DNA damage, protein oxidation, and mitochondrial impairment are all consequences of oxidative stress, which exacerbates synaptic abnormalities and neuronal cell damage.

A significant contributing factor to AD is neuronal inflammation. The aggregation of Aβ and the formation of NFTs stimulate the activation of microglial cells, the brain's defence cells. Proinflammatory cytokines generated by activated microglia have been connected to tau pathology, formation of Aβ, and neuronal cell injury. Neuroinflammation is exacerbated by the selection of peripheral immune cells, which intensifies the inflammatory response.

AD research has focused on new pathways and mechanisms. New research focuses on the gut-brain axis, mitochondrial dysfunction, vascular anomalies, and epigenetic mutations. Mitochondrial impairment increases the risk of oxidative stress and affects energy production. Reduced cerebral blood flow and metabolic issues are caused by vascular anomalies (Fig. **1**). Gene expression patterns are influenced by epigenetic changes, which may contribute to the progression of AD. The pathogenesis of dementia has been linked to the gut-brain axis, which involves a bilateral interaction between gut bacteria and the brain [16, 17].

Donepezil, galantamine, rivastigmine (cholinesterase inhibitors), memantine (glutamate regulators), suvorexant (orexin receptor antagonists), and aducanumab (anti-amyloid antibodies) are the Food and Drug Administration (FDA) approved drugs for the treatment of AD [18]. Their accessibility and possible adverse consequences restrict the use of these drugs. Therefore, producing a drug that is safe, effective, affordable, and efficient is crucial. Natural products may have therapeutic potential against AD, according to recent studies, and herbal medications are suggested as synthetic drug substitutes for AD management [19]. Effective use of these herbal medications will increase their anti-AD effectiveness by targeting several AD mechanisms with fewer adverse effects (Fig. **2**) [20]. Thus, it is beneficial to treat AD with herbal plants and their derivatives.

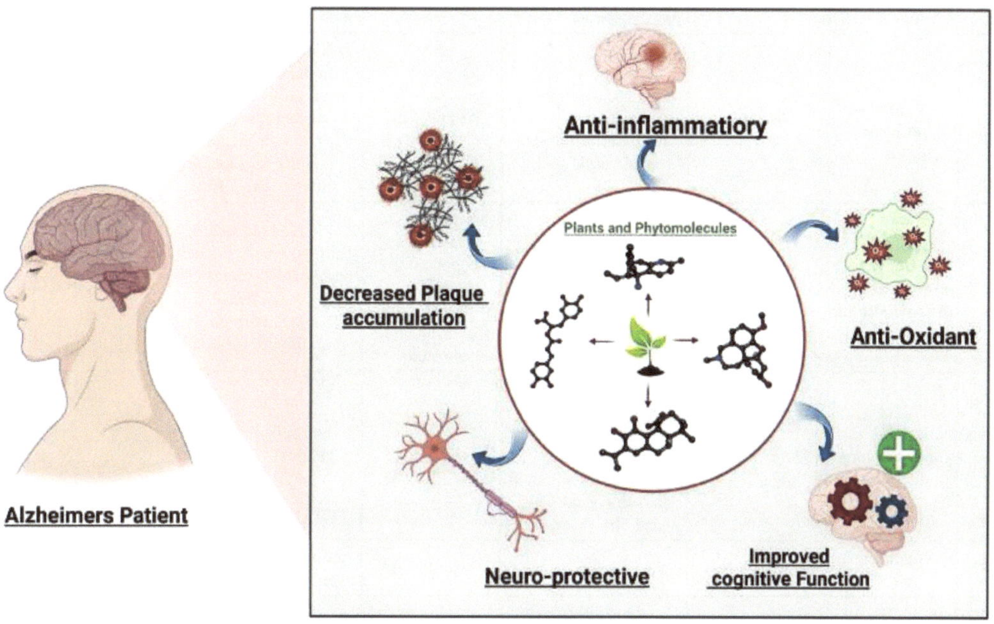

Fig. (2). Diagrammatic representative view of the mechanism of bioactive compounds on Alzheimer's Disease (Created by BioRender.com).

THE BENEFICIAL ROLE OF NATURAL PRODUCTS AND PHYTOCHEMICALS IN THE MANAGEMENT OF AD

In recent years, researchers have found a variety of active components in herbs, and these active ingredients have focused on the therapeutic use of herbal remedies for AD that have fewer side effects than traditional medications (Table **1**) [21]. The safety and efficacy of the plant-derived product in specific applications have been demonstrated following extensive standardisation (Fig. **3**).

Table 1. Plant sources with their bioactive compounds, along with their mechanism of action in the management of AD.

S.No.	Plant Sources	Family	Phytoconstituents	Major Constituents	Mechanism of Action	Chemical Structure	Reference
1.	***Bacopa monnieri***	*Scrophulariaceae*	alkaloids including brahmin, herpestine and nicotine, D-mannitol, glycosides, hersaponin, monnierin, and triterpenoids bisdesmosides, bacosides A and B, bacosaponins A, B, and C, bacopasaponins D, E, and F, betulic acid, jujubogenin, and sterols.	Bacosides A and B	Accelerating protein kinase activity, synapse restoration, and neuron production aid in nerve impulse transmission, antioxidant activity, and suppress lipoxygenase activity.		[22]
2.	***Centella asiatica***	*Apiaceae*	asiatic acid, asiaticoside, centella saponin, madecassoside, and madecassic acid.	Asiaticoside,	Anti-inflammatory, immunomodulatory, neuroprotective, nerve regenerative, anti-depressive, memory-enhancing, and antioxidative effects.	;	[39]
3.	***Convolvulus pluricaulis (Shankhpushpi)***	*Convolvulaceae*	anthocyanins, flavonoids, proanthocyanidins, steroids, and terpenes	kaempferol and quercetin, scopoletin and ayapanin, hydroxycinnamic acid, and β-sitosterol	Memory enhancing activity, brain stimulant, anti-neuroinflammatory, antidepressant, anticonvulsant, anxiolytic, sedative		[47]
4.	***Curcuma longa***	*Zingiberaceae*	Curcumin, phenol, saponin, and tannin, curcuminoids, cyclocurcumin, bisdemethoxycurcumin, demethoxycurcumin, and turmerone oil.	Curcumin	anti-inflammatory, antioxidant, anti-amyloidogenic		[51]
5.	***Galanthus nivalis***	*Amaryllidaceae*	Galanthamine	Galanthamine	anti-inflammatory, anti-cholinesterase activity,		[65]
6.	***Ginkgo biloba***	*Ginkgoaceae*	Ginkgolides, terpenoids, flavonoids, bilobalide, organic acids (vanillic, ascorbic acid, and shikimic acid), steroids (sitosterol and stigmasterol)	Ginkgolide B	anti-amyloidogenic, anti-oxidative, antiapoptotic, and scavenging pathways of free radicals		[69]
7.	***Glycyrrhiza glabra***	*Fabaceae*	chalcones, coumarins, flavonoids, gums, glycyrrhizin, isoflavonoids, lignins, saponins, triterpenoids, isoliquiritigenin, and volatile oils	Glycyrrhizin and glycyrrhitinic acid, isoliquiritigenin	anti-inflammatory, anti-epileptic, and antioxidant		[79]

(Table 1) cont.....

S.No.	Plant Sources	Family	Phytoconstituents	Major Constituents	Mechanism of Action	Chemical Structure	Reference
8.	***Huperzia serrata***	*Lycopodiaceae*	Huperzia A, huperzine A, caffeic acid, and ferulic acid	Huperzia A, huperzine A	anticonvulsant, anti-inflammatory, anti-nociceptive, anti-Alzheimer's, anti-schizophrenia, anti-apoptosis, antioxidant, protection from mitochondrial impairment, and prevention against poisoning with organophosphate and myasthenia gravis		[90]
9.	***Lepidium meyenii Walp.***	*Brassicaceae*	alkylbenzenes, amines, flavonoids, Phenols, saponins, and steroids	linoleic, palmitic, and oleic acid	Antioxidant, memory-enhancing activity		[91]
10.	***Melissa officinalis***	*Lamiaceae*	apigenin, luteolin, caffeic acid, cholinergic acid, rosmarinic acid, oleanolic acid, ursolic acid, and metrilic acid	Rosmarinic acid, ursolic acid	sedative-hypnotics, anti-anxiety effects, diaphoretic action, including antidepressants, anti-cholinesterase activity, antioxidant		[102]
11.	***Rosmarinus officinalis***	*Lamiaceae*	apigenin, carvacrol, eugenol, oleanolic acid, ursolic acid, rosmarinic acid.	Rosmarinic acid	anti-inflammatory and anti-oxidant, anti-acetylcholinesterase (AChE), and anti-butyrylcholinesterase (BChE) activity.		[106]
12.	***Salvia officinalis***	*Lamiaceae*	Caffeic acid, lithospermic acids, rosmarinic acid, sagernic acids, sage-coumarin, salvianolic acids, and Yunnaneic acid apigenin, hispidulin, kaempferol, luteolin, and quercetin α and β-thujone, α-humulene, β-caryophyllene, 1,8-cineole, camphor and viridiflorol tanshinones, carnosic acid, carnosol, and ursolic acid	Rosmarinic acid, carnosic acid	Anxiolytic and antidepressant activity, Antioxidants, Anticholinergic effect, Antiinflammatory		[113]
13.	***Tinospora cordifolia (Giloy)***	*Menispermaceae*	liquiritin, and Shoyuflavone A	Plastoquinone 3	Antioxidant, anticholinesterase activity		[116]

Bacopa monnieri

Bacopa monnieri is a nootropic herb that is mainly used to improve memory and thinking ability, anxiety, stress, attention deficit hyperactivity disorder (ADHD), and AD. The small perennial herbaceous plant, popularly referred to as "Brahmi," is a member of the Scrophulariaceae family. Phytochemical analysis of this plant extract revealed the presence of several bioactive phytoconstituents, including alcohols, triterpenoids, D-mannitol, glycosides, herasaponin, monierin, and alkaloids like brahmin, herpestine, and nicotine [22]. It also possesses an antioxidant effect due to the presence of several phytoconstituents (alkaloids, bisdesmosides, bacosides (A and B), bacosaponins (A, B, and C), bacopasaponins (D, E, and F), betulic acid, jujubogennin, sterols, and sulfhydryl components [23].

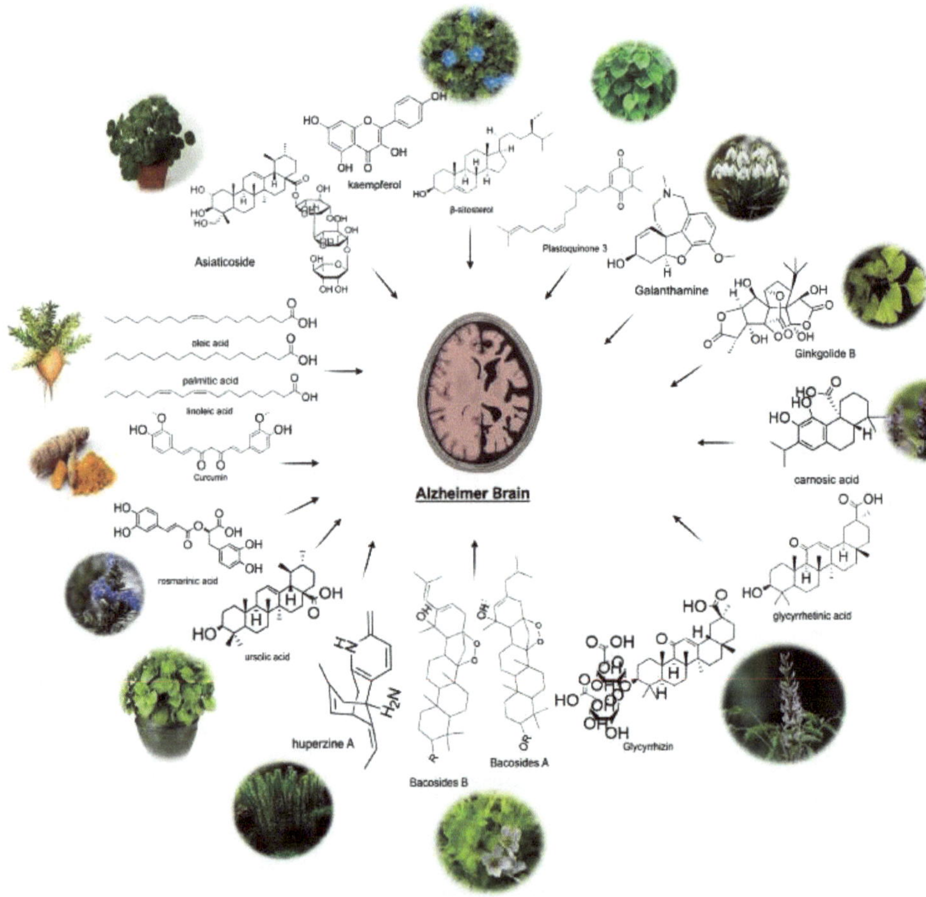

Fig. (3). Chemical structures and sources of natural products reported against Alzheimer's Disease (Created by BioRender.com)

The main constituents accountable for the neuroprotective activity are bacosides A and B [22]. These phytoconstituents stimulate the recovery of damaged neurons by enhancing protein kinase activity, promoting synapse restoration, and promoting neuron production. They also aid in nerve impulse transmission [24].

The antioxidant activity and neuroprotective effect of this plant make it an extremely intriguing herb for the treatment of AD [22].

This plant's antioxidant properties contribute to the upregulation of different antioxidant molecules (SOD and GSH), allowing it to mitigate H2O2-induced oxidative injury (in vitro & in vivo). Additionally, this herb reduces lipoxygenase (LOX) activity, which facilitates recovery from oxidative damage [22]. The diverse range of bioactive components present in this plant underscores its potential as a potent alternative treatment to enhance brain health and cognitive performance.

Centella asiatica (Gotu Kola)

For many years, gotu kola has been valued for its health advantages as a medicinal herb. It is mostly ingested as a green leafy vegetable [25]. It has been utilised for ages in Southeast Asian, Indian, and Sri Lankan Ayurvedic, Unani, and folk medicine. Additionally, it has also been observed in Chinese and African traditional medicine [26]. It is scientifically referred to as *Centella Asiatica*, belonging to the family *Apiaceae*.

According to Indian Ayurveda, C. asiatica is thought to have several beneficial effects [27]; its aqueous extract has been used to revive and rejuvenate neuronal cells, as well as being effective in conquering insomnia, and for thousands of years, Ayurveda has utilised this potential "cure-all herb" to treat both moderate and chronic disorders [28]. C. asiatica is believed to possess a broad range of therapeutic properties, such as anti-inflammatory, anticancer, antimicrobial, wound healing, immunomodulatory, neuroprotectant, anti-depressive, memory-enhancing, gastro-protective, cardio-protective, and anti-oxidative properties [26, 27, 29, 30].

The main phytoconstituent present in the *C.asiatica* is triterpenoid saponin glycosides, including asiatic acid, asiaticoside, centella saponin, madecassoside, and madecassic acid [31]. Asiaticoside, the main bioactive component of gotu kola, has neuroprotective and antioxidant properties [32]. *Centella asiatica's* neuroprotective properties can shield mitochondria from oxidative stress, which prevents caspase activity from initiating apoptosis [33].

C. asiatica is especially well-known for its ability to revitalise brain cells [34] and neurons and for its historical use as a memory enhancer [35]. This long-standing use is closely associated with its neuroprotective properties, suggesting that its bioactive constituents may be effective in treating neurological conditions common in elderly populations.

Furthermore, gotu kola may help cure AD, particularly in cases associated with Aβ poisoning, by inhibiting the synthesis of Aβ cells [36].

A study was conducted on trimethyltin (TMT) induced dementia rats by Yuliani to evaluate the effect of the extract of gotu kola on the caspase-3 expression in different regions of the hippocampus (CA-1, CA-2, and CA-3). The hippocampus is a crucial brain region involved in forming memories. It is the area of the CNS that is most vulnerable to oxidative stress. Oxidative stress in the hippocampus can lead to neuronal death, which is one factor in dementia development. It was discovered that administering gotu kola extract at varying dosages decreased caspase-3 expression. In the TMT-induced dementia paradigm in rats, it also demonstrated antiapoptotic activity in the pyramidal cells of the hippocampus [37].

Fresh Centella asiatica leaf extracts administered at a concentration of 250 mg/kg body weight significantly improved the rats' capacity for learning and memory retention, according to studies conducted on Wistar rats (an in vivo model) [38]. Additionally, it has demonstrated potential as a multifunctional treatment for some ailments, including depression, rheumatism, epilepsy, mental weakness, and stomachaches. Notably, it can lower oxidative stress responses and cure Aβ pathology, making it an essential health resource [39].

Convolvulus pluricaulis (Shankhpushpi)

Shankhpushpi is a perennial shrub, scientifically called *Convolvulus pluricaulis (C. pluricaulis),* and belongs to the family *Convolvulaceae.* It is well known for its memory-enhancing properties and is primarily used as a brain stimulant to manage various neurological disorders [40]. It has been traditionally used for a long time in India as a brain tonic (neural tonic) to enhance memory and cerebral function. The phytoconstituents and metabolites present in this crude herb possess several beneficial neuropharmacological activities (in vitro and in vivo) such as memory enhancer, anti-neuroinflammatory, antidepressant, anticonvulsant, anxiolytic, sedative, pain-relieving, and reversed the symptoms associated with AD [41]. Its cognitive benefits are being elucidated by earlier research, which indicates that aqueous, alcoholic, methanolic, and ethyl acetate extracts obtained from this plant may greatly enhance memory and learning capabilities [42].

The main phytoconstituents and metabolites of this plant include anthocyanins, flavonoids, kaempferol and quercetin, proanthocyanidins, steroids and terpenes, coumarins (scopoletin and ayapannin), phenolic acid (hydroxycinamic acid), and phytosterol (*βeta-sitosterol*) [43], which are solely accountable for the improvement in memory deficits and cognitive disability. By controlling the formation of stress hormones, these compounds are thought to foster nerve relaxation [44, 45].

Various studies on Shankhpushpi, conducted in rodents, have demonstrated a beneficial effect on memory by enhancing cerebral function. The ethanolic extract of *C. pluricaulis*, along with its ethyl acetate and aqueous fractions, showed memory-enhancing effects at doses of 100 and 200 mg/kg when administered orally [46]. Convolvine, a phytoconstituent of C. pluricaulis, decreased cognitive decline in AD and boosted the memory-improving properties of choline [47]. Administration of C. pluricaulis (160 mg/kg) for 3 months reduced acetylcholine esterase activity, further reducing free radical damage and preventing neuronal toxic effects [48]. In vitro, an alcohol extract of C. pluricaulis Choisy (leaves) reduced the formation of amyloid-β [49].

Based on network pharmacology evaluations, components of C. *pluricaulis* interact with serotonergic synapses, which are crucial for neurotransmission, AD, alcoholism, chronic depression, cognitive disorders, psychological conditions, and raising serotonin levels in the brain [50]. This research implies a mechanism linking the cholinergic system's influence and Shankhpushpi's memory-boosting benefits. As a result, the metabolites of Shankhpushpi exhibit nootropic and memory-boosting properties, thereby increasing their therapeutic relevance. The plant's potential as a memory-enhancing agent has been highlighted by numerous studies that have shown *C. pluricaulis* extracts to improve retention and spatial learning capacity in newborn rat pups and to enhance memory in older rats [42].

Curcuma longa

Curcuma longa (Turmeric) is a plant commonly cultivated in southern and southwestern tropical Asia. It is a member of the ginger (*Zingiberaceae*) family and is obtained from plant rhizomes. The significant turmeric phytoconstituents include curcumin (a water-insoluble, orange-yellow polyphenol), phenol, saponin, and tannin. Other bioactive molecules are curcuminoids (hydrophilic), cyclocurcumin, bisdemethoxycurcumin, demethoxycurcumin, and turmerone oil [51]. These constituents exhibit well-known anti-inflammatory, antioxidant, and anti-cancerous properties, and due to these properties, they play a significant role in the management and treatment of various diseases, including hyperglycaemia, neurological disorders, heart-related diseases, and other autoimmune disorders [52]. This naturally occurring polyphenol possesses several therapeutic properties and is frequently utilised in pharmaceuticals [53]. Due to its antioxidant, anti-inflammatory, and anti-amyloidogenic properties, it represents a practical therapeutic approach for managing AD [54, 55].

Specifically, curcumin has been related to a lower risk of AD [7]. It has been used to treat several kinds of disorders since ancient times, and research has linked its anti-inflammatory properties to a lower risk of Alzheimer's [56].

Curcumin has potent neuroprotective benefits and can be utilised to treat a range of neurological conditions due to its anti-inflammatory and antioxidant properties. It binds to Aβ proteins, which reduce inflammation and oxidative stress and improve cognition [57]. It is interesting to note that long-term use of low dosages of curcumin has proven more effective than large doses in treating AD. It has been discovered that a small amount of turmeric reduces proinflammatory cytokine levels, which regulate neuroinflammatory pathways, leading to the formation of neuritic plaques [58].

By raising intracellular cholesterol esters, hyperlipidemia and hypercholesterolemia accelerate the development of amyloid plaques. Curcumin is believed to have therapeutic benefits against Alzheimer's disease by reducing serum peroxides and inhibiting cholesterol synthesis [59]. Because of its numerous beneficial properties, curcumin is a natural substance that shows potential for controlling and possibly preventing AD.

Galanthus nivalis

A biennial herbaceous bulbous plant, *Galanthus nivalis,* is a member of the *Amaryllidaceae* family. It has been identified as one of the 20 most significant plants that produce alkaloids [60]. Due to its milk-white blooms, it is often referred to as a snowdrop. Its name comes from the Greek words "gala," meaning milk, and "nthos," meaning blossom. Since ancient Greek times, GN has been used as a folk remedy to relieve pain and neurological conditions. It has also achieved economic significance as a desirable plant [61].

Several key phytoconstituents, including terpenoids, flavonoids, phenolics, and an important class of alkaloids, were identified through phytochemical analyses of Galanthus nivalis. These substances possess numerous therapeutic properties, including acetylcholinesterase inhibitory, antiviral, anti-inflammatory, and antitumor properties [62, 63]. Galanthamine, an alkaloid of the isoquinoline class and effective against AD, is particularly significant. In the 1950s, Galanthamine was initially identified and isolated, and in 2000, both the US and the EU approved its synthetic form for the treatment of AD [64, 65]. Another study demonstrated the optimal outcomes of Galanthamine at a dose of 2mg/kg in different neurobehavioral parameters in rodents [62].

An *in-vitro* study was conducted in 2003, which revealed the inhibitory potential of the methanolic extract of Galanthus nivalis against AchE activity and established the neuroprotective potential of the drug [66].

These study results highlight the significance of Galanthus nivalis as an ethnobotanical plant that has potent neuroprotective properties and shows promise as an AD treatment.

Ginkgo biloba

Ginkgo biloba, a member of the *Ginkgoaceae* family, is a plant that is frequently used to treat AD and memory impairments. It possesses antioxidant and antiapoptotic activities [67, 68]. Terpenoids, flavonoids, bilobalide, organic acids (vanillic, ascorbic, and shikimic acid), steroids (sitosterol and stigmasterol), and ginkgolides are among the vital components found in the plant's leaves that are used to cure cognitive impairments [24]. Essential components of Ginkgo biloba include ginkgolides and terpenoids (diterpenes and sesquiterpene bilobalide) [69]. Approximately 22–27% of the plant extracts contain flavone glycosides, with terpene lactones accounting for the remaining 5-7% [70]. Isorhamnetin, kaempferol, and quercetin are examples of flavone glycosides, whereas bilobalide and ginkgolides A, B, and C are examples of terpene lactones.

Due to the presence of flavonoids, Ginkgo biloba has a significant role in the prevention of different forms of oxidative insult, which further leads to detrimental effects in the brains of individuals with AD [71]. The herb is believed to have therapeutic characteristics that help prevent and control AD through many mechanisms, including anti-amyloidogenic, anti-oxidative, antiapoptotic, and free radical scavenging pathways [72]. The extract of the plant impedes neurotoxicity induced by inhibiting amyloid-β (Aβ) by managing glutathione peroxidase (GPx) and superoxide dismutase (SOD) activity (in vitro), which limits the apoptosis in neurons, formation of ROS, absorption of glucose, mitochondrial impairment, and stimulation of the extracellular signal-regulated kinase (ERK) and c-Jun N-terminal kinase (JNK) pathways. GB has been demonstrated to improve cognition in Alzheimer's patients by enhancing oxygen flow and facilitating the body's ability to eliminate free radicals [24].

Several substances, including flavonoids, terpenoids, and organic acids, contribute to Ginkgo biloba's neuroprotective properties [73]. It may enhance cognition, particularly when taken in the early stages of AD, according to scientific reports [74].

Glycyrrhiza glabra

Liquorice is a herbaceous perennial with long, flexible, cylindrical roots and runners, and it is scientifically known as *Glycyrrhiza glabra.* It belongs to the family *Fabaceae*. Traditionally, the plant's roots and dried runners have been used therapeutically, as they possess various vital constituents, including amino acids, chalcones, coumarins, flavonoids, gums, glycyrrhizin, isoflavonoids, lignins, saponins, triterpenoids, and volatile oils. Glycyrrhizin and glycyrrhitinic acid are the main active constituents belonging to the triterpenoids, whereas the most prevalent isoflavone is glabridin [75].

Liquorice has been utilised for a long time in the treatment and management of various disorders such as asthma, stomach pain, sleeping disorders, peptic ulcer, malaria, viral infection, and psychiatric disorders [76]. It has anti-inflammatory, anticancer, antiviral, antihyperglycemic, and antimalarial activities, among other therapeutic effects [76].

Furthermore, research has suggested that liquorice may be utilised to treat neurological disorders like AD, dementia, or memory difficulties [77]. Research on rodents has demonstrated that the aqueous root extract of liquorice enhances learning and memory [78, 79]. The memory-boosting effect, which enhances neuronal function, is believed to be caused by its anti-inflammatory and antioxidant effects, which reduce exposure to oxidative stress and protect brain cells from damage. These studies suggest that liquorice might have a neuroprotective effect in preventing brain diseases.

Specifically, the active component, glycyrrhizin, has demonstrated neuroprotective properties against certain scopolamine-induced neurological disorders [80].

Liquorice extract also reduces neurological symptoms and has demonstrated promise as a neuroprotective therapy for acute ischaemic stroke patients [81]. It has also shown promising results in reducing neuropathic pain perceptions in rats [82]. Chalcones and flavonones, two phytoconstituents that have been isolated from liquorice, have shown potential in treating age-related cognitive disorders like Parkinson's and Alzheimer's [60]. The active components of liquorice root extracts, glycyrrhizin, glycyrrhetinic acid, and glycyrrhizic acid, have been shown to alleviate the symptoms of AD by inhibiting the death of neuronal cells caused by Aβ plaques [83]. The phytoconstituents present in liquorice extract have been shown to have beneficial effects on neuronal cell death, suggesting that they may be a neuroprotective agent that merits more research.

Thirty-five active components of liquorice stem and leaf flavonoids were evaluated in the network pharmacology study. These compounds activated 41 signalling pathways, including the PI3K-Akt pathway, and had a beneficial effect on AD through 67 targets. Additionally, behavioral data showed that the flavonoids in liquorice stem and leaf significantly reduced anxiety induced by inflammatory brain injury and enhanced learning and memory abilities in mice. Moreover, it was also demonstrated that the stem and leaf flavonoids of Glycyrrhiza glabra had a positive effect on oxidative stress. Additionally, results found from western blotting analysis reported the activation of the PI3K-Akt signalling pathway [84].

Moreover, liquorice flavonoids have been shown to have neuroprotective properties against seizure-induced neuronal cell death. Epileptogenesis is defined as a vigorous process that is triggered by injuries precipitated in the brain and leads to the progression of spontaneous recurrent seizures (epilepsy). A flavonoid found in liquorice, called isoliquiritigenin, has several pharmacological effects, including antioxidant and anti-inflammatory properties. Although the beneficial potential of isoliquiritigenin on cognitive deficits in epilepsy and its mechanism is not fully known. For a better understanding, a study was conducted by Zhu et al. in which a seizure model was established by administering kainic acid (intracerebroventricular injection) in rats. The changes in cognitive behaviour, synaptic plasticity, neuronal cell injury, and the glial cell number with the cytokines expression and nuclear factor-like (NRF2) signalling and (NLRP)3 inflammasome components were evaluated. It was found that isoliquiritigenin had a beneficial protective effect by regulating NRF2 signalling and the NLRP3 inflammasome pathway, thereby mitigating cognitive deficits induced by kainic acid in rats [85].

Huperzia serrata

Huperzia serrata is a perennial herb that grows extensively throughout the Americas, Oceania, and eastern and southern Asia. It is a member of the *Lycopodiaceae* family and is known scientifically as a toothed fern. The main component of the plant, lycopodium alkaloids, has demonstrated a potent cholinesterase inhibitory action [86]. This plant is also rich in other secondary metabolites, including phenolic acids, flavones, and triterpenes. For many years, it has been used to treat a variety of conditions, including rheumatism, colds, fevers, bruises, pain, strains, and contusions. Many pharmacological properties are also present, including anticonvulsant, anti-inflammatory, antinociceptive, anti-Alzheimer's, anti-schizophrenia, anti-apoptosis, antioxidant, protection against mitochondrial impairment, and prevention of organophosphate poisoning and Myasthenia Gravis (MG) [87].

Chinese phytochemists isolated huperzine A, a key alkaloid from H. serrata, which has been demonstrated to have anti-Alzheimer's effects [87]. *H. serrata* was found to enhance acetylcholinesterase (AchE) but not butyrylcholinesterase (BAChE), which may help prevent AD. Cognitive behaviour tests revealed that H. serrata reduced cognitive impairment in a model of scopolamine-induced cognitive impairment, suggesting that it could be used to treat dementia associated with AD. A dose of 30 mg/kg/day in mice can be used in preclinical studies to assess the efficacy and safety profile of H. serrata extract, and more specifically, its active ingredient, huperzine A [87, 88].

Using microwave technology, NSP01, a standardised green extract of H. serrata, was developed. Following phytochemical screening, the NSP01 extract's three primary constituents were ferulic acid, caffeic acid, and huperzine A. Additionally, it has been studied for its neuroprotective effects in primary neuronal cultures damaged by glutamate, as well as in another in vivo model involving cortical neurons. The NSP01 extract protected neurons and neurite networks from glutamate-induced damage, and improved neuron survival in subsequent studies [89].

Remarkably, the main ingredients—ferulic acid, caffeine, and huperzine A—showed a synergistic neuroprotective potential when combined, improving the neurite network and the survival of neurons. This research demonstrates H. serrata's potential as a neuroprotectant by demonstrating enhanced neuronal activity when injected into cultured nerve cells or in vivo paradigms of AD [90]. These study findings show that *H. serrata* has intriguing neuroprotective effects and may be used to treat AD.

Lepidium meyenii Walp

Lepidium meyenii is an edible herbaceous plant. It is indigenous to South America, namely the high Andes of Bolivia and a particular area of Peru. It is scientifically known as Peruvian ginseng or Maca and belongs to the family *Brassicaceae*. This plant has a long history of conventional herbal use. It is still used to treat a variety of diseases, such as depression, cancer, diminished memory, sexual disorders, and irregular menstruation. A detailed phytochemical study of its roots and leaves revealed a variety of secondary metabolites.

Using LC-Q-TOF analysis of the Maca leaf and root extract, Lee and Chang et al. (2009) found the biologically active secondary metabolism products, such as alkylbenzenes, amines, flavonoids, phenols, saponins, and steroids [62]. Remarkably, the amounts of flavonoids, phenol, and total saponin were higher in the leaves than in the roots. In vitro models utilising 2,2-diphenyl-1-picrylhydrazyl (DPPH) radical scavenging and FRAP tests to evaluate antioxidant effects revealed more vigorous activity in the leaves, suggesting that the leaves may have more beneficial therapeutic effects than the roots.

The well-known medicinal properties of Peruvian ginseng include antihyperglycemic, neuroprotective, dermatological, memory-boosting, fertility, energising, and antioxidant properties [91]. Studies on the methanolic extract of Maca, which was re-extracted in n-pentane, revealed neuroprotective effects. The highest effective dose with positive effects on neuronal function was found to be 3 mg/kg in in vivo studies conducted on adult rats affected by stroke. Another study was conducted on H2O2-exposed scampi neuronal cells for the induction of oxidative stress. Treatment at different doses (0.1, 0.3, 1.3, 10, and 30 µg/ml) of the pentane extract of Maca reported the ameliorative potential of L. mayenii [92, 93].

Macamides, another important class of secondary metabolites, were extracted from Maca L. mayenii Walp and exhibited neuroprotective properties. It was discovered that macamides mitigated the neurotoxic effects of Manganese (II) chloride (MnCl2) exposure on U-87 MG glioma cells. The CB1 receptors of macamides also played a role in this neuroprotection, according to an in vivo study on the receptors that caused the neuroprotective effects [94].

In general, L. mayenii Walp. It is a valuable medicinal herb that has potent neuroprotective properties. To confirm its benefits as a potent medication for managing various neurological conditions, thorough toxicity research employing multiple solvents for extract synthesis is essential.

Melissa officinalis

Melissa officinalis is a perennial herb that belongs to the family *Lamiaceae*. It has hairy leaves (2 to 8 cm in length) and is heart-shaped. The leaf has a rough, deeply veined surface. Due to its flavour and fragrance, it is commonly referred to as lemon balm [95]. There are several beneficial phytoconstituents, including apigenin, luteolin, caffeic, cholinergic, rosmarinic, oleanolic, ursolic, and metrilic acid present in the lemon balm extracts [96]. This plant exhibits several pharmacological effects, including sedative-hypnotic properties, anti-anxiety effects, and diaphoretic action, as well as antidepressant and antidiabetic properties [59, 97].

According to research, lemon balm may help cure AD since it has antioxidant properties and may inhibit acetylcholinesterase. Moreover, it directly binds to cholinergic receptors, modulates cholinergic activity, and exhibits anti-cholinesterase activity, which is responsible for its neuroprotective properties [33, 98]. Lemon balm contains some chemicals that interact with muscarinic and nicotinic acetylcholine receptors, lowering the activity of the AchE enzyme [99]. Furthermore, people with mild to moderate AD have shown improved cognitive function and decreased agitation after using lemon balm [100, 101].

Due to the presence of rosmarinic acid, lemon balm also has the benefit of destabilising the Aβ formation, which is a factor that complicates AD [102, 103]. It exhibits promise as a natural and therapeutic option for individuals with AD due to its several benefits for brain health.

Rosmarinus officinalis

Rosemary is a perennial shrub that has been used for centuries in traditional medicine to serve various purposes. It belongs to the family Lamiaceae. The main active constituents of this plant involve apigenin, carvacrol, eugenol, oleanolic acid, and ursolic acid, which are natural Cyclooxygenase-2 (COX-2) inhibitors [104]. The crucial extract present in rosemary is responsible for its antiseptic and other pharmacological activities [105]. Specifically, rosmarinic acid (a polyphenolic carboxylic acid) is responsible for various pharmacological activities, including anti-inflammatory and antioxidant effects, which further lead to neuroprotection. It has been demonstrated that rosemary also exhibits other beneficial properties, including antiviral, antibacterial, and anticancer effects, in addition to its neuroprotective effects.

Additionally, it has been shown that rosmarinic acid can reduce the production of nuclear factor kappa B (NF-κB) and Tumor Necrosis Factor alpha (TNF-α), which strongly suggests that it may play a role in lowering the risk of memory loss associated with Aβ [106]. According to in-vitro experiments, rosemary remarkably has potent protective effects against neurodegenerative diseases, including dementia and AD.

Moreover, rosemarinic acid and terpenes present in the rosemary as main constituents are also accountable for the anti-acetylcholinesterase (AChE) and anti-butyrylcholinesterase (BChE) activity. By raising the brain's choline levels, these cholinesterase enzyme inhibitors can effectively minimize memory deficits, anxiety, and other signs associated with dementia [107].

Moreover, a study reported the beneficial potential of plant extract (60% or 10% carnosic acid) and spearmint extract (5% rosmarinic acid) on memory and learning in mice [108]. Due to its numerous positive effects on mental health and cognitive function, rosemary can be utilized as a natural treatment for various neurological conditions.

Salvia officinalis

Salvia, a genus of plants in the *Lamiaceae* family, is used to treat a range of neurological conditions because of its neuroprotective effects. One of these, the fragrant herb S. officinalis, is well known for its medicinal benefits and for its capacity to cure neurological disorders such as AD [109]. This plant contains around 160 polyphenols, including various flavonoids and phenolic acids. Salvia species include a variety of phenolic chemicals, including lithospermic acids, rosmarinic acid, sagernic acids, sage-coumarin, salvianolic acids, and yunnaneic acid. Flavonoids such as luteolin, quercetin, kaempferol, hispidulin, and apigenin are present in it. Furthermore, terpenoids such as α and β-thujone, α-humulene, β-caryophyllene, 1,8-cineole, camphor, and viridiflorol are prevalent in the essential oils of Salvia species. Plants also contain considerable amounts of diterpenes and triterpenes, including ursolic acid, carnosic acid, carnosol, and tanshinones [110].

This plant has phytoconstituents that have a favourable effect on oxidative damage, inflammation, Aβ, and cholinergic function, as well as triggering multiple biochemical pathways associated with cognitive behaviour. It has been stated that this plant possesses anxiolytic and depressive properties [111]. Antioxidants, specifically rosmarinic and carnosic acid, found in *S. officinalis,* significantly shield the brain from oxidative injury [112]. The antioxidant activity of rosmarinic acid and other phytochemicals has also been reported in an *in vitro* paradigm [113]. Moreover, it is also accountable for activating the acetylcholine (Ach) receptor and binding the (muscarinic and nicotinic) receptors. It enhances memory retention by interacting with the cholinergic and muscarinic networks involved in memory retention [114]. Species of Salvia, particularly *S. officinalis*, have considerable potential as natural therapies for enhancing memory retention and mental well-being.

Tinospora cordifolia (Giloy)

It is well-renowned for its remarkable capacity to improve memory. It is effective for both healthy people and people with memory issues. Its effect on cognitive performance can be attributed mainly to choline supplementation, which boosts the production of acetylcholine, a neurotransmitter crucial for memory and learning processes, and fortifies the immune system. In the ancient Ayurvedic system, it was considered a potent enhancer of memory and learning. Due to its positive effects on memory enhancement, the stem of Guduchi (Tinospora cordifolia) is utilized as a Medhya treatment in Ayurvedic medicine. In 2023, Jamadagni et al. conducted a study to examine the Guduchi stem extract's Medhya abilities by investigating its ability to mitigate the neurotoxicity caused by aluminium chloride ($AlCl_3$) in rats used as a model of chronic memory loss. An aqueous extract of the Guduchi stem was administered to the Wistar rats treated with $AlCl_3$ for 42 days. The neurobehavioral (EPM) and biochemical parameters (oxidative stress, acetylcholinesterase inhibition assay) were measured. Rats treated with Guduchi extract exhibited substantially improved memory, accompanied by a decline in oxidative stress. These encouraging findings demonstrate the beneficial properties of Guduchi stem extract as a neuroprotectant against $AlCl_3$-induced neurotoxicity in rats [115].

Another study was conducted by Amrutha et al. to evaluate the mechanistic targets of the T. cordifolia protein associated with AD using network pharmacology and metabolomics. Using LC-MS/MS analysis, 1,186 metabolites were identified, including well-known physiologically active substances such as liquiritin, Plastoquinone 3, and Shoyoflavone A, among others. The network pharmacology study highlighted the relationship between metabolites and proteins by enriching 591 human proteins, including neurotransmitter receptors and other regulatory proteins. Pathway analysis demonstrated the enrichment of several neurotransmitter pathways (cholinergic, dopaminergic, serotonergic, and glutamatergic synapses), as well as apoptosis and signaling pathways (cAMP, mTOR, MAPK, and PI3K-Akt). According to the docking facts, metabolites of Guduchi may interact with BACE1 and MAO-B, two important targets for Alzheimer's disease, revealing a potential neuroprotective effect. These findings offer insight into the molecular mechanisms underlying T. cordifolia's therapeutic properties and lay the groundwork for future translational research on the plant [116].

Specifically, it has been shown that utilising an aqueous extract (200 mg/kg body weight) derived from Guduchi roots greatly improves verbal learning as well as logical memory [73]. These reports highlight the plant's historical significance in conventional healthcare as well as its potential as a valuable agent for cognitive enhancement [117].

THERAPEUTIC EFFECTS OF PHYTOCONSTITUENTS IN AD

Maintaining a healthy lifestyle requires a balanced, nutrient-dense diet, which is a significant defence against brain conditions like Alzheimer's. A diet high in bioactive substances has been linked to a decreased dementia risk [118, 119]. The extent to which the beneficial effects of bioactive compounds observed in in vitro and in vivo studies translate to neuroprotective advantages in humans in daily life remains unknown, as human interventional research is currently lacking.

Furthermore, it is unclear whether the specific chemical forms and amounts of nutrients included in meals effectively enhance the body's ability to utilize them. Nevertheless, research that clarifies the molecular processes behind these phytoconstituents' potential to prevent AD supports their beneficial effects. Certain bioactive substances, which fall under various chemical categories, including phenolic compounds, fat-soluble vitamins, essential omega-3 fatty acids, isothiocyanates, and carotenoids, possess several intriguing properties. These substances have considerable potential to enhance brain function and reduce the risk of AD.

Phenolic Compounds

A key component in supporting brain wellness is olive oil, which is a rich source of phenolic compounds such as oleuropein, hydroxytyrosol, and oleocanthal. Glycosylated seco-iridoid oleuropein has strong antioxidant potential and shields neuronal cells from apoptosis spurred by neurotoxins. Additionally, it can lower Aβ levels and inhibit the enzyme glutaminyl cyclase, which is implicated in Aβ formation. Oleuropein undergoes metabolic changes in the digestive tract to form hydroxytyrosol, another potent antioxidant found in olive oil that exhibits higher bioavailability [120, 121]. Hydroxytyrosol has been shown to activate phase II detoxifying enzymes and protect neuronal cells from damage caused by Aβ [122].

Furthermore, by blocking the cyclooxygenase (COX) enzyme, which is responsible for producing proinflammatory prostaglandins, oleocanthal, a phenolic component found in olive oil, exhibits beneficial anti-inflammatory properties. Oleocanthal has shown promise in preventing Aβ from aggregating and facilitating its removal from the brain [120]. According to in vitro research on tau protein produced from *E. coli,* oleocanthal may prevent protein accumulation, which implies that it could aid in the degenerative processes that cause dementia [120, 123].

Additionally, research has indicated that olive oil has neuroprotective properties, particularly when combined with the Mediterranean Diet (MD). MD, which is high in olive oil and other fat sources, has been linked to a lower risk of AD [124]. Observational studies show that a greater degree of adherence to MD ideas is linked to a slower rate of cognitive decline and a reduced risk of AD [125]. Additionally, studies have shown that having MD, along with olive oil, can enhance working memory, focus, and learning performance [126].

Genistein, an isoflavone present in soy products, could potentially prevent AD by shielding mitochondria and diminishing oxidative stress [127]. Diets high in genistein-containing soy products have been related to a decreased dementia risk. It also reduces inflammation by lowering TNF-α and NF-κB levels and inhibits caspase-3 activity to prevent apoptosis. By inhibiting β-secretase and promoting α-secretase, genistein can reduce the production of Aβ and the development of senile plaque [128].

Hop extracts were found to have an intriguing neuroprotective influence, along with other promising inhibitors of Aβ production. Hop extracts have been shown to inhibit γ-secretase activity, thereby reducing the synthesis of Aβ in cultured cells. They have also been demonstrated to reduce Aβ depositions and learning and memory impairment in mice. Garcinelliptone, a key active ingredient in hop extracts that initially originated from garcinia, was extracted by Sasaoka et al. This compound inhibits γ-secretase and may also have the ability to decrease the formation or accumulation of Aβ [129].

Anthocyanins are a distinct type of neuroprotective phenolic component responsible for providing various colours (red, blue, and violet) in several fruits and vegetables. These compounds, which possess antioxidant qualities, have been shown to protect the brain from oxidative stress by reducing the generation of free radicals [130]. Additionally, it has been discovered that anthocyanins inhibit the aggregation of Aβ peptides and reduce the brain's surplus production of reactive species [131]. An important source of anthocyanins, berries have been connected to delayed memory loss [118].

Another phenolic compound with neuroprotective properties is curcumin, which is found naturally in turmeric. Curcumin suppresses inflammation, inhibits Aβ from adhering together, and prevents the oxidation of protein [132]. Curry, which is high in curcumin, may help cognitive functions and prevent cognitive decline, according to observational studies [133]. Research has demonstrated that taking curcumin supplements can enhance memory and attention, decrease the aggregation of tau and Aβ proteins, and have a potent neuroprotective effect [134].

Fat-soluble Vitamins and Fatty Acids

Vitamins and essential fatty acids contain bioactive compounds that are vital for mitigating AD. Due to its high oxygen consumption and a high content of polyunsaturated fatty acids, the CNS is particularly susceptible to oxidative damage. It is believed that lipophilic antioxidants and vitamins can reduce oxidative damage and, hence, lower the incidence of this disease.

Fish are the primary source of docosahexaenoic acid (DHA), a polyunsaturated fatty acid that shows promise as a treatment for ageing-related CNS problems. It belongs to the identical long-chain omega-3 family as n-3 docosapentaenoic acid (DPA) and eicosapentaenoic acid (EPA). DHA makes up a substantial portion of the phospholipids that make up the structural membranes of brain cells and is the brain's most abundant n-3 fatty acid [135, 136]. As people age, their production of DHA decreases because the enzyme that is responsible for its synthesis becomes less active. Both dietary consumption and liver conversion from shorter-chain precursors affect the amount of DHA in the brain. DHA reduces inflammation in the brain by competing with arachidonic acid (precursor of prostaglandins) and preventing the production of these eicosanoids. Additionally, by increasing neuroprotectin D1 and brain-derived neurotrophic factor (BDNF), it can protect the nervous system by reducing neuronal injury and promoting neuronal regeneration. It also influences the metabolism of amyloid precursor protein (APP) [137, 138].

Another neuroprotective bioactive substance that can be acquired from the diet, such as fish, and exposed to sunlight is vitamin D. Due to its anti-inflammatory and anti-amyloid properties, it enhances the brain's ability to eliminate Aβ and regulates Aβ production and enzymatic breakdown. Additionally, it has an impact on calcium homeostasis, which is crucial in brain diseases [139, 140]. Research has demonstrated that in animal models, vitamin D therapy boosts cognition and stimulates neurogenesis [141]. Vitamin D is crucial for preserving mental health in older people at risk for dementia since low levels are linked to an increased risk of AD [142].

Tocopherols and tocotrienols, two of the eight components of vitamin E that are present in nuts, seeds, and vegetable oils, may also help prevent AD. The decline in glutathione and catalase levels is indicative of AD. Vitamin E inhibits COX and lowers NF-κB activity for use as a medication that prevents inflammation [143 - 145]. Studies based on observation show that the antioxidant properties of vitamin E help prevent dementia [146].

ISOTHIOCYANATES

One important class of bioactive compounds that may help prevent AD is isothiocyanates. These compounds are derived from glucosinolates found in cruciferous vegetables. Isomethiocyanates, especially those with an aromatic ring next to the thiocyanate group, have antioxidant properties that are attributed to the sulphur atom that is present in their molecules. They also have anti-inflammatory and COX-inhibiting effects. A study demonstrated that some isothiocyanates inhibit acetylcholinesterase activity, thereby increasing acetylcholine levels, a neurotransmitter that is typically decreased in individuals with Alzheimer's [147].

Sulforaphane is one of the extensively studied isothiocyanates, created from glucoraphanin during the processing of plant tissues. Sulforaphane enhances the antioxidant activity of glutathione peroxidase and glutaredoxin. In neurons and neuroglial cells, it enhances sulfiredoxin activity, which facilitates the regeneration of other antioxidant enzymes. Additionally, sulforaphane helps with clearance processes by regulating the activity of proteasomes in nerve cells [148].

The neuroprotective properties of isothiocyanates have been established through epidemiological investigations. Nurk et al. studied 2031 people between the ages of 70 and 74 to evaluate their nutritional consumption and cognitive function. Cognitive outcomes were greater for individuals who consumed more cruciferous vegetables, including broccoli, cauliflower, Brussels sprouts, and cabbage, compared to those who consumed them less frequently [149]. Additional studies in this field also produced encouraging findings [150].

The potential value of cruciferous vegetables in preserving cognitive function and possibly protecting against AD is highlighted by their isothiocyanate content. Further research is required to fully comprehend their mechanisms of action and therapeutic benefits in the context of neurodegenerative diseases.

CAROTENOIDS

Carotenoids are plant-based pigments that give a variety of fruits and vegetables their vivid yellow, orange, and red hues. Additionally, they are present in microalgae, which makes them a vital component of marine life's nutrition. Carotenoids prevent photooxidation and are necessary for photosynthesis. Astaxanthin is among the most beneficial compounds in the carotenoid family. It reduces oxidative stress, lipid peroxidation, and protein peroxidation by acting as a potent free radical scavenger. Astaxanthin also enhances the activity of antioxidant enzymes, including superoxide dismutase and catalase. Its neuroprotective effects include boosting neurogenesis and preventing the death of neuronal cells via altering mitogen-activated kinases and reducing caspase-3 activity [151].

In a clinical research with volunteers ages 45 to 64 who received an astaxanthin extract (6/12 mg/day) for 12 weeks, Katagiri et al. confirmed the neuroprotective benefits of astaxanthin. On cognitive and learning tests, both intervention groups outperformed the placebo group [151]. Zeaxanthin and lutein, two more carotenoids, have also been linked to psychological benefits. According to a study involving 2796 people over 60, increased consumption of lutein and zeaxanthin was associated with improved memory function of lutein and zeaxanthin was associated with improved memory function [152]. Other research conducted by Power et al. demonstrated that participants' cognitive ability and episodic memory were enhanced after taking supplements of lutein (10 mg), zeaxanthin (2 mg), and meso-zeaxanthin (10 mg) for twelve months [153].

Lycopene, another neuroprotective carotenoid, is a potent antioxidant that neutralises singlet oxygen, lowers lipid oxidation, and stops oxidative damage to DNA [154]. Hwang et al.'s in vitro study showed that it enhanced the survival of neuronal cells and decreased apoptosis in brain cells. Additionally, lycopene prevented amyloid-induced mitochondrial damage and reduced the level of free radicals [155]. According to these results, lycopene might help prevent neurological conditions.

Overall, carotenoids, phenolic compounds extracted from olive oil, anthocyanins, curcumin, and genistein, as well as bioactive molecules found in vitamins and essential fatty acids, hold great promise in preventing and minimizing AD by reducing oxidative stress, promoting neurogenesis, and regulating key processes involved in the progression of AD. Further research and clinical studies are needed to fully understand and exploit the therapeutic benefits of these bioactive compounds.

FUTURE PROSPECTIVES

Research on ethnomedicinal plants as treatments for Alzheimer's disease (AD) is up-and-coming due to their potential to tackle the disease's complex nature. However, essential knowledge gaps must be addressed to realise their full potential. To ensure the safety and efficacy of these plants, future studies should focus on identifying the active ingredients they contain and conducting thorough clinical trials to validate their effectiveness. Extract and dosage standardisation, along with comprehensive toxicity tests, are necessary for safe incorporation into treatments. In addition to considering hereditary characteristics that can influence therapy outcomes, a deeper understanding of pharmacokinetics, drug interactions, and mechanisms of action is also crucial. Understanding the sociocultural contexts from which these plants originate is vital to advancing this research. It will ensure that bioprospecting is conducted ethically and that traditional knowledge and scientific innovation are effectively integrated. By addressing these challenges, we may see advancements in AD treatment and greater recognition of the benefits of traditional medicine.

CONCLUSION

AD is spreading among the elderly at a startling rate around the globe. In most cases, AD and its comorbidities are fatal if left untreated. Therefore, many nations concentrate on efforts to conduct research and development to manage and treat AD efficiently. In the management and treatment of AD, natural anti-Alzheimer medicines extracted from plants may be a potential substitute for synthetic medications. Numerous bioactive substances found in these plants may have an impact on primary disease targets, including inflammation, metal dysregulation, oxidative insult, hyperphosphorylation of tau protein, Aβ aggregation, and cholinergic deficits. While tau phosphorylation, which is necessary for the development of neurofibrillary tangles, may be altered by some plant extracts, others may have anti-amyloidogenic properties that may inhibit amyloid plaques from proliferating or aggregating. Metal-chelating substances in these plants can treat the metal imbalance that contributes to the pathogenesis of AD, while antioxidant molecules present in herbal remedies can reduce oxidative stress. While anti-inflammatory drugs can reduce the neuroinflammation frequently seen in AD, compounds that enhance cholinergic transmission can help address the neurotransmitter deficits associated with cognitive loss. These substances, derived from plants, have demonstrated promise in enhancing cognitive performance and slowing the progression of the condition. However, thorough toxicological research and clinical trials are required to ensure the safety as well as efficiency of these plant-based constituents. More research is also needed to better understand the biological mechanisms underlying the positive neuroprotective activity of these natural phytochemicals, to develop more targeted and effective treatments. Future research will focus on investigating plant-derived compounds for the treatment of AD, with the ultimate goal of providing better care and management strategies for individuals affected by the disease.

LIST OF ABBREVIATIONS

CNS Central Nervous System

NDDs Neurodegenerative disorders

AD Alzheimer's disease

Aβ Amyloid beta

MAO Monoamine oxidase

NFTs Neurofibrillary tangles

AβPP Amyloid-β protein precursor

ROS Reactive Oxygen Species

SOD Superoxide dismutase

GP_x Glutathione peroxidase

ERK Extracellular signal-regulated kinase

JNKs c-JUN N-Terminal Kinase

ADHD Attention-Deficit Hyperactivity Disorder

COX-2 Cyclooxygenase-2

NF-κB Nuclear factor kappa B

TNF-α Tumor Necrosis Factor Alpha

Nrf2 Nuclear factor erythroid 2–related factor 2

NLRP3 Nucleotide-binding domain, leucine-rich–containing family, pyrin domain–containing-3

AchE Acetylcholinesterase

BAChE Butyrylcholinesterase

DPPH 2,2-diphenyl-1-picrylhydrazyl

$MnCl_2$ Manganese(II) chloride

CB1 Cannabinoid receptor 1

TMT Trimethyltin

$AlCl_3$ Aluminium chloride

COX Cyclooxygenase

MD Mediterranean diet

DHA Docosahexaenoic acid

DPA Docosapentaenoic acid

EPA Eicosapentaenoic acid

REFERENCES

[1] Goyal R, Mittal P, Gautam RK, *et al.* Natural products in the management of neurodegenerative diseases. Nutr Metab (Lond) 2024; 21(1): 26. [http://dx.doi.org/10.1186/s12986-024-00800-4] [PMID: 38755627]

[2] Tyagi S, Thakur AK. Beneficial role of capsaicin through modulation of mitochondrial functions in MPTP-injected mice. Neurosci Behav Physiol 2025; 55(1): 1-15. [http://dx.doi.org/10.1007/s11055-024-01560-7]

[3] Goyal R, Wilson K, Saharan A, *et al.* Insights on aspects of apoptosis in neurodegenerative disorders: a comprehensive review. Explor Med 2024; 5(1): 89-100. [http://dx.doi.org/10.37349/emed.2024.00208]

[4] Tyagi S, Thakur AK. Effect of capsaicin on 3-NP-induced neurotoxicity: a pre-clinical study. Neurochem Res 2024; 49(8): 2038-59. [http://dx.doi.org/10.1007/s11064-024-04158-0] [PMID: 38814358]

[5] Onyango IG, Jauregui GV, Čarná M, Bennett JP Jr, Stokin GB. Neuroinflammation in Alzheimer's disease. Biomedicines 2021; 9(5): 524. [http://dx.doi.org/10.3390/biomedicines9050524] [PMID: 34067173]

[6] Tahami Monfared AA, Byrnes MJ, White LA, Zhang Q. Alzheimer's disease: epidemiology and clinical progression. Neurol Ther 2022; 11(2): 553-69. [http://dx.doi.org/10.1007/s40120-022-00338-8] [PMID: 35286590]

[7] Gregory J, Vengalasetti YV, Bredesen DE, Rao RV. Neuroprotective herbs for the management of Alzheimer's disease. Biomolecules 2021; 11(4): 543. [http://dx.doi.org/10.3390/biom11040543] [PMID: 33917843]

[8] Tyagi S, Thakur AK. Neuropharmacological study on capsaicin in scopolamine-injected mice. Curr Alzheimer Res 2023; 20(9): 660-76. [http://dx.doi.org/10.2174/0115672050286225231230130613] [PMID: 38213170]

[9] Anwal L. A comprehensive review on Alzheimer's disease. World J Pharm Pharm Sci 2021; 10(7): 1170.

[10] Pandey J. Flavonoid-based derivatives for modulating various targets of Alzheimer's disease Natural Product-based Synthetic Drug Molecules in Alzheimer's Disease: Therapeutic & Theranostic Agents. Springer 2023; pp. 267-315. [http://dx.doi.org/10.1007/978-981-99-6038-5_11]

[11] Qiu C, Kivipelto M, von Strauss E. Epidemiology of Alzheimer's disease: occurrence, determinants, and strategies toward intervention. Dialogues Clin Neurosci 2009; 11(2): 111-28. [http://dx.doi.org/10.31887/DCNS.2009.11.2/cqiu] [PMID: 19585947]

[12] Zhang XX, Tian Y, Wang ZT, Ma YH, Tan L, Yu JT. The epidemiology of Alzheimer's disease

modifiable risk factors and prevention. J Prev Alzheimers Dis 2021; 8(3): 313-21. [http://dx.doi.org/10.14283/jpad.2021.15] [PMID: 34101789]

[13] Imbimbo BP, Lombard J, Pomara N. Pathophysiology of Alzheimer's disease. Neuroimaging clinics. 2005;15(4):727-53.

[14] Guo T, Zhang D, Zeng Y, Huang TY, Xu H, Zhao Y. Molecular and cellular mechanisms underlying the pathogenesis of Alzheimer's disease. Mol Neurodegener 2020; 15(1): 40. [http://dx.doi.org/10.1186/s13024-020-00391-7] [PMID: 32677986]

[15] Kumar Thakur A, Kamboj P, Goswami K, Ahuja K. Pathophysiology and management of alzheimer's disease: an overview. J Anal Pharm Res 2018; 7(2): 226-35. [http://dx.doi.org/10.15406/japlr.2018.07.00230]

[16] Heneka MT, Carson MJ, Khoury JE, *et al.* Neuroinflammation in Alzheimer's disease. Lancet Neurol 2015; 14(4): 388-405. [http://dx.doi.org/10.1016/S1474-4422(15)70016-5] [PMID: 25792098]

[17] Wong-Guerra M, Calfio C, Maccioni RB, Rojo LE. Revisiting the neuroinflammation hypothesis in Alzheimer's disease: a focus on the druggability of current targets. Front Pharmacol 2023; 141161850 [http://dx.doi.org/10.3389/fphar.2023.1161850] [PMID: 37361208]

[18] Wang JH, Wu YJ, Tee BL, Lo RY. Medical comorbidity in Alzheimer's disease: a nested case-control study. J Alzheimers Dis 2018; 63(2): 773-81. [http://dx.doi.org/10.3233/JAD-170786] [PMID: 29660933]

[19] Noori T, Dehpour AR, Sureda A, Sobarzo-Sanchez E, Shirooie S. Role of natural products for the treatment of Alzheimer's disease. Eur J Pharmacol 2021; 898173974 [http://dx.doi.org/10.1016/j.ejphar.2021.173974] [PMID: 33652057]

[20] Wang ZY, Liu JG, Li H, Yang HM. Pharmacological effects of active components of Chinese herbal medicine in the treatment of Alzheimer's disease: a review. Am J Chin Med 2016; 44(8): 1525-41. [http://dx.doi.org/10.1142/S0192415X16500853] [PMID: 27848250]

[21] Wu TY, Chen CP, Jinn TR. Traditional Chinese medicines and Alzheimer's disease. Taiwan J Obstet Gynecol 2011; 50(2): 131-5. [http://dx.doi.org/10.1016/j.tjog.2011.04.004] [PMID: 21791295]

[22] Dubey T, Chinnathambi S. Brahmi (Bacopa monnieri): An ayurvedic herb against the Alzheimer's disease. Arch Biochem Biophys 2019; 676108153 [http://dx.doi.org/10.1016/j.abb.2019.108153] [PMID: 31622587]

[23] Roy A. Role of medicinal plants against Alzheimer's disease. Int J Complement Altern Med 2018; 11(4) [http://dx.doi.org/10.15406/ijcam.2018.11.00398]

[24] Kumar GP, Khanum F. Neuroprotective potential of phytochemicals. Pharmacogn Rev 2012; 6(12): 81-90. [http://dx.doi.org/10.4103/0973-7847.99898] [PMID: 23055633]

[25] Sabaragamuwa R, Perera CO, Fedrizzi B. Centella asiatica (Gotu kola) as a neuroprotectant and its potential role in healthy ageing. Trends Food Sci Technol 2018; 79: 88-97. [http://dx.doi.org/10.1016/j.tifs.2018.07.024]

[26] Jahan R, Hossain S, Seraj S, Nasrin D, Khatun Z, Das PR, *et al.* Centella asiatica (L.) Urb.: Ethnomedicinal uses and their scientific validations. American-Eurasian Journal of Sustainable Agriculture 2012; 6(4): 261-70.

[27] Das AJ. Review on nutritional, medicinal and pharmacological properties of *Centella asiatica* (Indian pennywort). Journal of Biologically Active Products from Nature 2011; 1(4): 216-28. [http://dx.doi.org/10.1080/22311866.2011.10719089]

[28] Gohil K, Patel J, Gajjar A. Pharmacological review on Centella asiatica: A potential herbal cure-all.

Indian J Pharm Sci 2010; 72(5): 546-56.
[http://dx.doi.org/10.4103/0250-474X.78519] [PMID: 21694984]

[29] Bandara MS, Lee EL, Thomas JE. Gotu Kola (*Centella asiatica* L.): An under-utilized herb. Americas Journal of Plant Science and Biotechnology 2011; 5(2): 20-31.

[30] Roy DC, Barman SK, Shaik MM. Current updates on Centella asiatica: phytochemistry, pharmacology and traditional uses. Med Plant Res 2013; 3(4): 20-36.

[31] Pittella F, Dutra RC, Junior DD, Lopes MTP, Barbosa NR. Antioxidant and cytotoxic activities of Centella asiatica (L) Urb. Int J Mol Sci 2009; 10(9): 3713-21.
[http://dx.doi.org/10.3390/ijms10093713] [PMID: 19865514]

[32] Djuwita I, Rahminiwati M, Darusman LK, Sa'adiah S. Induksi Ekstrak Pegagan Secara in vitro terhadap Proliferasi dan Diferensiasi Sel-Sel Otak Besar Anak Tikus. J Vet 2013; 14(2): 138-44.

[33] Beheshti S, Shahmoradi B. Therapeutic effect of Melissa officinalis in an amyloid-β rat model of Alzheimer's disease. Journal of herbmed pharmacology. 2018;7(3):193-9.

[34] Orhan IE. Centella asiatica (L.) Urban: From traditional medicine to modern medicine with neuroprotective potential. Evid Based Complement Alternat Med 2012; 2012(1): 1-8.
[http://dx.doi.org/10.1155/2012/946259] [PMID: 22666298]

[35] Seevaratnam V, Banumathi P, Premalatha M, Sundaram S, Arumugam T. Functional properties of Centella asiatica (L.): A review. Int J Pharm Pharm Sci 2012; 4(5): 8-14.

[36] Singh AK, Rai SN, Maurya A, *et al.* Therapeutic potential of phytoconstituents in management of Alzheimer's disease. Evid Based Complement Alternat Med 2021; 2021(1): 1-19.
[http://dx.doi.org/10.1155/2021/5578574] [PMID: 34211570]

[37] Yuliani S, Linar N, Eds. Effect of Gotu Kola (Centella asiatica) Extract Toward Expression of Caspase 3 of Hippocampus Pyramidal Cells on Dementia Model Rats Induced by Trimethyltin. 2019 Ahmad Dahlan International Conference Series on Pharmacy and Health Science (ADICS-PHS 2019) 2019.
[http://dx.doi.org/10.2991/adics-phs-19.2019.12]

[38] Jusril NA, Mukhtar SM, Abu Bakar SI, Md Saad WM, Wen NK, Adenan MI. Nanoemulsion of Ethanolic Extract of *Centella asiatica* (NanoSECA) Ameliorates Learning and Memory Performance by Enhancing Cholinergic Activities, Increasing Antioxidative Levels, and Attenuating Oxidative Stress Markers in Adult Rats. Pharm Nanotechnol 2024; 12(4): 329-39.
[http://dx.doi.org/10.2174/2211738511666230831122844] [PMID: 37653639]

[39] Matthews DG, Caruso M, Murchison CF, *et al.* Centella asiatica improves memory and promotes antioxidative signaling in 5XFAD mice. Antioxidants 2019; 8(12): 630.
[http://dx.doi.org/10.3390/antiox8120630] [PMID: 31817977]

[40] Chandra P, Ali Z, Fatima N, *et al.* Shankhpushpi (*Convolvulus Pluricaulis*): exploring its cognitive enhancing mechanisms and therapeutic potential in neurodegenerative disorders. Curr Bioact Compd 2025; 21(2)e290424229510.
[http://dx.doi.org/10.2174/0115734072292339240416095600]

[41] Sharma R, Singla RK, Banerjee S, Sinha B, Shen B, Sharma R. Role of Shankhpushpi (Convolvulus pluricaulis) in neurological disorders: An umbrella review covering evidence from ethnopharmacology to clinical studies. Neurosci Biobehav Rev 2022; 140104795
[http://dx.doi.org/10.1016/j.neubiorev.2022.104795] [PMID: 35878793]

[42] Irshad S, Ali MA, Makbul SAA, Shamsi S. Effect of Convolvulus pluricaulis (Shankhpushpi) and Bacopa Monnieri (Brahmi) on aging-related diseases and memory deficits. Medicinal Plants for the Management of Neurodegenerative Diseases 2024; pp. 85-98.
[http://dx.doi.org/10.1201/9781003392941-7]

[43] Hannan MA, Sultana A, Rahman MH, *et al.* Protective mechanisms of nootropic herb Shankhpushpi (Convolvulus pluricaulis) against dementia: network pharmacology and computational approach. Evid Based Complement Alternat Med 2022; 2022(1): 1-18.

[http://dx.doi.org/10.1155/2022/1015310] [PMID: 36225186]

[44] Mukherjee PK, Kumar V, Kumar NS, Heinrich M. The Ayurvedic medicine Clitoria ternatea—From traditional use to scientific assessment. J Ethnopharmacol 2008; 120(3): 291-301. [http://dx.doi.org/10.1016/j.jep.2008.09.009] [PMID: 18926895]

[45] Sethiya NK, Nahata A, Mishra SH, Dixit VK. An update on Shankhpushpi, a cognition-boosting Ayurvedic medicine. J Chin Integr Med 2009; 7(11): 1001-22. [http://dx.doi.org/10.3736/jcim20091101] [PMID: 19912732]

[46] Asthana S, Greig NH, Holloway HW, *et al.* Clinical pharmacokinetics of arecoline in subjects with Alzheimer's disease. Clin Pharmacol Ther 1996; 60(3): 276-82. [http://dx.doi.org/10.1016/S0009-9236(96)90054-5] [PMID: 8841150]

[47] Bihaqi SW, Sharma M, Singh AP, Tiwari M. Neuroprotective role of Convolvulus pluricaulis on aluminium induced neurotoxicity in rat brain. J Ethnopharmacol 2009; 124(3): 409-15. [http://dx.doi.org/10.1016/j.jep.2009.05.038] [PMID: 19505562]

[48] Liu LF, Durairajan SSK, Lu JH, Koo I, Li M. In vitro screening on amyloid precursor protein modulation of plants used in Ayurvedic and Traditional Chinese medicine for memory improvement. J Ethnopharmacol 2012; 141(2): 754-60. [http://dx.doi.org/10.1016/j.jep.2011.08.065] [PMID: 21920424]

[49] Abd Rani NZ, Husain K, Kumolosasi E. Moringa genus: a review of phytochemistry and pharmacology. Front Pharmacol 2018; 9: 108. [http://dx.doi.org/10.3389/fphar.2018.00108] [PMID: 29503616]

[50] Wal P, Aziz N, Dash B, Tyagi S, Vinod YR. Neuro-nutraceuticals: Insights of experimental evidences and molecular mechanism in neurodegenerative disorders. Future J Pharm Sci 2023; 9(1): 31. [http://dx.doi.org/10.1186/s43094-023-00480-6]

[51] Fuloria S, Mehta J, Chandel A, *et al.* A comprehensive review on the therapeutic potential of Curcuma longa Linn. in relation to its major active constituent curcumin. Front Pharmacol 2022; 13820806 [http://dx.doi.org/10.3389/fphar.2022.820806] [PMID: 35401176]

[52] Jyotirmayee B, Nayak SS, Mohapatra N, Sahoo S, Mishra M, Mahalik G. Bioactive compounds and biological activities of turmeric (*Curcuma longa* L) Bioactive Compounds in the Storage Organs of Plants. Springer 2024; pp. 395-423. [http://dx.doi.org/10.1007/978-3-031-44746-4_20]

[53] Sharifi-Rad J, Rayess YE, Rizk AA, *et al.* Turmeric and its major compound curcumin on health: bioactive effects and safety profiles for food, pharmaceutical, biotechnological and medicinal applications. Front Pharmacol 2020; 11: 01021. [http://dx.doi.org/10.3389/fphar.2020.01021] [PMID: 33041781]

[54] Reddy PH, Manczak M, Yin X, *et al.* Protective effects of Indian spice curcumin against amyloid-β in Alzheimer's disease. J Alzheimers Dis 2018; 61(3): 843-66. [http://dx.doi.org/10.3233/JAD-170512] [PMID: 29332042]

[55] Abdul-Rahman T, Awuah WA, Mikhailova T, *et al.* Antioxidant, anti-inflammatory and epigenetic potential of curcumin in Alzheimer's disease. Biofactors 2024; 50(4): 693-708. [http://dx.doi.org/10.1002/biof.2039] [PMID: 38226733]

[56] Voulgaropoulou SD, van Amelsvoort TAMJ, Prickaerts J, Vingerhoets C. The effect of curcumin on cognition in Alzheimer's disease and healthy aging: A systematic review of pre-clinical and clinical studies. Brain Res 2019; 1725146476 [http://dx.doi.org/10.1016/j.brainres.2019.146476] [PMID: 31560864]

[57] Islam F, Khadija JF, Harun-Or-Rashid M, *et al.* Bioactive compounds and their derivatives: an insight into prospective phytotherapeutic approach against alzheimer's disease. Oxid Med Cell Longev 2022; 2022(1)5100904 [http://dx.doi.org/10.1155/2022/5100904] [PMID: 35450410]

[58] Rao RV, Descamps O, John V, Bredesen DE. Ayurvedic medicinal plants for Alzheimer's disease: a review. Alzheimers Res Ther 2012; 4(3): 22. [http://dx.doi.org/10.1186/alzrt125] [PMID: 22747839]

[59] Akram M, Nawaz A. Effects of medicinal plants on Alzheimer's disease and memory deficits. Neural Regen Res. 2017 Apr; 12(4): 660–670. [http://dx.doi.org/10.4103/1673-5374.205108]

[60] Ramalingam M, Kim H, Lee Y, Lee YI. Phytochemical and pharmacological role of liquiritigenin and isoliquiritigenin from radix glycyrrhizae in human health and disease models. Front Aging Neurosci 2018; 10: 348. [http://dx.doi.org/10.3389/fnagi.2018.00348] [PMID: 30443212]

[61] Munawar T, Bibi Y, Ahmad F. Ethnomedicinal study of plants used for neurodegenerative diseases: A review: Ethnomedicinal study of plants used for neurodegenerative diseases. Proceedings of the Pakistan Academy of Sciences: B Life and Environmental Sciences. 2020;57(3):13-26.

[62] Bordoloi S, Pathak K, Devi M, *et al.* Some promising medicinal plants used in Alzheimer's disease: an ethnopharmacological perspective. Discover Applied Sciences 2024; 6(5): 215. [http://dx.doi.org/10.1007/s42452-024-05811-7]

[63] Heinrich M, Lee Teoh H. Galanthamine from snowdrop—the development of a modern drug against Alzheimer's disease from local Caucasian knowledge. J Ethnopharmacol 2004; 92(2-3): 147-62. [http://dx.doi.org/10.1016/j.jep.2004.02.012] [PMID: 15137996]

[64] Ding Y, Qu D, Zhang KM, *et al.* Phytochemical and biological investigations of Amaryllidaceae alkaloids: a review. J Asian Nat Prod Res 2017; 19(1): 53-100. [http://dx.doi.org/10.1080/10286020.2016.1198332] [PMID: 27406068]

[65] Kong CK, Low LE, Siew WS, *et al.* Biological activities of snowdrop (Galanthus spp., Family Amaryllidaceae). Front Pharmacol 2021; 11552453 [http://dx.doi.org/10.3389/fphar.2020.552453] [PMID: 33679383]

[66] Rhee IK, Appels N, Luijendijk T, Irth H, Verpoorte R. Determining acetylcholinesterase inhibitory activity in plant extracts using a fluorimetric flow assay. Phytochem Anal 2003; 14(3): 145-9. [http://dx.doi.org/10.1002/pca.695] [PMID: 12793460]

[67] Bağdat RB, Coşge B. The essential oil of lemon balm (*Melissa officinalis L.*), its components and using fields. Anadolu Tarım Bilimleri Dergisi 2006; 21(1): 116-21.

[68] Canevelli M, Adali N, Kelaiditi E, Cantet C, Ousset PJ, Cesari M. Effects of Gingko biloba supplementation in Alzheimer's disease patients receiving cholinesterase inhibitors: Data from the ICTUS study. Phytomedicine 2014; 21(6): 888-92. [http://dx.doi.org/10.1016/j.phymed.2014.01.003] [PMID: 24548724]

[69] Nowak A, Kojder K, Zielonka-Brzezicka J, *et al.* The use of Ginkgo biloba L. as a neuroprotective agent in the Alzheimer's disease. Front Pharmacol 2021; 12775034 [http://dx.doi.org/10.3389/fphar.2021.775034] [PMID: 34803717]

[70] Xie L, Zhu Q, Lu J. Can we use Ginkgo biloba extract to treat Alzheimer's disease? Lessons from preclinical and clinical studies. Cells 2022; 11(3): 479. [http://dx.doi.org/10.3390/cells11030479] [PMID: 35159288]

[71] Oken BS, Storzbach DM, Kaye JA. The efficacy of Ginkgo biloba on cognitive function in Alzheimer disease. Arch Neurol 1998; 55(11): 1409-15. [http://dx.doi.org/10.1001/archneur.55.11.1409] [PMID: 9823823]

[72] Dash S. Ginkgo biloba in Alzheimer's disease. Austin J Clin Neurol 2015; 2(3): 1028.

[73] John OO, Amarachi IS, Chinazom AP, *et al.* Phytotherapy: A promising approach for the treatment of Alzheimer's disease. Pharmacol Res Mod Chin Med 2022; 2100030 [http://dx.doi.org/10.1016/j.prmcm.2021.100030]

[74] Singhal A, Bangar OP, Naithani V. Medicinal plants with a potential to treat Alzheimer and associated symptoms. Int J Nutr Pharmacol Neurol Dis 2012; 2(2): 84-91. [http://dx.doi.org/10.4103/2231-0738.95927]

[75] Wahab S, Annadurai S, Abullais SS, *et al.* Glycyrrhiza glabra (Licorice): A comprehensive review on its phytochemistry, biological activities, clinical evidence and toxicology. Plants 2021; 10(12): 2751. [http://dx.doi.org/10.3390/plants10122751] [PMID: 34961221]

[76] Pastorino G, Cornara L, Soares S, Rodrigues F, Oliveira MBPP. Liquorice (*GLYCYRRHIZA GLABRA*): A phytochemical and pharmacological review. Phytother Res 2018; 32(12): 2323-39. [http://dx.doi.org/10.1002/ptr.6178] [PMID: 30117204]

[77] Sarkar S, Shaw P, Singh P, Acharya Chowdhury A. Emerging neuroprotective potential of Liquorice: Mechanistic insights for neurological disorders. S Afr J Bot 2023; 154: 149-58. [http://dx.doi.org/10.1016/j.sajb.2023.01.029]

[78] Murray MT. Glycyrrhiza glabra (licorice). Textbook of natural medicine. 2020:641.

[79] Hasan MK, Ara I, Mondal MSA, Kabir Y. Phytochemistry, pharmacological activity, and potential health benefits of *Glycyrrhiza glabra.* Heliyon 2021; 7(6)e07240 [http://dx.doi.org/10.1016/j.heliyon.2021.e07240] [PMID: 34189299]

[80] Ban JY, Park HK, Kim SK. Effect of glycyrrhizic acid on scopolamine-induced cognitive impairment in mice. Int Neurourol J 2020; 24 (Suppl. 1): S48-55. [http://dx.doi.org/10.5213/inj.2040154.077] [PMID: 32482057]

[81] Paudel YN, Angelopoulou E, Semple B, Piperi C, Othman I, Shaikh MF. Potential neuroprotective effect of the HMGB1 inhibitor glycyrrhizin in neurological disorders. ACS Chem Neurosci 2020; 11(4): 485-500. [http://dx.doi.org/10.1021/acschemneuro.9b00640] [PMID: 31972087]

[82] Akintoye O, Fabunmi O, Ajibare A, Olaseinde O, Ayilara G, Owoyele B. Anti-Nociceptive Activity of Aqueous Licorice Root Extract on Neuropathic Pain and its Effect on some Selected Biochemical Parameters in Male Wistar Rats. W J Biomed Res 2019; (2): 31-37.

[83] Amit Joshi T, Tatke P. Neuroprotective effects of licorice extracts in tibial and sural transection induced neuropathic pain in rats. Indian Journal of Pharmaceutical Education and Research 2020; 54(2s): s285-94. [http://dx.doi.org/10.5530/ijper.54.2s.85]

[84] Pei H, He L, Shi M, *et al.* PI3K-Akt signaling pathway based on network pharmacology for the anti-Alzheimer's disease effect of licorice stem flavonoids. Aging (Albany NY) 2023; 15(9): 3381-93. [http://dx.doi.org/10.18632/aging.204536] [PMID: 37166431]

[85] Zhu X, Liu J, Huang S, *et al.* Neuroprotective effects of isoliquiritigenin against cognitive impairment via suppression of synaptic dysfunction, neuronal injury, and neuroinflammation in rats with kainic acid-induced seizures. Int Immunopharmacol 2019; 72: 358-66. [http://dx.doi.org/10.1016/j.intimp.2019.04.028] [PMID: 31030091]

[86] Jiang S, Gao BB, Ou YF, Zhao QS. Lycopodium alkaloids from Huperzia serrata and their cholinesterase inhibitory activities. Phytochemistry 2024; 223: 114114. [http://dx.doi.org/10.1016/j.phytochem.2024.114114] [PMID: 38697240]

[87] Kim JK, Park SU. Pharmacological aspects of galantamine for the treatment of Alzheimer's disease. EXCLI J 2017; 16: 35-9. [PMID: 28337117]

[88] Hussain G, Rasul A, Anwar H, *et al.* Role of plant derived alkaloids and their mechanism in neurodegenerative disorders. Int J Biol Sci 2018; 14(3): 341-57. [http://dx.doi.org/10.7150/ijbs.23247] [PMID: 29559851]

[89] Callizot N, Campanari ML, Rouvière L, *et al.* Huperzia serrata extract 'NSP01'with neuroprotective

effects-potential synergies of huperzine a and polyphenols. Front Pharmacol 2021; 12: 681532. [http://dx.doi.org/10.3389/fphar.2021.681532] [PMID: 34526893]

[90] Ohba T, Yoshino Y, Ishisaka M, *et al.* Japanese *Huperzia serrata* extract and the constituent, huperzine A, ameliorate the scopolamine-induced cognitive impairment in mice. Biosci Biotechnol Biochem 2015; 79(11): 1838-44. [http://dx.doi.org/10.1080/09168451.2015.1052773] [PMID: 26059088]

[91] Lee YK, Chang YH. Physicochemical and antioxidant properties of methanol extract from Maca (Lepidium meyenii Walp.) leaves and roots. Food Sci Technol (Campinas) 2019; 39 (Suppl. 1): 278-86. [http://dx.doi.org/10.1590/fst.03818]

[92] da Silva Leitão Peres N, Cabrera Parra Bortoluzzi L, Medeiros Marques LL, *et al.* Medicinal effects of Peruvian maca (*Lepidium meyenii*): a review. Food Funct 2020; 11(1): 83-92. [http://dx.doi.org/10.1039/C9FO02732G] [PMID: 31951246]

[93] Figueroa ME, Abdel-Wahab O, Lu C, *et al.* Leukemic IDH1 and IDH2 mutations result in a hypermethylation phenotype, disrupt TET2 function, and impair hematopoietic differentiation. Cancer Cell 2010; 18(6): 553-67. [http://dx.doi.org/10.1016/j.ccr.2010.11.015] [PMID: 21130701]

[94] Rahman MM, Islam MR, Supti FA, *et al.* Exploring the therapeutic effect of neurotrophins and neuropeptides in neurodegenerative diseases: at a glance. Mol Neurobiol 2023; 60(8): 4206-31. [http://dx.doi.org/10.1007/s12035-023-03328-5] [PMID: 37052791]

[95] Moradkhani H, Sargsyan E, Bibak H *et al. Melissa officinalis L.,* a valuable medicine plant. A review. J Med Plants Res 2010; 4: 2753-9.

[96] Zarei A, Changizi-Ashtiyani S, Taheri S, Hosseini N. A brief overview of the effects of Melissa officinalis L. extract on the function of various body organs. Zahedan J Res Med Sci 2015; 17(7) [http://dx.doi.org/10.17795/zjrms1007]

[97] Akhondzadeh S, Noroozian M, Mohammadi M, Ohadinia S, Jamshidi AH, Khani M. Melissa officinalis extract in the treatment of patients with mild to moderate Alzheimer's disease: a double blind, randomised, placebo controlled trial. J Neurol Neurosurg Psychiatry 2003; 74(7): 863-6. [http://dx.doi.org/10.1136/jnnp.74.7.863] [PMID: 12810768]

[98] Vladimir-Knežević S, Blažeković B, Kindl M, Vladić J, Lower-Nedza A, Brantner A. Acetylcholinesterase inhibitory, antioxidant and phytochemical properties of selected medicinal plants of the Lamiaceae family. Molecules 2014; 19(1): 767-82. [http://dx.doi.org/10.3390/molecules19010767] [PMID: 24413832]

[99] Stojanović NM, Mladenović MZ, Maslovarić A, Stojiljković NI, Randjelović PJ, Radulović NS. Lemon balm (Melissa officinalis L.) essential oil and citronellal modulate anxiety-related symptoms – In vitro and in vivo studies. J Ethnopharmacol 2022; 284114788 [http://dx.doi.org/10.1016/j.jep.2021.114788] [PMID: 34718102]

[100] Pratap G, Ashwini S, Manjula S. Alzheimer's disease: A challenge in managing with certain medicinal plants-A review. Int J Pharm Sci Res 2017; 12(8): 4960-72.

[101] Sharma R, Kuca K, Nepovimova E, Kabra A, Rao MM, Prajapati PK. Traditional Ayurvedic and herbal remedies for Alzheimer's disease: from bench to bedside. Expert Rev Neurother 2019; 19(5): 359-74. [http://dx.doi.org/10.1080/14737175.2019.1596803] [PMID: 30884983]

[102] Mahboubi M. *Melissa officinalis* and rosmarinic acid in management of memory functions and Alzheimer disease. Asian Pac J Trop Biomed 2019; 9(2): 47-52. [http://dx.doi.org/10.4103/2221-1691.250849]

[103] Hase T, Shishido S, Yamamoto S, *et al.* Rosmarinic acid suppresses Alzheimer's disease development by reducing amyloid β aggregation by increasing monoamine secretion. Sci Rep 2019; 9(1): 8711.

[http://dx.doi.org/10.1038/s41598-019-45168-1] [PMID: 31213631]

[104] Bui TT, Nguyen TH. Natural product for the treatment of Alzheimer's disease. J Basic Clin Physiol Pharmacol 2017; 28(5): 413-23.
[http://dx.doi.org/10.1515/jbcpp-2016-0147] [PMID: 28708573]

[105] Rafie H, Soheila H, Grant E. Rosmarinus officinalis (rosemary): a novel therapeutic agent for antioxidant, antimicrobial, anticancer, antidiabetic, antidepressant, neuroprotective, anti-inflammatory and anti-obesity treatment. J Herb Med 2017; 3(2): 8-14.

[106] Habtemariam S. The therapeutic potential of rosemary (Rosmarinus officinalis) diterpenes for Alzheimer's disease. Evid Based Complement Alternat Med 2016; 2016(1)2680409.
[http://dx.doi.org/10.1155/2016/2680409] [PMID: 26941822]

[107] Khatoon R, Alam MA, Sharma PK. A comprehensive study of pharmacological behaviors, nano-formulations, and applications of rosemary. Nat Prod J 2021; 11(5): 629-47.
[http://dx.doi.org/10.2174/2210315510999201104165058]

[108] Farr SA, Niehoff ML, Ceddia MA, *et al.* Effect of botanical extracts containing carnosic acid or rosmarinic acid on learning and memory in SAMP8 mice. Physiol Behav 2016; 165: 328-38.
[http://dx.doi.org/10.1016/j.physbeh.2016.08.013] [PMID: 27527000]

[109] Miraj S, Kiani S. A review study of therapeutic effects of Salvia officinalis L. Pharm Lett 2016; 8(6): 299-303.

[110] Lopresti AL. Salvia (Sage): A review of its potential cognitive-enhancing and protective effects. Drugs R D 2017; 17(1): 53-64.
[http://dx.doi.org/10.1007/s40268-016-0157-5] [PMID: 27888449]

[111] El Gabbas Z, Bezza K, Laadraoui J, Makbal R, Aboufatima R, Chait A. Salvia officinalis induces antidepressant-like effect, anxiolytic activity and learning improvement in hippocampal lesioned and intact adult rats. Bangladesh J Pharmacol 2018; 13(4): 367-78.
[http://dx.doi.org/10.3329/bjp.v13i4.38375]

[112] Alrashdi J, Albasher G, Alanazi MM, Al-Qahtani WS, Alanezi AA, Alasmari F. Effects of Rosmarinus officinalis L. extract on neurobehavioral and neurobiological changes in male rats with pentylenetetrazol-induced epilepsy. Toxics 2023; 11(10): 826.
[http://dx.doi.org/10.3390/toxics11100826] [PMID: 37888677]

[113] Datta S, Patil S. Evaluation of traditional herb extract Salvia officinalis in treatment of Alzheimers disease. Pharmacogn J 2020; 12(1): 131-43.
[http://dx.doi.org/10.5530/pj.2020.12.20]

[114] Ayoub IM, George MY, Menze ET, *et al.* Insights into the neuroprotective effects of *Salvia officinalis* L. and *Salvia microphylla* Kunth in the memory impairment rat model. Food Funct 2022; 13(4): 2253-68.
[http://dx.doi.org/10.1039/D1FO02988F] [PMID: 35137748]

[115] Jamadagni SB, Ghadge PM, Tambe MS, *et al.* Amelioration of AlCl3-induced memory loss in the rats by an aqueous extract of Guduchi, a Medhya Rasayana. Pharmacogn Mag 2023; 19(2): 231-43.
[http://dx.doi.org/10.1177/09731296221145063]

[116] Amrutha S, Parvaje R, Abhinand CS, Prasad TSK, Modi PK. Targeting Alzheimer's Disease with Tinospora cordifolia (Guduchi): A metabolomics and network pharmacology approach highlighting BACE1 and MAO-B as potential targets. 2024. Available from: https://doi.org/10.21203/rs.3.rs-5395345/v1

[117] Asiful Islam M, Saif Khandker S, Alam F, Ibrahim Khalil M, Amjad Kamal M, Hua Gan S. Alzheimer's disease and natural products: Future regimens emerging from nature. Curr Top Med Chem 2017; 17(12): 1408-28.
[http://dx.doi.org/10.2174/1568026617666170103163054] [PMID: 28049401]

[118] Grodzicki W, Dziendzikowska K. The role of selected bioactive compounds in the prevention of

Alzheimer's disease. Antioxidants 2020; 9(3): 229.
[http://dx.doi.org/10.3390/antiox9030229] [PMID: 32168776]

[119] Alzobaidi N, Quasimi H, Emad NA, Alhalmi A, Naqvi M. Bioactive compounds and traditional herbal medicine: Promising approaches for the treatment of dementia. Degener Neurol Neuromuscul Dis 2021; 11: 1-14.
[http://dx.doi.org/10.2147/DNND.S299589] [PMID: 33880073]

[120] Rodríguez-Morató J, Xicota L, Fitó M, Farré M, Dierssen M, De la Torre R. Potential role of olive oil phenolic compounds in the prevention of neurodegenerative diseases. Molecules 2015; 20(3): 4655-4680.
[http://dx.doi.org/10.3390/molecules20034655] [PMID: 25781069]

[121] Daccache A, Lion C, Sibille N, *et al.* Oleuropein and derivatives from olives as Tau aggregation inhibitors. Neurochem Int 2011; 58(6): 700-7.
[http://dx.doi.org/10.1016/j.neuint.2011.02.010] [PMID: 21333710]

[122] Martín-Peláez S, Covas MI, Fitó M, Kušar A, Pravst I. Health effects of olive oil polyphenols: Recent advances and possibilities for the use of health claims. Mol Nutr Food Res 2013; 57(5): 760-71.
[http://dx.doi.org/10.1002/mnfr.201200421] [PMID: 23450515]

[123] Li W, Sperry JB, Crowe A, Trojanowski JQ, Smith AB III, Lee VMY. Inhibition of tau fibrillization by oleocanthal via reaction with the amino groups of tau. J Neurochem 2009; 110(4): 1339-51.
[http://dx.doi.org/10.1111/j.1471-4159.2009.06224.x] [PMID: 19549281]

[124] Scarmeas N, Stern Y, Mayeux R, Luchsinger JA. Mediterranean diet, Alzheimer disease, and vascular mediation. Arch Neurol 2006; 63(12): 1709-17.
[http://dx.doi.org/10.1001/archneur.63.12.noc60109] [PMID: 17030648]

[125] Scarmeas N, Stern Y, Tang MX, Mayeux R, Luchsinger JA. Mediterranean diet and risk for Alzheimer's disease. Ann Neurol 2006; 59(6): 912-21.
[http://dx.doi.org/10.1002/ana.20854] [PMID: 16622828]

[126] Valls-Pedret C, Sala-Vila A, Serra-Mir M, *et al.* Mediterranean diet and age-related cognitive decline: a randomized clinical trial. JAMA Intern Med 2015; 175(7): 1094-103.
[http://dx.doi.org/10.1001/jamainternmed.2015.1668] [PMID: 25961184]

[127] Devi KP, Shanmuganathan B, Manayi A, Nabavi SF, Nabavi SM. Molecular and therapeutic targets of genistein in Alzheimer's disease. Mol Neurobiol 2017; 54(9): 7028-41.
[http://dx.doi.org/10.1007/s12035-016-0215-6] [PMID: 27796744]

[128] Cai B, Ye S, Wang T, *et al.* Genistein protects hippocampal neurons against injury by regulating calcium/calmodulin dependent protein kinase IV protein levels in Alzheimer's disease model rats. Neural Regen Res 2017; 12(9): 1479-84.
[http://dx.doi.org/10.4103/1673-5374.215260] [PMID: 29089994]

[129] Sasaoka N, Sakamoto M, Kanemori S, *et al.* Long-term oral administration of hop flower extracts mitigates Alzheimer phenotypes in mice. PLoS One 2014; 9(1)e87185
[http://dx.doi.org/10.1371/journal.pone.0087185] [PMID: 24489866]

[130] Li D, Wang P, Luo Y, Zhao M, Chen F. Health benefits of anthocyanins and molecular mechanisms: Update from recent decade. Crit Rev Food Sci Nutr 2017; 57(8): 1729-41.
[http://dx.doi.org/10.1080/10408398.2015.1030064] [PMID: 26192537]

[131] Yamakawa MY, Uchino K, Watanabe Y, *et al.* Anthocyanin suppresses the toxicity of Aβ deposits through diversion of molecular forms in *in vitro* and *in vivo* models of Alzheimer's disease. Nutr Neurosci 2016; 19(1): 32-42.
[http://dx.doi.org/10.1179/1476830515Y.0000000042] [PMID: 26304685]

[132] Münch G, Venigalla M, Sonego S, Gyengesi E. Curcumin and Apigenin - novel and promising therapeutics against chronic neuroinflammation in Alzheimer's disease. Neural Regen Res 2015; 10(8): 1181-5.

[http://dx.doi.org/10.4103/1673-5374.162686] [PMID: 26487830]

[133] Ng TP, Chiam PC, Lee T, Chua HC, Lim L, Kua EH. Curry consumption and cognitive function in the elderly. Am J Epidemiol 2006; 164(9): 898-906.
[http://dx.doi.org/10.1093/aje/kwj267] [PMID: 16870699]

[134] Small GW, Siddarth P, Li Z, *et al.* Memory and brain amyloid and tau effects of a bioavailable form of curcumin in non-demented adults: a double-blind, placebo-controlled 18-month trial. Am J Geriatr Psychiatry 2018; 26(3): 266-77.
[http://dx.doi.org/10.1016/j.jagp.2017.10.010] [PMID: 29246725]

[135] Freund Levi Y, Vedin I, Cederholm T, *et al.* Transfer of omega-3 fatty acids across the blood–brain barrier after dietary supplementation with a docosahexaenoic acid-rich omega-3 fatty acid preparation in patients with A lzheimer's disease: the O meg AD study. J Intern Med 2014; 275(4): 428-36.
[http://dx.doi.org/10.1111/joim.12166] [PMID: 24410954]

[136] Dyall SC. Long-chain omega-3 fatty acids and the brain: a review of the independent and shared effects of EPA, DPA and DHA. Front Aging Neurosci 2015; 7: 52.
[http://dx.doi.org/10.3389/fnagi.2015.00052] [PMID: 25954194]

[137] Cole GM, Frautschy SA. DHA may prevent age-related dementia. J Nutr 2010; 140(4): 869-74.
[http://dx.doi.org/10.3945/jn.109.113910] [PMID: 20181786]

[138] Green KN, Martinez-Coria H, Khashwji H, *et al.* Dietary docosahexaenoic acid and docosapentaenoic acid ameliorate amyloid-β and tau pathology via a mechanism involving presenilin 1 levels. J Neurosci 2007; 27(16): 4385-95.
[http://dx.doi.org/10.1523/JNEUROSCI.0055-07.2007] [PMID: 17442823]

[139] Kalueff AV, Eremin KO, Tuohimaa P. Mechanisms of neuroprotective action of vitamin D(3). Biochemistry (Mosc) 2004; 69(7): 738-41.
[http://dx.doi.org/10.1023/B:BIRY.0000040196.65686.2f] [PMID: 15310271]

[140] Landel V, Annweiler C, Millet P, Morello M, Féron F. Vitamin D, cognition and Alzheimer's disease: the therapeutic benefit is in the D-tails. J Alzheimers Dis 2016; 53(2): 419-44.
[http://dx.doi.org/10.3233/JAD-150943] [PMID: 27176073]

[141] Morello M, Landel V, Lacassagne E, *et al.* Vitamin D improves neurogenesis and cognition in a mouse model of Alzheimer's disease. Mol Neurobiol 2018; 55(8): 6463-79.
[http://dx.doi.org/10.1007/s12035-017-0839-1] [PMID: 29318446]

[142] Feart C, Helmer C, Merle B, *et al.* Associations of lower vitamin D concentrations with cognitive decline and long-term risk of dementia and Alzheimer's disease in older adults. Alzheimers Dement 2017; 13(11): 1207-16.
[http://dx.doi.org/10.1016/j.jalz.2017.03.003] [PMID: 28522216]

[143] Boccardi V, Baroni M, Mangialasche F, Mecocci P. Vitamin E family: Role in the pathogenesis and treatment of Alzheimer's disease. Alzheimers Dement (N Y) 2016; 2(3): 182-91.
[http://dx.doi.org/10.1016/j.trci.2016.08.002] [PMID: 29067305]

[144] Mocchegiani E, Costarelli L, Giacconi R, *et al.* Vitamin E–gene interactions in aging and inflammatory age-related diseases: Implications for treatment. A systematic review. Ageing Res Rev 2014; 14: 81-101.
[http://dx.doi.org/10.1016/j.arr.2014.01.001] [PMID: 24418256]

[145] Gorji N, Moeini R, Memariani Z. Almond, hazelnut and walnut, three nuts for neuroprotection in Alzheimer's disease: A neuropharmacological review of their bioactive constituents. Pharmacol Res 2018; 129: 115-27.
[http://dx.doi.org/10.1016/j.phrs.2017.12.003] [PMID: 29208493]

[146] Cherubini A, Martin A, Andres-Lacueva C, *et al.* Vitamin E levels, cognitive impairment and dementia in older persons: the InCHIANTI study. Neurobiol Aging 2005; 26(7): 987-94.
[http://dx.doi.org/10.1016/j.neurobiolaging.2004.09.002] [PMID: 15748776]

[147] Burčul F, Generalić Mekinić I, Radan M, Rollin P, Blažević I. Isothiocyanates: cholinesterase inhibiting, antioxidant, and anti-inflammatory activity. J Enzyme Inhib Med Chem 2018; 33(1): 577-82.
[http://dx.doi.org/10.1080/14756366.2018.1442832] [PMID: 29513045]

[148] Tarozzi A, Angeloni C, Malaguti M, Morroni F, Hrelia S, Hrelia P. Sulforaphane as a potential protective phytochemical against neurodegenerative diseases. Oxid Med Cell Longev 2013; 2013(1): 1-10.
[http://dx.doi.org/10.1155/2013/415078] [PMID: 23983898]

[149] Nurk E, Refsum H, Drevon CA, *et al.* Cognitive performance among the elderly in relation to the intake of plant foods. The Hordaland Health Study. Br J Nutr 2010; 104(8): 1190-201.
[http://dx.doi.org/10.1017/S0007114510001807] [PMID: 20550741]

[150] Lee S, Choi BR, Kim J, *et al.* Sulforaphane Upregulates the Heat Shock Protein Co-Chaperone CHIP and Clears Amyloid-β and Tau in a Mouse Model of Alzheimer's Disease. Mol Nutr Food Res 2018; 62(12)1800240
[http://dx.doi.org/10.1002/mnfr.201800240] [PMID: 29714053]

[151] Katagiri M, Satoh A, Tsuji S, Shirasawa T. Effects of astaxanthin-rich Haematococcus pluvialis extract on cognitive function: a randomised, double-blind, placebo-controlled study. J Clin Biochem Nutr 2012; 51(2): 102-7.
[http://dx.doi.org/10.3164/jcbn.D-11-00017] [PMID: 22962526]

[152] Christensen K, Gleason CE, Mares JA. Dietary carotenoids and cognitive function among US adults, NHANES 2011–2014. Nutr Neurosci 2020; 23(7): 554-62.
[http://dx.doi.org/10.1080/1028415X.2018.1533199] [PMID: 30326796]

[153] Power R, Coen RF, Beatty S, *et al.* Supplemental retinal carotenoids enhance memory in healthy individuals with low levels of macular pigment in a randomized, double-blind, placebo-controlled clinical trial. J Alzheimers Dis 2018; 61(3): 947-61.
[http://dx.doi.org/10.3233/JAD-170713] [PMID: 29332050]

[154] Obulesu M, Dowlathabad MR, Bramhachari PV. Carotenoids and Alzheimer's Disease: An insight into therapeutic role of retinoids in animal models. Neurochem Int 2011; 59(5): 535-41.
[http://dx.doi.org/10.1016/j.neuint.2011.04.004] [PMID: 21672580]

[155] Hwang S, Lim J, Kim H. Inhibitory effect of lycopene on amyloid-β-induced apoptosis in neuronal cells. Nutrients 2017; 9(8): 883.
[http://dx.doi.org/10.3390/nu9080883] [PMID: 28813012]

CHAPTER 7

Various Immunotherapy and Targeted Therapeutics for Managing Alzheimer's Disease

Mansi Verma[1,2,*] and **Niraj Kumar Singh**[1]

[1] *Division of Pharmacology, Institute of Pharmaceutical Research, GLA University, Mathura, Uttar Pradesh 281406, India*

[2] *Sunder Deep Pharmacy College, Ghaziabad, Uttar Pradesh 201002, India*

Abstract: Globally, Alzheimer's Disease (AD) is the most common neurological disorder that affects the older population. However, the intricacy of AD pathophysiology causes disparities in our knowledge of the disease and could be the primary cause of the medication development failure for AD. Thankfully, several preclinical and clinical research projects are still in progress, which will continue to pave the way for the discovery of disease processes and direct approaches to AD diagnosis and medication development. For instance, because of the exceptional preclinical outcomes, immunotherapeutic approaches targeting the tau and amyloid-β (Aβ) proteins were originally thought to be almost successful in clinical treatment. This strategy has been called into question due to the numerous failures in clinical studies involving vaccinations and humanized anti-tau monoclonal, anti-Aβ antibodies. However, immunotherapy targeting Aβ may still have potential, as evidenced by the United States Food and Drug Administration (USFDA) recent endorsement of Aducanumab, a novel anti-Aβ monoclonal antibody. Immunotherapies targeting various targets, including microglia, tau, and the gut-brain axis, are also being developed in the meantime. To improve the accuracy and efficacy of immunotherapeutic treatments, more inspection is required to clarify the targeted proteins' structures and epitopes. In this review, we concentrate on the mechanisms of action of immunotherapies focused on microglia, tau, and Aβ in AD. We also discuss current developments and prospects for immunotherapeutic approaches to AD.

Keywords: Alzheimer's disease, Antibody, Drug development, Immunotherapy, Vaccine.

INTRODUCTION

The leading cause of dementia in the elderly is Alzheimer's disease (AD). Its pathological hallmarks include neurofibrillary tangles with hyperphosphorylated

* **Corresponding author Mansi Verma:** Division of Pharmacology, Institute of Pharmaceutical Research, GLA University, Mathura, Uttar Pradesh 281406, India, Sunder Deep Pharmacy College, Ghaziabad, Uttar Pradesh 201002, India; E-mail: mansi7184@gmail.com

tau and extracellular amyloid-β (Aβ) plaques [1, 2]. Clinically, memory loss and cognitive decline are the primary indicators [3]. AD typically manifests after the age of 65, affecting 5–10% of adults in this age group, and rising to nearly 50% among individuals above 85 years [4]. Patients usually experience progressive dementia, mobility difficulties, and eventually succumb 5–12 years after onset [5].

Both controllable and non-modifiable risk factors contribute to AD pathology. Lifestyle-related disorders such as cardiovascular diseases, diabetes, and hypertension are among the modifiable ones [6 - 9]. Current treatments, namely cholinesterase inhibitors and N-methyl-D-aspartate receptor antagonists, provide only symptomatic relief and do not slow disease progression [10, 11]. Thus, disease-modifying strategies are urgently needed.

Considerable research has focused on the Aβ hypothesis, which suggests that reducing Aβ production, preventing aggregation, or enhancing its clearance could alter disease progression. Vaccination and passive immunization with antibodies have shown promise, while other approaches target tau protein and microglia-mediated immune pathways (*e.g.*, TLRs, TREM2) [12 - 15].

Both active and passive immunotherapies present distinct advantages and limitations. Active immunization (vaccines) stimulates cellular and humoral immune responses, producing endogenous antibodies for prolonged periods at relatively low cost and with fewer injections. However, older patients may respond less effectively. Passive immunotherapy, in contrast, involves direct administration of antibodies, allowing more controlled dosing and the option to halt treatment if adverse effects occur. Yet both approaches carry risks of excessive immune activation, including cerebral vasculitis [16].

Anti-Aβ immunotherapies remain at the forefront of clinical research. These therapies primarily function by decreasing Aβ synthesis, inhibiting its aggregation, and enhancing clearance [17 - 19]. Although many candidate agents have been tested for reducing Aβ burden, most have failed to demonstrate meaningful clinical benefit, raising doubts about whether Aβ remains the optimal therapeutic target.

Nevertheless, Aβ accumulation is still considered a central driver of neurodegeneration, and enhancing its clearance continues to be an area of interest [20 - 24]. Active immunization with Aβ or its fragments has been shown to trigger endogenous antibody production [25]. The landmark 1999 study demonstrated that full-length Aβ immunization reduced plaque deposition in PDAPP mice [26]. This led to the development of AN1792, the first Aβ vaccine tested in humans, but clinical trials were halted after 6% of participants developed T cell-mediated meningoencephalitis [27, 28].

Second-generation vaccines, such as CAD106, were designed without the T-lymphocyte epitope to minimize overactive immune responses. While active immunotherapy offers the benefit of generating long-lasting antibodies with short-term treatment, it also carries unpredictable risks, particularly in elderly populations.

Table **1** and Fig. (**1**) illustrate some of the recent developments in the creation of active immunotherapy for AD. Humanized monoclonal antibodies and polyclonal immunoglobulins have been employed in passive immunotherapy to enhance Aβ clearance because of the limited reactivity of vaccinations & the development of T cell-dependent adverse responses [29]. Although passive immunotherapy guarantees reasonably stable antibody titers, it is typically associated with numerous negative adverse consequences, like cerebral amyloid angiopathy with microhaemorrhages and vasogenic edema. Antibodies affect the secondary structure of monomers of Aβ, preventing them from aggregating into oligomers or fibrils. They help antigen-antibody complexes pass through the blood-brain barrier through interactions with Fc receptors; they promote opsonization of the antigen, which triggers complement activation and subsequent macrophage phagocytosis; and they lower peripheral Aβ levels, which may enhance the central nervous system's Aβ outflow. These are some of the well-known mechanisms of passive immunization [29 - 33].

Table 1. Treatment approaches that target tau in the creation of AD medications.

Therapeutic Strategy	Compound	Sponsor	Mechanism	Population	Route of Administration	Phase of development	Clinical trial identifier	Status
Active immunotherapy	AN1792	Janssen/Pfizer	Vaccination	mild to moderate AD	intramuscular	II	NCT00021723	discontinue
	amilomotide (CAD106)	Novartis	Vaccination	Individuals at risk of developing clinical symptoms of AD	intramuscular	II/III	NCT02565511	discontinue
	UB-311	United Neuroscience	Vaccination	mild AD mild AD	IM	II II	NCT02551809 NCT03531710	Completed discontinue
	ABvac40	Araclon Biotech	Vaccination	participants having extremely mild AD or amnestic mild cognitive impairment	SC	II	NCT03461276	discontinue

(Table 1) cont.....

Therapeutic Strategy	Compound	Sponsor	Mechanism	Population	Route of Administration	Phase of development	Clinical trial identifier	Status
Passive Immunotherapy Therapeutic Strategy Therapeutic Strategy	Solanezumab (LY2062430)	Eli Lily	Monoclonal antibodies	Mild to moderate AD AD that is mild to moderate AD that is mild to moderate Mild AD Prodromal AD People who are susceptible to memory loss	IM	III III III III	NCT00905372 NCT01900665 NCT02760602 NCT02008357	Completed Completed discontinue discontinue discontinue Active, not recruitment
	Gantenerumab Compound	Roche Sponsor	Monoclonal antibodies Mechanism	Prodromal AD mildAD prodromal to mild Population	IV Route of Administration	II/III Phase of development	NCT01224106 Clinical trial identifier	Completed Completed Recruiting Status
				At the beginning of AD			NCT03444870	Recruiting
				At the beginning of AD			NCT03443973	active, not Recruiting
				At the beginning of AD			NCT04339413	active, not Recruiting
	aducanumab (BIIB037)	Biogen	monoclonal antibody	At the beginning of AD	intravenous	III		discontinue
				In the early AD		III		discontinue
				In the early AD		III		active, not recruiting
	Crenezumab (RG7412) Compound	Roche/AC Immune SA Sponsor	monoclonal antibody Mechanism	prodromal to mild AD Population prodromal to mild AD prodromal to mild AD	intravenous Route of Administration	III Phase of development III III	Clinical trial identifier	discontinue Status discontinue discontinue
	Lecanemab (BAN2401)	Biogen/Eisai	Monoclonal antibodies	early AD Preclinical AD	intravenous	III	NCT03887455	recruiting
	donanemab (LY3002813)	Eli Lilly	Monoclonal antibodies	AD with early symptoms Preclinical AD	intravenous	III III	NCT04437511 NCT05026866	recruiting recruiting

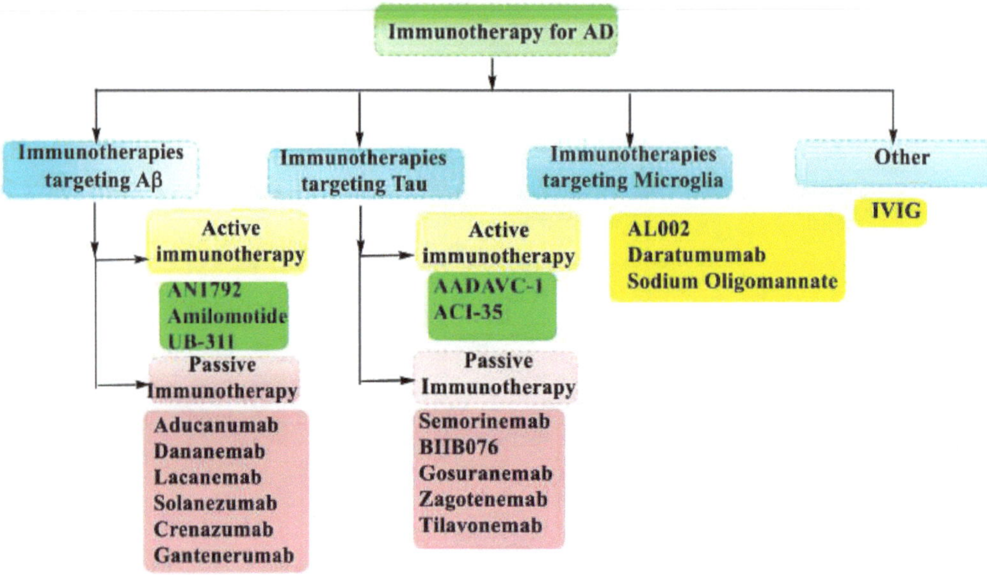

Fig. (1). Immunotherapies to treat AD (Drawing tool: Chemdraw).

The first *in vivo* Aβ antibody test was carried out by Bard F. *et al.* The findings showed that in an Alzheimer's disease animal model, peripheral administration of an antibody that targets the N-terminus of Aβ could cause microglia-mediated phagocytosis of Aβ, thereby lowering the pathology associated with Aβ [34].

The capacity of antibodies of Aβ to enhance Aβ clearance and stop cognitive deterioration in animal models has since been investigated. In addition to the N-terminus, these antibodies also target the mid-region and C-terminus. These days, antibodies that target a single, little Aβ peptide sequence have demonstrated some therapeutic benefits. Apart from aducanumab, which was recently authorised for commercialisation, Phase III trials for six monoclonal antibodies have begun, supported by Roche, Biogen, Eisai, and Eli Lilly. Furthermore, two vaccines: one from AC Immune SA and the other from Araclon Biotech, have advanced to phase II trials (Table **1** and Figure **1**). Phase I clinical trials are now being conducted on the monoclonal antibodies SHR-1707 from Jiangsu Hengrui Pharmaceuticals, RO7126209 from Roche, and LY3372993 from Eli Lilly.

IMMUNOTHERAPIES TARGETING AB

Active immunotherapy

AN1792

AN1792 was the first anti-Aβ vaccination to go through clinical trials. This vaccine, which includes the QS-21 adjuvant, is a synthetic version of the complete Aβ42 peptide. In a phase IIa clinical trial (NCT00021723) (19.7%), patients who had taken AN1792 had strong anti-AN1792 IgG titers, suggesting that they were antibody responders. Even though AN1792 treatment improved the neuropsychological test battery (NTB) score and levels of tau in cerebrospinal fluid (CSF) and minimized Aβ deposition, only the antibody responders showed this change. Six percent of patients who received the AN1792 injection developed T cell-mediated meningoencephalitis that led to clinical trial termination [35, 36]. Extended follow-up studies showed that antibody responders from the previous phase IIa clinical phase still had minimal but discernible levels of persistent anti-AN1792 antibody titers. This tenacity avoided functional degradation and resulted in long-term practical advantages [37].

Amilomotide

A vaccination known as Amilomotide (CAD106) uses Anti-Aβ antibodies that can be produced by the B-cell epitope N-terminal Aβ1-6 without triggering an Aβ-specific T-cell response [38]. According to the phase I study data (NCT00411580), CAD106 demonstrated a strong antibody response and an excellent safety profile [38]. According to phase II studies, tolerance & antibody response were suitably balanced in IIa (NCT00795418, NCT00956410, NCT01023685), NCT00733863 & IIb (NCT01097096).

CAD106, NCT02565511, the first vaccine to enroll in a phase II/III trial, resulted in unexpected alterations in brain volume, body weight, and cognitive performance, which prompted an early discontinuation of the study [38].

UB-311

UB-311 is composed of two synthetic Aβ1-14-targeting peptides, which are B-cell epitopes linked to several helper T-cell peptide epitopes (UBITh®). The Th2-biased delivery technique aims to lower the inflammatory reactivity of T cells and increase.

Immunogenicity [39]. According to the findings of a phase II trial (NCT02551809), UB-311 showed a 100% responder rate, considerable on-target immunogenicity, and the potential to enhance cognitive function in patients with early-to-mild AD [39].

According to the findings of a phase II trial (NCT02551809), UB-311 showed a 100% responder rate, considerable on-target immunogenicity, and the potential to enhance cognitive function in patients with early-to-mild AD.

Passive immunotherapy

Aducanumab

The N terminus of Aβ is permanently bound by the human IgG1 monoclonal antibody aducanumab (BIIB037) [40].

Aducanumab targets aggregates of Aβ, which involve insoluble fibrils and soluble oligomers. According to the phase Ib randomized study (NCT01677572), PRIME patients treated with aducanumab had a significantly lower composite (SUVRr) score, standard uptake value ratio of amyloid (PET), positron emission tomography, particularly those who are receiving 10 mg/kg of the medication over a 54-week period. The brain amyloid burden decreased in people with prodromal or moderate AD in a dose-and time-dependent manner. Furthermore, after aducanumab treatment, there was a delay in the Mini-Mental State Examination (MMSE) and Clinical Dementia Rating Sum of Boxes (CDR-SB) scores, indicating a positive effect on clinical development and cognitive function [41].

The favorable outcomes of phase I led to the elimination of the Phase II experiment. Two Phase III trials using the same strategy were EMERGE (NCT02484547) and ENGAGE (NCT02477800), both of which were stopped in March 2019 when futility evaluations showed that there was minimal possibility of therapeutic success [42].

In October 2019, a significant development occurred. A comprehensive study demonstrated that EMERGE had successfully achieved its primary objective after 78 weeks; the clinical drop in CDR-SB scores from baseline was statistically significantly reduced by 22% in patients receiving the high-dose treatment. The results of EMERGE were in line with data from patients receiving large doses of Aducanumab, despite the fact that ENGAGE failed to meet its main goal [43]. The question of whether EMERGE's positive outcomes are enough to prove validity in light of ENGAGE's poor outcomes is debatable.

Three members of the FDA's Peripheral and Central Nervous System Drugs Advisory Committee said aducanumab was an inadequate treatment for AD in a

March 2021 JAMA article [44].

Using an expedited approval process, the FDA formally approved aducanumab as a therapy for Alzheimer's disease in June 2021 based on the surrogate endpoint of removing amyloid plaques from the brain. The first new anti-AD medication approved by the FDA since memantine's 2003 launch requires the drug's safety; a new phase III clinical investigation (NCT04241068) is necessary. Many scientists and pharmaceutical developers believed that the FDA made a mistake with aducanumab's approval, which would eventually erode trust in the organization. In any case, the only medication that can currently target and improve Alzheimer's disease's basic disease mechanism is aducanumab. Meanwhile, people with In the early AD have found new hope since the approval of aducanumab and have offered valuable perspectives on research & innovation required to combat this condition.

Donanemab

Donanemab (LY3002813), a humanized monoclonal IgG1 antibody, binds specifically to the N-terminal pyroglutamate Aβ epitope, which is unique to deposited Aβ. In the phase II experiment TRAILBLAZER (NCT03367403), donanemab slowed the decline in the integrated AD rate scale score in people with AD at the beginning of AD, indicating less cognitive and functional loss, while secondary outcome results were not entirely consistent. Additionally, donanemab-treated individuals demonstrated a significant decrease in amyloid plaque at 76 weeks, according to 18F-florbetapir PET findings, and at the 52-week mark, 54.7% of the participants tested negative for amyloid.

Less cognitive deterioration & slower tau advancement at 24 weeks were strongly correlated with the increased reduction in amyloid plaque brought about by donanemab therapy. Within twelve weeks of initiating donanemab treatment, there was a notable drop in plasma P-tau217, a biomarker suggestive of AD pathology. Longer and more extensive studies are needed to evaluate the safety and efficacy of donanemab in AD because the incidence of ARIA-E was higher in the donanemab group than in the placebo group (26.7% vs. 0.8%).

A subsequent trial, called (NCT04640077) TRAILBLAZER-EXT, is currently in progress for patients receiving TRAILBLAZER.

Furthermore, a phase III study involving 1,500 participants, referred to as TRAILBLAZER-ALZ 2 (NCT04437511), is underway to ascertain whether donanemab can halt the clinical course of patients exhibiting pathological symptoms of AD who have not yet displayed clinical symptoms. Additionally, a second phase III trial, named TRAILBLAZER-ALZ 3 (NCT05026866), has commenced to assess the effectiveness and safety of donanemab in a cohort of approx 3,300 individuals with preclinical AD.

Lecanemab

One humanized IgG1 monoclonal antibody is lecanemab (BAN2401) that has efficacy against protofibrils, oligomers, and insoluble fibrils while preferring targeting soluble aggregated Aβ. The phase II trial BAN2401-G000-201 (NCT01767311) demonstrated a reduction in brain amyloid plaques, but the 12-month primary aim was not reached. At a maximal dose of 10 mg/kg every two weeks, specific clinical and biomarker endpoints demonstrated prolonged clinical remission. The amyloid PET SUVr value decreased in a dose-dependent manner over the course of 18 months, while the ADAS-Cog14 and Alzheimer's Disease Composite Score (ADCOMS) clinically declined. Over the course of 18 months, there was a dose-dependent decrease in the amyloid PET SUVr value and a clinical fall in the ADAS-Cog14 & Alzheimer's Disease Composite Score (ADCOMS)

The dosage of lecanemab (10 mg/kg every two weeks) decreased cognitive impairment more in APOE4 carriers than in non-carriers when compared to the placebo, according to the Bayesian sensitivity analysis. When compared to a placebo, CSF biomarker tests showed a decrease in p-tau and an increase in Aβ42; however, results on total tau varied from 12-18 months. In the overall population, 9.9% of people have ARIA-E at a dose of 10 mg/kg and 14.3% among APOE4 carriers, suggesting that lecanemab is well tolerated [46].

Clarity AD (NCT03887455), a phase III research, is being carried out to evaluate effectiveness, safety, and tolerability over an extended period. Lecanemab's safety and efficacy in preclinical AD patients are being assessed in AHEAD 3–45 (NCT04468659), another phase III trial. Determining the change in the Preclinical Alzheimer Cognitive Composite (PACC) 5 score from its baseline at 216 weeks of treatment is its main goal [47].

Solanezumab

By focusing on the mid domain of Aβ peptide (Aβ13–28), solanezumab (LY2062430), a humanized monoclonal antibody, raises Aβ clearance [47]. EXPEDITION 1 (NCT00905372) and EXPEDITION 2 (NCT00904683) two finished phase III clinical trials were unable to show that solanezumab was helpful in halting cognitive decline and enhancing functional ability in patients with mild-to-moderate AD [48].

Furthermore, the open-label extension study EXPEDITION EXT (NCT01127633), which followed EXPEDITION 1 and EXPEDITION 2, was halted because it failed to accomplish its main objective. Similarly, two additional phase III clinical studies, Expedition 3 (NCT01900665) and Expedition PRO (NCT02760602), were also discontinued as a result of their inability to stop cognitive deterioration. In a recent study (DIAN-TU, NCT01760005), solanezumab therapy was employed to activate its Aβ targets in order to assess its impact on patients with dominantly inherited AD; however, it did not lead to any improvement in cognitive function compared with the control group and, in fact, resulted in a significant deterioration [49].

Crenezumab

A humanized monoclonal antibody of the IgG1 class called crenezumab (RG7412) targets both clumps and monomers of Aβ [50]. Its affinity for oligomers is ten times greater [50]. Two Phase III clinical phase studies, CREAD (NCT02670083) and CREAD2 (NCT03114657), were halted after prearranged intermediate assessment found they were unlikely to meet their primary objective of increasing CDR SB scores. Furthermore, the clinical phase III CREAD OLE (NCT03491150) was terminated as a result of an interim analysis.

A clinical phase II research (NCT01998841) is now being conducted on preclinical AD patients with presenilin 1 (PSEN1) E280A autosomal dominant mutation to examine the safety and efficacy of crenezumab vs the placebo [50].

Gantenerumab

A human IgG1 monoclonal antibody called gantenerumab (RO4909832) has a high affinity for aggregated Aβ & helps the Fc receptor-mediated phagocytosis that removes Aβ [51, 52]. In February 2020, a phase II trial (NCT04623242, DIAN TU) comprising people with genetic AD disclosed that gantenerumab had not met the main objective. Following that, DIAN-TU-001 (NCT01760005), another phase II trial, was conducted in people who had mutations linked to early-onset AD. Treatment with gantenerumab mitigated the increases in neurofilament light chain and dramatically reduced CSF total tau, Aβ plaques, and phospho-tau181, but had no effect on cognitive tests. Oedema associated with Amyloid-associated imaging irregularities was seen in 19.2% of the participants [49].

Based on these findings, it was thought that a larger gantenerumab dosage was required for likely clinical success. Two parallel-group, placebo-controlled phase III trials, double-blind, randomized, GRADUATE 1 (NCT03444870) and GRADUATE 2 (NCT03443973), are being conducted to examine the effectiveness and gantenerumab safety in the larger population of AD patients whose condition is not directly brought on by gene mutations. To evaluate the safety and tolerability of long-term treatment, two open-label, multicenter, and rollover phase III investigations (NCT04339413 & NCT04374253) are also in progress [53].

Tau protein-based immunotherapies

Another important feature of AD is neurofibrillary tangles, which are made of tau (p-tau) protein that has been aberrantly phosphorylated. By attaching to tubulin during its polymerisation in normal conditions, the cytoplasmic protein tau can stabilise microtubules [54]. But in AD, hyperphosphorylated tau has a decreased capacity to bind microtubules, resulting in the creation of aggregates and neurofibrillary tangles [54]. Since tau protein seems to be more strongly connected with the degree of cognitive deterioration than Aβ, tau-targeting therapies may be helpful for AD patients [55 - 57].

Preventing aberrant tau phosphorylation, preventing tau aggregation, and encouraging the removal of tau aggregates are the three primary known anti-tau tactics.

Immunotherapies currently make up the majority of anti-tau drugs undergoing clinical studies, since the initial demonstration of tau immunotherapy effectiveness in the JNPL3 mice model (2007), several active vaccines, including AADvac1 and ACI-35. In addition, passive immunotherapeutic antibodies such as BIIB076, semorinemab, and gosuranemab have been created recently. In AD animal models, they all exhibit notable therapeutic effects [55], and most of them are currently being investigated in human environments. The specifics are displayed in Table **2**.

The pharmacological processes of active vaccinations stimulate the patient's immune system and cause the creation of natural anti-tau protein antibodies by administering phosphorylated or non-phosphorylated tau [55]. It has been demonstrated that immunogens of tau with moderate adjuvants can cause harmful levels of tau that don't result in significant immunological reactions [55]. For almost ten years, AADVac1 and ACI-35 have been the only two vaccines being studied in clinical settings; AADVac1 initiated the first phase II trial (NCT02579252) [56 - 60].

Because the purpose of passive antibodies is to target specific tau protein sites, they are a safer option than active vaccinations when it comes to lowering the possibility of negative immunological consequences. Furthermore, the benefits of passive immunization improved specificity for the epitopes that are being targeted. It has been demonstrated that passive antibodies can target intracellular tau proteins and enter neurons by receptors or bulk endocytosis [61]. Additionally, by stopping the spread of extracellular tau, anti-tau antibodies can slow the course of AD [61].

Table 2. Therapeutic strategies that target tau in the development of AD medications.

Therapeutic strategy	Compound	Sponsor	Mechanism	Population	Route of administration	Phase of development	Clinical trial identifier	Status
Active immunotherapy	ACI-35 AADvac1	Active vaccine Act ive vaccine	AC Immune SA, Janssen Axon Neuroscience SE	In early AD mild AD	intravenous intravenous	I/II II	NCT04445831 NCT02579252	Completed Recruiting
Passive immunotherapy	BIIB076 (NI-105)	Biogen	Monoclonal antibody	healthy and AD	intravenous	I	NCT03056729	completed
	RG7345 (RO6926496)	Discontinued	Monoclonal antibody	healthy	Intravenous	I	NCT02281786	completed
	Semorinemab (RO7105704)	Genentech Genentech	Monoclonal antibody	mild AD	intravenous	II	NCT03289143	completed
	Tilavonemab (ABBV-8E12)	Abbvie Abbvie	Monoclonal antibody	In early AD	intravenous	II	NCT02880956	completed Active, not recruiting
	Zagotenemab (LY3303560)	Eli Lilly	Monoclonal antibody	In early AD	intravenous	II	NCT03518073	Active, not recruiting
	Gosuranemab (BIIB092)	Biogen	Monoclonal antibody	mild AD	intravenous	II	NCT03352557	terminated
	PNT001	Pinteon Therapeutic	Monoclonal antibody	healthy	intravenous	I	NCT04096287	completed
	Lu AF87908	H.Lundbeck	Monoclonal antibody	healthy & AD	intravenous	I	NCT04149860	recruiting
	JNJ-63733657	Janssen	Monoclonal antibody	Early AD	intravenous	II	NCT04619420	recruiting
	E2814	Eisai	Monoclonal antibody	Mild to moderate AD	intravenous	I/II	NCT04971733	recruiting
	Bepranemab (UCB0107)	Hoffmann La Roche,UCB SA	Monoclonal antibody	mild cognitive impairment or mild AD	intravenous	II	NCT04867616	recruiting

Eleven antibodies have already begun clinical trials, and seven are currently undergoing them. The microtubule-binding domains (235–250, 386–408) and other related proteins, as well as proteins like pThr217, pThr231, and pSer422, are targeted by these antibodies. However, research on tau antibodies is not as advanced as that on Aβ, and none of the seven drugs have advanced to phase III trials. Since the medications must not only target the tau protein in neuron cells but also stop tau from diffusing outside of cells, research on tau antibodies is challenging. Maintaining this equilibrium is challenging, and following therapy, unfavorable inflammatory reactions could happen [62]. Nonetheless, since tau plays a significant role in the onset of AD, tau immunotherapy is still worth looking into.

Active Immunotherapy

AADvac1

Axon Neuroscience created the innovative active immunotherapy vaccine AADvac1, which targets the 12 amino acid KDNIKHVPGGGS sequence of the tau protein microtubule binding domain. It is safe, and neither the placebo group nor the immunised group experienced any uncommon side effects [63]. Furthermore, people with mild-to-moderate AD who received AADvac1 treatment experienced less cognitive impairment & brain atrophy in a phase I research (NCT02031198) [64]. Notably, AADvac1 markedly decreased the levels of p-tau181 and p-tau217, two CSF indicators of AD [63].

These encouraging outcomes provide credence to AADvac1's move to phase II testing. AADvac1 completed a phase II study on November 14, 2019, as per data (clinicaltrials.gov), but no more information has been released [65]. AADvac1's immunogenicity and effectiveness in reducing cognitive impairment in AD patients, safety, and tolerability of long-term AADvac1 therapy have been assessed in a recent phase II trial [66]. The findings indicated that whereas AADvac1 was safe and acceptable in 196 people, it had no effect on cognitive function.

ACI-35

Another active immunotherapy vaccine, ACI-35, was developed by AC Immune to target harmful types of hyperphosphorylated tau. Sixteen copies of synthetic tau that have been modified to recognize the aberrant tau phosphorylation sites S396 and S404 make up the vaccination. The safety and effectiveness of ACI-35 in patients with mild to moderate diseases are currently being assessed in Finland through double-blind, randomized phase I/II clinical trials. October 31, 2023, is when the trial is expected to end (NCT04445831) [67].

Passive immunotherapy

Semorinemab

With an immunoglobulin G4 isotype backbone, semorinemab (RO705705) is a humanized anti-tau monoclonal antibody that can bind all six human tau isoforms and shield neurons. Although AC Immune SA has released its safety profile, clinical studies have shown no signs of efficacy against AD [68]. It goes without saying that more research with a proper design is needed to assess its effectiveness. Up to October 2023, a semorinemab trial in individuals with mild AD is being conducted (NCT03828747) [69]. Semorinemab did not lessen AD symptoms, according to a January 2021 article in Nature Reviews Drug Discovery that also detailed the phase II trial's failure (NCT02754830) [70].

BIIB076 and Gosuranemab

Gosuranemab (BIIB092) and BIIB076 (NI-105) were both developed by Biogen. A monoclonal IgG1 that targets tau's mid-domain is called BIIB076. It is presently in the early stages of clinical development after completing a phase I study (NCT03056729) [71]. A humanised monoclonal antibody known as gosuranemab targets the extracellular N-terminal of tau in CSF & ISF that neurones produce. The unbound N-terminal tau in CSF was reduced by 98% in the gosuranemab group and increased by 11% in the placebo group according to a clinical phase II study in progressive supranuclear palsy; nevertheless, N-terminal tau neutralisation does not equate to clinical benefit (NCT03068468) [72].

Gosuranemab medication may be linked to glial reactions, such as the buildup of tau within astrocytic lysosomes, according to Kim and their colleagues recently verified the tissues of the brain of 3 patients undergoing treatment with the medication [73].

However, following the conclusion of the placebo-controlled period, Gosuranemab's clinical phase II trial in people with early AD was terminated due to its inability to reduce functional and cognitive disability [74].

ZAGOTENEMAB

Peter Davies' mouse monoclonal antibody against MCI-1, an early pathogenic form of tau, served as the basis for the humanized anti-tau antibody zagotenemab (LY3303560). So far, zagotenemab has successfully finished two phase I studies to investigate the safety of repeated dosages in both healthy individuals & AD patients (NCT02754830, NCT02754830) [75, 76]. In August 2019, 285 participants with at least six months of progressive memory loss participated in a phase II trial that ran until August 2021 and examined gradual memory loss

Lilly declared in October 2021 that the development of zagotenemab had been stopped since the trial had failed to meet its primary aim [77].

TILAVONEMAB

An antibody called tilavonemab (ABBV-8E12) attaches itself to the pathogenic tau N-terminus & recognizes its aggregated extracellular form. AbbVie and C2N Diagnostics created this medication, and a phase I study (NCT02880956) has shown its safety [78, 79]. However, a clinical phase II study that assessed the safety and efficacy of btilavonemab in 453 patients with early AD did not produce the expected results & as a result, tilavonemab is currently no longer used to treat AD (NCT02880956) [79]. Phase II experiment was completed on September 30, 2021, to assess tilavonemab's long-term safety and tolerability in 364 patients with early AD; however, the trial's final findings are not yet accessible (NCT03712787) [80].

Microglia-based immunotherapy

Microglia are thought to be essential for regulating inflammation and the neuro-immune response of the central nervous system. Two examples of chemical patterns linked to damage are Amyloid Beta (Aβ) and phosphorylated tau protein, which are believed to be picked up by receptors like TLR-4 on the surface of microglia [81, 82]. In reaction to this realization, inflammatory mediators are released by the brains of individuals suffering from AD. Consequently, these inflammatory chemicals accelerate the development of plaques of Aβ & neurofibrillary tangles, resulting in a harmful cycle that exacerbates the course of the illness. A recent study utilizing Positron Emission Tomography (PET) imaging with 130 participants revealed that the interaction between activated microglia and amyloid-beta (Aβ) affects the rate of tau propagation through the Braak stages. This discovery emphasizes the critical role that microglia & these pathogenic proteins play in the development of AD [83].

AL002

One important receptor that microglia in the brain specifically express is TREM2. It has been discovered that some TREM2 variations raise the chances of late-onset AD. It stimulates the TREM2 receptor's downstream signalling. A TREM2 agonistic antibody that attaches to the extracellular domain of TREM2 may increase microglia proliferation & reduce AD pathology in an animal model [84]. A few TREM2 antibody-based clinical studies have been started in light of the promising preclinical findings. Together, Alector and AbbVie created AL002, a humanized monoclonal IgG1 antibody that targets TREM2 exclusively. AL002 binds to the microglial receptor TREM2, activates TREM2 signaling, promotes microglia development, and raises the phosphorylation of TREM2 downstream effector Syk [85].

AL002 is presently being tested in a phase II study with 265 individuals who have early AD to determine its safety and effectiveness. This study will run for two years, ending in August 2023 [86].

Daratumumab

The FDA has authorised daratumumab (DARZALEX), a human antibody—called daratumumab (DARZALEX)—that targets CD38, to treat cancer of plasma cells. Daratumumab exerts immunomodulatory effects on nonplasma cells that express CD38 and is able to penetrate the blood–brain barrier [14]. Early AD patients have much higher levels of CD38+ CD8+ T-cells in their blood that can go to the central nervous system and have harmful consequences [87, 88]. With a projected completion date of June 2022, the clinical phase II study commenced to explore the potential benefits of administering daratumumab to individuals suffering from mild to moderate Alzheimer's [14].

Sodium oligomannate

Recent studies have highlighted the growing interest in gut microbiota due to its role in inflammation associated with AD and the activation of microglia. Due to its involvement in AD-related inflammation and microglia activation, recent research has brought attention to the growing interest in gut microbiota [89 - 92]. The Chinese National Medical Products Administration gave sodium oligomannate, a substance made of acidic linear oligosaccharides extracted from marine brown algae, provisional approval in November 2019 for the treatment of mild to severe Alzheimer's disease [93]. In December 2021, it was added to China's national medical insurance catalogue. By controlling the gut-brain axis and gut microbiota, sodium oligomannate delays cognitive decline by lowering microglial activation and neuroinflammation, while not being considered a conventional immunotherapeutic medication [94]. In Europe, a global phase III research study is currently testing sodium oligomannate among those suffering from mild to moderate AD [95, 96].

Intravenous Immunoglobulin (IVIG)

Intravenous immunoglobulin (IVIG) therapy is a form of passive immunotherapy that involves administering polyclonal antibodies intravenously. The plasma of thousands of healthy young donors is used to make these antibodies. IVIG has been utilized as a replacement treatment for a number of clinical diseases, including immunodeficiency syndromes, some forms of cancer, and haematological disorders. The primary component of IVIG is IgG antibodies, with only approximately 0.5% targeting amyloid-beta (Aβ) [97, 99].

The potential of IVIG in treating AD was first explored in 2002 when studies demonstrated that human pooled antibodies exhibited strong affinity to Aβ fibrils and neurotoxic oligomers were observed, whereas there was only a minimal interaction with its monomeric variant [99 - 101]. Beyond this, IVIG possesses immunomodulatory properties relevant to AD treatment. Initial trials indicated some success in slowing cognitive decline, encouraging further research [101 - 104].

A futility study of Gammagard IVIG (Baxter) in patients with mild to moderate AD revealed beneficial cognitive effects, and a phase 2 trial of IVIG in an open-label approach indicated symptomatic improvements. However, Baxter's phase 3 trial, funded by the US, did not meet its primary endpoint, leading to the trial's discontinuation [105].

Grifols conducted a pilot study where plasma was removed and replaced with Albutein in seven mild to moderate patients having AD, with the procedure performed biweekly over six months [106]. The company deemed this method a viable AD treatment strategy. Building on this, a phase 2 trial in 2017 involved 39 patients (20 sham group and 19 actively treated) and showed a distinct sawtooth pattern in plasma levels of Aβ40/Aβ42 in the treatment group, with adverse event rates being comparable between groups.

Over the course of 14 months, 496 mild to moderate patients of AD participated in the large-scale, double blind, placebo-controlled Alzheimer Management by (AMBAR) Albumin Replacement research by Grifols.

The study included three different treatment groups: (i) plasmapheresis with albumin replacement, (ii) modest doses of albumin and intravenous immunoglobulin in plasmapheresis (IVIG), (iii) plasmapheresis using IVIG and high-dose albumin. The control group got sham treatment. Variations in the Alzheimer's Disease Assessment Scale Cognitive Subscale and Alzheimer's Disease Cooperative Study Activities of Daily Living scores from the first assessment to the study end were used to measure the primary outcome [107].

According to the results, treated individuals saw a 42–70% decrease in ADCS-ADL scores and a 50–75% decrease in ADAS-Cog scores, worsening when compared to the control group. The average decline in treated patients' ADAS-Cog and ADCSADL scores was 66% and 52%, respectively, in comparison to the control group. Mildly AD patients progressed more slowly. Similar improvements were surprisingly seen in mild AD patients in the sham group.

Grifols also noted significant benefits in processing speed, memory, quality of life, and language performance in the group receiving high IVIG and high albumin treatment. Patients having moderate AD showed improvements in memory and life quality, whereas those who are suffering from mild AD demonstrated enhanced language capabilities and improved processing speed.

While the AMBAR study's findings are encouraging, several critical questions remain: the mechanisms behind the observed reduction in disease progression, the treatment's effectiveness in mild versus mild AD, the need for IVIG in the protocol, and how age, sex, and ApoE genotype affect therapy results [108].

Perspectives & Challenges of immunotherapy for AD

Aducanumab, a human monoclonal antibody that targets amyloid-beta, has been approved, marking a significant advancement in the treatment of Alzheimer's disease (AD). Encouraging the removal of amyloid plaques, a defining feature of AD pathogenesis, it presents the potential to slow down or perhaps stop neurodegeneration. Nevertheless, for immunotherapy to be regarded as a proven treatment for AD, several significant problems and ambiguities must be resolved. Some of the main issues are listed below [109].

1. Selecting suitable targets is a crucial stage in the creation of disease-modifying therapies for AD. For many years, Aβ and tau theories have been put out, and these two proteins are also linked to other postulated processes, including neuroinflammation. Nonetheless, there has been discussion on the connection between cognition and the production or aggregation of these two proteins. Postmortem studies indicate that certain Alzheimer's disease patients may lack Aβ deposits in their brains, while others who do have Aβ plaques may not show any signs of cognitive decline. Therefore, it is essential to examine additional important molecules that promote the course of AD or to clearly define tau and Aβ as useful targets for disease-modifying AD therapy. Clinical studies have examined a few unusual immunotherapy targets, including TREM2, CD38, and TNF-α [14, 95, 96, 105]. Recent developments in spatial transcriptome sequencing and single-cell RNA sequencing will help identify and target more promising medicines with immunotherapy.
2. Since several forms of Aβ have been found in the brains of individuals with AD, it is crucial to determine which type of form or forms of Aβ immunotherapy must be targeted in order to improve treatment outcomes. It has been demonstrated that Fibrillary aggregates are less dangerous than soluble Aβ oligomers and have a stronger association with the symptoms of AD [97, 98]. It has been proposed that microglia absorb and process the harmful Aβ oligomers to create a thick Aβ plaque that is comparatively less hazardous [99]. Thus, removing Aβ plaque instead of Aβ oligomers may not have much of an impact on improving AD patients' cognitive impairment.

Apart from Aβ1-40 & Aβ1-42, pyroglutamate Aβ3-42 & Aβ4-42 have also been identified as major Aβ isoforms that contribute significantly to AD neurodegeneration. Since there are no discernible differences between patients with sporadic and familial AD, these two isoforms may be viable targets for immunotherapy [100]. The Pyroglutamate Aβ monomers' N-terminal region has been confirmed to have a unique, pseudo β-hairpin shape, which provides fresh perspectives on active and passive AD vaccination [101].

Notably, a phase II clinical study found that Donanemab, a humanized antibody that targets the N-terminal Pyroglutamate Aβ epitope, reduces plaque burden and postpones cognitive decline in those with early-stage AD symptoms [45]. When combined, the mechanism of action against various Aβ types, epitopes, and isoforms may be revolutionary and need to be prioritised in AD immunotherapy.

3. Like Aβ immunotherapy, a number of variables might affect how well tau vaccinations or antibodies work.

The two most crucial factors among these may be the tau epitope and the place of action (intracellular or extracellular). Understanding the kind of tau that is found inside neuronal cells and the type that distributes to other cells to initiate tau pathology in AD is crucial. Anti-tau antibodies have the ability to enter neurones in the brain and target intracellular tau proteins. They can also bind extracellular tau to stop the progression of tau disease [61]. According to reports, patients with AD have increased CSF tau levels, which are mostly made up of tau fragments that range in length from 150-250 amino acids, suggesting that the distribution of extracellular tau may be stopped by antibodies that specifically target amino acids 150–250 [61, 101, 102].

4. An increasing body of research suggests that there may be variations in the pathophysiology and severity of individuals suffering from Alzheimer's Disease (AD), indicating that precise diagnosis and personalized treatment could yield better outcomes. It is widely recognized that in the brains affected by AD, the abnormal accumulation of amyloid-beta (Aβ) occurs prior to the development of tau pathology. Immunotherapies targeting Aβ, such as aducanumab, are likely to be more effective when administered at the onset or in the early stages of AD, when clinical symptoms are minimal due to the distinct roles and timing of tau & Aβ in the disease's progression. Conversely, strategies aimed at tau immunization may prove more beneficial in decelerating cognitive deterioration in patients with moderate to severe AD. The significance of early diagnosis and intervention in the treatment of AD is underscored by the fact that most participants in clinical trials for both tau and Aβ immunotherapies are individuals with moderate or early stage AD. However, the early identification

of AD has been hampered by the absence of a reliable biomarker or cognitive scale. Effective early diagnosis and treatment are critically needed.

5. Only a limited portion of medications may enter the brain, and both vaccinations and antibodies are provided peripherally. To guarantee adequate medication delivery into the brain, strategies to increase the effectiveness of vaccination or antibody administration across the blood–brain barrier must be devised.

Bioengineering and delivery technologies could be useful in relation to this problem. Biomimetic nanoparticles and other carriers would significantly increase the delivery efficiency.

6. In immunotherapy research, measures to lessen side effects should also be implemented. It's critical to prevent autoimmune T-cell activation with vaccinations.
7. For such a complicated condition, a medication that simply addresses one element of AD might not be sufficient to provide the intended therapeutic results. Therefore, it would be quite feasible to treat AD with an effective combination of several immunisation therapies. An active vaccination that targets both Aβ and tau concurrently, for instance, may be more therapeutically efficacious than vaccines that just target one target; nevertheless, the issue of elevated Th responses and antibody concentrations resulting from this type of vaccine should be considered

CONCLUSION AND FUTURE PERSPECTIVES

Following ten years of challenges in the effort to prevent AD *via* vaccination targeting Aβ, several vaccine candidates have advanced to phase 3 clinical trials, whereas alternative approaches are still undergoing preclinical evaluation. One major challenge with vaccination is timing. Research has shown that early intervention is crucial, as clearing plaques in later stages does not halt disease progression, likely due to ongoing tau aggregation. Consequently, eliminating amyloid during the pre-symptomatic phase may avert the clinical manifestation of Alzheimer's disease. Nevertheless, identifying the most effective time to get vaccinated is a significant challenge.

An additional concern with vaccination is achieving specific targeting of pathogenic Aβ and determining the appropriate antibody titer levels. Future research must focus on identifying the right vaccination window, ensuring the vaccine specifically targets harmful Aβ forms, and optimizing the immune response involving antibodies.

Compared to active vaccination, passive immunotherapy now exhibits greater promise. The FDA's recent, though controversial, acceptance of aducanumab highlights the potential of passive immunotherapy. One advantage of passive immunotherapy is the ability to adjust monoclonal antibody (mAb) doses and specificities. However, challenges like short-term antibody effects, limited cognitive improvement, and the prevalence of anomalies in imaging linked to amyloid (ARIA).

Given the intricate characteristics of Alzheimer's disease, forthcoming studies on both active and passive immunotherapy may necessitate the development of innovative approaches. Combination therapies that target both tau and Aβ proteins might offer benefits. Additionally, personalized treatments could be developed based on individual factors such as APOE genotype, immune profile, and Aβ strain variations. Tailoring therapies to these individual differences could enhance the effectiveness of immunotherapeutic approaches for preventing AD.

REFERENCES

[1] Ferrari C, Sorbi S. The complexity of Alzheimer's disease: an evolving puzzle. Physiol Rev 2021; 101(3): 1047-81.
[http://dx.doi.org/10.1152/physrev.00015.2020] [PMID: 33475022]

[2] Ballard C, Gauthier S, Corbett A, Brayne C, Aarsland D, Jones E. Alzheimer's disease. Lancet 2011; 377(9770): 1019-31.
[http://dx.doi.org/10.1016/S0140-6736(10)61349-9] [PMID: 21371747]

[3] Vermunt L, Sikkes SAM, van den Hout A, *et al.* Duration of preclinical, prodromal, and dementia stages of Alzheimer's disease in relation to age, sex, and *APOE* genotype. Alzheimers Dement 2019; 15(7): 888-98.
[http://dx.doi.org/10.1016/j.jalz.2019.04.001] [PMID: 31164314]

[4] Arvanitakis Z, Shah RC, Bennett DA. Diagnosis and management of dementia. JAMA 2019; 322(16): 1589-99.
[http://dx.doi.org/10.1001/jama.2019.4782] [PMID: 31638686]

[5] Long JM, Holtzman DM. Alzheimer disease: an update on pathobiology and treatment strategies. Cell 2019; 179(2): 312-39.
[http://dx.doi.org/10.1016/j.cell.2019.09.001] [PMID: 31564456]

[6] Schellenberg GD, Bird TD, Wijsman EM, *et al.* Genetic linkage evidence for a familial Alzheimer's disease locus on chromosome 14. Science 1992; 258(5082): 668-71.
[http://dx.doi.org/10.1126/science.1411576] [PMID: 1411576]

[7] Head E, Lott IT, Wilcock DM, Lemere CA. Aging in Down syndrome and the development of Alzheimer's disease neuropathology. Curr Alzheimer Res 2016; 13(1): 18-29.
[http://dx.doi.org/10.2174/1567205012666151020114607] [PMID: 26651341]

[8] Naj AC, Jun G, Reitz C, *et al.* Effects of multiple genetic loci on age at onset in late-onset Alzheimer disease: a genome-wide association study. JAMA Neurol 2014; 71(11): 1394-404.
[http://dx.doi.org/10.1001/jamaneurol.2014.1491] [PMID: 25199842]

[9] Holtzman DM, Herz J, Bu G. Apolipoprotein E and apolipoprotein E receptors: normal biology and roles in Alzheimer disease. Cold Spring Harb Perspect Med 2012; 2(3): a006312.
[http://dx.doi.org/10.1101/cshperspect.a006312] [PMID: 22393530]

[10] Nordberg A. Mechanisms behind the neuroprotective actions of cholinesterase inhibitors in Alzheimer disease. Alzheimer Dis Assoc Disord 2006; 20(2) (Suppl. 1): S12-8. [http://dx.doi.org/10.1097/01.wad.0000213804.59187.2d] [PMID: 16772751]

[11] Mangialasche F, Solomon A, Winblad B, Mecocci P, Kivipelto M. Alzheimer's disease: clinical trials and drug development. Lancet Neurol 2010; 9(7): 702-16. [http://dx.doi.org/10.1016/S1474-4422(10)70119-8] [PMID: 20610346]

[12] Agadjanyan MG, Petrovsky N, Ghochikyan A. A fresh perspective from immunologists and vaccine researchers: Active vaccination strategies to prevent and reverse Alzheimer's disease. Alzheimers Dement. 2015 Oct; 11(10): 1246–1259. [http://dx.doi.org/10.1016/j.jalz.2015.06.1884]

[13] Wang Y, Cella M, Mallinson K, *et al.* TREM2 lipid sensing sustains the microglial response in an Alzheimer's disease model. Cell 2015; 160(6): 1061-71. [http://dx.doi.org/10.1016/j.cell.2015.01.049] [PMID: 25728668]

[14] ClinicalTrials.gov. Study of Daratumumab in Patients With Mild to Moderate Alzheimer's Disease [Internet]. Bethesda (MD): National Library of Medicine (US); 2019 Aug 28. Identifier NCT04070378.

[15] Gambuzza M, Sofo V, Salmeri F, Soraci L, Marino S, Bramanti P. Toll-like receptors in Alzheimer's disease: a therapeutic perspective. CNS Neurol Disord Drug Targets 2014; 13(9): 1542-58. [http://dx.doi.org/10.2174/1871527313666140806124850] [PMID: 25106635]

[16] Lemere CA. Immunotherapy for Alzheimer's disease: hoops and hurdles. Mol Neurodegener 2013; 8(1): 36. [http://dx.doi.org/10.1186/1750-1326-8-36] [PMID: 24148220]

[17] Hardy J, Selkoe DJ. The amyloid hypothesis of Alzheimer's disease: progress and problems on the road to therapeutics. Science 2002; 297(5580): 353-6. [http://dx.doi.org/10.1126/science.1072994] [PMID: 12130773]

[18] Benilova I, Karran E, De Strooper B. The toxic Aβ oligomer and Alzheimer's disease: an emperor in need of clothes. Nat Neurosci 2012; 15(3): 349-57. [http://dx.doi.org/10.1038/nn.3028] [PMID: 22286176]

[19] Selkoe DJ, Hardy J. The amyloid hypothesis of Alzheimer's disease at 25 years. EMBO Mol Med 2016; 8(6): 595-608. [http://dx.doi.org/10.15252/emmm.201606210] [PMID: 27025652]

[20] Panza F, Lozupone M, Seripa D, Imbimbo BP. Amyloid-β immunotherapy for alzheimer disease: Is it now a long shot? Ann Neurol 2019; 85(3): 303-15. [http://dx.doi.org/10.1002/ana.25410] [PMID: 30635926]

[21] Liu YH, Giunta B, Zhou HD, Tan J, Wang YJ. Immunotherapy for Alzheimer disease—the challenge of adverse effects. Nat Rev Neurol 2012; 8(8): 465-9. [http://dx.doi.org/10.1038/nrneurol.2012.118] [PMID: 22751529]

[22] van Dyck CH. Anti-amyloid-β monoclonal antibodies for Alzheimer's disease: pitfalls and promise. Biol Psychiatry 2018; 83(4): 311-9. [http://dx.doi.org/10.1016/j.biopsych.2017.08.010] [PMID: 28967385]

[23] Sarazin M, Dorothée G, de Souza LC, Aucouturier P. Immunotherapy in Alzheimer's disease: do we have all the pieces of the puzzle? Biol Psychiatry 2013; 74(5): 329-32. [http://dx.doi.org/10.1016/j.biopsych.2013.04.011] [PMID: 23683656]

[24] Jack CR Jr, Knopman DS, Jagust WJ, *et al.* Hypothetical model of dynamic biomarkers of the Alzheimer's pathological cascade. Lancet Neurol 2010; 9(1): 119-28. [http://dx.doi.org/10.1016/S1474-4422(09)70299-6] [PMID: 20083042]

[25] Holmes C, Boche D, Wilkinson D, *et al.* Long-term effects of Aβ42 immunisation in Alzheimer's

disease: follow-up of a randomised, placebo-controlled phase I trial. Lancet 2008; 372(9634): 216-23. [http://dx.doi.org/10.1016/S0140-6736(08)61075-2] [PMID: 18640458]

[26] Schenk D, Barbour R, Dunn W, *et al.* Immunization with amyloid-β attenuates Alzheimer-disease-like pathology in the PDAPP mouse. Nature 1999; 400(6740): 173-7. [http://dx.doi.org/10.1038/22124] [PMID: 10408445]

[27] Morgan D. Immunotherapy for Alzheimer's disease. J Intern Med 2011; 269(1): 54-63. [http://dx.doi.org/10.1111/j.1365-2796.2010.02315.x] [PMID: 21158978]

[28] Gilman S, Koller M, Black RS, *et al.* Clinical effects of Aβ immunization (AN1792) in patients with AD in an interrupted trial. Neurology 2005; 64(9): 1553-62. [http://dx.doi.org/10.1212/01.WNL.0000159740.16984.3C] [PMID: 15883316]

[29] Wang A, Das P, Switzer RC III, Golde TE, Jankowsky JL. Robust amyloid clearance in a mouse model of Alzheimer's disease provides novel insights into the mechanism of amyloid-beta immunotherapy. J Neurosci 2011; 31(11): 4124-36. [http://dx.doi.org/10.1523/JNEUROSCI.5077-10.2011] [PMID: 21411653]

[30] DeMattos RB, Bales KR, Cummins DJ, Dodart JC, Paul SM, Holtzman DM. Peripheral anti-Aβ antibody alters CNS and plasma Aβ clearance and decreases brain Aβ burden in a mouse model of Alzheimer's disease. Proc Natl Acad Sci USA 2001; 98(15): 8850-5. [http://dx.doi.org/10.1073/pnas.151261398] [PMID: 11438712]

[31] Bates KA, Verdile G, Li Q-X, *et al.* Clearance mechanisms of Alzheimer's amyloid-β peptide: implications for therapeutic design and diagnostic tests. Mol Psychiatry 2009; 14(5): 469-86. [http://dx.doi.org/10.1038/mp.2008.96] [PMID: 18794889]

[32] Solomon B, Koppel R, Hanan E, Katzav T. Monoclonal antibodies inhibit *in vitro* fibrillar aggregation of the Alzheimer beta-amyloid peptide. Proc Natl Acad Sci USA 1996; 93(1): 452-5. [http://dx.doi.org/10.1073/pnas.93.1.452] [PMID: 8552659]

[33] Deane R, Sagare A, Hamm K, *et al.* IgG-assisted age-dependent clearance of Alzheimer's amyloid beta peptide by the blood-brain barrier neonatal Fc receptor. J Neurosci 2005; 25(50): 11495-503. [http://dx.doi.org/10.1523/JNEUROSCI.3697-05.2005] [PMID: 16354907]

[34] Bard F, Cannon C, Barbour R, *et al.* Peripherally administered antibodies against amyloid β-peptide enter the central nervous system and reduce pathology in a mouse model of Alzheimer disease. Nat Med 2000; 6(8): 916-9. [http://dx.doi.org/10.1038/78682] [PMID: 10932230]

[35] Tripathi SM, Sambhakar S, Mishra S, Malviya R, Sridhar SB. Cannabinoid receptor ligands with potential therapeutic applications and mechanisms of action: a versatile natural therapeutic agent. J Asian Nat Prod Res 2005; 1-20. [http://dx.doi.org/10.1080/10286020.2025.2522396]

[36] Ferrer I, Rovira MB, Guerra MLS, Rey MJ, Costa-Jussá F. Neuropathology and pathogenesis of encephalitis following amyloid-beta immunization in Alzheimer's disease. Brain Pathol 2004; 14(1): 11-20. [http://dx.doi.org/10.1111/j.1750-3639.2004.tb00493.x] [PMID: 14997933]

[37] Vellas B, Black R, Thal L, *et al.* Long-term follow-up of patients immunized with AN1792: reduced functional decline in antibody responders. Curr Alzheimer Res 2009; 6(2): 144-51. [http://dx.doi.org/10.2174/156720509787602852] [PMID: 19355849]

[38] Winblad B, Andreasen N, Minthon L, *et al.* Safety, tolerability, and antibody response of active Aβ immunotherapy with CAD106 in patients with Alzheimer's disease: randomised, double-blind, placebo-controlled, first-in-human study. Lancet Neurol 2012; 11(7): 597-604. [http://dx.doi.org/10.1016/S1474-4422(12)70140-0] [PMID: 22677258]

[39] Wang CY, Wang PN, Chiu MJ, *et al.* UB-311, a novel UBITh ® amyloid β peptide vaccine for mild Alzheimer's disease. Alzheimers Dement (N Y) 2017; 3(2): 262-72.

[http://dx.doi.org/10.1016/j.trci.2017.03.005] [PMID: 29067332]

[40] Arndt JW, Qian F, Smith BA, *et al.* Structural and kinetic basis for the selectivity of aducanumab for aggregated forms of amyloid-β. Sci Rep 2018; 8(1): 6412.
[http://dx.doi.org/10.1038/s41598-018-24501-0] [PMID: 29686315]

[41] Sevigny J, Chiao P, Bussière T, *et al.* The antibody aducanumab reduces Aβ plaques in Alzheimer's disease. Nature 2016; 537(7618): 50-6.
[http://dx.doi.org/10.1038/nature19323] [PMID: 27582220]

[42] Selkoe DJ. Alzheimer disease and aducanumab: adjusting our approach. Nat Rev Neurol 2019; 15(7): 365-6.
[http://dx.doi.org/10.1038/s41582-019-0205-1] [PMID: 31138932]

[43] Haeberlein SB, von Hehn C, Tian Y, *et al.* Emerge and Engage topline results: Phase 3 studies of aducanumab in early Alzheimer's disease. Alzheimers Dement 2020; 16(S9): e047259.
[http://dx.doi.org/10.1002/alz.047259]

[44] Alexander GC, Emerson S, Kesselheim AS. Evaluation of aducanumab for Alzheimer disease: scientific evidence and regulatory review involving efficacy, safety, and futility. JAMA 2021; 325(17): 1717-8.
[http://dx.doi.org/10.1001/jama.2021.3854] [PMID: 33783469]

[45] Mintun MA, Lo AC, Duggan Evans C, *et al.* Donanemab in early Alzheimer's disease. N Engl J Med 2021; 384(18): 1691-704.
[http://dx.doi.org/10.1056/NEJMoa2100708] [PMID: 33720637]

[46] Swanson CJ, Zhang Y, Dhadda S, *et al.* A randomized, double-blind, phase 2b proof-of-concept clinical trial in early Alzheimer's disease with lecanemab, an anti-Aβ protofibril antibody. Alzheimers Res Ther 2021; 13(1): 80.
[http://dx.doi.org/10.1186/s13195-021-00813-8] [PMID: 33865446]

[48] Doody RS, Thomas RG, Farlow M, *et al.* Phase 3 trials of solanezumab for mild-to-moderate Alzheimer's disease. N Engl J Med 2014; 370(4): 311-21.
[http://dx.doi.org/10.1056/NEJMoa1312889] [PMID: 24450890]

[49] Salloway S, Farlow M, McDade E, *et al.* A trial of gantenerumab or solanezumab in dominantly inherited Alzheimer's disease. Nat Med 2021; 27(7): 1187-96.
[http://dx.doi.org/10.1038/s41591-021-01369-8] [PMID: 34155411]

[50] Adolfsson O, Pihlgren M, Toni N, *et al.* An effector-reduced anti-β-amyloid (Aβ) antibody with unique aβ binding properties promotes neuroprotection and glial engulfment of Aβ. J Neurosci 2012; 32(28): 9677-89.
[http://dx.doi.org/10.1523/JNEUROSCI.4742-11.2012] [PMID: 22787053]

[51] Bohrmann B, Baumann K, Benz J, *et al.* Gantenerumab: a novel human anti-Aβ antibody demonstrates sustained cerebral amyloid-β binding and elicits cell-mediated removal of human amyloid-β. J Alzheimers Dis 2012; 28(1): 49-69.
[http://dx.doi.org/10.3233/JAD-2011-110977] [PMID: 21955818]

[52] Bateman RJ, Cummings J, Schobel S, *et al.* Gantenerumab: an anti-amyloid monoclonal antibody with potential disease-modifying effects in early Alzheimer's disease. Alzheimers Res Ther 2022; 14(1): 178.
[http://dx.doi.org/10.1186/s13195-022-01110-8]

[53] Klein G, Delmar P, Voyle N, *et al.* Gantenerumab reduces amyloid-β plaques in patients with prodromal to moderate Alzheimer's disease: a PET substudy interim analysis. Alzheimers Res Ther 2019; 11(1): 101.
[http://dx.doi.org/10.1186/s13195-019-0559-z] [PMID: 31831056]

[54] Ait-Bouziad N, Chiki A, Limorenko G, Xiao S, Eliezer D, Lashuel HA. Phosphorylation of the overlooked tyrosine 310 regulates the structure, aggregation, and microtubule- and lipid-binding

properties of Tau. J Biol Chem 2020; 295(23): 7905-22.
[http://dx.doi.org/10.1074/jbc.RA119.012517] [PMID: 32341125]

[55] Vaz M, Silvestre S. Alzheimer's disease: Recent treatment strategies. Eur J Pharmacol 2020; 887: 173554.
[http://dx.doi.org/10.1016/j.ejphar.2020.173554] [PMID: 32941929]

[56] Hawkes CA, Carare RO, Weller RO. Amyloid and tau in the brain in sporadic Alzheimer's disease: defining the chicken and the egg. Acta Neuropathol 2014; 127(4): 617-8.
[http://dx.doi.org/10.1007/s00401-014-1243-9] [PMID: 24452628]

[57] Crary JF, Trojanowski JQ, Schneider JA, *et al.* Primary age-related tauopathy (PART): a common pathology associated with human aging. Acta Neuropathol 2014; 128(6): 755-66.
[http://dx.doi.org/10.1007/s00401-014-1349-0] [PMID: 25348064]

[58] Asuni AA, Boutajangout A, Quartermain D, Sigurdsson EM. Immunotherapy targeting pathological tau conformers in a tangle mouse model reduces brain pathology with associated functional improvements. J Neurosci 2007; 27(34): 9115-29.
[http://dx.doi.org/10.1523/JNEUROSCI.2361-07.2007] [PMID: 17715348]

[59] ClinicalTrials.gov. 24 Months Safety and Efficacy Study of AADvac1 in Patients With Mild Alzheimer's Disease (ADAMANT) [Internet]. Bethesda (MD): National Library of Medicine (US); 2015 Oct 13. Identifier NCT02579252.

[60] Kontsekova E, Zilka N, Kovacech B, Novak P, Novak M. First-in-man tau vaccine targeting structural determinants essential for pathological tau–tau interaction reduces tau oligomerisation and neurofibrillary degeneration in an Alzheimer's disease model. Alzheimers Res Ther 2014; 6(4): 44.
[http://dx.doi.org/10.1186/alzrt278] [PMID: 25478017]

[61] Congdon EE, Sigurdsson EM. Tau-targeting therapies for Alzheimer disease. Nat Rev Neurol 2018; 14(7): 399-415.
[http://dx.doi.org/10.1038/s41582-018-0013-z] [PMID: 29895964]

[63] Novak P, Schmidt R, Kontsekova E, *et al.* FUNDAMANT: an interventional 72-week phase 1 follow-up study of AADvac1, an active immunotherapy against tau protein pathology in Alzheimer's disease. Alzheimers Res Ther 2018; 10(1): 108.
[http://dx.doi.org/10.1186/s13195-018-0436-1] [PMID: 30355322]

[64] Bittar A, Bhatt N, Kayed R. Advances and considerations in AD tau-targeted immunotherapy. Neurobiol Dis 2020; 134: 104707.
[http://dx.doi.org/10.1016/j.nbd.2019.104707] [PMID: 31841678]

[65] Panza F, Solfrizzi V, Seripa D, *et al.* Tau-based therapeutics for Alzheimer's disease: active and passive immunotherapy. Immunotherapy 2016; 8(9): 1119-34.
[http://dx.doi.org/10.2217/imt-2016-0019] [PMID: 27485083]

[66] Novak P, Kovacech B, Katina S, *et al.* ADAMANT: a placebo-controlled randomized phase 2 study of AADvac1, an active immunotherapy against pathological tau in Alzheimer's disease. Nat Aging 2021; 1(6): 521-34.
[http://dx.doi.org/10.1038/s43587-021-00070-2] [PMID: 37117834]

[67] ClinicalTrials.gov. A Study to Evaluate the Safety, Tolerability and Immunogenicity of Tau Targeted Vaccines in Participants With Early Alzheimer's Disease [Internet]. Bethesda (MD): National Library of Medicine (US); 2020 Jun 25. Identifier NCT04445831.

[68] Olsson B, Lautner R, Andreasson U, *et al.* CSF and blood biomarkers for the diagnosis of Alzheimer's disease: a systematic review and meta-analysis. Lancet Neurol 2016; 15(7): 673-84.
[http://dx.doi.org/10.1016/S1474-4422(16)00070-3] [PMID: 27068280]

[69] ClinicalTrials.gov. A Study of Semorinemab in Patients With Moderate Alzheimer's Disease (LAURIET) [Internet]. Bethesda (MD): National Library of Medicine (US); 2019 Feb 4. Identifier NCT03828747.

[70] Mullard A. Failure of first anti-tau antibody in Alzheimer disease highlights risks of history repeating. Nat Rev Drug Discov 2021; 20(1): 3-5.
[http://dx.doi.org/10.1038/d41573-020-00217-7] [PMID: 33303932]

[71] ClinicalTrials.gov. Single-Ascending-Dose Study of BIIB076 in Healthy Volunteers and Participants With Alzheimer's Disease [Internet]. Bethesda (MD): National Library of Medicine (US); 2017 Feb 21. Identifier NCT03056729.

[72] Dam T, Boxer AL, Golbe LI, *et al.* Safety and efficacy of anti-tau monoclonal antibody gosuranemab in progressive supranuclear palsy: a phase 2, randomized, placebo-controlled trial. Nat Med 2021; 27(8): 1451-7.
[http://dx.doi.org/10.1038/s41591-021-01455-x] [PMID: 34385707]

[73] Kim B, Mikytuck B, Suh E, *et al.* Tau immunotherapy is associated with glial responses in FTLD-tau. Acta Neuropathol 2021; 142(2): 243-57.
[http://dx.doi.org/10.1007/s00401-021-02318-y] [PMID: 33950293]

[74] ClinicalTrials.gov. Phase 2 Study of BIIB092 in Participants With Early Alzheimer's Disease (TANGO) [Internet]. Bethesda (MD): National Library of Medicine (US); 2017 Nov 24. Identifier NCT03352557.

[75] ClinicalTrials.gov. A Study of LY3303560 in Healthy Participants and Participants With Alzheimer's Disease (AD) [Internet]. Bethesda (MD): National Library of Medicine (US); 2016 Apr 28. Identifier NCT02754830.

[76] ClinicalTrials.gov. A Study of LY3303560 in Participants With Mild Cognitive Impairment or Alzheimer's Disease (TRAILBLAZER-ALZ) [Internet]. Bethesda (MD): National Library of Medicine (US); 2016 Apr 28. Identifier NCT02754830.

[77] ClinicalTrials.gov. A Study of LY3303560 in Participants With Early Symptomatic Alzheimer's Disease (TRAILBLAZER-ALZ 2) [Internet]. Bethesda (MD): National Library of Medicine (US); 2018 May 8. Identifier NCT03518073.

[78] West T, Hu Y, Verghese PB, *et al.* Preclinical and clinical development of ABBV-8E12, a humanized anti-Tau antibody, for treatment of Alzheimer's disease and other Tauopathies. J Prev Alzheimers Dis 2017; 4(4): 236-41.
[PMID: 29181488]

[79] ClinicalTrials.gov. An Extension Study of ABBV-8E12 in Early Alzheimer's Disease [Internet]. Bethesda (MD): National Library of Medicine (US); 2018 Oct 19. Identifier NCT03712787.

[80] Srivastava N, Upadhyay P, Srivastava A, Das S, Mishra S. Nanostructured drug carriers for nose-t--brain drug delivery in the treatment of neurological disorders. Apple Academic Press 2025; pp. 473-91.

[81] Ransohoff RM. How neuroinflammation contributes to neurodegeneration. Science 2016; 353(6301): 777-83.
[http://dx.doi.org/10.1126/science.aag2590] [PMID: 27540165]

[82] Cuello AC. Early and late CNS inflammation in Alzheimer's disease: two extremes of a continuum? Trends Pharmacol Sci 2017; 38(11): 956-66.
[http://dx.doi.org/10.1016/j.tips.2017.07.005] [PMID: 28867259]

[83] Pascoal TA, Benedet AL, Ashton NJ, *et al.* Microglial activation and tau propagate jointly across Braak stages. Nat Med 2021; 27(9): 1592-9.
[http://dx.doi.org/10.1038/s41591-021-01456-w] [PMID: 34446931]

[84] Wang S, Mustafa M, Yuede CM, *et al.* Anti-human TREM2 induces microglia proliferation and reduces pathology in an Alzheimer's disease model. J Exp Med 2020; 217(9): e20200785.
[http://dx.doi.org/10.1084/jem.20200785] [PMID: 32579671]

[85] Ewers M, Franzmeier N, Suárez-Calvet M, *et al.* Increased soluble TREM2 in cerebrospinal fluid is

associated with reduced cognitive and clinical decline in Alzheimer's disease. Sci Transl Med 2019; 11(507): eaav6221.
[http://dx.doi.org/10.1126/scitranslmed.aav6221] [PMID: 31462511]

[86] ClinicalTrials.gov. A Phase 2 Study to Evaluate Efficacy and Safety of AL002 in Participants With Early Alzheimer's Disease (INVOKE-2) [Internet]. Bethesda (MD): National Library of Medicine (US); 2020 Oct 19. Identifier NCT04592874.

[87] Gate D, Saligrama N, Leventhal O, *et al.* Clonally expanded CD8 T cells patrol the cerebrospinal fluid in Alzheimer's disease. Nature 2020; 577(7790): 399-404.
[http://dx.doi.org/10.1038/s41586-019-1895-7] [PMID: 31915375]

[88] Schindowski K, Eckert A, Peters J, *et al.* Increased T-cell reactivity and elevated levels of CD8+ memory T-cells in Alzheimer's disease-patients and T-cell hyporeactivity in an Alzheimer's disease-mouse model: implications for immunotherapy. Neuromolecular Med 2007; 9(4): 340-54.
[http://dx.doi.org/10.1007/s12017-007-8015-9] [PMID: 17963048]

[89] Cattaneo A, Cattane N, Galluzzi S, *et al.* Association of brain amyloidosis with pro-inflammatory gut bacterial taxa and peripheral inflammation markers in cognitively impaired elderly. Neurobiol Aging 2017; 49: 60-8.
[http://dx.doi.org/10.1016/j.neurobiolaging.2016.08.019] [PMID: 27776263]

[90] Vogt NM, Kerby RL, Dill-McFarland KA, *et al.* Gut microbiome alterations in Alzheimer's disease. Sci Rep 2017; 7(1): 13537.
[http://dx.doi.org/10.1038/s41598-017-13601-y] [PMID: 29051531]

[91] Minter MR, Zhang C, Leone V, *et al.* Antibiotic-induced perturbations in gut microbial diversity influences neuro-inflammation and amyloidosis in a murine model of Alzheimer's disease. Sci Rep 2016; 6(1): 30028.
[http://dx.doi.org/10.1038/srep30028] [PMID: 27443609]

[92] Zhuang ZQ, Shen LL, Li WW, *et al.* Gut microbiotais altered in patients with Alzheimer's disease. J Alzheimers Dis 2018; 63(4): 1337-46.
[http://dx.doi.org/10.3233/JAD-180176] [PMID: 29758946]

[93] Syed YY. Sodium oligomannate: first approval. Drugs 2020; 80(4): 441-4.
[http://dx.doi.org/10.1007/s40265-020-01268-1] [PMID: 32020555]

[94] Wang X, Sun G, Feng T, *et al.* Sodium oligomannate therapeutically remodels gut microbiota and suppresses gut bacterial amino acids-shaped neuroinflammation to inhibit Alzheimer's disease progression. Cell Res 2019; 29(10): 787-803.
[http://dx.doi.org/10.1038/s41422-019-0216-x] [PMID: 31488882]

[95] Shukla S, Tiwari S, Bhattacharya R, *et al.* Nanotechnology-Based Approaches for Nose-to-Brain Drug Delivery in Neurodegenerative Diseases. Lett Drug Des Discov 2024; 21(11): 1913-1921.
[http://dx.doi.org/10.2174/1570180820666230622120759]

[96] ClinicalTrials.gov. A Phase 2 Study to Evaluate Efficacy and Safety of AL002 in Participants With Early Alzheimer's Disease (INVOKE-2) [Internet]. Bethesda (MD): National Library of Medicine (US); 2020 Oct 19 [cited 2026 Feb 4]. Identifier NCT04592874.

[97] Walsh DM, Selkoe DJ. Aβ Oligomers – a decade of discovery. J Neurochem 2007; 101(5): 1172-84.
[http://dx.doi.org/10.1111/j.1471-4159.2006.04426.x] [PMID: 17286590]

[98] Kirkitadze MD, Bitan G, Teplow DB. Paradigm shifts in Alzheimer's disease and other neurodegenerative disorders: The emerging role of oligomeric assemblies. J Neurosci Res 2002; 69(5): 567-77.
[http://dx.doi.org/10.1002/jnr.10328] [PMID: 12210822]

[99] McLean CA, Cherny RA, Fraser FW, *et al.* Soluble pool of Aβ amyloid as a determinant of severity of neurodegeneration in Alzheimer's disease. Ann Neurol 1999; 46(6): 860-6.
[PMID: 10589538]

[100] Portelius E, Bogdanovic N, Gustavsson MK, *et al.* Mass spectrometric characterization of brain amyloid beta isoform signatures in familial and sporadic Alzheimer's disease. Acta Neuropathol 2010; 120(2): 185-93.
[http://dx.doi.org/10.1007/s00401-010-0690-1] [PMID: 20419305]

[101] Bakrania P, Hall G, Bouter Y, *et al.* Discovery of a novel pseudo β-hairpin structure of N-truncated amyloid-β for use as a vaccine against Alzheimer's disease. Mol Psychiatry 2022; 27: 840–848.
[http://dx.doi.org/10.1038/s41380-021-01385-7] [PMID: 34776512]

[102] Butchart J, Brook L, Hopkins V, *et al.* Etanercept in Alzheimer disease: a randomized, placebo-controlled, double-blind, phase 2 trial 2015; 84(21): 2161-2168.
[http://dx.doi.org/10.1212/WNL.0000000000001617]

[103] Schnabel J. Vaccines: Chasing the dream. Nature 2011; 475(7355): S18-9.
[http://dx.doi.org/10.1038/475S18a] [PMID: 21760578]

[104] Szabo P, Mujalli DM, Rotondi ML, *et al.* Measurement of anti-beta amyloid antibodies in human blood. J Neuroimmunol 2010; 227(1-2): 167-74.
[http://dx.doi.org/10.1016/j.jneuroim.2010.06.010] [PMID: 20638733]

[105] Relkin NR, Szabo P, Adamiak B, *et al.* 18-Month study of intravenous immunoglobulin for treatment of mild Alzheimer disease. Neurobiol Aging 2009; 30(11): 1728-36.
[http://dx.doi.org/10.1016/j.neurobiolaging.2007.12.021] [PMID: 18294736]

[106] Alves RPS, Yang MJ, Batista MT, Ferreira LCS. Alzheimer's disease: is a vaccine possible? Braz J Med Biol Res 2014; 47(6): 438-44.
[http://dx.doi.org/10.1590/1414-431X20143434] [PMID: 24878604]

[107] Usman MB, Bhardwaj S, Roychoudhury S, *et al.* Immunotherapy for Alzheimer's Disease: Current Scenario and Future Perspectives. J Prev Alzheimers Dis 2021; 8(4): 534-51.
[http://dx.doi.org/10.14283/jpad.2021.52] [PMID: 34585229]

[108] Loeffler DA. AMBAR, an Encouraging Alzheimer's Trial That Raises Questions. Front Neurol 2020; 11: 459.
[http://dx.doi.org/10.3389/fneur.2020.00459] [PMID: 32547478]

[109] Song C, Shi J, Zhang P, *et al.* Immunotherapy for Alzheimer's disease: targeting β-amyloid and beyond. Transl Neurodegener 2022; 11(1): 18.
[http://dx.doi.org/10.1186/s40035-022-00292-3] [PMID: 35300725]

CHAPTER 8

Gene Therapy and RNA-Based Innovations for Management of Alziemer's

Juhi Tiwari[1], **Piyush Anand**[1], **Deepak Kumar**[1] and **Shashi Kant Singh**[1,*]

[1] *Faculty of Pharmaceutical Sciences, Mahayogi Gorakhnath University, Gorakhpur, Uttar Pradesh 273007, India*

Abstract: Gene therapy is the process of transferring genetic material to treat a disease or, at the very least, to enhance a patient's clinical condition. Gene therapy can be implemented by transforming viruses into genetic carriers that deliver the desired gene to the target cells. Depending on the kind of genome, these vectors can be classified as either DNA-based or RNA-based viral vectors. Gene therapy holds promise for treating illnesses that traditional medicine cannot address. To apply it, a patient's cells must receive one or more nucleic acids or a faulty gene. Clinical gene therapy has advanced significantly during the past ten years. One can list several noteworthy achievements, including the development of medicines for diabetes, Alzheimer's, Parkinson's, cystic fibrosis, and several types of cancer. Long-term studies on the therapy of Alzheimer's disease have led to the development of various drugs, such as monoclonal antibodies to Aβ aggregation, cholinesterase inhibitors, and inhibitors of tau aggregation. Researchers continued to work on the disease even though the drugs did not stop its advancement, and as a result, gene therapy, a recently developed, state-of-the-art technique for delivering genes to specific locations where they can express the desired functionalities, is being introduced. For AD, gene therapy is a promising disease-modifying treatment. "Protein-coding and noncoding RNAs make up between 70 and 90 percent of the human genome. These RNAs are the primary determinants of biological variation in cells and populations. Among the different kinds of nucleic acids, RNA is more adaptable than DNA because of its single-stranded structure, direct protein-encoding capability, and high degree of modification for specific regulatory and therapeutic uses. Even with its bright future in biomedicine, RNA-based therapy still has several obstacles to overcome. Notably, delivering RNA effectively and precisely while reducing immunological reactions is one of the biggest technical challenges. Viral vectors, Virus-Like Particles (VLPs), lipid nanoparticles (LNPs), and Extracellular Vesicles (EVs) are some of the methods that have been created for delivering RNA.

Keywords: Biopolymers, Encoded molecules, Germline, Immune response, Innovative therapy, Lipid nanoparticles, RNA delivery systems, Somatic cells, Transcriptions, Viral vectors.

* **Corresponding author Shashi Kant Singh:** Faculty of Pharmaceutical Sciences, Mahayogi Gorakhnath University, Gorakhpur, Uttar Pradesh 273007, India; E-mail: principal.fps@mgug.ac.in

Prashant Kumar Dhakad, Raghav Mishra, Shweta Sharma, Sudhanshu Mishra & Nishant Gaur (Eds.)

INTRODUCTION

Through the continuous calls for eventual treatments targeting diseases at the molecular level, the gene therapy field has managed to demonstrate considerable relative successes. It is now a generally accepted notion that these early successes are indeed likely to pave the path for a more constant drive for better prospects through not only appreciation and comprehension of the various obstacles standing in the way of efficient in vivo gene delivery but also by the judicious development of ad hoc technologies [1, 2]. Furthermore, the therapeutic work in the field of gene therapy has triggered additional spin-offs, like the development of accurate in vivo imaging methods. Indeed, a significant number of innovations capitalizing on gene expression are envisaged [3, 4].

In the rapidly evolving field of antisense and gene repair technology, the gap that once existed has now significantly narrowed between the numerous in vitro successes and the actual therapeutic applications that employ either cell in vivo or whole living organisms. Recent advancements in RNA interference have further accelerated the research aimed at bespoke modulation of gene function, uncovering possibilities for targeted interventions. The prospects in these areas appear to be quite encouraging and promising [5]. Currently, the new generations of RNA interference inhibitors are still in their early stages of development, but as these innovative technologies emerge, the full extent of their therapeutic capabilities will likely expand considerably [6]. The gene delivery aspect of this technology might also reflect similar developments in its unique manner, given the intense level of dynamism that characterizes this field. Indeed, considering that we have merely begun to scratch the surface of understanding how much more efficient, selective, targeted, or localized gene delivery could become, there are numerous encouraging and compelling hypotheses in progress. These ideas are simply awaiting validation, and their potential impact on medical science may be profound as we continue to unveil new methodologies and technologies in this exciting domain of research [7].

OVERVIEW OF GENE THERAPY

Gene therapy refers to the delivery and stable introduction of new genetic information into a cell to obtain a therapeutic benefit. Therapy based on gene manipulation has the potential to radically change how we treat life-threatening diseases [8, 9]. Gene therapy focused on delivering a functioning copy of a defective gene to correct a monogenic disorder, by gene augmentation, is now well into the ongoing clinical phase for a variety of genetic disorders. In recent years, knowledge of disease-causing mutations has increased exponentially, offering the possibility of correlating mutations to certain protein functions that aid in the definition of the path of disease. For some of the many genetic diseases for which the gene and mutation cause is known, there is already a defined sequence of events of disease progression, which could provide a predictive roadmap for therapeutic intervention [10, 11].

One of the properties of gene therapy that makes it attractive is the potential for a single administration of the therapeutic gene and its product. The ideal goal of gene therapy is to deliver a therapeutic payload without the requirement for modifying the host's genome, ensuring the long-term presence of the correctable defect protein. For monogenic diseases, this would parallel the most advantageous feature of protein replacement therapy [12]. The actuality of current vector technology, however, is that the gene is not specifically targeted to the appropriate tissue to maximize effectiveness; the gene is often not integrated in an expression-efficient site, and long-term expression is largely not achieved. In an attempt to develop a positive profile record of gene therapy, somatic cell gene therapy trials have begun, and results are being reported. In another approach, strategies attempting to obtain a particular pharmacological effect rather than full correction of the disease by gene therapy are being pursued by isolating or inhibiting RNA transcription of relevant gene sequences. Optimal results in systemic genetic manipulation in humans, however, remain objectives rather than accomplishments. Fig. (**1**) depicts gene therapy-based methods and RNA-based innovations used in the diagnosis and treatment [13].

EVOLUTION OF RNA-BASED INNOVATIONS

Recently, nature has been exploited to achieve highly proficient tools capable of effectively interfacing with a variety of biological systems. The first part of this overview outlines the evolutionary and progressive increases in knowledge of a very promising group of biotechnological tools: RNA-based molecules. These naturally occurring molecules interact with proteins to carry out enzymatic as well as regulatory biological functions, providing a conceptual framework for the development of RNAs as biotechnological tools. The second part of this overview summarizes some pioneering applications of RNA technologies in biotechnological and biomedical fields [14]. Genomes are made up of DNA and RNA, and the study of the regulation of genome expression, as well as the functional properties of regulatory RNA-based molecules, has been pivotal in the development of efficient, easy, and flexible RNA-based tools [15]. These tools provide a powerful toolkit of biotechnological instruments that have proved to be optimized and straightforward, therefore having a wide impact in investigations related to functional genomics, systems biology, and evolution. Furthermore, they hold huge potential in providing innovative and personalized solutions to issues related to human health, through the development of improved RNA-based vaccines, RNA-based tools, and therapeutic agents for gene therapy, and RNA-based strategies suitable for studying and inactivating the expression of allelic variants with no haploinductive phenotype for target human haplolethal genes [16, 17].

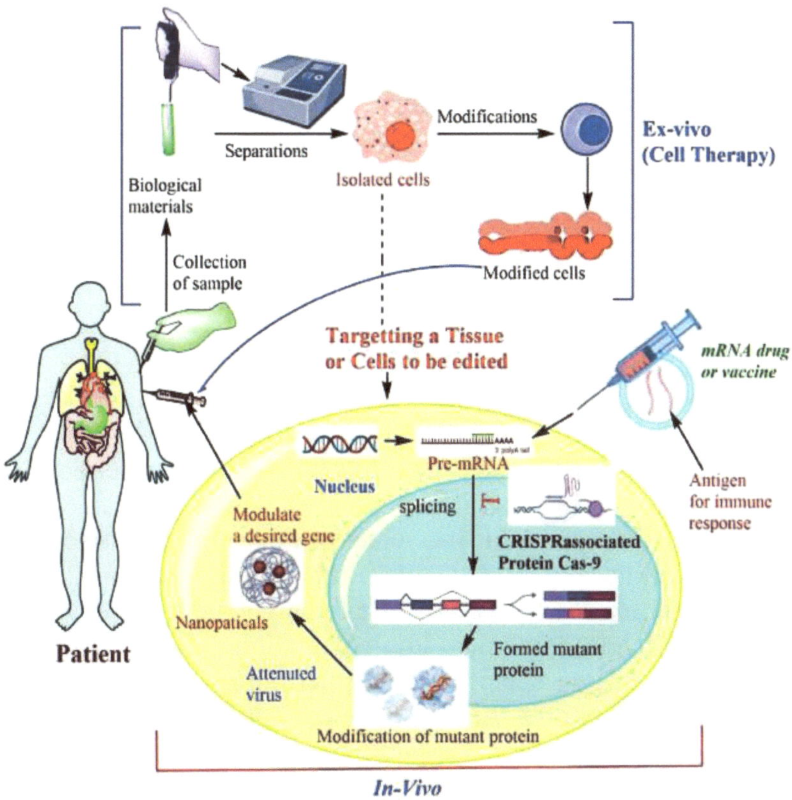

Fig. (1). Depicts gene therapy and RNA-based innovations used in the diagnosis and treatment.

Its Importance and Applications

The scientific and technological potential of gene therapy and RNA-based innovations has the potential to transform the treatment of human disease. The current trends and future prospects in this exciting field are changing our understanding of human biology and medicine. Research in gene therapy is focused on the development of therapeutic solutions for a large number of inherited or acquired diseases. Gene therapy has extended the range of therapeutic tools for addressing genetic diseases. The advances in biotechnological applications and the highly developed molecular and genetic tools available have driven innovation within the field of gene therapy. This fast evolution has provided new perspectives for gene therapy. The design and development of these new RNA technology molecules can offer new therapeutic interventions in areas where autologous cell or gene-based strategies are not suitable [18]. The field of RNA research has experienced great progress in the development of new RNA molecules. The design and development of these new RNA technology molecules can offer new therapeutic interventions in various biomedically compromised areas such as cancer, cardiovascular disease, infectious diseases, and neurodegenerative diseases. Pharmaceutical organizations are making large investments in this deep research field. Futuristic approaches such as the establishment of an RNA code, the emergence of RNA nanotechnology, and the construction of new therapies using modified RNA molecules emphasize the great potential of these genetic molecules. This evolutionary RNA research has led to a new era filled with personalized therapeutic interventions that were not possible before. RNA is the determinant of the functionality of the encoded molecules responsible for cell differentiation [19].

Fundamentals of Gene Therapy

Gene therapy allows selective, inheritable changes to the genome. It has the potential to cure otherwise incurable genetic diseases and can supplement or replace small-molecule drugs. This has made gene therapy an attractive option for treating such conditions [20, 21]. Cancer, hemophilia, and progressive retinal disease are just a few examples of diseases targeted by gene therapy. Despite great potential, many challenges to clinical implementation exist, including therapeutic vector delivery and editing specificity. Furthermore, ethical debates regarding germline editing affect not only the academic community but large sections of the world's population. This chapter presents an overview of current gene therapy clinical trials. The findings are synthesized to provide a current snapshot of gene therapy applications [22, 23].

Although gene therapy is the most intuitive method of gene transfer, the developments that made the method applicable and safe required decades of research and refinement. Gene therapy can be defined as the use of genetic material to modify or manipulate a person's cells to treat or cure a disease. Genetic material is introduced into a person's cells to replace or inactivate, deregulate, or amplify the functionality of a defective or wild-type gene [24, 25]. Alternatively, therapeutic genetic material is transferred to cells to alter cellular products or processes according to the principles of gene therapy. Gene therapy uses executing nucleic acids or genes that interrupt and interfere with nucleic acids. Successful gene therapy must always achieve the following key step: the expression of functional, active protein or ribozyme at therapeutically beneficial levels. As with the birth of molecular biology, this opening of a "brave new world" of gene therapy is based on the discoveries of a very wide variety of scientists in different fields [26, 27].

Mechanism of Gene Therapy

The goal of gene therapy is to correct the disease at the genetic level by either correcting the disease-causing gene at the mutation or adding therapeutic genes into patients' cells. Gene therapy is the use of DNA as a final product [28, 29]. This method is under intense investigation in numerous human afflictions. The discovered and understood genomic and genetic phenomena have reached a level that has made practical gene therapy methodology possible. The nature of genetic diseases, including the molecular defects that they result in, represents an optimal candidate for practical gene therapy application when the technologies are ready [30, 31].

Examples of the use of gene therapy technologies are: to repair or replace defective recessive housekeeping genes; to replace defective dominant negative genes; to add genes carrying therapeutic proteins to be expressed and naturally secreted at therapeutic levels; to add genes making therapeutic proteins; to restore a partially defective gene locus because of dosage effects; to enhance the level of expression of normal alleles of a gene; to change the expression of a gene under the inducer control of an administered substance; powerful gene therapy to manipulate the target to destroy abnormal cells; and to sensitize target cells for destruction by radiotherapy, chemotherapy, or immunotherapy agents [32].

Types of Gene Therapy

The development of gene transfer vectors based on adeno-associated viruses, nonpathogenic human viruses not associated with any human disease, and not naturally integrating, makes it possible to work on several types of monogenic diseases. The proof of concept comes from the approach of gene therapy in patients affected by different forms of congenital retinal dystrophy. The success of these tests opens new scenarios in the treatment of diseases linked to a single gene [33]. Gene therapy can correct the defect directly in the retina, avoiding the use of general gene therapy, so the genetic transfection can be better managed without concerning toxic effects due to the presence of viral capsids in sensitive tissues, *i.e.,* the liver, lungs, spleen, developing cells, and immune responses in general [34]. As these innovative therapies are becoming more and more care and intervention targets for the health of the future, the formation of an "innovative" doctor and the functioning of a new relationship between the pharmaceutical industry are significantly moving to the treatment, ideological planning, and management of the patient's whole life [35].

Scientific research on the potential of gene therapy in heteropathies with a simple transmission type can access resources, particularly in the case of orphan diseases, which are recognized by the scientific community as highly promising. This type of therapy works at the genetic level, correcting the specific genetic disease caused by the mutation. Gene therapy acts on the origin of the disease, unlike the previous approach of cure, and on the phenotype of the pathology, without offering a cure for the mutation that gives rise to the defective gene. This includes the introduction of a new gene, gene repair, excision of a disease gene, or the replacement of missing or non-functioning genes [35]. The identification of the genetic defect in many monogenic pathologies allows for the knowledge of the genetic mutation. The identification of the genetic mutation informs scientists of the person's genomic information to use to treat the disease. The possibility of replacing the sick gene with its counterparts by means of gene therapy opens up a way for the treated child to be able to spend a life without symptoms and without therapy or, in any case, as proposed by gene therapy clinical trials, less frequent and lighter treatments compared to current therapies [36].

Delivery Systems for Gene Therapy

Gene therapy involves the introduction of a gene and/or nucleic acid into an individual's cells for the purpose of repairing damaged or defective genes, and consequently to cure inherited diseases where no effective treatments exist. With accomplishments in recombinant DNA technology and the understanding of molecular biology, gene therapy gained momentum. Therefore, gene therapy has become a better strategy for gene modification and cell therapy [37]. Due to the development of various viral and non-viral vectors, random integration of gene therapy vectors is considered an unpredictable hazard of gene therapy. Since the 1970s, significant progress has been made in the development of gene therapy. Currently, gene therapy has started showing interesting advances in different clinical applications, including viral and non-viral vectors [38]. Today, viral carriers have shown good advances in in vivo strategies for monogenic diseases. In addition to the strategy of repairing diseases by somatic gene therapy, the potential of gene therapy comes mainly from the possibility of correcting hereditary diseases by germ cell therapy, especially for diseases that do not have effective treatment by classical methods [39].

Given the complexity of the treatment of some diseases, such as cancer, neurological, monogenic, and infectious diseases, the use of gene therapy technology has become the best solution for the treatment of these diseases. For use in gene therapy, viral vectors have been modified in two major areas: 1. Low pathogenicity or increased safety by changing the cell specificity or the dose of the virus, immediately followed by administration with the purpose of reducing viral cytotoxicity and a possible immune response. 2. Complex and more sophisticated non-viral vectors are utilized. In addition, the use of non-viral vectors is also attractive for their ability to be analyzed for off-target gene expression, most notably teratogenic responses to exogenous DNA [40]. This is due to the absence of viral particles that may or may not express genes that cause mutations in the genome. However, the use of non-viral vectors can promote weak gene expression, lower injection efficiency, and, in general, the number of vectors that remain associated with the genome or per cell is often less than one, compared to a plasmid or viral vectors that couple gene therapy [41, 42].

PRINCIPLES OF GENE THERAPY

Over the past decade, important advances in the understanding of gene regulation, gene delivery vectors, and gene expression have allowed the therapeutic use of at least two types of gene products. 1) Tumors can be treated directly with the use of tumor suppressor genes, pro-apoptotic genes, or even transferase genes in order to sensitize tumor cells to cytotoxic agents. 2) The objectives are to correct genetic disorders and acquire specific properties in cells by adding exogenous genes, a family of diseases today defined by up to 1,200 clinical entities. Gene therapy has the potential for treating cancer, a field called cancer gene therapy. Here, the therapeutic objective is usually to kill tumor cells or to sensitize them to antitumor treatments, such as chemotherapy and radiotherapy [43, 44].

Gene-based treatment offers new prospects for monogenic, multifactorial, and acquired diseases. Two types of genes can be used for therapy. The use of foreign DNAs is studied for therapeutic vaccines. The problem posed by these vectors concerns the coating protein's nature and the protein coat's immunogenicity [45]. The size and organization of cargo DNA are also critical. At present, these vectors are under study for surface vaccination, either as adjuvant therapy or as vaccination per se. Several studies evaluate retrospective loading of neoplastic proteins and, for this reason, are connected to cancer gene therapy. However, the principle on which these vectors are based is different, and specific problems occur. This will not be discussed here [46, 47].

Definition and Types of Gene Therapy

Genetics, the basic science of heredity, has a conceptual basis that is as ancient as human civilization itself. The main genetic message is transmitted, with slight variations, from cell to cell, going through species generation to human generation. These almost perfect and permanent laws are dedicated to the specific function of each protein, according to cell specialization. Insert a corrected gene into a cell, and release the cell to behave according to the new rules [48, 49].

Introduction This conceptually simple postulate, realizable only in the last two decades of the 20th century, can be seen to coincide with the beginning of investigations and trials of gene therapy. Therefore, the base of gene therapy is the increasingly refined ability to manipulate genetic substances of living organisms. The term gene therapy refers to therapies involving the transfer of genetic material to cells in vivo or in vitro. The concept of gene therapy is as old as the discovery of DNA in the early 1950s. In 1968, a pioneer of molecular biology asked if genetic mechanisms could be used not only to understand and experimentally determine the essence of those mechanisms but also to influence every aspect of genetic information [50, 51].

Gene therapy can be carried out in two ways: first, through somatic cells, and second, through germline cells. Somatic cells: These are the cells that form every tissue of the adult organism. The therapeutic application of gene therapy to somatic cells is defined as somatic gene therapy.

Embryonic or germline cells: In these cells, spermatozoa and oocytes meet, fertilize, and originate a new organism. Since the DNA of these cells is incorporated into the genetic message of a new organism, its modification is passed to the offspring. Due to this fact, the inheritance of therapeutic genes has been criticized as putting the genetic security of future generations in jeopardy. This fact is not a medical issue, but an ethical and philosophical question. The germline gene transfer affects human genetic heritage, being likely to bring changes that are transmitted from generation to generation, so that the risks can extend to future generations. There are many social, legal, and ethical implications of germ transgenesis. As far as the human species is concerned, germline therapy requires the intervention and acceptance of future adults. In practice, the extension and direction of potential gene therapy are, to a certain extent, limited by these factors, which should not be evaluated only based on the possible technical benefits or possible damages of potential gene therapy technology. In both cases, the patient can be submitted to ex vivo or in vivo gene therapy [52, 53].

Delivery Methods (Viral and Non-Viral Vectors)

The success of gene therapy, like traditional gene models for genetic diseases, depends heavily on the successful transfer of therapeutically useful genes that must also be able to demonstrate robust and sustained endogenous gene expression in their relevant target tissues. As an advanced delivery strategy, non-viral methods such as the direct injection of nucleic acids into specific tissues or cells, electroporation, and naked DNA are used together with physical and biochemical techniques to disrupt the membrane architecture of cells for gene delivery. The purpose of this review is to describe past and current advances in molecular therapy for all gene types. To this end, its aim is to provide a global understanding of the specific principles of gene-targeted cells that are responsible for gene delivery [54, 55].

The core content of most gene therapy approaches is a nucleic acid that encodes a therapeutic protein of interest. These therapeutic agents can be categorized into three major classes: endogenous genes, recombinant genes, and random integration of the gene into the genetic material of the host cell, and can be manipulated through viral or non-viral vectors. To date, there are historical gene vectors that are widely used for in vivo gene delivery, including retrovirus, adenovirus, adeno-associated virus, lentivirus, vesicular stomatitis virus, and herpes simplex virus [56]. Despite their potential, one formidable challenge for the advanced clinical application of viral gene vectors is cellular immunity, which is responsible for widespread clinical and preclinical evidence in multiple design vectors. Notably, the development of non-viral vectors appears to overcome such an obvious drawback, which are the second most significant delivery systems, as evidenced by the inception of RNA-based therapeutic exosomes and the frequent claims for outright gene delivery. For artificial exosomes, the presence of modified proteins in the tail resulted in altered characteristics, leading to enhanced reporting, like the ceRNA properties showing strong microRNA activity after isolating RNA from exosomes. In the cell, the conversion of exosomes into a ceaseless flux of therapeutic proteins would advance the use of RNA in situ [57, 58].

Challenges in Gene Therapy for Neurodegenerative Diseases

Challenges in Gene Therapy for Neurodegenerative Diseases Gene therapy is an innovative form of molecular medicine that involves the insertion of genetic material into the cells to compensate for abnormal genes or to make an effective gene product. This powerful tool has been utilized in the treatment of various human diseases, including different neurodegenerative diseases, by various administration routes. Viral and non-viral vectors can transport genes into the host genomes of neurons, where they are functional and stable to arrest any disease progression. Depending on the disease model, gene therapy could block neurotoxic proteins, replace them with neuroprotective growth factors, or change the inflammatory response. Despite significant improvements in the tools utilized for gene delivery into the central nervous system, challenges remain, including the development of an innovative delivery system and improvements in viral design. With the increasing knowledge about genome manipulation, the emerging technology of clustered regularly interspaced short palindromic repeats has potentially significant applications in the gene therapy of neurodegenerative diseases, which facilitates the reversible and precise manipulation of genes without leaving any genetic footprints behind [59, 60].

Gene Therapy Approaches in Alzheimer's Disease

Nearly 110 years after Alois Alzheimer first described the clinical and neuropathological features of his original patient, aggressive experimental genetic treatment and the expected transfer of some gene therapy techniques to the clinic were published [61]. In the shadow of this research breakthrough, this review describes the gene therapy approaches for Alzheimer's disease based on the new insights. Different genome editing solutions specifically eliminate polymorphic proteins or specific inclusion bodies, disaggregate amyloid, correct axonal repair, or extend the life of the model by combining these effects. Several options have been verified for stereotactic delivery, and related viral and non-viral vectors are also briefly outlined. In this work, we summarize Alzheimer's disease-oriented applications of gene therapy based on new gene editing techniques, including the treatment of inherited or sporadic forms of the disease [62, 63].

Gene therapy is ready to focus on the gene or cellular biological background of Alzheimer's disease, not only the cognitive enhancement of animal models that already have the capabilities. Therapy has already reached several technological challenges with solutions described to modify a specific part of Alzheimer's disease or to build a combined treatment, including stereotactic administration. According to optimistic predictions, effective and exclusive therapeutic directions can be verified in a small group of patients in a few years, in collaboration with state-of-the-art imaging and protein labeling techniques. These expeditions have already started, offering an intensive and fascinating research agenda for the near future [63].

Targeting Amyloid Beta Pathway

AD brain is massively damaged by Aβ peptides, such as parenchymal cortical and mainly limbic deposits of amyloid plaques. There has been significant interest in alternative approaches to Aβ immunotherapy and in employing the potential benefits of gene therapy for immunization against Aβ peptides. Some successful studies have also used DNA vectors to direct the expression of an Aβ-specific antibody in a number of animal models to evaluate the ability of a gene therapy approach in reducing Aβ burden [64, 65]. For gene immunotherapy, we used a plasmid vector and formulated the final products to generate powers for direct administration in the animals. With these systems, we showed for the first time in mice that a single administration of a plasmid formulated with cationic lipids and using low doses of DNA was able to attenuate AD pathology. Also, the wild-type mice were used as a model target for the antibody. The results showed how difficult it is to generate antibodies in a mouse model against a different species, such as a transgenic. However, although it has effectively delayed amyloid accumulation and has long-lived Aβ1-40/Aβ1-42 (reducing over 36 weeks), the development progress of this strategy has been slow. In conclusion, immunization with the gene gun, the single administration of dosages of 1.0 or 2.5 μg/ear, gave a significant reduction of APP processing products, the antibody, the Aβ1-40/Aβ1-42 levels, and the amyloid load [66, 67].

Modulation of Tau Protein Aggregation

The importance of the inhibition of tau protein hyperphosphorylation and beta-amyloid oligomers in improving memory loss and neurodegeneration as a consequence of their aggregation is recognized. These issues are related to disturbances of physiological processes. After the successful results that gene therapy in modulating the transcription of the genes of such proteins demonstrated, expectations for refraining from these disorders are high. The issue lies in the small number of genes or proteins, the location of the modification, the route and time for its intervention, and the evolution of possible side effects. Its use has extended, however, in silencing mRNAs into DNA structure as technology continues to increase these disappointments [68, 69].

Additional obstacles in the treatment arise from the drawbacks of constituents of delivery systems and the impaired distribution of non-viral and viral vectors for transferring genes to nervous system cells. Techniques providing an elongated half-life for delivery and distribution of genes to restricted neural tissues could improve gene therapy for reclaimed neuronal genes. Even more critical, however, are the limitations in preventing apoptotic changes that occur within nerves before their biological death, resulting from hyperphosphorylation of tau and aggregation of beta-amyloid. Gene therapy also fails to be symptomatic and recover any more neural tissue if the patient's tissue is completely degenerated. Commercial gene therapy of the future may be effective if applied to the prevention of the initiation of the disease, whose cause would be inhibited [70].

Enhancing Neuroprotection and Neurogenesis

We provide an overview of current and future innovative therapeutic approaches for neurological disorders. We present gene therapy and RNA-based strategies, as well as co-option tools to target neuroprotective and neurogenic factors. This review offers perspectives on the main problems that need to be addressed to develop a more effective therapeutic approach. Finally, we discuss patient-derived cellular models for therapy development and testing. Crucial issues like the delivery and efficiency of reprogramming factors to neuronal metabolomes are discussed [71 - 73].

Neuronal loss occurs in neurodegenerative conditions, mainly due to age-related pathologies such as Multiple Sclerosis, Alzheimer's Disease, Parkinson's Disease, and traumatic spinal cord injury. Further acute neurological damage can occur because of stroke, encephalitis, or chronic pathologies like epilepsy [74, 75]. Therefore, molecular reverb into ISR controllers, BDNF, NPY, APP, and BCL2 proteins must be examined. Impressively, an increase in IP3 and CREB is related to ISR suppression, IRE1/XBP1 axis reactivity, and normal neuronal functionality, finally connecting the metabolic status with neuronal reprogramming towards neuroprotective and neurogenic factors [76, 77].

Anti-inflammatory and Immune-based Strategies

Arthritis is a multifactorial joint disease in which the communication between the synovial membrane and other joint tissues is altered, giving rise to a state of chronic inflammation, thus destroying cartilage, bone, or tendons. This inflammation is characterized by the synovial hyperplasia produced mainly due to the formation of blood vessels and the infiltration of fibroblast-like synoviocytes and immune cells. Many transcription factors, kinases, and other proteins regulate inflammatory and anabolic reactions at the synovium. These molecules are the effectors and reinforce both the activation of the immune system and the synthesis of the extracellular matrix. Gene therapy offers the possibility of specifically knocking down a gene in a particular cell and organ without causing significant side effects, which can happen after chronic use of small synthetic chemical inhibitors. The progress is in genetic or chemical interventions that regulate transcription factors, kinases, and other molecular pathways that regulate synovial activity without causing a relevant immunological response at an effective concentration. An alternative is blocking the production of certain secreted molecules, such as cytokines, chemokines, or angiogenic factors. As anti-inflammatory gene therapy strategies, nucleic acid-based technologies that encode signaling inhibitors derived from Toll-like receptor adaptor proteins can be used. The administration of immunosuppressors remains a relevant protocol for both the pre- and post-operative maintenance of the organ transplant, as well as chronic diseases such as autoimmune arthritis. With the intention of reducing the side effects associated with long-term use and poor patient adherence, pre-clinical research is working on the surgical technique and gene therapy modifications to induce tolerance to the transplanted organ. The aim is to suppress the immune response through action on the T and B cells, transiently or for a prolonged time [78, 79].

RNA-Based Innovations

RNA therapies are increasing in popularity due to the low likelihood of causing cancer versus DNA therapies, minimizing the risk of altering future generations with unwanted genetic changes, and reducing the risk of unwanted immunologic reactions. RNA therapies may also require less sophisticated delivery technology than their DNA-based therapeutic cousins [80].With many variants of RNA molecule targets, such as mRNA, microRNA, and antisense RNA, the options for new tools to treat human disease appear nearly limitless [81]. These possibilities are limited largely by the weaknesses present in current RNA delivery technologies. Additionally, some potential RNA molecular therapies could fail due to an inability to control or regulate the duration of their therapeutic effects on patients. To address this shortcoming, our team and others study a major regulatory mechanism of RNA function—RNA decay [82, 83].

RNA decay pathways affect the function of diverse gene products, including extracellular factors, receptors, transcription factors, nuclear-localized regulatory RNAs like microRNAs and long noncoding RNAs, and ultimately all RNAs before their effects can be implemented. We and others have demonstrated that RNA decay has potent translational and disease-modifying effects that can be exploited to control the effectiveness and expression levels of therapeutic factors encoded within RNA transcripts [84]. We have also identified protein factors preferentially needed for the control of human RNA factors that are specific yet distinct from those controlling the decay of their respective endogenous RNA transcripts [85]. We are exploring species-specific differences in the function of these protein factors to enable the use of RNA-based therapeutic interventions to treat lethal human brain disease caused by the accumulation of nonfunctional protein products, without creating the lethal side effects that are observed when model systems engineer the accumulation of the same RNA products.

Collectively, the data generated by these studies advance the promise of RNA molecular therapy by increasing understanding of the mechanisms that control the expression of both endogenous and therapeutic human RNA components and providing new and specific factors for the control of human RNA decay that do not elicit the deleterious side effects of factors targeting the same genes in a model system [86, 87].

Types of RNA Therapeutics

The different types of RNA-based therapeutics can be broadly classified based on their functions: replacement or editing of genetic sequences. Each of these approaches has a different modified form, route of administration, and mode of action [88]. Depending on cellular uptake, mode of action, and the nature of the RNA therapeutic, it may require conjugation to various moieties. With a few exceptions, an effective delivery vehicle is needed for most RNA therapeutics to protect from nucleases and allow the cargo to enter cells and tissues to be transcribed or translated to the target [89, 90].

The first class of RNA therapeutics developed is small-interfering RNAs, which perform the natural regulation of gene expression. Synthetic siRNAs are used as a gene knockdown technology by inhibiting the production of functional protein from the encoded mRNA in the location with sufficient complementarity. New infections mediate RNA interference, but they can also cause hypercholesterolemia, viral infections, and cancer. As naked siRNA is rapidly degraded by ribonucleases and does not pass through cellular membranes due to its negative charge, delivery systems such as liposomes, neutral lipid emulsions, polymeric nanoparticles, or lipid nanoparticles have been developed. Second, microRNAs adjust post-transcriptional regulation of gene expression at the 3' UTR to varying degrees [91]. The overexpression of rvMiB has been investigated in some disease states, including viral infections, liver injury, and ocular diseases, using synthetic mimics or plasmid combination vectors. The introduction of miRNA modulators has been targeted to hepatic stellate cells for liver fibrosis therapy via liposome complexes. Third, long non-coding RNAs and antisense oligonucleotides play roles in gene regulation or serve as miRNA sponges for transcription in lncRNAs, ASOs, or similar RNA therapeutics. Due to the negative charge of the backbone of the RNA, non-covalent or covalent attachments to a variety of systems are necessary to enhance delivery into cells. The purists with native lncRNAs are more biocompatible with endogenous protein binding in a complex with naked RNA or after traveling to various cellular locations. In contrast, ASOs consist of DNA, which presents a low production problem in vivo and a long half-life, but also a major disadvantage, including viral infections and cancer [92, 93].

Finally, the sequence is inserted into DNA or the intent of CRISPR directly into cellular gene-editing effects RNA duplex fragments through guide RNA, unlike gene therapy and effective RNA interference-mediated gene editing, base modification, or light regulation. CRISPR-Cas9 RNA therapeutics are edited into the genome or knock out inefficient transcripts that can be fused to appropriate gene delivery systems to prevent cell delivery for valid therapeutic applications. To enhance the performance of treatment delivery accuracy and to decrease toxicity, progress has been made. These improvement strategies include the design of customized RNA therapeutics, modifications to the genome, RNPs, and the development of gene delivery or internal packaging agents. Collectively, treatment using these RNA-based therapeutics is known to be effective and brings the important benefits of longer half-lives [93, 94].

mRNA Vaccines and Their Impacts

Over the past few years, one of the biggest breakthroughs in gene therapy has been accomplished with the consistent and efficient usage of modified RNA and protein replacement. The rapid development of RNA and lipid technologies, as well as advances in the control of the translation process of those vaccines, has gained great attention in independent research, pharmaceutical companies, health departments, and vaccine knowledge circles around the world. One of the main aspects that has been important for the significant progress of this product is related to the use of technology called RNA-Seq, where the genome of a patient's tissue is mapped in order to identify genetic mutations that may cause diseases, and nanoparticles made by sensors capable of targeting those identified errors and correcting them. This study elucidates some of the possible changes that mRNA vaccines can present for the principles of classic vaccine protocol; these vaccines demonstrate the possibility of changing the way modern vaccines are thought of and administered [95]. These innovations allow developments for the most diverse vaccine categories, from dealing with globally recognized diseases such as dengue and Zika, to the internal administration of a vaccine in cases of cancer and the identification of disease mutations. As the global scientific community aligns and develops strategies for combating this unprecedented pandemic virus, traditional regulatory and bottom-line control measures are also forced to evolve at such a rapid pace. This breakthrough represents a significant advance in the real-time evaluation of RNA-based vaccine requirements, and projections regarding the intensive and successful distribution of these new vaccines are key in accelerating the return to normal family, social life, and economy [96, 97].

Challenges in RNA Stability and Delivery

RNA is a key player in many, or perhaps all, of the deadly human diseases that are currently treated or that could be potentially treated using gene therapy. A gene therapy that would harness key types of RNA to provide therapeutic benefit in vivo could be significantly simpler and less expensive than plasmid-based gene therapies. However, the viability of targeting endogenous mRNAs in a therapeutic setting necessitates careful attention to the stability of the RNA. That is, delivering an RNA improvement to provide some therapeutic benefit does not have the long-term impact of modulating mRNA levels as do gene therapies currently approved. The transient nature of RNA raises the question of time limit and level for therapeutic benefit. Furthermore, RNA biopharmaceutics faces a formidable physiological delivery barrier in vivo as well as the inherently fragile nature of RNA due to its established suite of intracellular RNA protection and turnover mechanisms [97].

mRNA, siRNAs, and miRNAs are all defined either primarily or uniquely by their potential to affect the transcriptome of a human cell. While the therapeutic target might be a protein, the number one delivery barrier is often the fragility of the nucleic acid [98, 99]. All three of the above-mentioned RNA types are, however, fragile, including exogenous polyA mRNAs in vaccines. RNA has many self-defense mechanisms, ranging from protein protection and chaperoning systems to prevent naked nucleic acids from being rapidly degraded following administration to being recognized by endogenous RNAs and RNA-derived species, including siRNAs, that, when present, can target them for destruction. While the protein protection for each type of RNA can be circumvented using proprietary solutions that include attachment of a protein partner or an artificially attached 5' cap, the biological turnover mechanisms designed to clear and thus potentially recycle naturally produced RNA after the current generation is used will continue to apply to each type [100, 101].

GENETIC BASIS OF ALZHEIMER'S DISEASE

Genes associated with both early-onset and late-onset Alzheimer's disease have been discovered and are well documented. Among the familial early-onset Alzheimer's disease genes, mutations in APP, PSEN1, and PSEN2 greatly contribute to the pathology of the disease. APP encodes the amyloid β protein. PSEN1 and PSEN2 code for proteins presenilin 1 and presenilin 2, respectively, which are involved in the proteolysis of APP, including γ-secretase-mediated cleavage of APP-C99. Alzheimer's disease-associated APP mutations, which include V717I, V717F, V717G, E718G, and K724N, result in APP proteolysis, either cleavage by α- or β-secretases, and aggregation of Aβ plaques. Mutations in PSEN1, such as E120G, E120K, E120Q, G122D, F105L, F105S, F105V, L166P, and M139V, result in the increased production of a more hydrophobic and aggregation-capable Aβ [102, 103]. The PSEN2 mutations, M239V, T122R, S130I, N141I, and P117L, are also associated with the same pathological mechanisms of Aβ production, cleavage, and deposition. The apolipoprotein and cholesterol pathway is recognized as having a regulatory influence on β- and γ-secretase activity, and a large number of Alzheimer's disease-associated loci, of many genes involved in the cholesterol metabolism process, are well characterized for their role in the Alzheimer's disease processes [104, 105].

Key Genes Involved (APP, PSEN1, PSEN2, APOE)

Alzheimer's disease is a chronic neurodegenerative disease characterized by extensive synaptic loss and profound neuron death, with insidious clinical onset and progressive memory decline over time. The lack of effective drugs or therapies in the clinic is another critical issue. Accumulation of Aβ is the primary cause of early-onset Alzheimer's disease. In brain cells, APP expression appears to be an early event preceding Aβ deposition. The expression and metabolism level of APP are of great significance for controlling the brain burden of Aβ and the treatment of Alzheimer's disease [106]. The main role of presenilin is to make γ-secretase activity, after which the enzymatic cleavage of other peptidases to produce Aβ will occur. More than 100 types of mutations are distributed in more than 4000 reports so far. The APOE gene has a significant relationship with the occurrence of late-onset Alzheimer's disease. Although the brain concentration is very low, specifically expressed in astrocytes and near neurons and synapses, neuron-released human apolipoprotein is of great significance for the overall brain repair of Alzheimer's disease by repairing damaged or dying neurons, synapses, and non-neural cells [107, 108].

Genes involved include APP, PSEN1, PSEN2, and APOE. Their changes may cause early or late-onset Alzheimer's disease. Among them, any mutation in APP, PSEN1, or PSEN2 can lead to the production of a small protein, possibly causing Alzheimer's disease. The most common form of Alzheimer's disease is late-onset, and ApoE4 is the most common risk factor gene associated with it. The copies of ApoE4 will affect the age at which Alzheimer's disease becomes irreversible and the course of disease progression. ApoE4 has been found in the brains of patients with Alzheimer's disease, mainly in the cells surrounding beta-amyloid plaques. Additionally, in the brains of individuals with normal aging, ApoE2/3/4 has been widely expressed. It is necessary to study the correlation between brain damage and apolipoprotein, which is important for treating Alzheimer's disease and future drug development, as well as for providing support to patients to live independently for longer. Table **1** describes different gene targets, their role in physiology, functions, therapeutic approaches and current status for the treatment of AD through gene therapy [109, 110].

Table 1. Describes different gene targets for the treatment of AD through gene therapy.

S.no	Gene	Function	Role in AD Pathology	Therapeutic Approach	Current Research Status	Reference
1.	APOE	Lipid metabolism, neuronal repair	APOE4 variant increases AD risk	CRISPR gene editing to convert APOE4 → APOE2	Preclinical studies	[105, 111, 112]
2.	BACE1	β-secretase enzyme	Initiates Aβ production	siRNA, ASO to silence BACE1 expression	Clinical trials	[113, 114]
3.	APP	Amyloid precursor protein	Mutations increase Aβ plaques	Gene silencing (RNAi), CRISPR correction	Preclinical studies	[115, 116]
4.	PSEN1	γ-secretase complex	Mutations increase toxic Aβ production	CRISPR gene correction	Early research	[117, 118]
5.	PSEN2	γ-secretase complex	Similar to PSEN1 in Aβ processing	CRISPR gene correction	Early research	[118, 119]
6.	TREM2	Microglial activation, Aβ clearance	Loss-of-function variants impair immune response	Gene therapy to enhance TREM2 function	Preclinical studies	[120, 121]
7.	PGRN (GRN)	Neuroinflammation, lysosomal function	Deficiency leads to neurodegeneration	Gene replacement therapy	Early research	[122, 123]
8.	MAPT (Tau gene)	Microtubule stability	Hyperphosphorylation causes Tau tangles	ASO therapy to reduce Tau levels	Clinical trials	[124, 125]
9.	CLU (Clusterin)	Lipid metabolism, neuroprotection	Risk variant increases Aβ aggregation	Gene therapy to enhance protective variants	Early research	[126, 127]
10.	BIN1	Endocytosis, Tau regulation	Alters Tau pathology, linked to cognitive decline	Gene silencing approaches	Early research	[128, 129]

(Table 1) cont.....

S.no	Gene	Function	Role in AD Pathology	Therapeutic Approach	Current Research Status	Reference
11.	CD33	Microglial activation	Risk variant inhibits Aβ clearance	Gene knockdown to improve Aβ clearance	Preclinical studies	[130, 131]
12.	SORL1	APP trafficking	Mutations increase Aβ accumulation	Gene therapy to enhance function	Preclinical studies	[132, 133]
13.	ABCA7	Lipid transport, Aβ clearance	Risk variant reduces Aβ clearance	Gene therapy for functional restoration	Early research	[134, 135]
14.	PLCG2	Immune response, microglial function	Protective variant reduces AD risk	Gene activation therapy	Early research	[136]
15.	HIF1A	Hypoxia response	Regulates neuroinflammation and Aβ	Gene therapy for neuroprotection	Preclinical studies	[137]

Genetic and Sporadic Forms of Alzheimer's

Alzheimer's Disease (AD) exists as a hereditary syndrome, known as Familial AD (FAD), accounting for 5-10% of all AD cases. FAD always develops before the age of 60 and often in the forties, specifically in its early onset form. It occurs due to mutations in disease-causing genes. To date, more than 230 FAD mutations and duplications in these genes have been described, increasing age-related expression or specific misfolded generation [138, 139]. Therefore, ongoing clinical trials focus on precursor protein proteolytic cleavage enzymes via traditional small-molecule inhibitors or gene-targeted genome-editing approaches, which directly intervene in mutagenesis or gene silencing. A major RNA biocatalytic revelation can facilitate the development of RNA-based innovations or prevent age-related hemisynchronous increases of FAD foci and sporadic AD, regulating noncoding RNAs at the locations, compatible stoichiometry of RNA structures, or adaptable local RNA basis compositions. These tools are easy to improve, easy to make, or easy to synthesize; thus, sharing economy liberalization and personalized care are attainable to reduce people's disaster risks and enhance disaster resilience [140, 141].

Alzheimer's Disease (AD) exists as a rare hereditary syndrome, known as familial AD or FAD, and accounts for approximately 5–10% of all AD cases. In general, it develops before age 60 and often in the forties, specifically in its early onset form. Mutations in disease-causing genes cause syndrome development, and more than 230 different FAD mutations and duplications in these genes have been described

[142]. It is widely proposed that syndrome development is due to increased expression or even the increased production of a proteolytically generated toxic product—specifically, a misfolded peptide. Therefore, ongoing clinical trials largely focus on how to inhibit the specific mutually regulated proteolytic cleavage enzymes via isoform-specific traditional small-molecule inhibitors or the newest type of gene-editing methods [143, 144].

Pathophysiological Mechanisms

Despite successful recombinant DNA technology, it may still take years to develop large-scale human gene therapeutic strategies. The reason: increasingly complex endocrine, oncologic, neurogenetic, or hereditary diseases. Hereditary disorders are the most likely candidates for gene therapy trials because they are some of the most complex, costly, and emotionally difficult chronic or lethal diseases that afflict our species. Basic concepts from cell and molecular biology have driven the design of therapeutic gene delivery strategies derived from in vitro and animal models of genetic alteration. These isolated examples are needed to develop innovative strategies for millennia of human diseases, but until now, this has been only a prelude to the age of clinical medicine using recombinant DNA techniques. Future prospects will be considered within the context of existing laws and standards [145].

Several viruses have the potential to carry genetic material into mammalian host cells. Numerous types of tissue culture cells, indeed, fibroblasts, epithelial cells, lymphocytes, hepatocytes, spermatogenic, and bone-marrow-derived cells are consistently susceptible to infection with experimental live vaccines [146]. Wild-type viruses and nonviral particles could infect and lyse selective host cells in the process, particularly those of the immune system. Cytotoxicity from viral infection was initially thought to be advantageous since it enhanced the replication of exogenously applied gene transfer vectors. However, scientists learned early in the development of gene therapy techniques that most human subjects would suffer additional pathology from such viruses [147, 148].

Clinical Applications and Case Studies

To date, nearly 3,000 clinical trials, including gene-based therapies or therapeutics, have been registered, and several hundred patients have received approved exponential gene therapy applications. The specific impact of gene therapy technologies is indeed demonstrated by over 10 billion US dollars invested in gene therapy and gene-editing technologies in the 2011-2017 period. We can associate with the molecular character of genomic medicine in terms of reviewing any application in genome diagnostics and personalized therapy, as well as factors directing the necessary transition into realization. With the progress in the last decade, gene therapy has gained a novel status in its role to revolutionize the treatment of diverse categories of selected diseases [149]. Still, the current gene therapy potential appears to be rooted in the identification and application of specific products of intact functional gene bodies, for which reported human gene replacements can be applied for participative treatment of genetic defects themselves or associated monogenic diseases. In this regard, gene therapy becomes a specific personalized treatment, which indeed becomes the progressive, reliable, and cost-efficient approach predominantly in life-saving strategies, like bone marrow transplantation and gene addition. In its more technically straightforward form, gene therapy represents the approach of gene "addition." Traditionally ineffective monogenic Duchenne myopathy has been primarily targeted, and the initial but encouraging perspectives have emerged after the urgent first approach of adeno-associated viral-mediated dystrophin-encoding miniaturized gene delivery. Disease-relevant activity studies in the "immunodeficient" dog model definitely point out the beneficial potential for the future treatment of young patients [150, 151].

Gene Therapy used for Rare Genetic Disorders

Gene therapy involves the introduction of foreign nucleic acids into human cells for the correction of various gene disorders. This technique provides easier and direct access to rigorous intracellular targets, resulting in highly specific and potent therapeutics compared to any other conventional approaches. Gene therapy is now widely used in various genetic disorders, such as inherited and acquired conditions that are life-threatening and caused by the mutation of a specific gene. As postulated in the concepts of predictive and personalized medicine, gene therapy is increasingly being considered as the future direction towards the "curative" or at least "disease-altering" therapy, not only for classical gene disorders but also for complex genetic and epigenetic disorders, such as cancer, neurodegenerative disorders, and infectious diseases. The current trend of gene therapy throughout the world, recent successes, failures, and future perspectives of this field covering various genetic and epigenetic disorders are summarized in this report. Table **2** describes different delivery methods and mechanisms involved for RNA-based treatment of AD using gene therapy [151].

Table 2. Describes the delivery methods and mechanisms for RNA-based treatment of AD.

S. No.	Method	Delivery Mechanism	Advantage	Limitations	Reference
1.	Adeno-Associated Virus (AAV)	AAV binds to specific receptors on neurons, enters the cell, and integrates its genetic material into the host cell's nucleus	Long-term Expressions,	Limited Gene and Costly Production	[152]
2.	Lentivirus (LV)	Integrating the Genetic material into the Genome	Stable, long-lasting gene	Risk of the Genome causing mutations	[153]
3.	Herpes Simplex Virus (HSV-1)	It infects the neuron and provides a large genetic payload	High Efficiency	Neurotoxicity	[154]
4.	Lipid Nanoparticles (LNPs)	Encapsulate siRNAs and effectively deliver them across the BBB	Safer, most widely used	Low BBB Penetration	[155]
5.	Polymeric Nanoparticles	Receptor-mediated endocytosis	Easily customizable and Biodegradable	Less Efficient in viral terms	[156]
6.	Exosome-Based Delivery	Use Natural extracellular vesicles to deliver RNA	High BBB Penetration	Difficult to produce large quantities	[157]
7.	Peptide-Based Delivery Systems	"Peptide binds to specific receptors on the endothelial cells of the BBB, allowing the drug to be transported into the brain parenchyma."	High Tunable	Short life	[158]
8.	CRISPR/Cas9-Based Nanoparticles	"Designed to deliver the CRISPR-Cas9 gene editing system directly to affected brain cells"	Precisely Gene Correction	Low Efficiency	[159]
9.	Aptamer-Conjugated Delivery Systems	Uses the DNA and RNA Aptamers for gene transfer	High specificity for Neurons	Require optimization for BBB Transfer	[160]
10.	Magnetoception (Magnetic Nanoparticles)	Uses the Magnetic Field to guide the Genetic material into the cells	Enhanced targeted delivery	Required external magnetic fields	[161]

(Table 2) cont.....

S. No.	Method	Delivery Mechanism	Advantage	Limitations	Reference
11.	Ultrasound-Mediated BBB Disruption	"combined with microbubbles, are used to temporarily open the BBB, allowing therapeutic drugs to directly access the brain tissue and potentially target the underlying pathology, like amyloid beta plaques, by delivering treatment directly to the affected areas"	Non-invasive	Need Precise control, can damage BBB	[162]
12.	mRNA-LNP Therapy	Delivers mRNA encoding therapeutic protein	Fast Transient	Some Stability issues, Need Repeated Dosing	[72]
13.	Hydrogel-Based Delivery Systems	"The mechanism involves the hydrogel forming a gel-like matrix once injected or applied, which slowly releases the therapeutic agent over time, maximizing drug bioavailability and minimizing side effects."	Controlled release over time	Still experimental in AD	[163]
14.	Dendrimers	"Leveraging their unique structure to efficiently transport therapeutic agents across the Blood-Brain Barrier (BBB) and directly deliver them to affected brain regions"	High Stability	Potential Toxicity	[164]
15.	Galactose-Functionalized Nanoparticles	It targets microglia and astrocytes for the AD Therapy	Highly Specific	Still in Preclinical Development	[165]

Gene Therapy Used for Rare Genetic Disorders.

Currently, gene addition is mostly used in patients with single gene mutations, which are usually hereditary diseases where another copy of the gene is not present, or it is not possible to replace. In gene therapy, an additional copy of the normal gene is introduced into the genome, along with appropriate regulatory elements, where it is used to produce the protein that is lacking in the patient. The viral vectors used for gene addition express ectopically delivered genes from their integrated proviruses for over a patient's lifetime, without the need for complex in vitro maturation or laborious cell selection procedures. The most significant success of ex vivo gene addition was the treatment of patients with X-linked severe combined immune deficiency. The WASP gene is located on the X chromosome, and its mutation leads to a deficient proliferation of immunocompetent cells. Although related to an increased risk of cancer and leukemias, long-term follow-up studies have been recently reported. The use of genetically modified hematopoietic progenitor cells expressing the corrected gene is thus capable of permanently rescuing the patients from the disease [166, 167]. The effect of the treatment is permanent; the treated patients are still healthy 13 years after therapy. Other applications of lentiviral gene addition were mainly for inherited blood disorders, including hemoglobinopathies such as beta-thalassemia, sickle cell anemia, and/or Fanconi anemia. Drawbacks related to the use of integrating vectors remain, but the successful clinical treatments performed so far raise hope that gene addition will be automated soon for other disease indications. In addition to safety concerns, other obstacles to overcome are related to the cost of the therapy and the supply of sufficient numbers of vector-producing cells for each treatment. Random insertion and non-negligible levels of insertional mutagenesis, or in the case of lentiviral vectors, rare but recurrent break-back formation, have raised accompanying safety and supply concerns. Still, aside from regulatory issues related to above-threshold marketing, the field of gene therapy has not experienced unexpected trials related to such issues, and gene addition has made substantial progress and met every clinical target posed so far [168, 169].

RNA Therapies for Cancer Treatment

Cancer is a complex genetic disorder characterized by cell proliferation that occurs in the presence of mutations in designated genes that promote or inhibit the cell cycle. The aim of cancer gene therapy is to explore the underlying molecular pathways related to the development and cellular progression of cancer, with the goal of treating the disease by using single genes, gene products, and genetic modification [170, 171]. While initially all gene therapies were based on genomic genes, resulting in single-gene therapies, the introduction of RNA fragments was a major breakthrough in the field and has enabled multiple-gene therapeutic approaches. This was highlighted by the use of small interfering RNA for silencing in both cancer expression of specific genes and using mRNA for the transfer of anti-tumorigenic protein-coding regions. Such cancer treatments have recently proven to be successful in basic discovery research, and even some recipients in clinical trials have exhibited promising findings [172, 173].

Interest in RNA-based therapies has since expanded into the non-gene therapy field, a point that evolves some of the gene therapy techniques. It became known that RNA fragments are not equal; instead, they could act as signaling molecules or be subjected to modifications that further enhance their therapeutic and diagnostic qualities. Sodium and pH resistance and modified nucleotide sequences are two significant RNA characteristics that are considered crucial for hindering the body's immune response to RNA fragments and promoting longer half-lives. However, there are some difficulties in developing RNA-based therapies, such as the inability of the RNA fragments to go through the cell membrane without assistance, the side effects caused by stimulating the immune response of the body, and the potential cytotoxicity due to high or continuous dosage administrations. Technologies have been developed to address these challenges, such as designing special lipids or exosomes that function as artificial RNA carriers, and creating RNA molecules that are resistant to sodium and pH and that encode immunogenic resistant sequences. These advances have further expanded the repertoire of therapeutic abilities of RNA therapies [174, 175].

CHALLENGES AND ETHICAL CONSIDERATIONS

As a final section, it is imperative to highlight some of the contemporary challenges that gene therapy faces, to foster discussion and address potential threats to translational momentum. With a focus on RNA-based gene products, we also delve into some of the foreseeable ethical considerations that relate to their consumption and lifestyle [176].

The ethical considerations for gene therapy include demonstration of the safety and efficacy, as well as informed consent for a novel, minimally tested therapy. Importantly, the therapy must be able to improve the health of the population that consumes it, either directly or indirectly, by improving the health of society as a whole. The high costs associated with such a treatment must also be overcome to allow access to a wide range of individuals [177, 178].

In conclusion, RNA-based gene products are a promising therapy and are poised to alter the landscape of inborn errors and drug treatments. While the investigation is just beginning and presents several challenges, the rewards of the ultimate treatment of these inherited disorders warrant the difficult efforts that lie ahead. The continued revolution in biology and computational ability seeks to stoke these drug therapies, enabling individuals to sidestep their defective genes and to allow resolution [178].

Safety and Off-Target Effects

Special attention has been paid to the design of gene editing technologies that create minimal off-target mutations. As described above, CRISPR and TALEN technologies can each be designed with high target specificity, with each based on molecularly rigid guide RNAs or protein-DNA structures, respectively. Studies have begun to investigate the genome-wide off-target effects of various targeted gene editing approaches for both CRISPR and TALEN. For all editing platforms, it is critical to understand the activity of reagents prior to translational use. Indeed, due to the low fidelity of the Cas9 nuclease, it may be necessary to screen mouse or patient-derived cell lines for off-target mutations. Future design of CRISPR editing nucleases will also likely incorporate in silico quantitative algorithms to predict off-target activity and create libraries of high-fidelity mutants. However, many gene editing off-target studies conducted to date have been of low-throughput quality and thus require more thorough investigation, especially in the in vivo setting [179, 180].

Regulatory and Approval Pathways

Procedures for gene and cell therapy require a comparison of inherent features of the approaches versus more established biopharmaceuticals. The regulatory process for pharmacodynamic properties of gene or cell pharmaceuticals is more complex than those associated with production, manufacturing, standardization, and safety that are required for products such as blood derivatives, including recombinant blood factors, therapeutic monoclonal antibodies, and recombinant viral vaccines. This section reviews the current regulatory philosophies delineated in U.S. legislation and their impact on the design of clinical protocols and pilot U.S. activities in translational gene therapy. Regulations vary in different countries and may be complemented in the future with renewed international deliberations and harmonization. Guidelines and procedures for the quality of information, especially when dealing with genetic manipulations, are still evolving, as demonstrated by regional differences in European regulations. Biological information required to assess potential risks and advantages of gene and cell therapy, particularly to anticipate long-term evolutionary changes, must be matched with ethical concerns [181, 182].

The approval process required for a cell or gene therapy under the current U.S. Act requires a formal application consisting of descriptive documents to support safety and efficacy data from adequate and well-controlled studies that fulfill international guidelines. Facilities must conform to current Good Manufacturing Practice for biological products, which ensures traceability of therapeutic agents, principally by maintaining extensive records and by frequent and documented inspections. In general, clinical investigations to support efficacy from adequate and well-controlled studies will require substantial patient exposure at significant doses of therapeutic agents, chosen to support statistically valid conclusions of clinical superiority [183].

Ethical Debates for Genetic Engineering

The development of recombinant DNA technology has led to a rapid increase in interest, funding, and outlets, both academically and commercially, for studies on the modification of the mammalian genetic heritage. However, they have also raised important ethical questions and critical concerns. In the face of available technological capabilities for altering human germline cells, the moral questions are complex and multifaceted for many reasons: partly because of the novel issues that arise, partly because of the novelty of the concept of tampering with human nature, and partly because of citizens' fears. If certain technologies are in place, is the installation of biological modification hypothetical, desired, collisional, or too risky for the generation of undesired human products? Given the threats of scientific abuse, it is important to keep public concerns in mind through a legitimate ethical evaluation of these issues by expert technicians [184, 185].

In this direction, the first international summits organized and conducted provided guidelines. The Working Group sought to bifurcate the Human Genome Project and to publicly debate the wisdom of introducing germline genetic modifications for treatment. During the IV international summit on Gene Therapy, there was a large consensus that gene therapy should be off-limits, but also transmitted recommendations to that effect. The safety issues of gene therapies were critical; standards in gene therapy were to be kept at the highest levels, able to control the phase of trial, evaluate, and manage the risks associated with artificial genetic modifications to human beings. There is also more affinity for other types of clinics for suffering children. The second ethical issue relates to the concept of genetically modifying offspring. Thanks to stem cell treatments, transplantations, and in vitro studies of fertilization, there are ongoing debates on shared aspects. People whose genomes are edited will not often be able to report truly informed consent. The third ethical issue of the alteration of germline cells has to do with the consequences of scientific research in animals. The last ethical issue has to do with international cooperation. Therefore, the main ethical issues are safety, the long-term and transgenerational effects, and the effects of genetic modifications on the welfare of the people who will be introduced into the riskiest trials. A comprehensive strategy is attempted. Given the great interest of the media, it administered and disseminated its indications and proposals. The researchers also expressed themselves, permitting the transmission of knowledge on the topic. Working for the good of the people is the cornerstone of gene therapy. The evolution of the discussants was incredible, capable of overcoming one hundred myths and prejudices [186, 187].

FUTURE PERSPECTIVES AND EMERGING TRENDS

Keyseukeu et al. published a comprehensive and detailed review on the current trends and future prospects in gene therapy. They summarize four major categories of current trends: molecular advances in gene therapy, the contribution of gene therapy in rare and severe diseases, the first gene therapy marketed, and factors that limit gene therapy market access. They conclude that gene therapy has moved from conception to promising results in single-gene inborn conditions and are now testing it in acquired diseases like cancer, Alzheimer's, or diabetes. The current gene therapy market contains molecules with moderate to strong efficacy in rare and severe diseases and is expected to grow [187].

Despite the significant challenges of vector development, toxicities, and manufacturing guidelines, an increasing number of biopharma companies are becoming aware of the revenue potential. Regarding future prospects, advances and new perspectives of DNA/RNA vaccines in cancer therapy are discussed. It is described that cancer-antigen DNA/RNA vaccine approaches operate by "reprogramming" tumor cells or patient dendritic cells by expressing and exposing the antigen to the immune system, and that preclinical and clinical results against infectious disease and cancer predict success in the future. The challenges faced by DNA/RNA technologies are highlighted, including their low cost and easy scalability, regulatory flexibility, such as clinical trials initiated during the preclinical stage, stability during syringe handling, no need for cold chain, adjuvant-free, no risk of oncogenesis, and multiantigen combined vaccination. Additionally, personalized RNA cancer vaccines, consisting of an intranodal injection of patient-specific synthetic long peptides coupled to an mRNA encoding CD40L, CD70, and a constitutively active Toll-like receptor 4, can be produced within 5 weeks after the selection of neoepitopes, based on their capacity to enhance cytotoxic T cell and Tregs activity. Vaccine approaches like these will guide new treatments in the irradiation, precision medicine, and cancer immunotherapy areas [188, 189].

Next-Generation Gene Editing Technologies

In the last few years, a number of groundbreaking technologies have been developed that will significantly advance the field of gene editing in attempts to modify the genome. These technologies, which generally use endonucleases that require recognition of DNA sequences and double-strand break functions, include zinc-finger nucleases, transcription activator-like effector nucleases, and meganucleases. More recent promising new developments are CRISPR/Cas9 systems based on prokaryotic type II clustered regularly interspaced short palindromic repeats systems and composed of CRISPR-associated proteins, such as Cas9, which has become one of the most convenient and notable in terms of versatility. CRISPR/Cas9 systems encode a CRISPR RNA, which is essential for target recognition, and transactivating crRNA and form a single guide RNA by combining crRNA and tracrRNA [190, 191].

In recent years, a number of significant findings have been made showing that new-generation gene editing technology is not only improving in efficacy and range of frequently identified alleles, but also contributing new basic knowledge about gene editing DSB repair in response to CRISPR/Cas9-mediated DSBs [192]. Specific trends and results include the success of gene therapy using CRISPR/Cas9, which has been tested to reduce the impact of various diseases caused by diverse genetic mutations regulated by different systems, and the delivery of a set of two sgRNAs and Cas9, followed by gene correction. In addition, orthogonal CRISPR/Cas9 have been engineered for selective cleavage of a single target or double targets in a single genomic locus [193, 194].

RNA-Based Drug Development

The discovery and validation of RNA drug targets are leading to novel, innovative RNA molecular design approaches with clinical potential. These developments are addressing notable needs ranging from a shortage of RNA therapeutics in various disease areas to RNA sequence-induced immunoactivities [195]. Current clinical validation and ultimate development success should further substantiate the utility of interfering RNA for therapeutic intervention. Open research and development questions in the areas, including comprehensive RNA sequence activity and potential adverse effect characterization, sequence delivery challenges, and molecular design strategies, should continue to fuel academic research efforts as well as potential collaborative interest within the pharma sector [196, 197].

Innovation around RNA-based molecule design and development has been intensifying since the realization that short interfering RNAs could be synthesized and transfected into cells as a post-transcriptional level gene silencing strategy or lead to suppression of developing diseases. Passive RNA functions, two nucleotides comprising both of its substrate molecules and an exclusive, limited set of sugar, base, and phosphodiester molecular units, are ripe for systematic design. The pure diversity of RNA sequence-based reactivity provides an infinite solution space and search that could easily handle the complexity of an element like microRNA to both gain a basic understanding and define RNA intervention strategies [198].

Personalized Medicine and Precision Therapies

New frontiers in biotechnology are opening up the exploitation of our genetic code to attain more effective and safer treatments for a vast number of diseases. Gene therapy and RNA-based drugs have made it possible for researchers to start realizing the dream of personalized medicine. These promising biopharmaceuticals are expected not only to merit the development of an industry for personalized medicines but also to help realize the healthcare model of precision medicine and its engineering protocols based on molecular-based refined disease classification systems, reinvent the pharmaceutical industry and therapeutics by providing customizable, effective, minimally invasive therapeutics and maintain the long-term sustainability of healthcare systems. Recent advances have made it possible not only to slow down the progression of a genetic disease but also to cure it. However, gene therapy's main limitation—its specificity coupled with the complexity involved in identifying the genes responsible for multifactorial diseases—has until now been a serious handicap. In recent years, new systems have been discovered for increasing the therapeutic impact of gene therapy vectors and oligonucleotide-system drugs, such as RNA interference, the exploitation of genome editing guided nucleases, or the new generation of catalytically and structurally optimized AONs that increase the efficiency, overcoming the previous limitations imposed by current RNA-based drugs. All these caveats make the progress of this area toward personalized therapeutic systems complementary to the benefits stemming from the recent advances in high-throughput systems, coupled with sophisticated data analysis methods that have led to the discovery of new direct and indirect relationships connecting the genotype to the phenotype [199].

CONCLUSION

In conclusion, gene therapy is on the verge of reaching the stage of product licensing. The current trends and fast pace of developments in the field predict that gene therapy will mature as a successful and sustainable pharmaceutical intervention. Successful gene therapy will necessitate ties between various disciplines, aligning them with future requirements in affordability and in tune with the needs of diverse patient populations. Any invention that addresses these issues is bound to succeed. Furthermore, the inventiveness of scientists in the field, the capacity of society to embrace this technology, and the economic stamina of nations that invest in the success of such great achievements will only dictate how quickly gene therapy turns into a successful modality. Finally, RNA intervention has progressed at a fast pace and holds great promise to alleviate a variety of pathological conditions, including those for which other interventions fall short.

RNA-based therapy is still in a phase of infancy, but outcomes from the interventions that have been completed with significant success present a rosy picture of RNA's power as a therapeutic agent. Of particular note, double-strand RNA-based targeted interventions that employ RNAi hold tremendous potential as a potent class of therapeutics, the reach of which extends to the vast majority of pathological conditions. As basic research uncovers RNA's roles as a versatile therapeutic agent, practical applications are likely to flourish, taking RNA-based therapy to a successful conclusion. Central to the success, as with any therapeutic agent, is the economic feasibility of RNA-based therapy. There exists a significant potential in RNA mimicking natural RNA for it to fail in clinical trials due to the immune response generated by RNA. Such an immune response could perhaps be overcome by a synthetic technological invention that lends RNA a chance to become immune unresponsive. Naturally, technological advancements will predict the success of future RNA inventions. These inventions should also be coupled with risk–benefit evaluations that are socially and economically acceptable. A scientific society that can envision the promise and advantages of selecting RNA as a therapeutic agent and harnesses its attributes will be favorably making headway toward the eventual development of RNA-based therapeutics.

AUTHORS' CONTRIBUTIONS

Conceptualization: All the authors contributed equally to the development of this book's chapter framework and concept. This collaborative effort ensured that a well-rounded perspective was preserved throughout the study.

LIST OF ABBREVIATIONS

RNA	Ribonucleic acid
DNA	Deoxyribonucleic acid
CRISPR	Clustered Regularly Interspaced Short Palindromic Repeats
LNP	Lipid Nanoparticle
siRNAs	small interfering RNAs
AAV	Adeno-Associated Virus
TALEN	Transcription Activator-Like Effector Nucleases
mLNP	Multi-Layered Nanoparticles

APOE Apolipoprotein E

HSV Herpes Simplex Virus

ACKNOWLEDGEMENTS

The authors are thankful to Mahayogi Gorakhnath University, Gorakhpur, for providing the necessary facilities to execute this manuscript.

REFERENCES

[1] Cring MR, Sheffield VC. Gene therapy and gene correction: targets, progress, and challenges for treating human diseases. Gene Ther 2022; 29(1-2): 3-12. [http://dx.doi.org/10.1038/s41434-020-00197-8] [PMID: 33037407]

[2] Sayed N, Allawadhi P, Khurana A, *et al.* Gene therapy: Comprehensive overview and therapeutic applications. Life Sci 2022; 294: 120375. [http://dx.doi.org/10.1016/j.lfs.2022.120375] [PMID: 35123997]

[3] Bulcha JT, Wang Y, Ma H, Tai PWL, Gao G. Viral vector platforms within the gene therapy landscape. Signal Transduct Target Ther 2021; 6(1): 53. [http://dx.doi.org/10.1038/s41392-021-00487-6] [PMID: 33558455]

[4] Mead RJ, Shan N, Reiser HJ, Marshall F, Shaw PJ. Amyotrophic lateral sclerosis: a neurodegenerative disorder poised for successful therapeutic translation. Nat Rev Drug Discov 2023; 22(3): 185-212. [http://dx.doi.org/10.1038/s41573-022-00612-2] [PMID: 36543887]

[5] Perelló-Trias MT, Serrano-Muñoz AJ, Rodríguez-Fernández A, Segura-Sampedro JJ, Ramis JM, Monjo M. Intraperitoneal drug delivery systems for peritoneal carcinomatosis: Bridging the gap between research and clinical implementation. J Control Release 2024; 373: 70-92. [http://dx.doi.org/10.1016/j.jconrel.2024.07.017] [PMID: 38986910]

[6] Nissanka MC, Dilhari A, Wijesinghe GK, Weerasekera MM. Advances in experimental bladder models: bridging the gap between in vitro and in vivo approaches for investigating urinary tract infections. BMC Urol 2024; 24(1): 206. [http://dx.doi.org/10.1186/s12894-024-01590-w] [PMID: 39313789]

[7] Munda M, Velnar T. Stem cell therapy for degenerative disc disease: Bridging the gap between preclinical promise and clinical potential. Biomol Biomed 2024; 24(2): 210-8. [http://dx.doi.org/10.17305/bb.2023.9518] [PMID: 37669102]

[8] Khair K, Steadman L, Chaplin S, Holland M, Jenner K, Fletcher S. Parental perspectives on gene therapy for children with haemophilia: The Exigency study. Haemophilia 2021; 27(1): 120-8. [http://dx.doi.org/10.1111/hae.14188] [PMID: 33216422]

[9] Papaioannou I, Owen JS, Yáñez-Muñoz RJ. Clinical applications of gene therapy for rare diseases: A review. Int J Exp Pathol 2023; 104(4): 154-76. [http://dx.doi.org/10.1111/iep.12478] [PMID: 37177842]

[10] Herzeg A, Almeida-Porada G, Charo RA, *et al.* Prenatal somatic cell gene therapies: charting a path toward clinical applications (proceedings of the CERSI-FDA meeting). J Clin Pharmacol 2022; 62(S1) (Suppl. 1): S36-52. [http://dx.doi.org/10.1002/jcph.2127] [PMID: 36106778]

[11] Califano I, Smulever A, Jerkovich F, Pitoia F. Advances in the management of anaplastic thyroid carcinoma: transforming a life-threatening condition into a potentially treatable disease. Rev Endocr Metab Disord 2024; 25(1): 123-47. [http://dx.doi.org/10.1007/s11154-023-09833-1] [PMID: 37648897]

[12] Dogbey DM, Torres VES, Fajemisin E, *et al.* Technological advances in the use of viral and non-viral

vectors for delivering genetic and non-genetic cargos for cancer therapy. Drug Deliv Transl Res 2023; 13(11): 2719-38.
[http://dx.doi.org/10.1007/s13346-023-01362-3] [PMID: 37301780]

[13] Shchaslyvyi AY, Antonenko SV, Tesliuk MG, Telegeev GD. Current state of human gene therapy: approved products and vectors. Pharmaceuticals (Basel) 2023; 16(10): 1416.
[http://dx.doi.org/10.3390/ph16101416] [PMID: 37895887]

[14] Cao X, Zhang Y, Ding Y, Wan Y. Identification of RNA structures and their roles in RNA functions. Nat Rev Mol Cell Biol 2024; 25(10): 784-801.
[http://dx.doi.org/10.1038/s41580-024-00748-6] [PMID: 38926530]

[15] Deogharia M, Gurha P. The "guiding" principles of noncoding RNA function. Wiley Interdiscip Rev RNA 2022; 13(4): e1704.
[http://dx.doi.org/10.1002/wrna.1704] [PMID: 34856642]

[16] Zhu Y, Zhu L, Wang X, Jin H. RNA-based therapeutics: an overview and prospectus. Cell Death Dis 2022; 13(7): 644.
[http://dx.doi.org/10.1038/s41419-022-05075-2] [PMID: 35871216]

[17] Yao ZT, Yang YM, Sun MM, *et al.* New insights into the interplay between long non-coding RNAs and RNA-binding proteins in cancer. Cancer Commun (Lond) 2022; 42(2): 117-40.
[http://dx.doi.org/10.1002/cac2.12254] [PMID: 35019235]

[18] Shahryari A, Burtscher I, Nazari Z, Lickert H. Engineering gene therapy: advances and barriers. Adv Ther (Weinh) 2021; 4(9): 2100040..
[http://dx.doi.org/10.1002/adtp.202100040]

[19] Sun J, Roy S. Gene-based therapies for neurodegenerative diseases. Nat Neurosci 2021; 24(3): 297-311.
[http://dx.doi.org/10.1038/s41593-020-00778-1] [PMID: 33526943]

[20] Papanikolaou E, Bosio A. The promise and the hope of gene therapy. Front Genome Ed. 2021; 3: 618346.
[http://dx.doi.org/10.3389/fgeed.2021.618346]

[21] Puranik N, Yadav D, Chauhan PS, Kwak M, Jin JO. Exploring the role of gene therapy for neurological disorders. Curr Gene Ther 2021; 21(1): 11-22.
[http://dx.doi.org/10.2174/1566523220999200917114101] [PMID: 32940177]

[22] Zabaleta N, Unzu C, Weber ND, Gonzalez-Aseguinolaza G. Gene therapy for liver diseases — progress and challenges. Nat Rev Gastroenterol Hepatol 2023; 20(5): 288-305.
[http://dx.doi.org/10.1038/s41575-022-00729-0] [PMID: 36646909]

[23] Zhou K, Han J, Wang Y, Xu Y, Zhang Y, Zhu C. The therapeutic potential of bone marrow-derived macrophages in neurological diseases. CNS Neurosci Ther 2022; 28(12): 1942-52.
[http://dx.doi.org/10.1111/cns.13964] [PMID: 36066198]

[24] Zhang G, Han S, Wang L, Yao Y, Chen K, Chen S. A ternary synergistic eNOS gene delivery system based on calcium ion and L-arginine for accelerating angiogenesis by maximizing NO production. Int J Nanomedicine 2022; 17: 1987-2000.
[http://dx.doi.org/10.2147/IJN.S363168] [PMID: 35530975]

[25] De Silva SR, Moore AT. Optogenetic approaches to therapy for inherited retinal degenerations. J Physiol 2022; 600(21): 4623-32.
[http://dx.doi.org/10.1113/JP282076] [PMID: 35908243]

[26] Palmieri M, Pozzer D, Landsberger N. Advanced genetic therapies for the treatment of Rett syndrome: state of the art and future perspectives. Front Neurosci 2023; 17: : 1172805..
[http://dx.doi.org/10.3389/fnins.2023.1172805] [PMID: 37304036]

[27] Sevin C, Deiva K. Clinical trials for gene therapy in lysosomal diseases with CNS involvement. Front Mol Biosci 2021; 8: : 624988..

[http://dx.doi.org/10.3389/fmolb.2021.624988] [PMID: 34604300]

[28] Zhang B. CRISPR/Cas gene therapy. J Cell Physiol 2021; 236(4): 2459-81. [http://dx.doi.org/10.1002/jcp.30064] [PMID: 32959897]

[29] Botto C, Rucli M, Tekinsoy MD, Pulman J, Sahel JA, Dalkara D. Early and late stage gene therapy interventions for inherited retinal degenerations. Prog Retin Eye Res 2022; 86: : 100975.. [http://dx.doi.org/10.1016/j.preteyeres.2021.100975] [PMID: 34058340]

[30] Gurevich I, Agarwal P, Zhang P, *et al.* In vivo topical gene therapy for recessive dystrophic epidermolysis bullosa: a phase 1 and 2 trial. Nat Med 2022; 28(4): 780-8. [http://dx.doi.org/10.1038/s41591-022-01737-y] [PMID: 35347281]

[31] Mendell JR, Al-Zaidy SA, Rodino-Klapac LR, *et al.* Current clinical applications of in vivo gene therapy with AAVs. Mol Ther 2021; 29(2): 464-88. [http://dx.doi.org/10.1016/j.ymthe.2020.12.007] [PMID: 33309881]

[32] Subramaniam KS, Antoniou MN, McGrath JA, Lwin SM. The potential of gene therapy for recessive dystrophic epidermolysis bullosa. Br J Dermatol 2022; 186(4): 609-19. [http://dx.doi.org/10.1111/bjd.20910] [PMID: 34862606]

[33] Wagner HJ, Weber W, Fussenegger M. Synthetic biology: emerging concepts to design and advance adeno-associated viral vectors for gene therapy. Adv Sci (Weinh) 2021; 8(9): 2004018. [http://dx.doi.org/10.1002/advs.202004018] [PMID: 33977059]

[34] Wang JH, Gessler DJ, Zhan W, Gallagher TL, Gao G. Adeno-associated virus as a delivery vector for gene therapy of human diseases. Signal Transduct Target Ther 2024; 9(1): 78. [http://dx.doi.org/10.1038/s41392-024-01780-w] [PMID: 38565561]

[35] Liu J, Koay TW, Maiakovska O, Zayas M, Grimm D. Progress in bioengineering of myotropic adeno-associated viral gene therapy vectors. Hum Gene Ther 2023; 34(9-10): 350-64. [http://dx.doi.org/10.1089/hum.2023.057] [PMID: 37082964]

[36] Jensen TL, Gøtzsche CR, Woldbye DPD. Current and future prospects for gene therapy for rare genetic diseases affecting the brain and spinal cord. Front Mol Neurosci 2021; 14: 695937.. [http://dx.doi.org/10.3389/fnmol.2021.695937] [PMID: 34690692]

[37] Kulkarni JA, Witzigmann D, Thomson SB, *et al.* The current landscape of nucleic acid therapeutics. Nat Nanotechnol 2021; 16(6): 630-43. [http://dx.doi.org/10.1038/s41565-021-00898-0] [PMID: 34059811]

[38] Sun D, Lu ZR. Structure and function of cationic and ionizable lipids for nucleic acid delivery. Pharm Res 2023; 40(1): 27-46. [http://dx.doi.org/10.1007/s11095-022-03460-2] [PMID: 36600047]

[39] Zhang Y, Liu Q, Zhang X, *et al.* Recent advances in exosome-mediated nucleic acid delivery for cancer therapy. J Nanobiotechnology 2022; 20(1): 279. [http://dx.doi.org/10.1186/s12951-022-01472-z] [PMID: 35701788]

[40] Maestro S, Weber ND, Zabaleta N, Aldabe R, Gonzalez-Aseguinolaza G. Novel vectors and approaches for gene therapy in liver diseases. JHEP Rep Innov Hepatol 2021; 3(4): 100300.. [http://dx.doi.org/10.1016/j.jhepr.2021.100300] [PMID: 34159305]

[41] Belete TM. The current status of gene therapy for the treatment of cancer. Biologics 2021; 15: 67-77. [PMID: 33776419]

[42] Yu C, Li L, Hu P, *et al.* Recent advances in stimulus-responsive nanocarriers for gene therapy. Adv Sci (Weinh) 2021; 8(14): : 2100540.. [http://dx.doi.org/10.1002/advs.202100540] [PMID: 34306980]

[43] Nawaz W, Huang B, Xu S, *et al.* AAV-mediated *in vivo* CAR gene therapy for targeting human T-cell leukemia. Blood Cancer J 2021; 11(6): 119. [http://dx.doi.org/10.1038/s41408-021-00508-1] [PMID: 34162832]

[44] Pan X, Veroniaina H, Su N, Sha K, Jiang F, Wu Z, Qi X. Applications and developments of gene therapy drug delivery systems for genetic diseases. Asian J Pharm Sci. 2021 Nov; 16(6): 687–703. [http://dx.doi.org/10.1016/j.ajps.2021.05.003]

[45] Vickram AS, Manikandan S, Richard T, Lakshmi SV, Chopra H. Targeted Gene Therapy: Promises and Challenges in Disease Management. J BioX Res 2024; 7(02): 0007. [http://dx.doi.org/10.34133/jbioxresearch.0007]

[46] Cetin B, Erendor F, Eksi YE, Sanlioglu AD, Sanlioglu S. Gene and cell therapy of human genetic diseases: Recent advances and future directions. J Cell Mol Med 2024; 28(17): e70056. [http://dx.doi.org/10.1111/jcmm.70056] [PMID: 39245805]

[47] Wang Y, Hu LF, Zhou TJ, *et al.* Gene therapy strategies for rare monogenic disorders with nuclear or mitochondrial gene mutations. Biomaterials 2021; 277: : 121108.. [http://dx.doi.org/10.1016/j.biomaterials.2021.121108] [PMID: 34478929]

[48] Rouches MV, Xu Y, Cortes LBG, Lambert G. A plasmid system with tunable copy number. Nat Commun 2022; 13(1): 3908. [http://dx.doi.org/10.1038/s41467-022-31422-0] [PMID: 35798738]

[49] Yilmaz M, Yalcin E, Presumey J, *et al.* Overexpression of schizophrenia susceptibility factor human complement C4A promotes excessive synaptic loss and behavioral changes in mice. Nat Neurosci 2021; 24(2): 214-24. [http://dx.doi.org/10.1038/s41593-020-00763-8] [PMID: 33353966]

[50] Bashor CJ, Hilton IB, Bandukwala H, Smith DM, Veiseh O. Engineering the next generation of cell-based therapeutics. Nat Rev Drug Discov 2022; 21(9): 655-75. [http://dx.doi.org/10.1038/s41573-022-00476-6] [PMID: 35637318]

[51] Parambi DGT, Alharbi KS, Kumar R, *et al.* Gene therapy approach with an emphasis on growth factors: theoretical and clinical outcomes in neurodegenerative diseases. Mol Neurobiol 2022; 59(1): 191-233. [http://dx.doi.org/10.1007/s12035-021-02555-y] [PMID: 34655056]

[52] Cecchin R, Troyer Z, Witwer K, Morris KV. Extracellular vesicles: The next generation in gene therapy delivery. Mol Ther 2023; 31(5): 1225-30. [http://dx.doi.org/10.1016/j.ymthe.2023.01.021] [PMID: 36698310]

[53] Philips JG, Martin-Avila E, Robold AV. Horizontal gene transfer from genetically modified plants - Regulatory considerations. Front Bioeng Biotechnol 2022; 10: 971402. [http://dx.doi.org/10.3389/fbioe.2022.971402] [PMID: 36118580]

[54] Gorovits B, Azadeh M, Buchlis G, *et al.* Evaluation of cellular immune response to adeno-associated virus-based gene therapy. AAPS J 2023; 25(3): 47. [http://dx.doi.org/10.1208/s12248-023-00814-5] [PMID: 37101079]

[55] Wang X, Han JN, Zhang X, *et al.* Reversible thermal regulation for bifunctional dynamic control of gene expression in Escherichia coli. Nat Commun 2021; 12(1): 1411. [http://dx.doi.org/10.1038/s41467-021-21654-x] [PMID: 33658500]

[56] Zu H, Gao D. Non-viral vectors in gene therapy: recent development, challenges, and prospects. AAPS J 2021; 23(4): 78. [http://dx.doi.org/10.1208/s12248-021-00608-7] [PMID: 34076797]

[57] Li Z, Zhao P, Zhang Y, *et al.* Exosome-based *Ldlr* gene therapy for familial hypercholesterolemia in a mouse model. Theranostics 2021; 11(6): 2953-65. [http://dx.doi.org/10.7150/thno.49874] [PMID: 33456582]

[58] Lin N, Ouyang Y, Qin Y, *et al.* Spatially localized entropy-driven evolution of nucleic acid-based constitutional dynamic networks for intracellular imaging and spatiotemporal programmable gene therapy. J Am Chem Soc 2024; 146(30): 20685-99. [http://dx.doi.org/10.1021/jacs.4c03651] [PMID: 39012486]

[59] Amado DA, Davidson BL. Gene therapy for ALS: A review. Mol Ther 2021; 29(12): 3345-58. [http://dx.doi.org/10.1016/j.ymthe.2021.04.008] [PMID: 33839324]

[60] Byun S, Lee M, Kim M. Gene therapy for Huntington's disease: the final strategy for a cure? J Mov Disord 2022; 15(1): 15-20. [http://dx.doi.org/10.14802/jmd.21006] [PMID: 34781633]

[61] Bhardwaj S, Agrawal P, Kumar S, *et al.* CRISPR/Cas9 gene editing: New hope for Alzheimer's disease. J adv Res 2022; 40: 207-221.

[62] Wu J, Yang Y, Wan Y, *et al.* New insights into the role and mechanisms of ginsenoside Rg1 in the management of Alzheimer's disease. Biomed Pharmacother 2022; 152: : 113207.. [http://dx.doi.org/10.1016/j.biopha.2022.113207] [PMID: 35667236]

[63] Chopade P, Chopade N, Zhao Z, Mitragotri S, Liao R, Chandran Suja V. Alzheimer's and Parkinson's disease therapies in the clinic. Bioeng Transl Med 2023; 8(1): e10367. [http://dx.doi.org/10.1002/btm2.10367] [PMID: 36684083]

[64] Usman MB, Bhardwaj S, Roychoudhury S, *et al.* Immunotherapy for Alzheimer's disease: current scenario and future perspectives. J Prev Alzheimers Dis 2021; 8(4): 534-51. [http://dx.doi.org/10.14283/jpad.2021.52] [PMID: 34585229]

[65] Yi LX, Tan EK, Zhou ZD. Passive immunotherapy for Alzheimer's disease: challenges & future directions. J Transl Med 2024; 22(1): 430. [http://dx.doi.org/10.1186/s12967-024-05248-x] [PMID: 38715084]

[66] Islam MR, Rabbi MA, Hossain T, Sultana S, Uddin S. Mechanistic Approach to Immunity and Immunotherapy of Alzheimer's Disease: A Review. ACS Chem Neurosci 2024; 15(20): 3589–3614. [http://dx.doi.org/10.1021/acschemneuro.4c00360] [PMID: 39173186]

[67] Parrocha CMT, Nowick JS. Current peptide vaccine and immunotherapy approaches against Alzheimer's disease. Pept Sci (Hoboken) 2023; 115(1): e24289.. [http://dx.doi.org/10.1002/pep2.24289] [PMID: 36778914]

[68] Garbuz DG, Zatsepina OG, Evgen'ev MB. Beta amyloid, tau protein, and neuroinflammation: an attempt to integrate different hypotheses of Alzheimer's disease pathogenesis. Mol Biol (Mosk) 2021; 55(5): 734-47. [PMID: 34671002]

[69] Ashrafian H, Zadeh EH, Khan RH. Review on Alzheimer's disease: Inhibition of amyloid beta and tau tangle formation. Int J Biol Macromol 2021; 167: 382-94. [http://dx.doi.org/10.1016/j.ijbiomac.2020.11.192] [PMID: 33278431]

[70] Romaniyanto , Mahyudin F, Sigit Prakoeswa CR, *et al.* An update of current therapeutic approach for Intervertebral Disc Degeneration: A review article. Ann Med Surg (Lond) 2022; 77: : 103619.. [http://dx.doi.org/10.1016/j.amsu.2022.103619] [PMID: 35638079]

[71] Morris G, Schorge S. Gene therapy for neurological disease: State of the art and opportunities for next-generation approaches. Neuroscience 2022; 490: 309-14. [http://dx.doi.org/10.1016/j.neuroscience.2022.03.010] [PMID: 35304290]

[72] Tsakiri M, Zivko C, Demetzos C, Mahairaki V. Lipid-based nanoparticles and RNA as innovative neuro-therapeutics. Front Pharmacol 2022; 13: : 900610.. [http://dx.doi.org/10.3389/fphar.2022.900610] [PMID: 36016560]

[73] Khorkova O, Hsiao J, Wahlestedt C. Nucleic acid–based therapeutics in orphan neurological disorders: Recent developments. Front Mol Biosci 2021; 8: : 643681.. [http://dx.doi.org/10.3389/fmolb.2021.643681] [PMID: 33996898]

[74] Azam S, Haque ME, Balakrishnan R, Kim IS, Choi DK. The ageing brain: molecular and cellular basis of neurodegeneration. Front Cell Dev Biol 2021; 9: : 683459.. [http://dx.doi.org/10.3389/fcell.2021.683459] [PMID: 34485280]

[75] Pandya VA, Patani R. Region-specific vulnerability in neurodegeneration: lessons from normal ageing. Ageing Res Rev 2021; 67: 101311.
[http://dx.doi.org/10.1016/j.arr.2021.101311] [PMID: 33639280]

[76] Lee J, Kim HJ. Normal aging induces changes in the brain and neurodegeneration progress: review of the structural, biochemical, metabolic, cellular, and molecular changes. Front Aging Neurosci 2022; 14: 931536.
[http://dx.doi.org/10.3389/fnagi.2022.931536] [PMID: 35847660]

[77] Traxler L, Lucciola R, Herdy JR, Jones JR, Mertens J, Gage FH. Neural cell state shifts and fate loss in ageing and age-related diseases. Nat Rev Neurol 2023; 19(7): 434-43.
[http://dx.doi.org/10.1038/s41582-023-00815-0] [PMID: 37268723]

[78] Palanki R, Peranteau WH, Mitchell MJ. Delivery technologies for in utero gene therapy. Adv Drug Deliv Rev 2021; 169: 51-62.
[http://dx.doi.org/10.1016/j.addr.2020.11.002] [PMID: 33181188]

[79] Charlesworth CT, Hsu I, Wilkinson AC, Nakauchi H. Immunological barriers to haematopoietic stem cell gene therapy. Nat Rev Immunol 2022; 22(12): 719-33.
[http://dx.doi.org/10.1038/s41577-022-00698-0] [PMID: 35301483]

[80] Chehelgerdi M, Chehelgerdi M. The use of RNA-based treatments in the field of cancer immunotherapy. Mol Cancer 2023; 22(1): 106.
[http://dx.doi.org/10.1186/s12943-023-01807-w] [PMID: 37420174]

[81] Tian Z, Liang G, Cui K, *et al.* Insight into the prospects for RNAi therapy of cancer. Front Pharmacol 2021; 12: 644718.
[http://dx.doi.org/10.3389/fphar.2021.644718] [PMID: 33796026]

[82] Toden S, Zumwalt TJ, Goel A. Non-coding RNAs and potential therapeutic targeting in cancer. Biochim Biophys Acta Rev Cancer. 2021 Jan; 1875(1): 188491.

[83] Huang D, Ma N, Li X, *et al.* Advances in single-cell RNA sequencing and its applications in cancer research. J Hematol Oncol 2023; 16(1): 98.
[http://dx.doi.org/10.1186/s13045-023-01494-6] [PMID: 37612741]

[84] Chen X, Cubillos-Ruiz JR. Endoplasmic reticulum stress signals in the tumour and its microenvironment. Nat Rev Cancer 2021; 21(2): 71-88.
[http://dx.doi.org/10.1038/s41568-020-00312-2] [PMID: 33214692]

[85] O'Grady T, Njock MS, Lion M, *et al.* Sorting and packaging of RNA into extracellular vesicles shape intracellular transcript levels. BMC Biol 2022; 20(1): 72.
[http://dx.doi.org/10.1186/s12915-022-01277-4] [PMID: 35331218]

[86] Rani V, Sengar RS. Biogenesis and mechanisms of microRNA-mediated gene regulation. Biotechnol Bioeng 2022; 119(3): 685-92.
[http://dx.doi.org/10.1002/bit.28029] [PMID: 34979040]

[87] Hwang HJ, Ha H, Lee BS, Kim BH, Song HK, Kim YK. LC3B is an RNA-binding protein to trigger rapid mRNA degradation during autophagy. Nat Commun 2022; 13(1): 1436.
[http://dx.doi.org/10.1038/s41467-022-29139-1] [PMID: 35302060]

[88] Kim YK. RNA therapy: rich history, various applications and unlimited future prospects. Exp Mol Med 2022; 54(4): 455-65.
[http://dx.doi.org/10.1038/s12276-022-00757-5] [PMID: 35440755]

[89] Winkle M, El-Daly SM, Fabbri M, Calin GA. Noncoding RNA therapeutics — challenges and potential solutions. Nat Rev Drug Discov 2021; 20(8): 629-51.
[http://dx.doi.org/10.1038/s41573-021-00219-z] [PMID: 34145432]

[90] Tani H. Recent advances and prospects in RNA drug development. Int J Mol Sci 2024; 25(22): 12284.
[http://dx.doi.org/10.3390/ijms252212284] [PMID: 39596348]

[91] Ranasinghe P, Addison ML, Dear JW, Webb DJ. Small interfering RNA: Discovery, pharmacology and clinical development—An introductory review. Br J Pharmacol 2023; 180(21): 2697-720. [http://dx.doi.org/10.1111/bph.15972] [PMID: 36250252]

[92] Smith ES, Whitty E, Yoo B, Moore A, Sempere LF, Medarova Z. Clinical applications of short non-coding RNA-based therapies in the era of precision medicine. Cancers (Basel) 2022; 14(6): 1588. [http://dx.doi.org/10.3390/cancers14061588] [PMID: 35326738]

[93] Damase TR, Sukhovershin R, Boada C, Taraballi F, Pettigrew RI, Cooke JP. The limitless future of RNA therapeutics. Front Bioeng Biotechnol 2021; 9: : 628137.. [http://dx.doi.org/10.3389/fbioe.2021.628137] [PMID: 33816449]

[94] Robinson EL, Port JD. Utilization and potential of RNA-based therapies in cardiovascular disease. JACC Basic Transl Sci 2022; 7(9): 956-69. [http://dx.doi.org/10.1016/j.jacbts.2022.02.003] [PMID: 36317129]

[95] Webb C, Ip S, Bathula NV, *et al.* Current status and future perspectives on MRNA drug manufacturing. Mol Pharm 2022; 19(4): 1047-58. [http://dx.doi.org/10.1021/acs.molpharmaceut.2c00010] [PMID: 35238565]

[96] Karmacharya P, Patil BR, Kim JO. Recent advancements in lipid–mRNA nanoparticles as a treatment option for cancer immunotherapy. J Pharm Investig 2022; 52(4): 415-26. [http://dx.doi.org/10.1007/s40005-022-00569-9] [PMID: 35369363]

[97] Li D, Liu C, Li Y, *et al.* Messenger RNA-based therapeutics and vaccines: what's beyond COVID-19? ACS Pharmacol Transl Sci 2023; 6(7): 943-69. [http://dx.doi.org/10.1021/acsptsci.3c00047] [PMID: 37470024]

[98] Wang P, Zhou Y, Richards AM. Effective tools for RNA-derived therapeutics: siRNA interference or miRNA mimicry. Theranostics 2021; 11(18): 8771-96. [http://dx.doi.org/10.7150/thno.62642] [PMID: 34522211]

[99] Ma B, Wang S, Wu W, *et al.* Mechanisms of circRNA/lncRNA-miRNA interactions and applications in disease and drug research. Biomed Pharmacother 2023; 162: 114672. [http://dx.doi.org/10.1016/j.biopha.2023.114672] [PMID: 37060662]

[100] Liu Y, Zhou Z, Guo S, *et al.* Transcriptome analysis reveals key miRNA–mRNA pathways in ovarian tissues of Yunshang black goats with different kidding numbers. Front Endocrinol (Lausanne) 2022; 13: 883663. [http://dx.doi.org/10.3389/fendo.2022.883663] [PMID: 35663314]

[101] Nguyen LD, Wei Z, Silva MC, *et al.* Small molecule regulators of microRNAs identified by high-throughput screen coupled with high-throughput sequencing. Nat Commun 2023; 14(1): 7575. [http://dx.doi.org/10.1038/s41467-023-43293-0] [PMID: 37989753]

[102] Yu W, Yu W, Yang Y, Lü Y. Exploring the key genes and identification of potential diagnosis biomarkers in Alzheimer's disease using bioinformatics analysis. Front Aging Neurosci 2021; 13: 602781. [http://dx.doi.org/10.3389/fnagi.2021.602781] [PMID: 34194312]

[103] Sirkis DW, Bonham LW, Johnson TP, La Joie R, Yokoyama JS. Dissecting the clinical heterogeneity of early-onset Alzheimer's disease. Mol Psychiatry 2022; 27(6): 2674-88. [http://dx.doi.org/10.1038/s41380-022-01531-9] [PMID: 35393555]

[104] Schwartzentruber J, Cooper S, Liu JZ, *et al.* Genome-wide meta-analysis, fine-mapping and integrative prioritization implicate new Alzheimer's disease risk genes. Nat Genet 2021; 53(3): 392-402. [http://dx.doi.org/10.1038/s41588-020-00776-w] [PMID: 33589840]

[105] Raulin AC, Doss SV, Trottier ZA, Ikezu TC, Bu G, Liu CC. ApoE in Alzheimer's disease: pathophysiology and therapeutic strategies. Mol Neurodegener 2022; 17(1): 72. [http://dx.doi.org/10.1186/s13024-022-00574-4] [PMID: 36348357]

[106] Cheng Z, Li M, Dey R, Chen Y. Nanomaterials for cancer therapy: current progress and perspectives. J Hematol Oncol 2021; 14(1): 85.
[http://dx.doi.org/10.1186/s13045-021-01096-0] [PMID: 34059100]

[107] Garin N, Sole N, Lucas B, *et al.* Drug related problems in clinical practice: a cross-sectional study on their prevalence, risk factors and associated pharmaceutical interventions. Sci Rep 2021; 11(1): 883.
[http://dx.doi.org/10.1038/s41598-020-80560-2] [PMID: 33441854]

[108] Huang R, Zhou PK. DNA damage repair: historical perspectives, mechanistic pathways and clinical translation for targeted cancer therapy. Signal Transduct Target Ther 2021; 6(1): 254.
[http://dx.doi.org/10.1038/s41392-021-00648-7] [PMID: 34238917]

[109] Fan K, Francis L, Aslam MM, *et al.* Investigation of the independent role of a rare APOE variant (L28P; APOE*4Pittsburgh) in late-onset Alzheimer disease. Neurobiol Aging 2023; 122: 107-11.
[http://dx.doi.org/10.1016/j.neurobiolaging.2022.11.007] [PMID: 36528961]

[110] Blumenfeld J, Yip O, Kim MJ, Huang Y. Cell type-specific roles of APOE4 in Alzheimer disease. Nat Rev Neurosci 2024; 25(2): 91-110.
[http://dx.doi.org/10.1038/s41583-023-00776-9] [PMID: 38191720]

[111] Huang Y, Mahley RW. Apolipoprotein E: Structure and function in lipid metabolism, neurobiology, and Alzheimer's diseases. Neurobiol Dis 2014; 72(Pt A): 3-12.
[http://dx.doi.org/10.1016/j.nbd.2014.08.025] [PMID: 25173806]

[112] Williams T, Borchelt DR, Chakrabarty P. Therapeutic approaches targeting Apolipoprotein E function in Alzheimer's disease. Mol Neurodegener 2020; 15(1): 8.
[http://dx.doi.org/10.1186/s13024-020-0358-9] [PMID: 32005122]

[113] Taylor HA, Przemylska L, Clavane EM, Meakin PJ. BACE1: More than just a β-secretase. Obes Rev 2022; 23(7): e13430.
[http://dx.doi.org/10.1111/obr.13430] [PMID: 35119166]

[114] Evin G, Hince C. BACE1 as a therapeutic target in Alzheimer's disease: rationale and current status. Drugs Aging 2013; 30(10): 755-64.
[http://dx.doi.org/10.1007/s40266-013-0099-3] [PMID: 23842796]

[115] Dawkins E, Small DH. Insights into the physiological function of the β-amyloid precursor protein: beyond Alzheimer's disease. J Neurochem 2014; 129(5): 756-69.
[http://dx.doi.org/10.1111/jnc.12675] [PMID: 24517464]

[116] Chang JL, Hinrich AJ, Roman B, *et al.* Targeting amyloid-β precursor protein, APP, splicing with antisense oligonucleotides reduces toxic amyloid-β production. Mol Ther 2018; 26(6): 1539-51.
[http://dx.doi.org/10.1016/j.ymthe.2018.02.029] [PMID: 29628304]

[117] Bagaria J, Bagyinszky E, An SSA. Genetics, functions, and clinical impact of presenilin-1 (PSEN1) gene. Int J Mol Sci 2022; 23(18): 10970.
[http://dx.doi.org/10.3390/ijms231810970] [PMID: 36142879]

[118] Moore B, Sharma A, Goulet M, *et al.* Developing a Gene Therapy for the Treatment of Autosomal Dominant Alzheimer's Disease. Hum Gene Ther 2023; 34(19-20): 1049-63.
[http://dx.doi.org/10.1089/hum.2023.092] [PMID: 37578141]

[119] Cai Y, An SSA, Kim S. Mutations in presenilin 2 and its implications in Alzheimer's disease and other dementia-associated disorders. Clin Interv Aging 2015; 10: 1163-72.
[PMID: 26203236]

[120] Yaghmoor F, Noorsaeed A, Alsaggaf S, *et al.* The role of TREM2 in Alzheimer's disease and other neurological disorders. J Alzheimers Dis Parkinsonism 2014; 4(5): 160.
[http://dx.doi.org/10.4172/2161-0460.1000160] [PMID: 25664220]

[121] Wolf EM, Fingleton B, Hasty AH. The therapeutic potential of TREM2 in cancer. Front Oncol 2022; 12: : 984193..

[http://dx.doi.org/10.3389/fonc.2022.984193] [PMID: 36119485]

[122] Townley RA, Boeve BF, Benarroch EE. Progranulin. Neurology 2018; 90(3): 118-25. [http://dx.doi.org/10.1212/WNL.0000000000004840] [PMID: 29263224]

[123] Boylan MA, Pincetic A, Romano G, Tatton N, Kenkare-Mitra S, Rosenthal A. Targeting progranulin as an immuno-neurology therapeutic approach. Int J Mol Sci 2023; 24(21): 15946. [http://dx.doi.org/10.3390/ijms242115946] [PMID: 37958929]

[124] Zhang CC, Xing A, Tan MS, Tan L, Yu JT. The role of MAPT in neurodegenerative diseases: genetics, mechanisms and therapy. Mol Neurobiol 2016; 53(7): 4893-904. [http://dx.doi.org/10.1007/s12035-015-9415-8] [PMID: 26363795]

[125] VandeVrede L, Boxer AL, Polydoro M. Targeting tau: Clinical trials and novel therapeutic approaches. Neurosci Lett 2020; 731: : 134919.. [http://dx.doi.org/10.1016/j.neulet.2020.134919] [PMID: 32380145]

[126] Foster EM, Dangla-Valls A, Lovestone S, Ribe EM, Buckley NJ. Clusterin in Alzheimer's disease: mechanisms, genetics, and lessons from other pathologies. Front Neurosci 2019; 13: 164. [http://dx.doi.org/10.3389/fnins.2019.00164] [PMID: 30872998]

[127] Mamun MAA, Zheng YC, Wang N, *et al.* Decoding CLU (Clusterin): Conquering cancer treatment resistance and immunological barriers. Int Immunopharmacol 2024; 137: 112355. [http://dx.doi.org/10.1016/j.intimp.2024.112355] [PMID: 38851158]

[128] Tan MS, Yu JT, Tan L. Bridging integrator 1 (BIN1): form, function, and Alzheimer's disease. Trends Mol Med 2013; 19(10): 594-603. [http://dx.doi.org/10.1016/j.molmed.2013.06.004] [PMID: 23871436]

[129] Thomas S, Mercogliano G, Prendergast GC. Bin1 targeted immunotherapy alters the status of the enteric neurons and the microbiome during ulcerative colitis treatment. PLoS One 2022; 17(11): e0276910.. [http://dx.doi.org/10.1371/journal.pone.0276910] [PMID: 36322599]

[130] Griciuc A, Serrano-Pozo A, Parrado AR, *et al.* Alzheimer's disease risk gene CD33 inhibits microglial uptake of amyloid beta. Neuron 2013; 78(4): 631-43. [http://dx.doi.org/10.1016/j.neuron.2013.04.014] [PMID: 23623698]

[131] Jiang T, Yu JT, Hu N, Tan MS, Zhu XC, Tan L. CD33 in Alzheimer's disease. Mol Neurobiol 2014; 49(1): 529-35. [http://dx.doi.org/10.1007/s12035-013-8536-1] [PMID: 23982747]

[132] Yin RH, Yu JT, Tan L. The role of SORL1 in Alzheimer's disease. Mol Neurobiol 2015; 51(3): 909-18. [http://dx.doi.org/10.1007/s12035-014-8742-5] [PMID: 24833601]

[133] Al-Akhrass H, Pietilä M, Lilja J, *et al.* Sortilin-related receptor is a druggable therapeutic target in breast cancer. Mol Oncol 2022; 16(1): 116-29. [http://dx.doi.org/10.1002/1878-0261.13106] [PMID: 34564954]

[134] De Roeck A, Van Broeckhoven C, Sleegers K. The role of ABCA7 in Alzheimer's disease: evidence from genomics, transcriptomics and methylomics. Acta Neuropathol 2019; 138(2): 201-20. [http://dx.doi.org/10.1007/s00401-019-01994-1] [PMID: 30903345]

[135] Dib S, Pahnke J, Gosselet F. Role of ABCA7 in Human Health and in Alzheimer's Disease. Int J Mol Sci 2021; 22(9): 4603. [http://dx.doi.org/10.3390/ijms22094603] [PMID: 33925691]

[136] Baysac K, Sun G, Nakano H, *et al.* PLCG2-associated immune dysregulation (PLAID) comprises broad and distinct clinical presentations related to functional classes of genetic variants. J Allergy Clin Immunol 2024; 153(1): 230-42. [http://dx.doi.org/10.1016/j.jaci.2023.08.036] [PMID: 37769878]

[137] Mitroshina EV, Vedunova MV. The role of oxygen homeostasis and the HIF-1 factor in the development of neurodegeneration. Int J Mol Sci 2024; 25(9): 4581. [http://dx.doi.org/10.3390/ijms25094581] [PMID: 38731800]

[138] Bolognini D, Halgren A, Lou RN, *et al.* Recurrent evolution and selection shape structural diversity at the amylase locus. Nature 2024; 634(8034): 617-25. [http://dx.doi.org/10.1038/s41586-024-07911-1] [PMID: 39232174]

[139] Monroig Ó, Shu-Chien AC, Kabeya N, Tocher DR, Castro LFC. Desaturases and elongases involved in long-chain polyunsaturated fatty acid biosynthesis in aquatic animals: From genes to functions. Prog Lipid Res 2022; 86: 101157. [http://dx.doi.org/10.1016/j.plipres.2022.101157] [PMID: 35104467]

[140] Luquette LJ, Miller MB, Zhou Z, *et al.* Single-cell genome sequencing of human neurons identifies somatic point mutation and indel enrichment in regulatory elements. Nat Genet 2022; 54(10): 1564-71. [http://dx.doi.org/10.1038/s41588-022-01180-2] [PMID: 36163278]

[141] Olesen MA, Villavicencio-Tejo F, Quintanilla RA. The use of fibroblasts as a valuable strategy for studying mitochondrial impairment in neurological disorders. Transl Neurodegener 2022; 11(1): 36. [http://dx.doi.org/10.1186/s40035-022-00308-y] [PMID: 35787292]

[142] Gustavsson A, Norton N, Fast T, *et al.* Global estimates on the number of persons across the Alzheimer's disease continuum. Alzheimers Dement 2023; 19(2): 658-70. [http://dx.doi.org/10.1002/alz.12694] [PMID: 35652476]

[143] Reitz C, Pericak-Vance MA, Foroud T, Mayeux R. A global view of the genetic basis of Alzheimer disease. Nat Rev Neurol 2023; 19(5): 261-77. [http://dx.doi.org/10.1038/s41582-023-00789-z] [PMID: 37024647]

[144] Wightman DP, Jansen IE, Savage JE, *et al.* A genome-wide association study with 1,126,563 individuals identifies new risk loci for Alzheimer's disease. Nat Genet 2021; 53(9): 1276-82. [http://dx.doi.org/10.1038/s41588-021-00921-z] [PMID: 34493870]

[145] Georgiou M, Fujinami K, Michaelides M. Inherited retinal diseases: Therapeutics, clinical trials and end points—A review. Clin Exp Ophthalmol 2021; 49(3): 270-88. [http://dx.doi.org/10.1111/ceo.13917] [PMID: 33686777]

[146] Baggen J, Vanstreels E, Jansen S, Daelemans D. Cellular host factors for SARS-CoV-2 infection. Nat Microbiol 2021; 6(10): 1219-32. [http://dx.doi.org/10.1038/s41564-021-00958-0] [PMID: 34471255]

[147] Nooraei S, Bahrulolum H, Hoseini ZS, *et al.* Virus-like particles: preparation, immunogenicity and their roles as nanovaccines and drug nanocarriers. J Nanobiotechnology 2021; 19(1): 59. [http://dx.doi.org/10.1186/s12951-021-00806-7] [PMID: 33632278]

[148] Hendel SJ, Shoulders MD. Directed evolution in mammalian cells. Nat Methods 2021; 18(4): 346-57. [http://dx.doi.org/10.1038/s41592-021-01090-x] [PMID: 33828274]

[149] Di Minno G, Miesbach W, Castaman G, Peyvandi F. Next-generation strategies to improve safety and efficacy of adeno-associated virus-based gene therapy for hemophilia: lessons from clinical trials in other gene therapies. Haematologica 2024; 109(12): 3879-91. [http://dx.doi.org/10.3324/haematol.2023.284622] [PMID: 38450517]

[150] Saute JAM, Giugliani R, Baldo G, Pasqualim G, Souza CFM, Matte US. Clinical trials to gene therapy development and production in Brazil: a review. Lancet Reg Health Am. 2025; 43: 100994.

[151] Britten-Jones AC, Jin R, Gocuk SA, *et al.* The safety and efficacy of gene therapy treatment for monogenic retinal and optic nerve diseases: A systematic review. Genet Med 2022; 24(3): 521-34. [http://dx.doi.org/10.1016/j.gim.2021.10.013] [PMID: 34906485]

[152] Naso MF, Tomkowicz B, Perry WL III, Strohl WR. Adeno-associated virus (AAV) as a vector for gene therapy. BioDrugs 2017; 31(4): 317-34.

[http://dx.doi.org/10.1007/s40259-017-0234-5] [PMID: 28669112]

[153] Zeng CY, Yang TT, Zhou HJ, *et al.* Lentiviral vector–mediated overexpression of Klotho in the brain improves Alzheimer's disease–like pathology and cognitive deficits in mice. Neurobiol Aging 2019; 78: 18-28.
[http://dx.doi.org/10.1016/j.neurobiolaging.2019.02.003] [PMID: 30851437]

[154] Harris SA, Harris EA. Molecular mechanisms for herpes simplex virus type 1 pathogenesis in Alzheimer's disease. Front Aging Neurosci 2018; 10: 48.
[http://dx.doi.org/10.3389/fnagi.2018.00048] [PMID: 29559905]

[155] Ahmed T. Lipid nanoparticle mediated small interfering RNA delivery as a potential therapy for Alzheimer's disease. Eur J Neurosci 2024; 59(11): 2915-54.
[http://dx.doi.org/10.1111/ejn.16336] [PMID: 38622050]

[156] La Barbera L, Mauri E, D'Amelio M, Gori M. Functionalization strategies of polymeric nanoparticles for drug delivery in Alzheimer's disease: Current trends and future perspectives. Front Neurosci 2022; 16: 939855.
[http://dx.doi.org/10.3389/fnins.2022.939855] [PMID: 35992936]

[157] Sun M, Chen Z. Unveiling the Complex Role of Exosomes in Alzheimer's Disease. J Inflamm Res 2024; 17: 3921-48.
[http://dx.doi.org/10.2147/JIR.S466821] [PMID: 38911990]

[158] Al-azzawi S, Masheta D, Guildford A, Phillips G, Santin M. A peptide-based nanocarrier for an enhanced delivery and targeting of flurbiprofen into the brain for the treatment of Alzheimer's disease: An in vitro study. Nanomaterials (Basel) 2020; 10(8): 1590.
[http://dx.doi.org/10.3390/nano10081590] [PMID: 32823499]

[159] Lu L, Yu X, Cai Y, Sun M, Yang H. Application of CRISPR/Cas9 in Alzheimer's Disease. Front Neurosci 2021; 15: : 803894..
[http://dx.doi.org/10.3389/fnins.2021.803894] [PMID: 34992519]

[160] Murakami K, Izuo N, Bitan G. Aptamers targeting amyloidogenic proteins and their emerging role in neurodegenerative diseases. J Biol Chem 2022; 298(1): 101478..
[http://dx.doi.org/10.1016/j.jbc.2021.101478] [PMID: 34896392]

[161] Chaparro CIP, Simões BT, Borges JP, Castanho MARB, Soares PIP, Neves V. A promising approach: magnetic nanosystems for Alzheimer's disease theranostics. Pharmaceutics 2023; 15(9): 2316.
[http://dx.doi.org/10.3390/pharmaceutics15092316] [PMID: 37765284]

[162] Gandhi K, Barzegar-Fallah A, Banstola A, Rizwan SB, Reynolds JNJ. Ultrasound-mediated blood–brain barrier disruption for drug delivery: a systematic review of protocols, efficacy, and safety outcomes from preclinical and clinical studies. Pharmaceutics 2022; 14(4): 833.
[http://dx.doi.org/10.3390/pharmaceutics14040833] [PMID: 35456667]

[163] Gong B, Cheng W, Ji W, *et al.* Hydrogel: A promising new technique for treating Alzheimer's disease. J Transl Int Med 2022; 10(3): 188-90.
[http://dx.doi.org/10.2478/jtim-2022-0008] [PMID: 35702184]

[164] Mroziak M, Kozłowski G, Kołodziejczyk W, *et al.* Dendrimers—Novel Therapeutic Approaches for Alzheimer's Disease. Biomedicines 2024; 12(8): 1899.
[http://dx.doi.org/10.3390/biomedicines12081899] [PMID: 39200363]

[165] Muolokwu CE, Chaulagain B, Gothwal A, *et al.* Functionalized nanoparticles to deliver nucleic acids to the brain for the treatment of Alzheimer's disease. Front Pharmacol 2024; 15: : 1405423..
[http://dx.doi.org/10.3389/fphar.2024.1405423] [PMID: 38855744]

[166] Mendiratta G, Ke E, Aziz M, Liarakos D, Tong M, Stites EC. Cancer gene mutation frequencies for the U.S. population. Nat Commun 2021; 12(1): 5961.
[http://dx.doi.org/10.1038/s41467-021-26213-y] [PMID: 34645806]

[167] Long J, Wang D, Wang A, *et al.* A mutation-based gene set predicts survival benefit after

immunotherapy across multiple cancers and reveals the immune response landscape. Genome Med 2022; 14(1): 20.
[http://dx.doi.org/10.1186/s13073-022-01024-y] [PMID: 35197093]

[168] Arriola E, Bernabé R, Campelo RG, *et al.* Cost-effectiveness of next-generation sequencing versus single-gene testing for the molecular diagnosis of patients with metastatic non–small-cell lung cancer from the perspective of Spanish reference centers. JCO Precis Oncol 2023; 7(7): e2200546.
[http://dx.doi.org/10.1200/PO.22.00546] [PMID: 36862967]

[169] Hu H, Ye Z, Qin Y, *et al.* Mutations in key driver genes of pancreatic cancer: molecularly targeted therapies and other clinical implications. Acta Pharmacol Sin 2021; 42(11): 1725-41.
[http://dx.doi.org/10.1038/s41401-020-00584-2] [PMID: 33574569]

[170] Khorkova O, Stahl J, Joji A, Volmar CH, Wahlestedt C. Amplifying gene expression with RNA-targeted therapeutics. Nat Rev Drug Discov 2023; 22(7): 539-61.
[http://dx.doi.org/10.1038/s41573-023-00704-7] [PMID: 37253858]

[171] Ma Y, Liao J, Cheng H, Yang Q, Yang H. Advanced gene therapy system for the treatment of solid tumour: A review. Mater Today Bio 2024; 27: : 101138..
[http://dx.doi.org/10.1016/j.mtbio.2024.101138] [PMID: 39027677]

[172] Lei T, Wang Y, Zhang Y, *et al.* Leveraging CRISPR gene editing technology to optimize the efficacy, safety and accessibility of CAR T-cell therapy. Leukemia 2024; 38(12): 2517-43.
[http://dx.doi.org/10.1038/s41375-024-02444-y] [PMID: 39455854]

[173] Zhang H, Qin C, An C, *et al.* Application of the CRISPR/Cas9-based gene editing technique in basic research, diagnosis, and therapy of cancer. Mol Cancer 2021; 20(1): 126.
[http://dx.doi.org/10.1186/s12943-021-01431-6] [PMID: 34598686]

[174] Helm M, Schmidt-Dengler MC, Weber M, Motorin Y. General principles for the detection of modified nucleotides in RNA by specific reagents. Adv Biol 2021; 5(10): 2100866.
[http://dx.doi.org/10.1002/adbi.202100866] [PMID: 34535986]

[175] Ye L, Gribling-Burrer AS, Bohn P, *et al.* Short- and long-range interactions in the HIV-1 5′ UTR regulate genome dimerization and packaging. Nat Struct Mol Biol 2022; 29(4): 306-19.
[http://dx.doi.org/10.1038/s41594-022-00746-2] [PMID: 35347312]

[176] Iyer AA, Saade D, Bharucha-Goebel D, *et al.* Ethical challenges for a new generation of early-phase pediatric gene therapy trials. Genet Med 2021; 23(11): 2057-66.
[http://dx.doi.org/10.1038/s41436-021-01245-3] [PMID: 34234300]

[177] Karunarathna I, *et al.* The ethics of genetic editing: Navigating the future of science. Proceedings of the Uva Clinical Research 2024.

[178] Newman NJ, Yu-Wai-Man P, Carelli V, *et al.* Efficacy and safety of intravitreal gene therapy for Leber hereditary optic neuropathy treated within 6 months of disease onset. Ophthalmology 2021; 128(5): 649-60.
[http://dx.doi.org/10.1016/j.ophtha.2020.12.012] [PMID: 33451738]

[179] Bessoltane N, Charlot F, Guyon-Debast A, *et al.* Genome-wide specificity of plant genome editing by both CRISPR–Cas9 and TALEN. Sci Rep 2022; 12(1): 9330.
[http://dx.doi.org/10.1038/s41598-022-13034-2] [PMID: 35665758]

[180] Houghton BC, Panchal N, Haas SA, *et al.* Genome editing with TALEN, CRISPR-Cas9 and CRISPR-Cas12a in combination with AAV6 homology donor restores T cell function for XLP. Front Genome Editing 2022; 4: 828489.
[http://dx.doi.org/10.3389/fgeed.2022.828489] [PMID: 35677600]

[181] Haider R. Pharmaceutical and biopharmaceuticals industries: Revolutionizing healthcare. Asian Journal of Natural Sciences 2023; 2(2): 69-80.
[http://dx.doi.org/10.55927/ajns.v2i2.4179]

[182] Madhavan A, Arun KB, Sindhu R, *et al.* Customized yeast cell factories for biopharmaceuticals: from

cell engineering to process scale up. Microb Cell Fact 2021; 20(1): 124. [http://dx.doi.org/10.1186/s12934-021-01617-z] [PMID: 34193127]

[183] Han X, Alu A, Liu H, *et al.* Biomaterial-assisted biotherapy: A brief review of biomaterials used in drug delivery, vaccine development, gene therapy, and stem cell therapy. Bioact Mater 2022; 17: 29-48.
[http://dx.doi.org/10.1016/j.bioactmat.2022.01.011] [PMID: 35386442]

[184] Robinson J, Barker DJ, Marsh SGE. 25 years of the IPD-IMGT/HLA D atabase. HLA 2024; 103(6): e15549.
[http://dx.doi.org/10.1111/tan.15549] [PMID: 38936817]

[185] Ebrahimi SB, Samanta D. Engineering protein-based therapeutics through structural and chemical design. Nat Commun 2023; 14(1): 2411.
[http://dx.doi.org/10.1038/s41467-023-38039-x] [PMID: 37105998]

[186] Hayenhjelm M, Nordlund C. Positions on the Germline Boundary: Stop, Wait, or Go? In: The Risks and Ethics of Human Gene Editing. Technikzukünfte, Wissenschaft und Gesellschaft / Futures of Technology, Science and Society. Wiesbaden: Springer VS; 2025. p. 45–68.
[http://dx.doi.org/10.1007/978-3-658-46979-5_3]

[187] Hayenhjelm M, Nordlund C. The Risks and Ethics of Human Gene Editing: A Philosophical Guide to the Arguments. Springer Nature 2025.
[http://dx.doi.org/10.1007/978-3-658-46979-5]

[188] Fan T, Zhang M, Yang J, Zhu Z, Cao W, Dong C. Therapeutic cancer vaccines: advancements, challenges and prospects. Signal Transduct Target Ther 2023; 8(1): 450.
[http://dx.doi.org/10.1038/s41392-023-01674-3] [PMID: 38086815]

[189] Gupta M, Wahi A, Sharma P, *et al.* Recent advances in cancer vaccines: challenges, achievements, and futuristic prospects. Vaccines (Basel) 2022; 10(12): 2011.
[http://dx.doi.org/10.3390/vaccines10122011] [PMID: 36560420]

[190] Ferreira P, Choupina AB. CRISPR/Cas9 a simple, inexpensive and effective technique for gene editing. Mol Biol Rep 2022; 49(7): 7079-86.
[http://dx.doi.org/10.1007/s11033-022-07442-w] [PMID: 35716290]

[191] Aljabali AAA, El-Tanani M, Tambuwala MM. Principles of CRISPR-Cas9 technology: Advancements in genome editing and emerging trends in drug delivery. J Drug Deliv Sci Technol 2024; 92: : 105338..
[http://dx.doi.org/10.1016/j.jddst.2024.105338]

[192] Ali G, Tariq MA, Shahid K, Ahmad FJ, Akram J. Advances in genome editing: the technology of choice for precise and efficient β-thalassemia treatment. Gene Ther 2021; 28(1-2): 6-15.
[http://dx.doi.org/10.1038/s41434-020-0153-9] [PMID: 32355226]

[193] Baloglu MC, Celik Altunoglu Y, Baloglu P, Yildiz AB, Türkölmez N, Özden Çiftçi Y. Gene-editing technologies and applications in legumes: Progress, evolution, and future prospects. Front Genet 2022; 13: : 859437..
[http://dx.doi.org/10.3389/fgene.2022.859437] [PMID: 35836569]

[194] Sun L, Lai M, Ghouri F, *et al.* Modern plant breeding techniques in crop improvement and genetic diversity: from molecular markers and gene editing to artificial intelligence—A critical review. Plants 2024; 13(19): 2676.
[http://dx.doi.org/10.3390/plants13192676] [PMID: 39409546]

[195] Zhang J, Fei Y, Sun L, Zhang QC. Advances and opportunities in RNA structure experimental determination and computational modeling. Nat Methods 2022; 19(10): 1193-207.
[http://dx.doi.org/10.1038/s41592-022-01623-y] [PMID: 36203019]

[196] Falese JP, Donlic A, Hargrove AE. Targeting RNA with small molecules: from fundamental principles towards the clinic. Chem Soc Rev 2021; 50(4): 2224-43.

[http://dx.doi.org/10.1039/D0CS01261K] [PMID: 33458725]

[197] Gurung AB, Ali MA, Lee J, Farah MA, Al-Anazi KM. An updated review of computer-aided drug design and its application to COVID-19. BioMed Res Int 2021; 2021(1): 8853056. [http://dx.doi.org/10.1155/2021/8853056] [PMID: 34258282]

[198] Murray A, Hearn J, Turner S. The emerging landscape of RNA-targeted small molecules. Nat Rev Drug Discov. 2021 May; 20(5): 334.

[199] Lee MJ, Lee I, Wang K. Recent advances in RNA therapy and its carriers to treat the single-gene neurological disorders. Biomedicines 2022; 10(1): 158. [http://dx.doi.org/10.3390/biomedicines10010158] [PMID: 35052837]

CHAPTER 9

Comorbidities and their Impact on Disease Progression: Focused on Alzheimer's Disease

Ganesh Sonawane[1,*], **Kajal Pansare**[1], **Deepak Sonawane**[1], **Chandrashekhar Patil**[1] and **Sunil Mahajan**[1]

[1] *Divine College of Pharmacy, Satana, Nashik, Maharashtra 423301, India*

Abstract: Alzheimer's Disease (AD) is a progressive neurodegenerative disorder that primarily affects cognitive and functional abilities. While much study has focused on the basic pathology of AD, the role of comorbidities in influencing its progression has gained increasing attention. Comorbidities, such as cardiovascular diseases, diabetes, depression, and sleep disorders, are highly prevalent in individuals with AD and significantly impact its trajectory, outcomes, and management. This chapter focuses on the intricate interplay between Alzheimer's disease and common comorbid conditions, highlighting how these comorbidities contribute to faster cognitive decline, worsened behavioral symptoms, and increased caregiver burden. Key mechanisms linking comorbidities to AD progression, including chronic inflammation, vascular dysfunction, metabolic dysregulation, and oxidative stress, are discussed in detail. These processes increase neuronal damage and can accelerate the formation of amyloid-beta and tau proteins, indicators of AD pathology. The chapter also addresses the clinical challenges posed by comorbidities in AD patients, such as diagnostic complexities, polypharmacy, and the lack of tailored management guidelines. Practical strategies for effective management are outlined, including comprehensive assessments, personalized care plans, and the merging of pharmacological and non-pharmacological treatments. The role of multidisciplinary teams in optimizing patient outcomes is emphasized, along with emerging research into novel therapeutic targets and biomarkers for early detection and intervention. This chapter aims to provide a significant resource for healthcare practitioners, researchers, and caregivers, promoting a more complete approach to managing Alzheimer's disease in the presence of comorbidities.

Keywords: Alzheimer's disease, Biomarkers, Cognitive decline, Comorbidities, Integrated care, Metabolic disorders, Vascular dysfunction.

* **Corresponding author Ganesh Sonawane:** Divine College of Pharmacy, Satana, Nashik, Maharashtra 423301, India; E-mail: gbsonawane8@gmail.com

Prashant Kumar Dhakad, Raghav Mishra, Shweta Sharma, Sudhanshu Mishra & Nishant Gaur (Eds.)

INTRODUCTION

Comorbidities are defined as the coexistence of one or more chronic medical conditions alongside a primary disease, and they play a critical role in determining the course, management, and prognosis of patients [1, 2]. Their presence often complicates therapeutic strategies, as the interaction between multiple disease processes can produce synergistic or antagonistic effects, leading to increased morbidity and reduced treatment effectiveness.

In the context of Alzheimer's Disease (AD), a progressive neurodegenerative disorder primarily driven by the accumulation of amyloid-beta plaques and neurofibrillary tangles in the brain [3], comorbidities are especially significant. AD is the leading cause of dementia worldwide, with prevalence rising sharply with advancing age, and is associated with profound cognitive deficits, behavioral alterations, and impaired ability to perform activities of daily living [4, 5]. These clinical manifestations not only compromise the quality of life of patients but also place a substantial burden on caregivers and healthcare systems.

Due to the aging process and the systemic effects of neurodegeneration, patients with AD are particularly vulnerable to comorbid conditions [6]. Common examples include metabolic disorders such as diabetes mellitus, cardiovascular diseases such as hypertension and stroke, and neuropsychiatric illnesses including depression and anxiety [7]. These comorbidities do not occur in isolation but often interact with the underlying neurodegenerative pathology, thereby accelerating cognitive decline, worsening functional impairment, and complicating pharmacological as well as non-pharmacological management.

Importance of Understanding the Interplay between Comorbidities and AD Progression

AD is rarely an isolated condition. Most patients with AD have multiple comorbidities that can either contribute to the onset of dementia or exacerbate cognitive and functional decline. The presence of these conditions complicates disease management in several ways:

- **Acceleration of Cognitive Decline:** Most comorbidities, including diabetes, hypertension, and depression, contribute to a high risk of the development of dementia and accelerated progression of cognitive impairment. Chronic systemic inflammation, oxidative stress, and vascular dysfunction are key mechanisms linking these conditions to AD pathology.
- **Increased Burden of Symptoms:** Comorbidities can worsen behavioral and psychological symptoms of dementia like agitation, depression, and sleep

disturbances, which result in a higher requirement for medical procedures and social services.

- **Complications in Diagnosis and Management:** The presence of overlapping symptoms between AD and comorbid conditions poses challenges in diagnosis. For instance, cognitive impairment in AD may be worsened by uncontrolled diabetes, making it more difficult to determine the primary cause of cognitive dysfunction. Additionally, polypharmacy—commonly required to manage multiple chronic diseases—raises the risk of adverse drug interactions and further cognitive deterioration.
- **Increased Mortality and Morbidity:** Most of the studies suggest that AD patients with multiple comorbidities have a greater mortality rate than those without. Cardiovascular diseases, infections, and metabolic disorders contribute to poorer survival outcomes, highlighting the need for a holistic treatment approach.
- **Impact on Caregivers and Healthcare Systems:** The presence of comorbidities increases the complexity of care, leading to higher caregiver burden and healthcare costs. The need for frequent medical visits, hospitalizations, and specialized care can overwhelm both family caregivers and healthcare professionals [8 - 10].

Objectives and Scope of the Chapter

The main aim of this chapter is to provide a comprehensive knowledge of the role of comorbidities in AD progression by exploring their prevalence, underlying mechanisms, and impact on cognitive and functional decline. It seeks to define and classify common comorbid conditions, including metabolic, cardiovascular, neurological, infectious, and psychiatric disorders, while examining how these conditions accelerate AD pathology through mechanisms such as systemic inflammation, oxidative stress, vascular dysfunction, and metabolic dysregulation. Additionally, the chapter will analyze the consequences of comorbidities on disease outcomes, including faster cognitive deterioration, increased behavioral symptoms, reduced treatment efficacy, higher caregiver burden, and increased mortality rates. A critical focus will be on the challenges in diagnosing and managing comorbidities in AD due to symptom overlap, polypharmacy risks, and the lack of standardized treatment guidelines. Also, the chapter will offer effective management strategies, emphasizing personalized care plans, pharmacological and non-pharmacological interventions, and the role of multidisciplinary healthcare teams in providing integrated care. Finally, the chapter points out future research, such as advances in biomarker identification and new therapeutic strategies, to counteract the effects of comorbid conditions on AD progression. Through these discussions, the chapter seeks to inform healthcare providers, researchers, and caregivers to embrace a comprehensive, patient-centered strategy to treat Alzheimer's disease and its comorbidities.

OVERVIEW OF ALZHEIMER'S DISEASE

AD is a chronic, progressive neurodegenerative disorder that is characterized by a decrease in cognitive ability, behavioral disturbances, and functional impairment in activities of daily living. It is the leading cause of dementia, responsible for 60–70% of all dementia cases globally. AD mostly occurs in older adults, and it has enormous healthcare and socio-economic impacts [11].

Pathophysiology and Progression of AD

The pathophysiology of AD is complex and involves several cellular and molecular pathways. The two characteristic pathological hallmarks of AD are amyloid-beta (Aβ) plaques and neurofibrillary tangles (NFTs), which are involved in causing neuronal injury and cognitive impairment [12 - 14].

- **Amyloid Plaque Formation:** Amyloid plaques are extracellular deposits of Aβ peptide fragments, resulting from the abnormal cleavage of the Amyloid Precursor Protein (APP). Aβ peptides aggregate in the brain, causing synaptic dysfunction, neuroinflammation, and neuronal loss. The disruption implicated in the imbalance between Aβ production and clearance is a central driver of disease progression [15].
- **Neurofibrillary Tangles (NFTs):** NFTs are intracellular accumulations of hyperphosphorylated tau protein, which interfere with the normal functioning of microtubules in neurons. This leads to impaired neuronal transport, synaptic dysfunction, and eventually cell death [16].
- **Neuroinflammation:** Chronic activation of glial cells (microglia and astrocytes) in response to Aβ and tau pathology contributes to neuroinflammation, releasing pro-inflammatory cytokines that exacerbate neuronal damage [17].
- **Neurotransmitter Dysregulation:** The cholinergic system, crucial for memory and learning, is severely affected in AD, leading to a deficit in acetylcholine. Other neurotransmitter systems, including glutamate, serotonin, and dopamine, are also disrupted, contributing to cognitive and behavioral symptoms [18].
- **Vascular Contributions to AD:** Impaired cerebral blood flow, Blood-Brain Barrier (BBB) dysfunction, and vascular disease (*e.g.*, small vessel disease, stroke) underlie cognitive impairment and hasten disease progression [19].

Stages of AD Progression

AD progresses through three general stages: preclinical, Mild Cognitive Impairment (MCI) due to AD, and Alzheimer's dementia (mild, moderate, and severe). Each stage is characterized by a slow deterioration in cognitive function, changes in behavior, and caregiver dependence. AD stages and key characteristics are explained in detail in Table **1**.

Table 1. Alzheimer's disease stages and key characteristics.

Stage	Key Symptoms	Cognitive Decline	Functional Impact
Preclinical Stage	No noticeable memory issues; subtle brain changes begin	No significant behavioral changes	No functional decline; individuals maintain daily activities
Mild Cognitive Impairment (MCI) Due to AD	Mild memory lapses, difficulty with word recall, trouble with complex tasks	Mild anxiety or depression, occasional confusion	Independent in daily activities but may need reminders
Mild AD	Memory loss, difficulty managing finances, trouble organizing daily activities, and recalling recent events	Mild impairment in problem-solving and attention	Able to perform daily activities independently but with noticeable difficulties
Moderate AD	Pronounced confusion, personality changes, difficulty recognizing familiar people, wandering behavior	Significant memory loss, difficulty with language and reasoning	Needs assistance with ADLs such as dressing and hygiene
Severe AD	Loss of ability to communicate, severe cognitive decline, physical deterioration, difficulty swallowing	Completely dependent, unable to recognize family members	Requires full-time care, bedridden, loss of bladder and bowel control

- **Preclinical Stage:** This stage may last for years or even decades before clinical symptoms appear. Aβ accumulation and tau pathology begin, but cognitive function remains intact. Biomarkers such as cerebrospinal fluid (CSF) Aβ42, tau levels, and PET imaging can detect early changes.
- **Mild Cognitive Impairment (MCI) Due to AD:** Patients experience subtle memory lapses, difficulty in word-finding, and challenges in complex tasks. Functional independence is largely preserved.
- **Alzheimer's Dementia (Mild, Moderate, and Severe Stages):** As AD progresses, symptoms become increasingly severe. In the mild stage, people have worsening memory loss, with difficulty in handling money, keeping track of daily routines, and remembering recent events. As the disease advances to the moderate stage, confusion becomes more pronounced, accompanied by noticeable personality changes, difficulty recognizing familiar people, and a tendency to wander. In the severe stage, patients lose the ability to communicate effectively, require full-time care, and suffer from significant physical decline,

including difficulties in swallowing and mobility, ultimately leading to complete dependence on caregivers [20 - 23].

Symptoms of AD

AD leads to cognitive impairments, behavioral disturbances, and impaired daily activities. The symptoms are illustrated in Fig. (**1**).

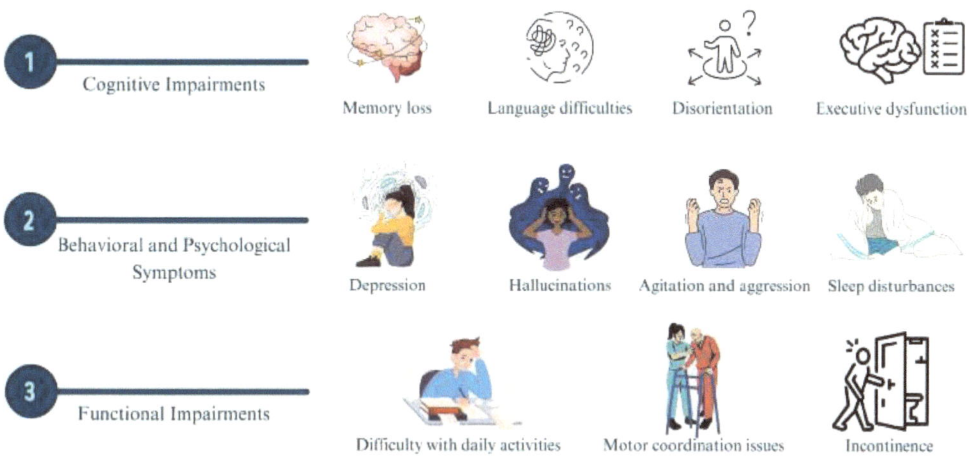

Fig. (1). Symptoms of Alzheimer's disease.

- **Cognitive Impairments:** AD most notably targets cognitive abilities, resulting in progressive impairment of memory, reasoning, and problem-solving. Loss of memory is the most common and initial symptom, with short-term memory being specifically affected, rendering it hard for patients to remember recent events or conversations. Difficulty with language also occurs, which results in word-finding issues, decreased vocabulary, and comprehension problems with conversation. As the disease advances, disorientation becomes evident, causing confusion about time, place, and familiar surroundings. Executive function deficits, such as difficulties with problem-solving, multitasking, and decision-making, further impact daily activities. Additionally, impaired reasoning and judgment become apparent, leading to poor financial decisions, an inability to recognize danger, and difficulty planning tasks effectively [24].
- **Behavioral and Psychological Symptoms:** Apart from cognitive decline, AD is accompanied by numerous behavioral and psychological symptoms that have a great impact on patients and caregivers. Depression and apathy are prevalent and lead to withdrawal from social relationships, loss of interest in activities once enjoyed, and ongoing states of sadness. Agitation and aggression may also

develop, manifesting as verbal or physical outbursts, often triggered by frustration or confusion. Hallucinations and delusions are frequent in later stages, with individuals experiencing false perceptions or paranoid beliefs. Sleep disturbances, including insomnia, nighttime wandering, and excessive daytime sleepiness, further exacerbate the burden of the disease. These symptoms contribute to increased caregiver stress and complicate disease management [25].

- **Functional Impairments:** With the progression of AD, patients lose their independence in performing activities of daily living. Simple Activities of Daily Living (ADLs), including dressing, grooming, eating, and personal hygiene, become challenging. Motor coordination deteriorates over time, leading to difficulty in walking, increased risk of falls, and muscle stiffness, particularly in the advanced stages of the disease. Additionally, loss of bladder and bowel control becomes common, contributing to incontinence issues that require ongoing care and supervision. These functional impairments significantly reduce the quality of life for individuals with AD and necessitate continuous assistance from caregivers [26, 27].

Risk Factors and Epidemiology of AD

AD is affected by a mix of non-modifiable and modifiable risk factors. Of the non-modifiable risk factors, age is the most powerful determinant, with AD prevalence doubling about every five years after the age of 65. Genetic susceptibility is also important, especially the presence of the APOE ε4 allele, which enhances the risk of AD. Moreover, uncommon mutations in the APP, PSEN1, and PSEN2 genes are also responsible for causing early-onset familial Alzheimer's. A positive family history, especially having a first-degree relative with AD, also increases the risk of developing the disease. Sex differences also play a role, with women being at greater risk than men, possibly due to hormonal differences and increased life expectancy.

Aside from age and genetic factors, various modifiable risk determinants impact the development and exacerbation of AD. Cardiovascular diseases like hypertension, atherosclerosis, atrial fibrillation, and stroke are highly correlated with a higher risk for cognitive impairment. Likewise, metabolic diseases like type 2 diabetes and obesity also aid in neurodegeneration due to insulin resistance, sustained inflammation, and metabolic dysregulation. Such lifestyle factors as smoking and heavy drinking aggravate oxidative stress and neuronal injury, further increasing AD risk. Sedentary living, low-quality diet—particularly those diets with high levels of saturated fats and carbohydrates, and a lack of social and mental challenge, also impair cognitive functioning. By contrast, following a Mediterranean diet and mental and social stimulation are linked with a lower AD risk. Sleep disruptions, especially chronic insomnia and sleep apnea, are responsible for enhanced amyloid-beta deposition in the brain, which hastens disease worsening. Furthermore, individuals with recurrent Traumatic Brain Injuries (TBIs) also have an increased risk of suffering from neurodegenerative disorders, such as AD. Awareness of these risk factors can aid in deferring the onset and worsening of Alzheimer's disease [28 - 30].

Global and Regional Epidemiology

AD impacts more than 55 million individuals globally, with the number of cases predicted to triple by 2050 as a result of aging populations. The incidence is regionally variable, with the highest incidence rates found in North America and Western Europe and increasing cases in Asia and Latin America as a result of demographic changes. The disease poses a significant economic burden, with healthcare costs projected to exceed $1 trillion annually in the coming decades [31 - 33].

COMMON COMORBIDITIES IN AD

AD often coexists with multiple comorbid conditions that influence disease progression, symptom severity, and overall patient outcomes. The common comorbidities associated with Alzheimer's disease are illustrated in Fig. (**2**). These comorbidities can accelerate cognitive decline, increase functional impairments, and complicate disease management. Understanding these conditions and their interactions with AD is crucial for optimizing treatment strategies and improving patient care.

Metabolic Disorders

- **Diabetes Mellitus:** Diabetes mellitus, particularly type 2 diabetes, is a significant risk factor for Alzheimer's disease. Insulin resistance and chronic hyperglycemia contribute to neurodegenerative changes by increasing amyloid-beta (Aβ) accumulation and tau phosphorylation. Elevated blood sugar levels also promote oxidative stress and neuroinflammation, exacerbating neuronal damage. Moreover, vascular complications associated with diabetes, such as microvascular dysfunction, impair cerebral blood flow, further worsening cognitive deficits. Studies suggest that antidiabetic medications like metformin may have neuroprotective effects, highlighting the need for careful metabolic management in AD patients.
- **Obesity:** Obesity, particularly in midlife, is linked to an increased risk of AD due to its association with systemic inflammation, insulin resistance, and metabolic syndrome. Adipose tissue secretes pro-inflammatory cytokines that contribute to neuroinflammation, a key pathological feature of AD. Furthermore,

obesity-related vascular dysfunction increases the risk of cerebral hypoperfusion, further exacerbating cognitive impairment. Weight management and lifestyle modifications, including diet and exercise, are essential for reducing AD risk and slowing disease progression.

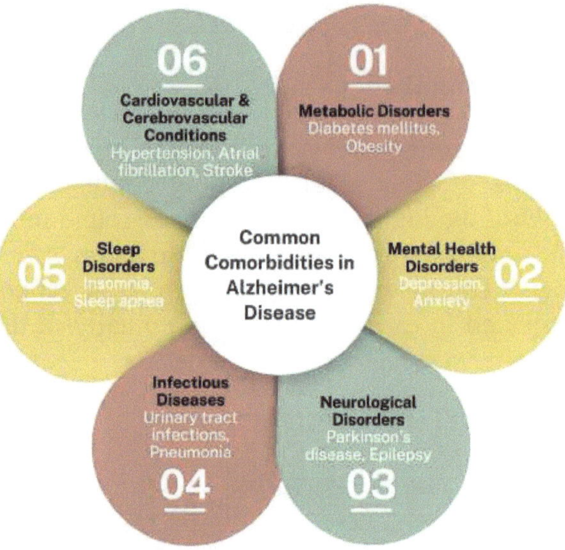

Fig. (2). Common comorbidities in Alzheimer's disease.

Mental Health Disorders

- **Depression:** Depression is a common comorbidity in AD and is related to accelerated cognitive decline and increased functional impairment. Neurobiological mechanisms linking depression and AD include alterations in hippocampal volume, dysregulation of neurotransmitters (serotonin and dopamine), and increased cortisol levels leading to neuronal damage. Identification of depression and treatment in AD patients can enhance the quality of life and decelerate cognitive deterioration. Selective Serotonin Reuptake Inhibitors (SSRIs) are widely used, but their effects on cognitive outcomes require further research.
- **Anxiety:** Anxiety disorders are prevalent among AD patients and can worsen cognitive and behavioral symptoms. Anxiety is often linked to increased agitation, restlessness, and sleep disturbances, further complicating disease management. Neurochemical imbalances, including heightened stress hormone levels, may contribute to both anxiety and AD progression. Non-pharmacological interventions, such as Cognitive-Behavioral Therapy (CBT) and mindfulness practices, can be beneficial in managing anxiety symptoms in AD patients [34, 35].

Other Neurological Disorders

- **Parkinson's Disease:** Both AD and Parkinson's Disease (PD) have similar pathological processes involving neuroinflammation, oxidative stress, and synaptic impairment. Although PD essentially involves motor function, most of its patients will eventually develop dementia and cognitive impairment, which is referred to as Parkinson's Disease Dementia (PDD). A combination of PD and AD in the patient results in increased cognitive and motor impairment and requires an individualized treatment protocol. Dopaminergic therapies may improve motor symptoms but must be carefully managed to avoid worsening cognitive dysfunction.
- **Epilepsy:** Epilepsy is being recognized more and more as a comorbidity in AD, especially toward the end stages of the disorder. Patients with AD who have epilepsy show faster cognitive deterioration and increased neuronal loss. The presence of recurrent seizures may exacerbate synaptic dysfunction and increase neurodegeneration. Antiepileptic drugs (AEDs) are often required for seizure control, but caution is necessary, as some AEDs can cause cognitive side effects.

Infectious Diseases

- **Urinary Tract Infections (UTIs):** UTIs are common in AD patients, particularly in those with incontinence or reduced mobility. Infections can lead to sudden cognitive deterioration, increased agitation, and delirium. The immune response to infections may trigger systemic inflammation, worsening neuroinflammation and accelerating disease progression. Prompt diagnosis and treatment of UTIs are essential to prevent complications in AD patients.
- **Pneumonia:** Pneumonia is a leading cause of mortality in AD patients, often due to aspiration resulting from swallowing difficulties in advanced disease stages. Respiratory infections can cause acute cognitive decline, confusion, and increased hospitalization rates. Vaccination, good oral hygiene, and swallowing assessments can help reduce the risk of pneumonia in AD patients.

Sleep Disorders

- **Insomnia:** Sleep disturbances, particularly insomnia, are prevalent in AD patients and contribute to cognitive decline, increased agitation, and reduced quality of life. Disruptions in the sleep-wake cycle can impair memory consolidation and exacerbate neurodegenerative processes. Melatonin supplements and behavioral sleep interventions—such as maintaining a consistent sleep schedule—can help manage insomnia in AD patients.
- **Sleep Apnea**: Obstructive Sleep Apnea (OSA) is linked to a higher risk of AD due to intermittent hypoxia and fragmentation of sleep, which favor neurodegeneration. OSA can hasten cognitive decline by limiting oxygen

delivery to the brain and enhancing amyloid-beta deposition. Continuous Positive Airway Pressure (CPAP) therapy has been demonstrated to enhance cognitive function and retard disease progression in AD patients with sleep apnea.

Cardiovascular and Cerebrovascular Conditions

- **Hypertension:** Hypertension is a firmly established risk factor for both Alzheimer's disease and vascular dementia. High blood pressure over the long term causes damage to cerebral blood vessels, resulting in decreased cerebral perfusion and greater vulnerability to neurodegenerative alterations. Hypertension also enhances Blood-Brain Barrier (BBB) dysfunction, which allows the entry of neurotoxic substances into the brain. Successful blood pressure control with antihypertensive drugs, including ACE inhibitors and calcium channel blockers, may delay cognitive impairment in hypertensive AD patients.
- **Atrial Fibrillation:** Atrial Fibrillation (AF) carries a higher risk of stroke and cognitive disturbance in AD patients. The irregular heart rhythm in AF can lead to embolic strokes, causing brain damage that worsens cognitive deficits. Additionally, AF-related hypoperfusion may contribute to chronic cerebral ischemia, further accelerating neurodegeneration. Anticoagulant therapy and effective control of AF can decrease the risk of cognitive impairment and enhance global brain health.
- **Stroke:** Stroke significantly increases the likelihood of developing AD and can worsen cognitive impairment in existing cases. Post-stroke dementia often results from ischemic damage to brain regions critical for memory and executive function. Additionally, stroke-related inflammation and BBB dysfunction may exacerbate AD pathology. Stroke prevention strategies, including lifestyle modifications and appropriate anticoagulation therapy, are essential for minimizing AD-related complications [36].

MECHANISMS LINKING COMORBIDITIES TO ALZHEIMER'S DISEASE PROGRESSION

AD is a multifaceted neurodegenerative disease that involves progressive cognitive deterioration, neuronal degeneration, and pathological features like Amyloid-Beta (Aβ) plaques and tau tangles. Current evidence indicates that several comorbidities drive AD progression *via* several interlinked biological pathways. These include chronic inflammation, oxidative stress, vascular impairment, Blood-Brain Barrier (BBB) disruption, metabolic dysregulation, and disrupted amyloid and tau protein dynamics. Recognizing these pathways is important to identifying targeted interventions for slowing the disease and enhancing outcomes for patients.

Chronic Inflammation and Neuroinflammation

Chronic systemic inflammation and neuroinflammation play a central role in AD pathology. Many comorbid conditions, such as diabetes, obesity, cardiovascular disease, infections, and autoimmune disorders, contribute to persistent low-grade inflammation, which exacerbates neurodegeneration.

- **Mechanisms:** Microglial activation is a key component of the neuroinflammatory response in AD. Microglia, the CNS resident immune cells, are hyperactivated due to systemic inflammation from comorbid conditions like diabetes, infections, and cardiovascular diseases. This chronic activation results in overproduction of pro-inflammatory cytokines such as Tumor Necrosis Factor-Alpha (TNF-α), interleukin-1β (IL-1β), and interleukin-6 (IL-6), which induce synaptic dysfunction and neuronal injury. In addition, astrocytes, another key glial cell, become reactive under inflammation and secrete neurotoxic compounds that are responsible for neuronal damage and cognitive deficits. Peripheral inflammatory signals from systemic conditions can cross the Blood-Brain Barrier (BBB), further triggering neuroinflammatory cascades in the brain, exacerbating neuronal damage, and accelerating the progression of AD pathology.
- **Impact on AD Progression:** The sustained neuroinflammatory response driven by microglial and astrocytic activation leads to several detrimental effects in AD. Chronic inflammation promotes the aggregation of amyloid-beta (Aβ) plaques by impairing clearance mechanisms, thereby increasing plaque burden. Furthermore, neuroinflammation enhances tau hyperphosphorylation, which leads to the formation of neurofibrillary tangles, further compromising neuronal integrity. The persistent inflammatory environment also disrupts synaptic function, contributing to neuronal loss and cognitive deficits. As a result, AD patients with heightened neuroinflammation experience a faster rate of cognitive decline, worsening behavioral symptoms, and reduced responsiveness to therapeutic interventions [37].

Oxidative Stress and Its Impact on Neuronal Damage

Oxidative stress occurs because of an imbalance in the production of Reactive Oxygen Species (ROS) and the antioxidant defense systems of the brain. It is a major contributor to neurodegeneration and is strongly associated with comorbidities such as diabetes, cardiovascular disease, and obesity.

- **Mechanisms:** Mitochondrial dysfunction is a key feature of AD pathogenesis, since mitochondria are extremely vulnerable to oxidative damage. Damaged mitochondria generate high levels of Reactive Oxygen Species (ROS), which induce oxidative damage to neuronal DNA, proteins, and lipids, culminating in neuronal cell death. ROS-induced lipid peroxidation also interferes with the integrity of neuronal membranes, impairing synaptic function and interneuronal communication. Another important mechanism is protein misfolding and aggregation, where oxidative stress accelerates amyloid-beta (Aβ) and tau protein misfolding. This facilitates the generation of amyloid plaques and neurofibrillary tangles, which are characteristic pathological features of AD. The collective effect of these mechanisms leads to neurodegeneration and cognitive impairment.
- **Impact on AD Progression:** The recurring oxidative stress and mitochondrial malfunction in AD cause extensive neuronal apoptosis and synaptic loss, further worsening cognitive impairments. Enhanced oxidative damage speeds up amyloid-beta aggregation and tau hyperphosphorylation, culminating in the accelerated development of amyloid plaque and neurofibrillary tangles. This, in turn, undermines cognitive functions like memory, learning, and executive processing, resulting in exacerbation of symptoms in AD patients. As oxidative stress persists in injuring neural structures, the person undergoes an increased decline in cognitive functions, decreased neuronal plasticity, and increased susceptibility to neurodegenerative processes.

Vascular Contributions to Cognitive Impairment and Dementia (VCID)

Vascular dysfunction is a major contributor to cognitive impairment and a key link between comorbid conditions (*e.g.*, hypertension, atrial fibrillation, diabetes, and stroke) and AD progression. Vascular Cognitive Impairment and Dementia (VCID) often coexist with AD, leading to more severe cognitive deficits.

- **Mechanisms:** Reduced cerebral perfusion is a major contributor to AD progression, as chronic hypertension and atherosclerosis lead to the narrowing of cerebral blood vessels. This restricts the oxygen and nutrient supply to neurons, impairing their function and increasing susceptibility to neurodegeneration. Additionally, small vessel disease and silent strokes cause white matter lesions and microinfarcts, which contribute to white matter degeneration and cognitive decline. Another critical mechanism is hypoxia and ischemia, where reduced blood flow leads to neuronal oxygen deprivation, triggering oxidative stress and inflammatory responses. These pathological changes create an environment conducive to neurodegeneration and further exacerbate AD pathology.

- **Impact on AD Progression:** Vascular impairments significantly accelerate the progression of AD by disrupting essential clearance mechanisms and promoting toxic protein accumulation. Impaired blood flow hinders the glymphatic system's ability to clear amyloid-beta, leading to its accumulation and increased plaque formation. Additionally, hypoxia-induced stress responses contribute to tau hyperphosphorylation, which further exacerbates neurofibrillary tangle formation. These changes collectively result in worsened cognitive decline, with significant deficits in executive function and memory. The combination of vascular dysfunction and AD pathology creates a vicious cycle that accelerates disease progression and leads to greater functional impairments.

Blood-Brain Barrier (BBB) Dysfunction

The BBB is a selective barrier that regulates the exchange of molecules between the bloodstream and the brain. Many comorbid conditions, including diabetes, hypertension, and infections, disrupt BBB integrity, allowing harmful substances to enter the brain and contribute to AD progression.

- **Mechanisms:** Dysfunction of the BBB is an essential element in AD progression. Ongoing inflammation makes endothelial tight junctions vulnerable, enhancing BBB permeability and entry of inflammatory cytokines, neurotoxic proteins, and peripheral immune cells into the brain. This contributes to neuroinflammation and promotes neuronal loss. Moreover, BBB is also essential in clearing amyloid-beta out of the brain through transport proteins like low-density Lipoprotein Receptor-Related Protein 1 (LRP1). When there is compromised integrity of the BBB, clearance of amyloid-beta becomes ineffective, with the result being its accumulation and the development of toxic plaques. Another key mechanism is neurovascular uncoupling, where the disruption of normal neurovascular function impairs the brain's ability to regulate blood flow. This results in localized ischemia and oxidative stress, further promoting neurodegeneration.
- **Impact on AD Progression:** BBB dysfunction significantly accelerates AD progression by promoting amyloid-beta accumulation and persistent neuroinflammation. Increased permeability allows for the excessive deposition of amyloid plaques, which further drives neurotoxicity and synaptic loss. Additionally, the infiltration of peripheral immune cells into the brain enhances inflammatory responses, worsening neuronal injury and impairing cognitive function. The combined effects of reduced amyloid clearance, heightened neuroinflammation, and disrupted neurovascular function contribute to rapid cognitive decline, increasing memory deficits and impairing executive function in AD patients [38].

Role of Metabolic Dysregulation (Insulin Resistance, Glucose Toxicity)

Metabolic disorders, particularly diabetes and obesity, are strongly linked to AD due to their effects on insulin signaling and glucose metabolism in the brain.

- **Mechanisms:** Metabolic dysregulation, particularly insulin resistance in the brain, plays a significant role in AD. Insulin is crucial for neuronal survival, synaptic plasticity, and memory function. However, insulin resistance leads to impaired glucose uptake, reduced energy production, and increased oxidative stress, contributing to neurodegeneration. Chronic hyperglycemia further exacerbates AD pathology through glucose toxicity, where excessive glucose leads to the formation of advanced glycation end-products (AGEs). These AGEs promote neuronal damage and enhance amyloid-beta aggregation, accelerating plaque formation. Additionally, mitochondrial dysfunction arises from metabolic disturbances, leading to excessive oxidative stress and neuronal apoptosis, further contributing to cognitive decline.
- **Impact on AD Progression:** Metabolic dysregulation accelerates AD progression by increasing amyloid-beta production while reducing its clearance, leading to the formation of toxic plaques. Furthermore, insulin resistance and glucose toxicity contribute to greater tau hyperphosphorylation, promoting the development of neurofibrillary tangles, which disrupt neuronal function. The combined effects of impaired energy metabolism, oxidative stress, and neuronal damage result in faster cognitive decline and worsening functional impairments, ultimately leading to severe memory deficits and loss of daily living abilities in AD patients.

Amyloid and Tau Protein Dynamics in the Presence of Comorbid Conditions

Amyloid-beta plaques and tau neurofibrillary tangles are the hallmark pathological features of AD. Comorbid conditions can influence their accumulation, aggregation, and clearance, exacerbating disease progression.

- **Mechanisms:** Chronic inflammation plays a central role in AD by disrupting amyloid-beta clearance pathways, leading to an increased plaque burden in the brain. Inflammatory cytokines interfere with the normal degradation and removal of amyloid-beta, exacerbating its accumulation. Oxidative stress further contributes to AD pathology by promoting tau hyperphosphorylation, which leads to the formation of neurofibrillary tangles and disrupts neuronal function. Additionally, vascular dysfunction reduces cerebral blood flow, impairing the clearance of amyloid-beta through the glymphatic and vascular systems. This combination of inflammation, oxidative stress, and vascular impairment accelerates neuronal damage and cognitive decline.

- **Impact on AD Progression:** The persistence of these pathological processes enhances neurodegeneration and synaptic loss, leading to widespread brain atrophy. As a result, AD patients experience greater cognitive decline, memory impairment, and behavioral disturbances such as agitation and mood changes. Furthermore, chronic inflammation and vascular dysfunction reduce the effectiveness of therapeutic interventions aimed at targeting amyloid-beta and tau, complicating disease management and limiting treatment success [39].

IMPACT OF COMORBIDITIES ON DISEASE OUTCOMES

AD is a degenerative neurologic disorder that most notably impacts cognition, behavior, and functional ability. The occurrence of comorbid conditions such as cardiovascular diseases, diabetes, metabolic disorders, and psychiatric diseases has a profound impact on the course and severity of AD (Table **2**). Comorbidities precipitate accelerated cognitive decline, greater neuropsychiatric symptoms, poorer therapeutic response, increased caregiver burden, and higher mortality. Knowledge of the effect of these comorbidities is critical in maximizing treatment options and enhancing patient outcomes.

Table 2. Impact of comorbidities on disease outcomes.

Comorbidity	Cognitive Decline	Behavioral & Psychological Symptoms	Response to Treatment	Mortality Risk
Diabetes Mellitus	Faster memory impairment	Increased depression and apathy	Reduced response to cholinesterase inhibitors	Higher due to cardiovascular complications
Hypertension	Worsened executive function	Increased agitation and aggression	Poor response to AD medications due to vascular damage	Elevated due to stroke risk
Depression	Accelerated cognitive decline	Greater risk of anxiety, hallucinations	Poor adherence to AD treatments	Increased due to suicide risk
Sleep Apnea	Impaired attention and memory	Increased confusion and irritability	Limited efficacy of AD drugs	Higher due to cardiovascular effects
Stroke	Severe decline in all cognitive functions	Worsened paranoia and delusions	Reduced ability to participate in therapies	Markedly increased due to disability and secondary strokes

Faster Cognitive Decline

Comorbid conditions such as diabetes, hypertension, stroke, and metabolic syndrome accelerate the progression of AD by exacerbating neuroinflammation, oxidative stress, and vascular dysfunction. Insulin resistance and chronic hyperglycemia impair glucose metabolism in the brain, leading to energy deficits and neuronal dysfunction. Similarly, hypertension and atherosclerosis reduce cerebral perfusion, increasing the risk of silent strokes and white matter lesions that further contribute to cognitive impairment. Patients with multiple comorbidities experience a steeper decline in memory, executive function, and attention compared to those with AD alone.

Increased Behavioral and Psychological Symptoms

Neuropsychiatric symptoms such as agitation, depression, anxiety, and psychosis are common in AD and are often exacerbated by comorbid conditions. Depression, which frequently coexists with AD, is associated with an increased risk of faster cognitive decline and reduced motivation for social and cognitive engagement. Cardiovascular diseases and metabolic disorders contribute to systemic inflammation, which is linked to worsening mood disturbances and agitation. Patients with cerebrovascular disease are also more prone to experiencing hallucinations and delusions, further complicating disease management [40].

Reduced Response to Pharmacological and Non-Pharmacological Therapies

Comorbidities can complicate the effectiveness of AD pharmacological and non-pharmacological treatments. Cardiovascular diseases, for example, disrupt drug metabolism and distribution and hence compromise the efficacy of AD management using cholinesterase inhibitors and NMDA receptor antagonists. Polypharmacy in elderly subjects with more than one medical illness raises the risks of drug-drug interactions, side effects, and nonadherence to drugs. Non-pharmacological interventions, including cognitive training, exercise, and diet changes, could also be less successful in persons with more serious comorbid conditions because they have less active participation and physical impairments.

Higher Caregiver Burden and Reduced Quality of Life

Caregivers of individuals with AD often experience significant physical, emotional, and financial stress, which is further exacerbated by the presence of comorbidities. Patients with multiple health conditions require more intensive medical care, frequent hospitalizations, and complex medication regimens, increasing the burden on caregivers. Behavioral symptoms such as aggression, wandering, and sleep disturbances can also strain caregivers, leading to burnout and reduced quality of life. Support systems and healthcare interventions must be tailored to address the needs of both patients and caregivers to improve overall well-being [41].

Increased Mortality Rates

The presence of comorbidities significantly increases the risk of mortality in AD patients. Cardiovascular diseases, diabetes, infections (such as pneumonia and urinary tract infections), and frailty contribute to higher hospitalization rates and complications that can lead to premature death. Respiratory infections are particularly concerning, as aspiration pneumonia is a common cause of death in advanced AD. Additionally, poor nutritional status and immobility increase the risk of pressure ulcers, sepsis, and other fatal complications. Managing comorbid conditions effectively can help slow disease progression and improve survival rates in AD patients [42].

CHALLENGES IN MANAGING COMORBIDITIES IN ALZHEIMER'S DISEASE

Managing comorbidities in AD presents unique challenges due to the progressive cognitive decline and complex medical needs of affected individuals. Patients with AD often have multiple chronic conditions, including cardiovascular diseases, diabetes, metabolic disorders, and psychiatric illnesses, which complicate diagnosis and treatment. Additionally, age-related physiological changes, polypharmacy, and the lack of specific clinical guidelines for managing comorbidities in dementia patients further hinder effective care. Fig. (**3**) explores the key challenges in managing comorbid conditions in AD.

Diagnostic Complexities Due to Overlapping Symptoms

Diagnosing comorbid conditions in AD patients is particularly challenging because many symptoms of systemic illnesses overlap with the cognitive and behavioral symptoms of dementia.

- **Cognitive Impairment Masking Symptoms:** Patients with AD may have difficulty communicating symptoms such as pain, fatigue, or discomfort, leading to delays in diagnosis and treatment of underlying conditions.

- **Neuropsychiatric Symptom Overlap:** Depression, anxiety, and sleep disturbances are common in both AD and conditions such as Parkinson's disease, hypothyroidism, and metabolic disorders, making it difficult to attribute symptoms to a specific cause.
- **Vascular and Metabolic Contributions to Cognitive Decline:** Conditions such as diabetes, hypertension, and stroke can mimic or exacerbate AD-related cognitive deficits, complicating differential diagnosis.
- **Altered Pain Perception:** AD patients may have a reduced ability to perceive and express pain, leading to underdiagnosis of conditions such as arthritis, neuropathy, or gastrointestinal disorders.

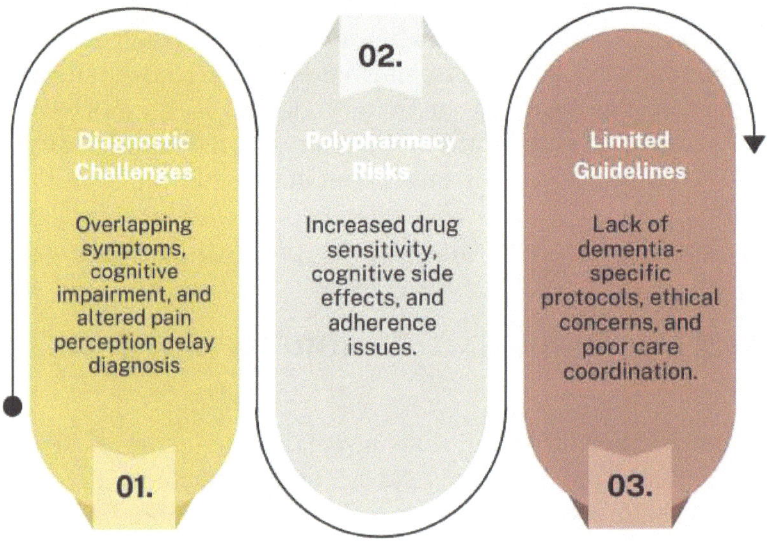

Fig. (3). Challenges in managing comorbidities in AD.

To address these challenges, clinicians must use comprehensive assessment tools, obtain collateral histories from caregivers, and employ biomarker-based diagnostics to differentiate between AD-related symptoms and comorbid conditions [43].

Polypharmacy and Drug-Drug Interactions

Polypharmacy is a major concern in AD patients, as they often require multiple medications to manage cognitive symptoms, behavioral disturbances, and comorbid conditions. However, excessive medication use increases the risk of drug-drug interactions, adverse effects, and treatment complications.

- **Higher Sensitivity to Medications:** Older adults with AD have altered pharmacokinetics and pharmacodynamics, making them more susceptible to medication side effects, such as sedation, confusion, dizziness, and falls.
- **Cognitive Side Effects of Commonly Used Drugs:** Many medications used to treat comorbid conditions (*e.g.*, antihistamines, benzodiazepines, antipsychotics, and certain pain relievers) have anticholinergic properties that can worsen cognitive impairment in AD patients.
- **Increased Risk of Adverse Events:** Drug interactions may lead to serious complications such as hypotension, renal dysfunction, gastrointestinal bleeding, and increased susceptibility to infections.
- **Challenges in Adherence:** Cognitive impairment makes it difficult for AD patients to manage complex medication regimens, leading to missed doses, overdosing, or medication mismanagement.

A structured approach to medication management, including regular medication reviews, deprescribing of unnecessary drugs, and the use of simplified dosing regimens, is essential for reducing polypharmacy risks in AD patients [44].

Limited Clinical Guidelines for Managing Comorbidities in Dementia Patients

Despite the high prevalence of comorbidities in AD, clinical guidelines for managing these conditions in dementia patients remain limited. Most guidelines for conditions such as diabetes, hypertension, and psychiatric disorders are based on studies conducted in cognitively intact individuals and may not fully address the unique needs of AD patients.

- **Lack of Dementia-Specific Treatment Protocols:** Current guidelines for conditions like hypertension and diabetes do not always consider how aggressive treatment should be in dementia patients, particularly in late-stage AD, where the benefits of strict disease control may be limited.
- **Uncertainty in Treatment Goals:** In advanced AD, there is debate over whether to prioritize symptom management and quality of life over disease modification for comorbid conditions.
- **Ethical Considerations in Care Decisions:** Decisions about invasive procedures, aggressive treatments, and end-of-life care are complex in AD patients, requiring discussions with caregivers and ethical considerations regarding patient autonomy and quality of life.
- **Limited Integration of Multidisciplinary Care:** Effective management of comorbidities requires coordination between neurologists, primary care physicians, geriatricians, psychiatrists, and caregivers, but healthcare systems often lack the infrastructure for integrated dementia care.

Developing dementia-specific guidelines for common comorbidities, incorporating personalized treatment plans, and enhancing interdisciplinary collaboration can improve the management of comorbid conditions in AD patients [45].

STRATEGIES FOR EFFECTIVE MANAGEMENT OF COMORBIDITIES IN AD

Managing comorbidities in AD requires a holistic and multidimensional approach. Given the progressive nature of AD and its impact on cognition, functional ability, and behavior, the presence of additional chronic conditions further complicates patient care. Effective management strategies must integrate comprehensive assessments, personalized treatment plans, appropriate pharmacological adjustments, non-pharmacological interventions, and a multidisciplinary approach to optimize patient outcomes and improve quality of life. Strategies for managing comorbidities in AD, including cardiovascular care, neuropsychiatric management, and lifestyle interventions, are summarized in Table **3**.

Table 3. Strategies for managing comorbidities in AD.

Management Strategy	Key Interventions	Expected Benefit
Comprehensive Assessment	Cognitive, functional, and comorbidity evaluations	Early identification of risk factors and tailored interventions
Personalized Care Plans	Integrated treatment of AD and comorbidities	Improved patient outcomes and quality of life
Pharmacological Management	Adjusted drug regimens, minimizing polypharmacy	Reduced side effects, improved medication adherence
Non-Pharmacological Approaches	Exercise, dietary modifications, and cognitive stimulation	Slowed disease progression, enhanced overall well-being
Multidisciplinary Team Approach	Collaboration among neurologists, psychiatrists, geriatricians, and caregivers	Holistic and coordinated care, reduced caregiver burden

Comprehensive Assessment

A thorough evaluation of AD patients is essential to identify cognitive deficits, functional impairments, and coexisting medical conditions. The assessment should include:

Cognitive Evaluation

- Standardized cognitive (*e.g.*, Mini-Mental State Examination [MMSE], and Montreal Cognitive Assessment [MoCA]) to assess the severity of cognitive impairment.
- Monitoring for changes in memory, executive function, language, and spatial abilities over time.

Functional Assessment

- Assessment of activities of daily living (*e.g.*, bathing, dressing, eating) with scales like the Activities of Daily Living (ADL) and Instrumental Activities of Daily Living (IADL).
- Identification of functional decline that may indicate worsening of AD or the presence of comorbid conditions.

Comorbidity Screening

- Regular monitoring for common conditions such as hypertension, diabetes, cardiovascular disease, depression, and osteoporosis.
- Laboratory tests (*e.g.*, blood glucose, lipid profiles, renal function tests) to detect metabolic disorders.
- Imaging studies (*e.g.*, MRI, CT scan) to assess cerebrovascular contributions to cognitive impairment.
- Comprehensive assessments allow for early detection of comorbidities, guiding tailored treatment strategies and improving overall patient care [46].

Personalized Care Plans

Given the heterogeneity of AD progression and the presence of comorbidities, individualized treatment plans are essential to balance the management of AD symptoms and other medical conditions.

Balancing AD and Comorbidity Treatment

- Prioritization of interventions based on the severity of AD and the impact of comorbid conditions.
- Adjusting treatment goals based on disease stage (*e.g.*, aggressive management in early AD vs. symptom control in advanced AD).

Care Plan Adaptations Over Time

- Regular reassessment to modify treatment strategies as AD progresses.

- Incorporation of palliative care principles in late-stage AD to focus on comfort and quality of life.
- A personalized approach ensures that interventions align with the patient's cognitive, functional, and overall health status.

Pharmacological Considerations

Medication management in AD patients is complex due to polypharmacy, altered drug metabolism, and increased susceptibility to adverse effects.

Adjustments for Polypharmacy

- Regular medication reviews to minimize unnecessary drugs and reduce the risk of drug-drug interactions.
- Avoidance of medications with strong anticholinergic properties, as they can worsen cognitive impairment.
- Consideration of dose adjustments based on renal and hepatic function in elderly AD patients.
- Minimizing Adverse Effects
- Selection of medications with the lowest risk of sedation, dizziness, and falls.
- Use of simplified medication regimens and caregiver-assisted administration to improve adherence.
- A judicious pharmacological approach enhances safety while maintaining therapeutic benefits for both AD symptoms and comorbid conditions.

Non-Pharmacological Approaches

In addition to pharmacological management, lifestyle modifications and behavioral interventions play a crucial role in improving AD patient outcomes.

Lifestyle Interventions

- **Diet:** A Mediterranean or DASH diet that is high in antioxidants, omega-3 fatty acids, and vitamins can promote brain health and prevent metabolic disorders.
- **Exercise:** Physical exercise enhances cardiovascular well-being, lowers the risk of vascular dementia, and improves mood.
- **Cognitive Stimulation:** Activities such as puzzles, reading, music therapy, and social engagement assist in maintaining cognitive function and slowing cognitive decline.

Caregiver Support and Education

- Providing education on disease progression, symptom management, and emergency care planning.
- Offering support groups and respite care services to reduce caregiver burden.
- Encouraging behavioral strategies to manage neuropsychiatric symptoms such as agitation, aggression, and sleep disturbances.
- Non-pharmacological approaches complement medical treatment and improve both patient and caregiver well-being [47].

Role of Multidisciplinary Teams

Effective management of AD and its comorbidities requires a coordinated effort among healthcare professionals, caregivers, and support services.

Key Contributors to Multidisciplinary Care

- **Neurologists:** Oversee cognitive assessments and AD-specific treatments.
- **Psychiatrists:** Manage neuropsychiatric symptoms, including depression and anxiety.
- **Geriatricians:** Provide comprehensive care tailored to the aging population.
- **Primary Care Physicians:** Monitor chronic conditions, conduct routine screenings, and coordinate overall patient care.
- **Nurses and Caregivers:** Assist with daily activities, medication adherence, and behavioral support.

Enhancing Team-Based Collaboration

- Regular interdisciplinary meetings to review patient progress and adjust treatment plans.
- Use of electronic health records for seamless communication between specialists.
- Integration of palliative care services for advanced AD patients to ensure a focus on comfort and dignity.
- A well-coordinated multidisciplinary approach enhances patient outcomes and supports caregivers in delivering high-quality care [48].

EMERGING RESEARCH AND FUTURE DIRECTIONS IN ALZHEIMER'S DISEASE AND COMORBIDITIES

AD is a multifactorial neurodegenerative condition that is typified by progressive cognitive impairment and neuronal dysfunction. The complexity of AD is also augmented by the occurrence of comorbidities, such as metabolic disorders, cardiovascular disease, and chronic inflammatory diseases. Current research has aimed at elucidating the complex pathophysiological relationships between AD and these comorbid conditions, which has led to new developments in potential therapeutic interventions and the use of biomarkers in early diagnosis and management of the disease. As scientific advancements continue to unfold, the exploration of novel therapeutic targets and precision medicine approaches offers hope for improving patient outcomes [49].

Pathophysiological Links Between AD and Comorbidities

Recent studies have demonstrated that systemic diseases such as diabetes, cardiovascular disorders, and gut dysbiosis contribute significantly to AD progression. Insulin resistance, commonly associated with Type 2 diabetes, impairs glucose metabolism in the brain and promotes amyloid-beta accumulation, a hallmark of AD pathology. This metabolic dysfunction has led researchers to conceptualize AD as "Type 3 Diabetes," highlighting the critical role of insulin signaling in neurodegeneration.

Similarly, cardiovascular diseases, including hypertension and atherosclerosis, have been linked to impaired cerebral blood flow, leading to reduced amyloid clearance and increased neuroinflammation. The disruption of the blood-brain barrier in cardiovascular conditions exacerbates the entry of peripheral inflammatory molecules into the brain, accelerating neurodegeneration.

Emerging evidence also underscores the influence of the gut-brain axis in AD pathophysiology. The gut microbiome modulates neuroinflammatory processes through microbial metabolites, which can either protect against or exacerbate AD pathology. An imbalance in gut bacteria, known as dysbiosis, has been associated with increased production of pro-inflammatory cytokines, blood-brain barrier dysfunction, and enhanced tau pathology. These findings emphasize the need to consider systemic health as an integral factor in AD progression and management [50].

Potential Therapeutic Targets Addressing AD and Comorbid Conditions

Given the growing recognition of shared pathological mechanisms between AD and comorbidities, researchers are focusing on therapeutic strategies that target both conditions simultaneously. One promising approach is the use of metabolic and insulin-sensitizing therapies, such as Glucagon-Like Peptide-1 (GLP-1) receptor agonists (*e.g.*, liraglutide and semaglutide). Originally developed for diabetes treatment, these drugs have shown neuroprotective effects by enhancing insulin sensitivity and reducing amyloid-beta deposition. Similarly, metformin, a widely prescribed antidiabetic drug, has been found to lower AD risk by improving mitochondrial function and reducing oxidative stress.

Anti-inflammatory and immunomodulatory approaches are also being explored, given the role of chronic inflammation in AD pathogenesis. Non-Steroidal Anti-Inflammatory Drugs (NSAIDs) have been investigated for their potential to mitigate neuroinflammation, although clinical trials have yielded mixed results. Monoclonal antibodies targeting inflammatory pathways, such as Canakinumab and TNF-alpha inhibitors, are being studied for their ability to reduce neuroinflammatory damage in AD patients with comorbid inflammatory conditions.

In addition to metabolic and inflammatory interventions, vascular-targeted therapies are gaining traction. Antihypertensive medications, including Angiotensin-Converting Enzyme (ACE) inhibitors and angiotensin II receptor blockers, have been found to improve cerebral perfusion and slow cognitive decline in hypertensive AD patients. Lipid-lowering agents, such as statins, offer neuroprotection through anti-inflammatory and antioxidant mechanisms, while antithrombotic agents like aspirin and clopidogrel help prevent microvascular thrombosis, thereby reducing the risk of vascular dementia.

Another emerging avenue of research focuses on gut microbiome modulation as a potential intervention for AD. Probiotics and prebiotics are being studied for their ability to restore microbial balance and reduce systemic inflammation. Experimental studies on Fecal Microbiota Transplantation (FMT) suggest that it may help modulate gut-brain interactions and improve cognitive outcomes in neurodegenerative diseases [51].

Role of Biomarkers in Early Detection and Management of AD

The advent of biomarker research has revolutionized the early diagnosis and monitoring of AD, enabling more precise detection and targeted interventions. Blood-based biomarkers, such as plasma amyloid-beta (Aβ42/40 ratio), phosphorylated tau (p-Tau 181, p-Tau 217, p-Tau 231), and neurofilament light chain (NfL), offer non-invasive methods for identifying AD pathology and monitoring disease progression. Additionally, inflammatory markers, including IL-6, TNF-alpha, and C-Reactive Protein (CRP), provide insights into systemic inflammation and its impact on neurodegeneration.

Cerebrospinal fluid (CSF) biomarkers, including amyloid-beta, total tau, and phosphorylated tau, continue to be the gold standard for diagnosing AD and differentiating it from other dementias. Recent research is also exploring lipid and metabolic biomarkers in CSF, which may reflect underlying metabolic dysfunction linked to both AD and diabetes.

Advancements in neuroimaging biomarkers have further improved diagnostic accuracy. Positron emission tomography (PET) imaging allows for direct visualization of amyloid plaques and tau neurofibrillary tangles, providing valuable insights into disease staging. Magnetic Resonance Imaging (MRI) aids in identifying brain atrophy, white matter changes, and vascular contributions to cognitive decline. Emerging research on gut microbiome-derived biomarkers holds promise for identifying individuals at risk for AD and tailoring personalized interventions [52, 53].

FUTURE DIRECTIONS IN AD RESEARCH

As our understanding of the complex interplay between AD and comorbidities deepens, future research should focus on developing precision medicine approaches that integrate genetic, metabolic, and inflammatory profiles to tailor individualized treatment strategies. The identification of novel therapeutic targets, including those addressing systemic inflammation, metabolic dysfunction, and vascular contributions, is crucial for advancing AD treatment. Additionally, refining biomarker panels for early detection and monitoring disease progression will enhance diagnostic accuracy and treatment efficacy.

Moreover, interdisciplinary collaboration among neurologists, geriatricians, cardiologists, and microbiome researchers will be essential in developing comprehensive management strategies for AD patients with comorbid conditions. Clinical trials should aim to evaluate combination therapies that address multiple pathological pathways simultaneously, rather than focusing solely on amyloid and tau [54].

Emerging research highlights the necessity of a holistic approach to AD management, considering the significant impact of comorbidities on disease progression and treatment response. By integrating advances in pathophysiology, therapeutic targets, and biomarker discovery, researchers and clinicians can move toward more effective interventions that enhance quality of life and alter the trajectory of AD progression [55]. The future of AD research lies in a multidimensional approach that not only targets neurodegeneration but also addresses systemic health factors that influence disease outcomes.

CONCLUSION

Alzheimer's disease is a multifaceted neurodegenerative disorder that is significantly influenced by comorbid conditions such as diabetes, cardiovascular diseases, metabolic syndromes, and chronic inflammation. The complex interplay between these conditions accelerates cognitive decline, worsens disease pathology, and complicates treatment strategies. Emerging research highlights the importance of targeting shared mechanisms, including insulin resistance, vascular dysfunction, neuroinflammation, and oxidative stress, to develop more effective therapeutic approaches. Advances in biomarker discovery are revolutionizing early detection, enabling timely intervention and personalized treatment strategies. Future directions in AD research emphasize the need for precision medicine, combination therapies, and an integrative approach that addresses both neurodegeneration and systemic health. By bridging the gap between clinical and translational research, a comprehensive and multidisciplinary strategy can enhance patient outcomes, improve quality of life, and ultimately pave the way for more effective prevention and management of Alzheimer's disease.

REFERENCES

[1] Scheltens P, De Strooper B, Kivipelto M, *et al.* Alzheimer's disease. Lancet 2021; 397(10284): 1577-90.
[http://dx.doi.org/10.1016/S0140-6736(20)32205-4] [PMID: 33667416]

[2] Scheltens P, Blennow K, Breteler MMB, *et al.* Alzheimer's disease. Lancet 2016; 388(10043): 505-17.
[http://dx.doi.org/10.1016/S0140-6736(15)01124-1] [PMID: 26921134]

[3] Ballard C, Gauthier S, Corbett A, Brayne C, Aarsland D, Jones E. Alzheimer's disease. the Lancet. 2011 Mar 19;377(9770):1019-31.

[4] Newcombe EA, Camats-Perna J, Silva ML, Valmas N, Huat TJ, Medeiros R. Inflammation: the link between comorbidities, genetics, and Alzheimer's disease. J Neuroinflammation 2018; 15(1): 276.
[http://dx.doi.org/10.1186/s12974-018-1313-3] [PMID: 30249283]

[5] Santiago JA, Potashkin JA. The impact of disease comorbidities in Alzheimer's disease. Front Aging Neurosci 2021; 13: 631770.
[http://dx.doi.org/10.3389/fnagi.2021.631770] [PMID: 33643025]

[6] Doraiswamy PM, Leon J, Cummings JL, Marin D, Neumann PJ. Prevalence and impact of medical comorbidity in Alzheimer's disease. J Gerontol A Biol Sci Med Sci 2002; 57(3): M173-7.
[http://dx.doi.org/10.1093/gerona/57.3.M173] [PMID: 11867654]

[7] Maciejewska K, Czarnecka K, Szymański P. A review of the mechanisms underlying selected comorbidities in Alzheimer's disease. Pharmacol Rep 2021; 73(6): 1565-81.
[http://dx.doi.org/10.1007/s43440-021-00293-5] [PMID: 34121170]

[8] Matej R, Tesar A, Rusina R. Alzheimer's disease and other neurodegenerative dementias in comorbidity: A clinical and neuropathological overview. Clin Biochem 2019; 73: 26-31.
[http://dx.doi.org/10.1016/j.clinbiochem.2019.08.005] [PMID: 31400306]

[9] Calderón-Larrañaga A, Vetrano DL, Ferrucci L, *et al.* Multimorbidity and functional impairment–bidirectional interplay, synergistic effects and common pathways. J Intern Med 2019; 285(3): 255-71.
[http://dx.doi.org/10.1111/joim.12843] [PMID: 30357990]

[10] Nelson PT, Braak H, Markesbery WR. Neuropathology and cognitive impairment in Alzheimer disease: a complex but coherent relationship. J Neuropathol Exp Neurol 2009; 68(1): 1-14.
[http://dx.doi.org/10.1097/NEN.0b013e3181919a48] [PMID: 19104448]

[11] Wang SS, Good TA. An overview of Alzheimer's disease. Journal of the Chinese Institute of

Chemical Engineers 2005; 36(6): 533-59.

[12] Kumar A, Singh A, Ekavali . A review on Alzheimer's disease pathophysiology and its management: an update. Pharmacol Rep 2015; 67(2): 195-203. [http://dx.doi.org/10.1016/j.pharep.2014.09.004] [PMID: 25712639]

[13] Abubakar MB, Sanusi KO, Ugusman A, *et al.* Alzheimer's disease: an update and insights into pathophysiology. Front Aging Neurosci 2022; 14742408 [http://dx.doi.org/10.3389/fnagi.2022.742408] [PMID: 35431894]

[14] Yarns BC, Holiday KA, Carlson DM, Cosgrove CK, Melrose RJ. Pathophysiology of Alzheimer's disease. Psychiatr Clin North Am 2022; 45(4): 663-76. [http://dx.doi.org/10.1016/j.psc.2022.07.003] [PMID: 36396271]

[15] Takahashi RH, Nagao T, Gouras GK. Plaque formation and the intraneuronal accumulation of β□amyloid in Alzheimer's disease. Pathol Int 2017; 67(4): 185-93. [http://dx.doi.org/10.1111/pin.12520] [PMID: 28261941]

[16] Brion JP. Neurofibrillary tangles and Alzheimer's disease. Eur Neurol 1998; 40(3): 130-40. [http://dx.doi.org/10.1159/000007969] [PMID: 9748670]

[17] Obulesu M, Jhansilakshmi M. Neuroinflammation in Alzheimer's disease: an understanding of physiology and pathology. Int J Neurosci 2014; 124(4): 227-35. [http://dx.doi.org/10.3109/00207454.2013.831852] [PMID: 23919560]

[18] Snowden SG, Ebshiana AA, Hye A, *et al.* Neurotransmitter imbalance in the brain and Alzheimer's disease pathology. J Alzheimers Dis 2019; 72(1): 35-43. [http://dx.doi.org/10.3233/JAD-190577] [PMID: 31561368]

[19] Kalaria RN, Akinyemi R, Ihara M. Does vascular pathology contribute to Alzheimer changes? J Neurol Sci 2012; 322(1-2): 141-7. [http://dx.doi.org/10.1016/j.jns.2012.07.032] [PMID: 22884479]

[20] Parnetti L, Chipi E, Salvadori N, D'Andrea K, Eusebi P. Prevalence and risk of progression of preclinical Alzheimer's disease stages: a systematic review and meta-analysis. Alzheimers Res Ther 2019; 11(1): 7. [http://dx.doi.org/10.1186/s13195-018-0459-7] [PMID: 30646955]

[21] Davis M, O'Connell T, Johnson S, *et al.* Estimating Alzheimer's disease progression rates from normal cognition through mild cognitive impairment and stages of dementia. Curr Alzheimer Res 2018; 15(8): 777-88. [http://dx.doi.org/10.2174/1567205015666180119092427] [PMID: 29357799]

[22] Komarova NL, Thalhauser CJ. High degree of heterogeneity in Alzheimer's disease progression patterns. PLOS Comput Biol 2011; 7(11)e1002251 [http://dx.doi.org/10.1371/journal.pcbi.1002251] [PMID: 22072952]

[23] Galasko D, Corey-Bloom J, Thal LJ. Monitoring progression in Alzheimer's disease. J Am Geriatr Soc 1991; 39(9): 932-41. [http://dx.doi.org/10.1111/j.1532-5415.1991.tb04463.x] [PMID: 1842209]

[24] Werner P. Knowledge about symptoms of Alzheimer's disease: correlates and relationship to help□seeking behavior. Int J Geriatr Psychiatry 2003; 18(11): 1029-36. [http://dx.doi.org/10.1002/gps.1011] [PMID: 14618555]

[25] Lanctôt KL, Amatniek J, Ancoli-Israel S, *et al.* Neuropsychiatric signs and symptoms of Alzheimer's disease: New treatment paradigms. Alzheimers Dement (N Y) 2017; 3(3): 440-9. [http://dx.doi.org/10.1016/j.trci.2017.07.001] [PMID: 29067350]

[26] Bature F, Guinn B, Pang D, Pappas Y. Signs and symptoms preceding the diagnosis of Alzheimer's disease: a systematic scoping review of literature from 1937 to 2016. BMJ Open 2017; 7(8)e015746 [http://dx.doi.org/10.1136/bmjopen-2016-015746] [PMID: 28851777]

[27] Lyketsos CG, Carrillo MC, Ryan JM, *et al.* Neuropsychiatric symptoms in Alzheimer's disease. Alzheimers Dement 2011; 7(5): 532-9.
[http://dx.doi.org/10.1016/j.jalz.2011.05.2410] [PMID: 21889116]

[28] Zhang XX, Tian Y, Wang ZT, Ma YH, Tan L, Yu JT. The epidemiology of Alzheimer's disease modifiable risk factors and prevention. J Prev Alzheimers Dis 2021; 8(3): 313-21.
[http://dx.doi.org/10.14283/jpad.2021.15] [PMID: 34101789]

[29] Reitz C, Mayeux R. Alzheimer disease: Epidemiology, diagnostic criteria, risk factors and biomarkers. Biochem Pharmacol 2014; 88(4): 640-51.
[http://dx.doi.org/10.1016/j.bcp.2013.12.024] [PMID: 24398425]

[30] Samieri C. Epidemiology and risk factors of Alzheimer's disease: a focus on diet. Biomarkers for Preclinical Alzheimer's Disease 2018; pp. 15-42.

[31] Li X, Feng X, Sun X, Hou N, Han F, Liu Y. Global, regional, and national burden of Alzheimer's disease and other dementias, 1990–2019. Front Aging Neurosci 2022; 14937486
[http://dx.doi.org/10.3389/fnagi.2022.937486] [PMID: 36299608]

[32] Javaid SF, Giebel C, Khan MAB, Hashim MJ. Epidemiology of Alzheimer's disease and other dementias: rising global burden and forecasted trends. F1000 Res 2021; 10: 425.
[http://dx.doi.org/10.12688/f1000research.50786.1]

[33] Tahami Monfared AA, Byrnes MJ, White LA, Zhang Q. Alzheimer's disease: epidemiology and clinical progression. Neurol Ther 2022; 11(2): 553-69.
[http://dx.doi.org/10.1007/s40120-022-00338-8] [PMID: 35286590]

[34] Beach TG, Malek-Ahmadi M. Alzheimer's disease neuropathological comorbidities are common in the younger-old. J Alzheimers Dis 2021; 79(1): 389-400.
[http://dx.doi.org/10.3233/JAD-201213] [PMID: 33285640]

[35] Silverberg JI. Comorbidities and the impact of atopic dermatitis. Ann Allergy Asthma Immunol 2019; 123(2): 144-51.
[http://dx.doi.org/10.1016/j.anai.2019.04.020] [PMID: 31034875]

[36] Nday CM, Eleftheriadou D, Jackson G. Shared pathological pathways of Alzheimer's disease with specific comorbidities: current perspectives and interventions. J Neurochem 2018; 144(4): 360-89.
[http://dx.doi.org/10.1111/jnc.14256] [PMID: 29164610]

[37] Falsetti L, Viticchi G, Zaccone V, *et al.* Shared molecular mechanisms among Alzheimer's disease, neurovascular unit dysfunction and vascular risk factors: a narrative review. Biomedicines 2022; 10(2): 439.
[http://dx.doi.org/10.3390/biomedicines10020439] [PMID: 35203654]

[38] Demetrius LA, Driver J. Alzheimer's as a metabolic disease. Biogerontology 2013; 14(6): 641-9.
[http://dx.doi.org/10.1007/s10522-013-9479-7] [PMID: 24249045]

[39] Duric V, Clayton S, Leong ML, Yuan LL. Comorbidity factors and brain mechanisms linking chronic stress and systemic illness. Neural Plast 2016; 2016(1): 1-16.
[http://dx.doi.org/10.1155/2016/5460732] [PMID: 26977323]

[40] Piccirillo JF, Costas I. The impact of comorbidity on outcomes. ORL J Otorhinolaryngol Relat Spec 2004; 66(4): 180-5.
[http://dx.doi.org/10.1159/000079875] [PMID: 15467342]

[41] Burgel PR, Escamilla R, Perez T, *et al.* Impact of comorbidities on COPD-specific health-related quality of life. Respir Med 2013; 107(2): 233-41.
[http://dx.doi.org/10.1016/j.rmed.2012.10.002] [PMID: 23098687]

[42] Sarfati D, Koczwara B, Jackson C. The impact of comorbidity on cancer and its treatment. CA Cancer J Clin 2016; 66(4): 337-50.
[http://dx.doi.org/10.3322/caac.21342] [PMID: 26891458]

[43] Gauthier S, Cummings J, Ballard C, *et al.* Management of behavioral problems in Alzheimer's disease. Int Psychogeriatr 2010; 22(3): 346-72.
[http://dx.doi.org/10.1017/S1041610209991505] [PMID: 20096151]

[44] Bharati K. Challenges in Management of Elderly and Frail Alzheimer's Disease Patients Perspective 2024; 2(1) 000105.

[45] Prasher V, Mahmood H, Mitra M. Challenges faced in managing dementia in Alzheimer's disease in patients with Down syndrome. Degener Neurol Neuromuscul Dis 2016; 6: 85-94.
[http://dx.doi.org/10.2147/DNND.S91754] [PMID: 30050371]

[46] Clodomiro A, Gareri P, Puccio G, *et al.* Somatic comorbidities and Alzheimer's disease treatment. Neurol Sci 2013; 34(9): 1581-9.
[http://dx.doi.org/10.1007/s10072-013-1290-3] [PMID: 23370896]

[47] Grossberg GT, Desai AK. Management of Alzheimer's disease. J Gerontol A Biol Sci Med Sci 2003; 58(4): M331-53.
[http://dx.doi.org/10.1093/gerona/58.4.M331] [PMID: 12663697]

[48] Urrestarazu E, Iriarte J. Clinical management of sleep disturbances in Alzheimer's disease: current and emerging strategies. Nat Sci Sleep 2016; 8: 21-33.
[http://dx.doi.org/10.2147/NSS.S76706] [PMID: 26834500]

[49] Dams-O'Connor K, Guetta G, Hahn-Ketter AE, Fedor A. Traumatic brain injury as a risk factor for Alzheimer's disease: current knowledge and future directions. Neurodegener Dis Manag 2016; 6(5): 417-29.
[http://dx.doi.org/10.2217/nmt-2016-0017] [PMID: 27599555]

[50] Popa-Wagner A, Sandu RE, Buga AM, Uzoni A, Petcu EB. Neuroinflammation and comorbidities are frequently ignored factors in CNS pathology. Neural Regen Res 2015; 10(9): 1349-55.
[http://dx.doi.org/10.4103/1673-5374.165208] [PMID: 26604877]

[51] Miranda O, Jiang C, Qi X, Kofler J, Sweet RA, Wang L. Exploring Potential Medications for Alzheimer's Disease with Psychosis by Integrating Drug Target Information into Deep Learning Models: A Data-Driven Approach. Int J Mol Sci 2025; 26(4): 1617.
[http://dx.doi.org/10.3390/ijms26041617] [PMID: 40004081]

[52] Frisoni GB, Boccardi M, Barkhof F, *et al.* Strategic roadmap for an early diagnosis of Alzheimer's disease based on biomarkers. Lancet Neurol 2017; 16(8): 661-76.
[http://dx.doi.org/10.1016/S1474-4422(17)30159-X] [PMID: 28721928]

[53] Vrahatis AG, Skolariki K, Krokidis MG, Lazaros K, Exarchos TP, Vlamos P. Revolutionizing the early detection of Alzheimer's disease through non-invasive biomarkers: the role of artificial intelligence and deep learning. Sensors (Basel) 2023; 23(9): 4184.
[http://dx.doi.org/10.3390/s23094184] [PMID: 37177386]

[54] Calsolaro V, Edison P. Neuroinflammation in Alzheimer's disease: Current evidence and future directions. Alzheimers Dement 2016; 12(6): 719-32.
[http://dx.doi.org/10.1016/j.jalz.2016.02.010] [PMID: 27179961]

[55] Liu H, Zhang B, Mouftah H, Shen X, Ma J. Opportunistic routing for wireless ad hoc and sensor networks: Present and future directions. IEEE Commun Mag 2009; 47(12): 103-9.
[http://dx.doi.org/10.1109/MCOM.2009.5350376]

CHAPTER 10

Emerging Pharmacological Innovation in Alzheimer's Treatment

Arvind Kumar Patel[1,*], **Manish Singh[1]**, **Shivam Kumar[1]**, **Shankul Kumar[1]**, **Neha Singh[2]**, **Neetu Sachan[2]** and **Phool Chandra[2]**

[1] *Narayan Institute of Pharmacy, Gopal Narayan Singh University, Jamuhar, Bihar 821305, India*

[2] *Department of Pharmaceutical Sciences, School of Pharmaceutical & Biological Sciences, Harcourt Butler Technical University, Kanpur, Uttar Pradesh, 208002, India*

Abstract: Alzheimer's disease is a complicated neurological disease characterized by behavioural defects, memory loss, and progressive cognitive deterioration. Recent developments in pharmaceutical innovation have demonstrated promise in tackling fundamental pathogenic pathways, such as neuroinflammation, synaptic dysfunction, mitochondrial impairments, amyloid-\u03B2 and more. Notable strategies include synapse-restoring medicines, tau aggregation inhibitors, neuroinflammatory modulators, and monoclonal antibodies that target amyloid-β2 aggregates. Therapeutic possibilities are further enhanced by emerging methods such as gene editing and RNA-based therapeutics, as well as biomarkers for personalized medicine. These developments provide fresh optimism for altering the course of the disease and enhancing patient outcomes, even though issues like safety profiles and efficient delivery methods still exist. While tau aggregation inhibitors and anti-tau antibodies are being studied to reduce the formation of neurofibrillary tangles, monoclonal antibodies such as aducanumab and lecanemab have shown potential in reducing the amyloid plaque burden. Neuroinflammation is a key player in the widespread pathology of Alzheimer's disease, and drugs that alter microglial activity or stop pro-inflammatory cytokines such as TNF-α and IL-1β are used to combat it. Due to growing understanding of the meaning of mitochondrial dysfunction, mitophagy activators and mitochondrial protectants have been developed to preserve the energy balance of neurons. Gene and RNA-based therapies may hold promise for precision medicine, including CRISPR/Cas9 editing and antisense oligonucleotide platforms that allow us to correct protein expression and genetic factors to more precisely address genetic and phenotypic variables. Moreover, advances in imaging and fluid biomarkers are enabling more personalized treatment plans and improved early detection. Though challenges still remain, including in managing side effects and delivering medications across the blood-brain barrier, these developments represent a significant breakthrough

[*] **Corresponding author Arvind Kumar Patel:** Narayan Institute of Pharmacy, Gopal Narayan Singh University, Jamuhar, Bihar 821305, India; E-mail: akpatel4393@gmail.com

Prashant Kumar Dhakad, Raghav Mishra, Shweta Sharma, Sudhanshu Mishra & Nishant Gaur (Eds.)

in the fight against Alzheimer's. Future efforts must prioritize cross-disciplinary approaches integrating lifestyle shifts with pharmaceutical advancements to enable complete illness remediation.

Keywords: Alzheimer's disease, Diagnosis, Emerging innovation, Gene therapy, Tau aggregation inhibitors, Treatment.

INTRODUCTION

Alzheimer's disease is a neurological illness and the most prevalent kind of dementia. Currently, an estimated 50 million individuals globally experience some types of dementia; nevertheless, as the lifespan rate increases, this figure is predicted to rise by 2050. Alzheimer's Disease (AD) entails deficits in memory, mental problems, changes in behaviour and loss of individual autonomy. AD is characterized by both histopathological features, such as abnormally phosphorylated tau protein, which contributes to the development of Neurofibrillary Tangled Structures (NFTs) in the cortex of the brain and subcortical gray matter, and outside the cell aggregates of β-amyloid peptides (Aβ) fibrils, resulting in neuritic plaques [1, 2]. Endogenous "damage signals" like Aβ oligomers may activate microglial cells, thereby leading to the secretion of cytokines that are pro-inflammatory. It, in turn, causes signaling pathways in neurons, leading to hyperphosphorylation and Tau protein aggregation. As neurons die, this protein develops, which stimulates the microglial cells and leads to neurodegeneration. Alzheimer's disease has been classified as either sporadic or familial. Among them, familial is Early-Onset Alzheimer's Disease (EOAD) in people below 65 years of age. This class is marked by mutations in certain genes, including the presenilin 1 gene (PSEN1, 14q24.2), as well as the presenilin 2 gene (PSEN2, 1q 42.13), and the amyloid precursor protein gene (APP, 21q 21.3), which has been found to be present in up to 70% of cases of familial Alzheimer disease [3 - 5]. However, on the other hand, sporadic AD occurs in people older than 65 years of age. Stress, depression, environmental exposure, social exclusion, poor academic achievement, traumatic brain injury, and metabolic syndrome have been identified as important risk factors for sporadic AD [6].

By 2050, the total number of instances of dementia is predicted to reach two to three times, with Alzheimer's disease accounting for the vast majority. In Latin America, as in other regions, with low incomes, the challenge is not only a spike in the percentage of people suffering from dementia, but also a shortage of investment in health professionals and training and epidemiological studies, which strengthens chronic barriers related to resources, culture and stigma [7]. Genetics was at the forefront of many advances that have assisted scientists in comprehending AD. Conversely, advancement in developing medications for managing Alzheimer's disease has yet to be achieved. All authorized medications treat symptoms, and there is no reliable proof that any may delay the progression of illness [8].

PATHOPHYSIOLOGY OF ALZHEIMER'S DISEASE

Amyloid β and neurofibrillary tangles in the hippocampus are two of the most noticeable morphological characteristics of Alzheimer's disease. Given the abundance of accounts of idiopathic and familial AD, this condition can be generally classified as either Familial AD (FAD) or Sporadic AD (SAD), with the build-up of mostly neurofibrillary tangles and amyloid β in the hippocampus. In FAD, it is linked to alterations in PSEN1, PSEN2 and APP, while in SAD, it is linked to age, genetics, metabolism, and environment. A detailed understanding of the pathophysiology of AD is extremely important in the use of therapeutic regimens. Different hypotheses have been investigated, and a multitude of mechanisms have been considered to provide a better insight into the pathophysiology of AD [9].

There are different types of genes that influence Alzheimer's disease progression by the development of amyloid plaque, tau pathology and neuroinflammation. These genes and their molecular mechanism for the progression of Alzheimer's disease are given in Table **1** [8, 9].

Table 1. Genes involved in the progression of Alzheimer's disease.

Gene	Pharmacological function	Molecular pathways in AD	References
ADAM10 (15q21.3)	It is the primary secretase in the brain, contributing to the non-amyloidogenic route of APP metabolism.	Changes in APP metabolism (by the non-amyloidogenic route), synaptic plasticity and hippocampal neurogenesis.	[10, 11]
BIN1 (2q14.3)	Participates in immunological reactions, calcium balance, apoptosis, synaptic vesicle endocytosis, and plasma membrane dynamics	Assists in amyloid (by secretase activity) and Tau disease, and is linked to inflammatory processes, apoptosis and calcium equilibrium.	[12, 13]
EPHA1 (7q34)	It contributes to synaptic formation and plasticity by regulating cell migration, angiogenesis and death.	Mutations in immunological pathways, endocytosis and breakdown of the blood-brain barrier.	[14]
CD33 (19q 13.3)	It inhibits immune cell activity and cytokine synthesis	Influences the activation of microglial (neuroinflammation) and Aβ elimination *via* microglial cells	[15]
TOMM40 (19q13.32)	Helps stabilize the mitochondrial membrane respiratory chain	Impairment of the mitochondrial membrane leads to oxidative stress	[16]

(Table 1) cont.....

Gene	Pharmacological function	Molecular pathways in AD	References
INPP50 (2q37.1)	Engages in modulating microglial expression of genes	Microglia with deficient phagocytic capability leading to increased Aβ accumulation	[17]
BCHE (3q26.1)	Assist in the breakdown and stimulation of hazardous organophosphate esters by acetylcholine	Associated with NFT and NPF	[18]
ECHDC3 (10p14)	This involves fatty acid production in mitochondria and responsiveness to insulin.	Changes in the metabolism process of lipid and cholesterol can affect cognitive processes.	[19]

The amyloid beta (Aβ) hypothesis

The amyloid beta (Aβ) hypothesis has been one of the more influential ideas driving Alzheimer's research. This hypothesis holds that, in addition to other pathological aspects, abnormal deposition of Aβ through the formation of neurofibrillary tangles ultimately leads to cell death, vascular damage, and dementia [20]. It is thought that the aberrant cleavage of a bigger protein called Amyloid Precursor Protein (APP) results in the formation of the Aβ peptide. The transmembrane protein known as APP is normally broken down by specific enzymes called secretases in different places, each of which creates a peptide fragment crucial to neural processes. When the N-terminal side of APP is broken by the B-site APP cleavage enzyme, or BACE, the carboxy terminus CTF99/89 fragment is left behind, and the sAPPβ peptide fragment is released. Following the cleavage of the CTF99/89 fragment by γ-secretase, many Aβ peptides are produced, the two most prominent of which are Aβ40 and Aβ42. The longer Aβ42 is more likely to aggregate than the shorter forms [21]. Aβ oligomers, which are small clusters of aggregated forms of Aβ peptides, are thought to be especially toxic to neurons and synapses [22]. The exact mechanism for Aβ aggregation is unclear. The Aβ theory is nevertheless supported by mutations in APP and secretases linked to uncommon instances of early-onset familial AD by aiding the accumulation of longer, more aggregative Aβ species that hasten the disease course [23].

Aβ peptides, particularly the longer and more amyloidogenic ones, can be overproduced as a result of changes in chromosome 21's APP gene, increasing the risk that Aβ will aggregate and form amyloid plaque in the brain. It has been determined that approximately 25 mutations are harmful and can cause autosomal dominant AD. Due to differences resulting from the cleavage motif, these mutations change the Aβ42/Aβ40 ratio in a way that produces variable aggregation and morphological characteristics [24]. The D7H mutation on APP boosts the oligomerization of Aβ42 [25]. Aβ synthesis is increased as a result of aberrant processing of the APP protein caused by PSEN1 and PSEN2 mutations on chromosomes 14 and 1, respectively [26]. The apolipoprotein E (APOE) gene, located on chromosome 19, is another gene linked to the onset of AD. ε2, ε3, and ε4 are the three main alleles of the APOE gene. The most prevalent kind of AD is late-onset AD, for which the APOE ε4 allele is the biggest genetic risk factor. A further elevated risk of developing AD is linked to the presence of one or two copies of the ε4 allele [27]. It should be mentioned that this mutation causes a large buildup of Aβ plaque since it is linked to increased Aβ aggregation and impaired Aβ clearance. In comparison to noncarriers, more neuroinflammation and neuronal damage are common among APOE ε4 carriers [28].

It is believed that the buildup of Aβ peptides sets off a series of processes that ultimately result in neurodegeneration. By activating kinases, including GSKβ and CDK-5, Aβ aggregation may generally lead to tau phosphorylation (Figure **1**) [29]. Tau pathology and increased GSKβ activity are caused by the APOE ε4 allele. Amylin receptors are likely the mechanism by which Aβ peptides activate microglial cells, which in turn triggers the NLRP3 inflammasome pathway and results in neuroinflammation [30]. Massive changes in mitochondria are brought on by the buildup of mitochondria-targeted Aβ1-42, which ultimately leads to mitochondrial malfunction and the activation of apoptotic pathways. By altering NMDA receptor signaling, they also harm synapses. Excitatory synapses are eliminated when the oligomers bind at the synaptic density (Table **2**) [31].

Table 2. Key aspects of the amyloid cascade hypothesis.

Aspect	Description	References
Genetic Evidence	Mutations in APP, PSEN1, and PSEN2 linked to familial AD, supporting Aβ's role.	[32]
Pathological Correlations	Aβ plaques correlate with cognitive decline in AD patients.	[33]
Experimental Models	Transgenic mice with APP mutations develop Aβ plaques and cognitive deficits.	[34]
Shift to Oligomers	Soluble Aβ oligomers, like ADDLs, are toxic and cause synaptic dysfunction.	[32, 35]
Intracellular Aβ	Intracellular Aβ accumulation, particularly in lysosomes, contributes to pathology.	[33, 36]
Tau Pathology	Aβ promotes tau pathology, contributing to neurodegeneration.	[34]
Vascular Mechanisms	Impaired Aβ clearance *via* the brain's microcirculatory system.	[35]
Therapeutic Strategies	Immunotherapies and small-molecule inhibitors targeting Aβ.	[37]
Multimodal Therapies	Combining Aβ-targeted therapies with interventions for other pathologies.	[38]

(Table 2) cont.....

Aspect	Description	References
Early Intervention	Importance of early intervention based on Aβ accumulation preceding symptoms.	[39]

Tau protein and neurofibrillary tangles

Another main pathological characteristic of AD, besides the accumulation of amyloid-beta plaques, is the existence of neurofibrillary tangles, which are abnormal tau protein aggregates. A microtubule-related protein called tau plays important roles in maintaining and stabilizing the structure of neuronal microtubules, a cellular function fundamental to the working exchange of materials and nutrient supply through neurons [40]. In healthy neurons, the tau protein is under very tight control of regulation and phosphorylation, hence allowing binding to tubulin and polymerization of microtubules. Tau gets modified post-translationally in AD in abnormal ways, such as hyperphosphorylation, affecting its normal functionality [41].

Self-aggregation is the nature of hyperphosphorylated tau molecules that eventually become insoluble Paired Helical Filaments (PHFs) and Straight Filaments (SFs). In the brains of AD patients, aggregated tau structures are the main components of NFTs [42]. Tau pathology also includes amyloid-beta fibrils. The formation and accumulation of NFTs contribute to neuronal dysfunction and progressive neurodegeneration through diverse means. The first of them is that tau's association with microtubules is destroyed by a number of posttranslational alterations that take place in its microtubule-binding domain, including lysine succinylation, lysine acetylation, and tyrosine phosphorylation. Neuronal NFT buildup results in a decreased amount of acetylated alpha-tubulin [43]. Inflammations in the brain are the second thing that NFT buildup causes. In the NFT-rich areas of the brain, microglia begin to hypertrophy and react to NFTs more than amyloid plaques [44]. Contrary research, however, showed that S100B overexpressed astrocytes and IL-1 overexpressed microglia had NFTs of proportionate amounts and formation stages for microglial density. Such a constant inflammatory status aggravates neuronal injury and advances the propagation of tau pathology further [45]. According to more recent research, tau aggregates can propagate to the brain because of their prion-like characteristics. Abnormal tau forms are hypothesized to promote the gradual spread of tau in many brain areas, which are thought to encourage the misfolding of regular tau in nearby neurons. It is believed that this spreading process plays a role in AD and the subsequent development of cognitive decline in people who are affected. Therefore, one promising avenue for disease-modifying therapy is to target aberrant tau aggregation and its consequences (Fig. **1**) [46].

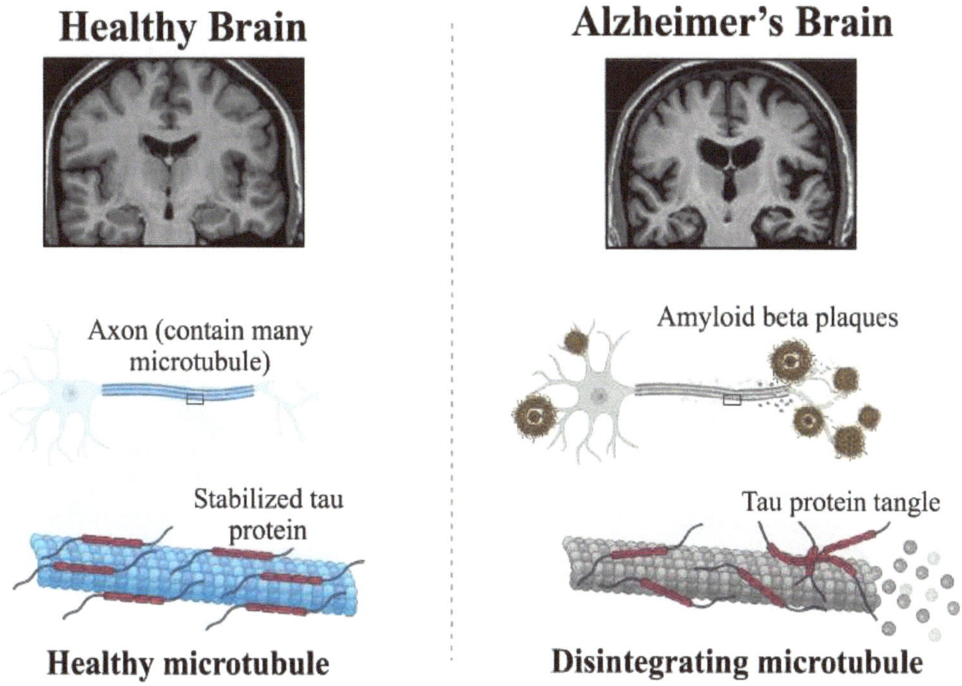

Fig. (1). Pathophysiology of Alzheimer's disease due to tau protein tangle and formation of amyloid beta plaques.

Neuroinflammation

The role of inflammatory processes in the progression of AD is substantial, as one of the defining characteristics of AD patients is persistent brain inflammation, or neuroinflammation, which is linked to the disease's progression [47]. In addition to the release of pro-inflammatory chemicals, including cytokines, chemokines, and reactive oxygen species, the activation of immune cells such as astrocytes and microglia is indicative of the neuroinflammatory response in AD. Short-term immune responses like these may be useful in defending the brain and getting rid of harmful substances, but long-term, unchecked inflammation might harm neurons and worsen the course of the illness [48]. Aβ plaques in AD activate a range of cell surface receptors, such as CD14, TLR4-TLR6, CD36-CD47-α6-1-integrin complex, and scavenger receptors, which in turn activate microglia. Cardiolipin, which is similarly released by microglia, promotes the absorption of Aβ plaques [49]. The initial reaction to Aβ plaques is to remove them, but prolonged exposure causes chronic activation primarily of the M1 phenotype and the release of pro-inflammatory cytokines, such as interleukin 1β and TNF-α [50]. Because ApoE4 binds to microglia's TREM2 receptors, activating and secreting inflammatory cytokines such as IL-1, IL-6, TNF-α, and IFN-γ, which all contribute to neuronal damage, it plays a crucial role in the development of FAD [51]. Furthermore, microglia's absorption of amyloid beta is enhanced when APOE and APOJ bind to TREM2. When monocyte chemoattractant protein-1 (MCP-1) is present, astrocytes, another inflammatory cell in AD, migrate near Aβ plaques and break them down. Additionally, by secreting glypican-4, which encourages tau hyperphosphorylation, they aid in the accumulation of neuroinflammation with microglia [52]. The problem is made worse by the involvement of peripheral mononuclear cells in the NFT-induced C3 activation on astrocytes, which results in neuroinflammation and neuronal damage [53]. Peripheral mononuclear cell involvement exacerbates AD pathology because AD shares pathophysiological characteristics with peripheral inflammation and systemic illnesses. Neuroinflammation is made worse by chronic systemic inflammation, which is seen in conditions like diabetes, obesity, and cardiovascular disorders. This, in turn, raises the risk of AD. Systemic inflammation frequently destroys the blood-brain barrier, which makes it possible for inflammatory chemicals and peripheral immune cells to enter the brain and enhance neuroinflammation [54].

Oxidative stress

An imbalance between the generation of ROS and the body's capacity to counteract them through antioxidant mechanisms is known as oxidative stress [55]. Oxidative stress has a significant impact on the development and course of Alzheimer's disease. Because of their high rates of metabolism and the fact that ROS generation is an inevitable result of oxidative phosphorylation, mitochondria, the cell's powerhouses, are especially susceptible to oxidative damage [56]. Disrupted energy metabolism, changed dynamics, inhibited antioxidant capacity, and elevated ROS are the foundations of mitochondrial dysfunctions seen in both human AD brains and animal models. Moreover, AD oxidative stress is made worse by the breakdown of cellular antioxidant defense systems. The activities of glutathione peroxidase, catalase, and superoxide dismutase (SOD) are essential for eliminating free ROS. However, excessive ROS generation in AD frequently disrupts antioxidant mechanisms, increasing oxidative stress [57]. In AD, oxidative stress and mitochondrial dysfunction are caused by a variety of factors. First, the buildup and aggregation of Aβ peptides, which have been demonstrated to directly affect mitochondrial functioning, is one of the characteristics of AD [58].

Aβ peptides can build up and interfere with mitochondrial processes in the intermembrane space of the mitochondria, which lowers ATP synthesis and increases ROS production. Furthermore, a vicious loop is created by Aβ-induced mitochondrial damage since more dysfunctional mitochondrial activity might lead to increased Aβ synthesis and aggregation. Secondly, the idea of tau-linked mitochondrial dysfunction is a key mechanism that contributes to AD pathogenesis. Tau hyperphosphorylation and aggregation obstruct mitochondrial transport *via* neural axons, perhaps leading to oxidative stress and localized energy shortages [59]. This is supported by the fact that tau disease alters mitochondrial dynamics, compromising the fission and fusion mechanisms that are essential for preserving a robust mitochondrial network. On the other hand, tau aggregation is brought on by oxidative stress and mitochondrial malfunction [60]. They worsen the progression of the illness process by damaging and killing neurons, disrupting synaptic function and plasticity, and causing neuroinflammation. In addition to accelerating neurodegenerative processes, oxidative stress may also play a role in the development and buildup of Aβ plaques and neurofibrillary tangles [61]. Targeting these mechanisms has been viewed as an appealing possible treatment approach because of the roles that oxidative stress and mitochondrial dysfunction played in AD. While dietary or pharmacological interventions are designed to restore redox balance and reduce oxidative damage, antioxidant therapies encompass a wide range of approaches to accomplish these aims. Compounds targeting mitochondrial function and dynamics also appear to be able to preserve neuronal health and therefore alleviate symptoms associated with AD [62].

Vascular abnormalities

The blood vessels of the brain are responsible for the continual energy supply to the brain in order to deliver oxygen and nutrients critical for the entire functioning of neurons. When vascular function is impaired, it imposes itself on the delivery of that supply, resulting in hypoperfusion with decreased delivery of oxygen and nutrients to the brain. Vascular aging stands out as a prominent contribution to the activation of the AD pathology. More precisely, conditions such as cerebral atherosclerosis, arteriosclerosis, and cardiovascular diseases put individuals at a higher risk of being afflicted with AD [63]. Those who are chronically hypoperfused in the brain die off neuronally. This prolonged state of ischemia and hypoxia in the brain may induce the deposition of beta-amyloid peptides, characteristic of Alzheimer's disease. The reduction in cerebral perfusion also creates problems with the clearance of Aβ, ultimately contributing to a greater burden of that protein and aggregation [64]. High blood pressure, diabetes, high cholesterol, smoking, and other chronic vascular risk factors all enhance the chance of AD. These risk factors encourage a series of biological processes that impair cerebral blood flow and vascular health by causing endothelial dysfunction, oxidative stress, inflammation, and atherosclerosis. Vascular pathology, which includes cerebral infarcts, white matter lesions, and microhaemorrhages, can result from cerebrovascular disorders caused by atherosclerosis and small vessel disease. These vascular lesions accelerate AD-related neurodegeneration and place a cognitive strain on the brain [65]. Another vascular condition that has been closely linked to AD is cerebral amyloid angiopathy. In this case, Aβ is deposited in the walls of the brain's blood vessels, usually in capillaries and tiny veins. Increased neuroinflammation and neuronal damage can result from CAAs' disruption of blood flow regulation and further damage to the blood-brain barrier. Free radical production and mitochondrial failure are further effects of hypoperfusion [66].

Synaptic dysfunction and neurotransmitter alterations

The pathophysiology of Alzheimer's Disease (AD) is characterized by changes in neurotransmitter systems and synaptic dysfunction. Neurons communicate with one another at synapses, which are critical for cognitive processes including memory and learning [67]. Early-onset synaptic disruption in AD is linked to the cognitive deterioration of those who are afflicted. Aβ peptide buildup and the development of neurofibrillary tangles, which are caused by hyperphosphorylated tau protein and result in synaptic loss and damage, are the causes of this [68]. Excitotoxicity, in which excessive stimulation of glutamatergic receptors results in cell death and neuronal injury, is linked to glutamate neurotransmission dysfunction. Dopaminergic, serotonergic, and noradrenergic system alterations that impact mood, behavior, and cognitive abilities are also observed in AD [69, 70].

BARRIERS TO EFFICIENT TREATMENT OF ALZHEIMER'S DISEASE

Diagnosis at a very late stage

The early diagnosis of Alzheimer's Disease (AD) becomes indispensable in the effective treatment and retardation of disease progression. Many therapeutic trials for AD were either partly or wholly unsuccessful because treatments were often given too late, thus allowing irreversible brain damage to occur. More effective therapy and better treatment outcomes require diagnoses at the preclinical stage of AD, prior to major neurodegeneration [71].

Limited knowledge of pathogenesis mechanisms involved in Alzheimer's disease

In several publications, the picture of the complexity of AD in terms of its pathogenesis is put forward. Seo and Holtzman (2024) state that "the exact pathophysiology of the disease is not completely clear," right after saying that some much-needed progress in treating AD with anti-Aβ antibody treatments has surfaced over time. Numerous hypotheses were proposed to explain these failures, including unclear knowledge about the disease pathways and processes. However, because the process has never been fully figured out, so many clinical trials failed [72].

Blood-brain barrier

This barrier constitutes one of the key protective barriers of the blood-brain interface; it is mainly responsible for governing substance transfer between blood and the CNS. It serves as a protective barrier between blood and the CNS, effectively screening off toxins and pathogens; therefore, it hampers the delivery of drugs to treat various CNS disorders [73].

Complex pathology

The nature of AD pathology is complex, owing to processes beyond Aβ and tau aggregation, including inflammation, oxidative stress, and mitochondrial damage, coupled with an imbalance of neurotransmitters [74]. Usually, AD also co-exists with other diseases associated with aging, including cerebrovascular lesions, Lewy bodies, and TDP-43 pathology, which further complicate the clinical picture and the progression of the illness. With the heaviness of AD pathology in consideration, there is a growing consensus that multipronged and holistic approaches in both diagnosis and treatment are warranted. This includes combination therapies, multifunctional modulators targeting various pathological factors, and multiplexed biomarker detection for more reliable diagnostics [75].

Lack of effective drugs

Needless to say, the present treatments are largely symptomatic in the case of diseases such as Alzheimer's disease, schizophrenia, Huntington's disease, or even neurodegenerative diseases like Parkinson's, which aggravate the necessity of finding new therapeutics that at least delay, halt, or reverse the progression of such diseases. For instance, in Alzheimer's disease, none of the current drugs can cure or even halt disease progression but merely alleviate symptoms for a small time [76].

Clinical trial challenges

Drug development for AD faces significant hurdles, having reached an astonishing 99% failure rate in clinical trials without displaying any drug-placebo difference. Such a high rate of failure risks delaying such treatments from being available for patients and discourages investment into AD drug development. A main hurdle in AD clinical trials is the recruitment of appropriate participants. Low rates of participation impose a major hindrance to advancing research and care for AD [77].

Genetic and environmental factors

AD is a highly challenging, complex neurodegenerative disorder where treatment strategies are influenced, among many other conditions, by genetic predisposition and environmental factors. The interaction of these factors, in conjunction with the presence of genetic variants, has aided in forming a multi-factorial gamut of AD that includes contributing environmental exposures on disease risk, onset, and progression [78].

Aging population

Generally, the rapidly increasing number of aged individuals across the world is very likely to lead to the development of Alzheimer's disease and dementias that were unthinkable before. This means that dementia is a real burden that is increasingly consuming health resources around the world. From 2013, the number of people suffering from dementia is expected to rise to 7.8 million and will exceed 27 million in 2050 in Latin America. Worldwide, in the population aged 60 years or older, the prevalence of dementia due to AD in community settings was estimated at 40.2 per 1000 persons [17, 62].

Costs of care

Alzheimer's Disease (AD) places a huge economic burden on patients, caregivers, and healthcare systems. The total healthcare costs for AD treatment in the United States were estimated at $305 billion in 2020, while those were forecasted to be greater than $1 trillion by 2050. The major proportion of these costs has been attributed to skilled nursing care, home healthcare, and hospice care. In Europe, annual care costs per patient were estimated to lie between €6,614 and €64,426 in 2005 [17, 79].

Stigma and awareness of Alzheimer's

In AD, stigma and a lack of awareness have a great influence on seeking help and participating in a clinical trial. Dementia-associated social stigma brings a delay in diagnosing and treating it, which makes people unwilling to engage in research [80].

ADVANCEMENTS IN DRUG DELIVERY SYSTEMS

Nanotechnology in the treatment of Alzheimer's

In the treatment of Alzheimer's Disease (AD), nanotechnology has shown great promise, especially when it comes to improving drug delivery methods. Delivery of therapeutic drugs to the brain is largely hampered by the Blood-Brain Barrier (BBB), although nanotechnology offers creative ways to get around this problem. Several nanoparticle-based systems have been created that cross the BBB, target specific brain regions, and provide sustained release of therapeutic compounds to improve the delivery and efficacy of drugs intended to treat Alzheimer's disease. Major advances in this field are listed below [81].

Lipid nanosystems

Therapeutic treatments such as galantamine and curcumin can be delivered across the blood-brain barrier using lipid-based nanoparticles, increasing their bioavailability and targeted efficacy [82].

Polymer-based nanocarrier

Drugs such as huperzine A and curcumin have been delivered *via* PLGA nanoparticles, which are often modified with specific ligands to enhance their solubility and BBB penetration [83, 84].

Inorganic nanocarriers

Systems with multifunctional features, such as AgAuSe quantum dots and magnetic iron oxide nanoparticles, have been used for both drug delivery and diagnostic applications [85, 86].

Multifunctional and targeted approaches

Hybrid nano-system

Using ingredients such as curcumin and cerium oxide nanoparticles, these systems have been designed to simultaneously target multiple Alzheimer's diseases, including oxidative stress and β-amyloid aggregation [87, 88].

Site-specific delivery

Techniques involving surface modification and ligand conjugation improve the targeting of nanoparticles to specific areas of the brain, thereby reducing side effects and improving therapeutic outcomes [89, 90].

Clinical implications and challenges

Nanoparticle systems have shown promise in clinical settings by reducing AD pathology and improving cognitive performance in preclinical models [91]. Although the outcomes are promising, there are significant challenges regarding safety, scalability, and clinical applicability. More research needs to be done to make these systems as user-friendly [92].

Although nanotechnology has the potential to significantly enhance AD treatment, it also raises concerns about toxicity and long-term safety. Essentially, more studies and clinical testing must be conducted for nanoparticle-based drug delivery systems to fully utilize their potential in treating Alzheimer's disease effectively [93].

Strategies for bypassing the blood-brain barrier

As the Blood-Brain Barrier (BBB) prevents therapeutic drugs from accessing the brain, developing strategies for overcoming the BBB is an essential part of successfully treating Alzheimer's disease (AD). Recent improvements have been dedicated towards innovative drug-targeting methods as well as counter-external environment approaches to bypass this barrier. Among other things, these approaches include the use of gene therapy, nanoparticles, and targeted ultrasound to aid the delivery of treatments across the Blood-Brain Barrier [94].

Nanoformulation delivery technique

Development of Liposomes, Exosomes, and Nanoparticles for Drug Delivery across the Blood-Brain-Barrier. These carriers can improve absorption and decrease systemic side effects by encapsulating drugs. AD with antioxidant-based nanomedicine. Neuroinflammation and oxidative stress can be targeted, for example, with curcumin-loaded silica nanoparticles and resveratrol-loaded selenium nanoparticles. After that, BACE1 RNAi-loaded Prussian Blue nanocomplex-mediated transduction was applied to AD. In particular, as was previously demonstrated, the model improves cognitive skills and increases BBB permeation while also decreasing amyloid β production and accumulation [95].

Targeted ultrasound

Targeted ultrasound is a non-invasive technique to deliver drugs specifically by temporarily disrupting the blood-brain barrier. When combined with monoclonal antibodies like aducanumab, it has been employed to lower amyloid beta concentrations in AD patients. MRI-guided FUS holds promise for maximizing drug delivery to the central nervous system due to the precision afforded, as well as the availability for real-time imaging [96].

Genome therapy and CRISPR-Cas9

Gene therapy and CRISPR-Cas9 techniques are being explored to treat the genetic abnormalities linked to AD. These approaches may modify disease progression by targeting specific genetic pathways [97]. While these approaches seem to hold potential, clinical use of them must still overcome obstacles. There continue to be concerns that the BBB is complex, requires precise targeting, and could have safety implications. Furthermore, while early treatments using these techniques may slow disease progression, they have not yet been effective in treating AD at an advanced stage. AD cannot be reversed. To improve these methods and prevent AD, further research and clinical studies are needed to increase their efficacy and accessibility for the treatment of AD [98].

SIGNIFICANCE OF BIOMARKERS IN PHARMACEUTICAL INNOVATION

Biomarkers for diagnosis and prediction of Alzheimer's disease

Finding predictive and diagnostic biomarkers for Alzheimer's disease (AD) is essential for successful treatment plans and early intervention. Recent studies have highlighted the importance of biomarkers in AD diagnosis, especially in the early stages when disease-modifying treatments may be most successful. This overview will address the diversity of biomarkers, their clinical use, and the challenges in using them [99].

Types of Biomarkers

Neuroimaging Biomarkers

Methods such as MRI and PET are crucial for visualizing the alterations in the brain linked to AD, which aids in the diagnosis confirmation.

Cerebrospinal Fluid (CSF) Biomarkers

Proteins, including tau and amyloid-beta (Aβ), are utilized in therapeutic settings and are essential for comprehending AD pathogenesis [100].

Blood-Based Biomarkers

With excellent predictive values, emerging blood assays, namely p-tau217, show promise in both diagnosing AD and distinguishing it from other dementias [101].

Clinical Applications

Due to the fact that biomarkers inform clinicians about the most effective treatment strategies, particularly as new therapies targeting AD pathology become available, they allow for the early diagnosis of AD before clinical symptoms appear, allowing for timely intervention and treatment guidance (Fig. **2**) [102].

Challenges

While blood-based biomarkers provide a less intrusive option, there are still issues with their general clinical adoption. In contrast, the interpretation of biomarker results can be complex and involves consideration of specific patient characteristics [103].

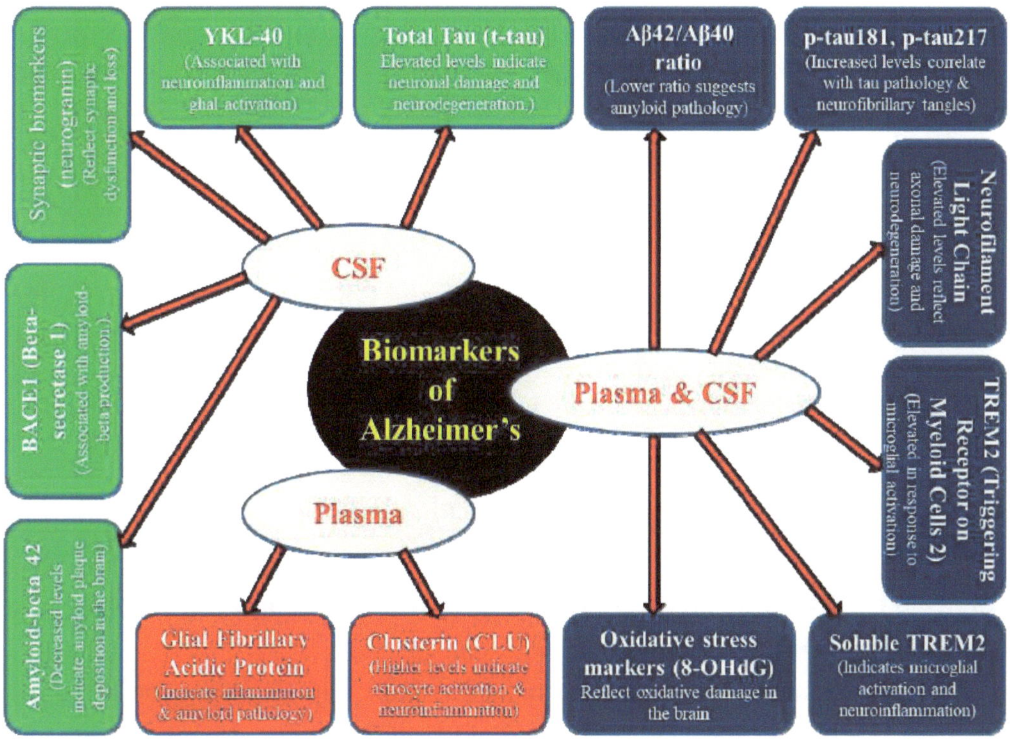

Fig. (2). Biomarkers for the diagnosis of Alzheimer's disease.

To guarantee their successful application in clinical practice, additional validation and standardization are still required, notwithstanding the progress made in biomarker research. The therapy of Alzheimer's disease may be greatly improved by incorporating these indicators into standard care [104].

Biomarker-guided drug development

For Alzheimer's disease (AD), biomarker-guided medication development is a game-changing strategy that improves early diagnosis and treatment effectiveness. Researchers can target interventions at preclinical stages and potentially change the course of disease by finding particular biomarkers. To guarantee that treatments are customized to each patient's unique profile, this approach places a strong emphasis on incorporating biomarkers into clinical trials and routine evaluations [105].

Role in Clinical Trials

In clinical trials, biomarkers are crucial for patient selection and monitoring, especially for medications that target amyloid-β. By standardizing biomarker nomenclature, the FDA-NIH BEST framework improves communication amongst drug development stakeholders [106].

Advancements in Biomarker Technology

New blood tests have become widely available, and diagnostic tools are available at reasonable prices, leading to increased use of biomarkers. Biomarker discovery is optimized by machine learning frameworks such as BRAIN, which uncover complex interdependencies that can guide drug development strategies [107]. Although there are many possibilities to improve AD treatment by focusing on biomarkers, there are still challenges in ensuring the accuracy and utility of these diagnostic tools, especially in different populations [108].

CLINICAL TRIAL AND REGULATORY FRAMEWORK

With the introduction of disease-modifying therapies (DMTs) and the emphasis on amyloid-targeting monoclonal antibodies, the clinical trials and regulatory environment surrounding AD treatment have changed significantly. Emerging data focusing on the need for combination drugs and imaginative trial designs to tackle the complicated pathophysiology of AD suggest that the tide is slowly turning. Regional regulatory standards are adapting to these advancements, fostering a collaborative method of drug development [109].

Clinical trials framework

Recently, a variety of amyloid-targeting monoclonal antibodies have been tested in clinical trials, including aducanumab, lecanemab, and donanemab. These antibodies appear to have the potential to decrease cognitive decline through amyloid plaque reduction in the brain [110, 111]. However, to show the potential impact of these treatments, studies focus on early-stage intervention and use biomarkers for the selection of patients and monitoring of efficacy (Fig. **3**) [112,

113]. It is hoped that combination therapies that target multiple pathways, including tau and amyloid proteins, will improve treatment outcomes [77].

Current Alzheimer's disease clinical trials are investigating several new treatment approaches, indicating a trend toward more targeted interventions. These studies address the complex nature of AD using a variety of approaches, such as pharmacological treatments and brain stimulation techniques [114].

Hydralazine Trial: A triple blind randomized trial with 424 participants is evaluating the efficacy of hydralazine in mild to moderate AD, with an emphasis on cognitive performance over a 12-month period. Changes in the Alzheimer's Disease Rating Scale are one of the main outcomes [115].

Amyloid-targeting therapies: A major breakthrough in AD therapy options, recently approved monoclonal antibodies that target amyloid plaques have demonstrated promise in halting cognitive deterioration [116, 117].

Novel strategies: Studies are looking into non-invasive techniques that have the potential to improve cognitive function with little adverse effects, such as transcranial direct current stimulation (tDCS) and repetitive transcranial magnetic stimulation (rTMS) [118].

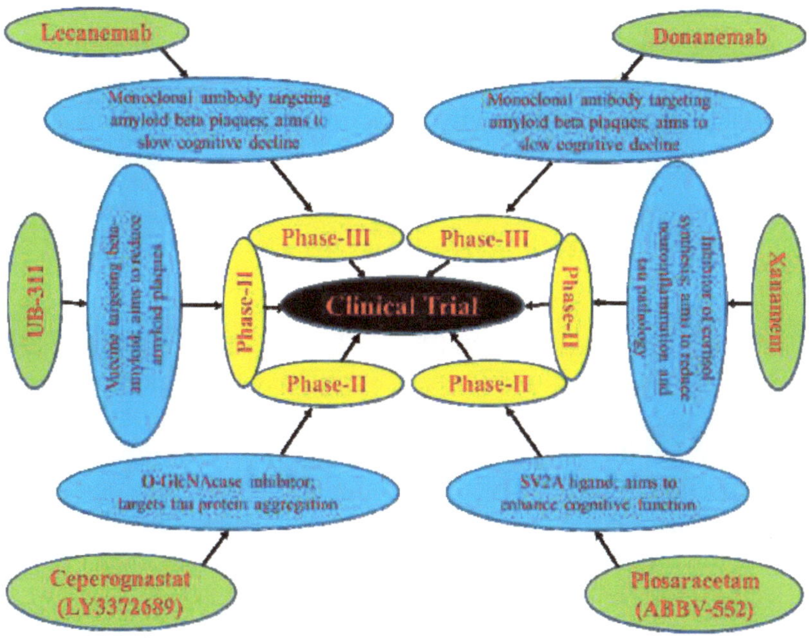

Fig. (3). Drugs ongoing in clinical trials for Alzheimer's treatment.

Regulatory landscape

Through fast routes, the FDA has approved a number of amyloid-targeting treatments; nonetheless, post-marketing studies are necessary to verify therapeutic benefits [119]. In order to facilitate worldwide drug development, regulatory guidelines from the US, EU, and Japan emphasize the importance of strong clinical data, especially in predementia stages, and promote multiregional clinical studies [120]. The necessity for strong proof of efficacy and safety is shown by CMS's decision to only fund amyloid-targeting treatments in approved clinical trials [121].

Even so, there are still disagreements over the therapeutic significance of the advantages seen in trials, with some speculating as to whether gains in cognitive scores result in appreciable enhancements in patients' quality of life [122]. Safety issues, like anomalies in imaging caused by amyloid, call for close observation and emphasize the necessity of thorough safety evaluations in clinical trials [123, 124]. In order to address concerns about equality in treatment access and guarantee the generalizability of results, it is imperative that studies include a variety of groups [125].

The complexity of Alzheimer's disease means that more effective treatments may require a multidisciplinary approach that addresses multiple pathogenetic pathways, although the development of drugs targeting amyloid represents an important milestone. In response to these advances, the regulatory environment is fostering collaboration and creativity to address unmet needs in AD treatment [126].

ARTIFICIAL INTELLIGENCE IN DRUG DEVELOPMENT FOR ALZHEIMER'S

Artificial intelligence (AI) is essential for the advancement of AD drug discovery by improving knowledge of disease mechanisms and accelerating the identification of potential treatment targets. AI-driven strategies facilitate the creation of new molecules, maximize drug repurposing, and increase the effectiveness of clinical trials. Given the complex pathophysiology of AD and the shortcomings of existing treatments, these advancements are essential. In addition to accelerating the process, integrating AI into drug research increases the chances of discovering effective treatments for AD. Below are key features of AI in new medical innovations for Alzheimer's [127].

AI-enabled drug design

Predictive models and generative algorithms are two examples of AI technologies used to create nontoxic inhibitors that target specific enzymes, such as DYRK1A, that are linked to the course of Alzheimer's disease. These models aid in the development of strong inhibitors with anti-inflammatory and antioxidant qualities by forecasting chemical affinities and evaluating toxicity [128].

Machine learning in drug discovery

Machine learning (ML) makes it possible to quickly analyse large datasets and find new drug targets and candidates. Navigating the multifactorial nature of AD, which includes a number of pathophysiological theories beyond amyloid-beta build-up, such as aberrant tau protein and mitochondrial dysfunction, is made easier with this method [129].

Genomic and real-world data integration

When paired with genetic data, AI and ML techniques aid in the discovery of druggable targets and repurposable medications. To illustrate the promise of AI in customizing treatments based on genetic and real-world patient data, studies have identified particular genetic variations and pharmacological targets, such as EPHX2, that are linked to a lower risk of AD [130].

Personalized treatment and prognostic modelling

By tracking patient progress and forecasting individual responses to medicines, AI improves individualized treatment plans. Furthermore, it helps in prognostic modelling, which predicts the course of diseases and optimizes care regimens using multimodal data.

While AI has greatly advanced the discovery of new drugs for AD, challenges such as data privacy, ethical issues, and clinical workflow integration remain. To unlock the potential of AI to transform Alzheimer's care and deliver improved patient outcomes, these hurdles will need to be removed [131].

FUTURE DIRECTIONS

The future of AD therapeutic pharmacology aims at an efficacious approach that encompasses not only multi-targeted therapy, but also novel drug delivery technologies and precision medicine within the setting of pathological heterogeneity in AD. By addressing different underlying mechanisms of AD, such as tau protein tangles, beta amyloid plaques, and neuroinflammation, these methods hope to target the multi-weighted pathology of the disease. The integration of genetic and immunological features with sophisticated drug discovery methods is enabling more efficient and individualized treatments [132].

Customized immunotherapy and precision medicine

Personalized immunotherapy, based on immunological and genetic profiles, is a promising approach to improve the effectiveness of the therapy and reduce adverse drug reactions in AD patients. Patient classification and identification of specific biomarkers for accurate treatment has become an urgent task through this method [133]. These include precision medicine technologies that tailor treatment to an individual's genetic profiles and biomarkers to provide a more individualized therapeutic framework [134].

Combination and multitarget therapy

To combat the multifactorial nature of AD, strategies involving the development of multi-target therapeutics are under consideration. Such strategies include multi-target drug development able to simultaneously interfere with at least two pathways implicated in AD, including tau, amyloid beta and neuroinflammation [135, 136]. The effectiveness of combination drugs in precision medicine was further underscored by their ability to reverse disease-associated gene networks in a cell-type–specific way, such as letrozole and irinotecan [137].

Advanced drug delivery systems

Novel drug delivery technologies such as nanoparticles and nanomaterials are thus in development to enhance the delivery and efficacy of therapies. These systems attempt to increase drug absorption and blood-brain barrier crossing, while minimizing systemic adverse effects [138].

Cutting-edge therapeutic targets

Investigators are seeking new therapeutic targets beyond tau and amyloid, including inflammation, neuroprotection, and synaptic plasticity. These targets are involved in the development of disease-modifying treatments that can slow or stop the progression of the disease [139].

Aside from the potential of these new technologies, there are still barriers to implementing them in clinical settings. To improve these tactics and guarantee their effectiveness and safety, scientists, physicians, and pharmaceutical corporations must continue their study and work together due to the intricacy of AD pathology. Furthermore, combining lifestyle modifications with early detection techniques may improve treatment results and AD patients' quality of life [140].

Emerging technologies and innovation pipelines

A deeper comprehension of the intricate biology of Alzheimer's disease and the need for efficient therapies are driving the rapid evolution of emerging technologies and innovation pipelines for its treatment. A wide range of treatment options, such as disease-modifying medications, novel drug delivery systems, and personalized medicine techniques, have been made possible by recent developments. Key features of these new technologies and their implications for treating AD are described in the sections that follow [141].

Disease-modifying therapies

Current treatments target neuroinflammation, tau tangles, and amyloid-beta plaques. One hundred twenty-three distinct candidates have been found in late-stage trials. Different types of interventions, Small molecule and biologic medicines, account for 54% and 30% of studies, respectively, suggesting a high priority for these modalities [142].

Innovative drug delivery systems

Drug distribution across the blood-brain barrier is improved by nanotechnology techniques like liposomes, exosomes, and nanoparticles, which lessen systemic side effects. The potential of heterocyclic compounds to target oxidative stress and mitochondrial dysfunction factors that are critical in AD pathology is being investigated [143].

Personalized medicine and biomarkers

By using customized approaches, creating individualized treatment programs that incorporate biomarkers and neuroimaging can boost the effectiveness of therapy. Genetic Interventions Gene therapy and CRISPR-Cas9 are two promising strategies to address the genetic factors associated with AD.

Although these developments raise the prospect of more potent treatments for AD, obstacles still exist, including as the disease's complexity and the requirement for additional study to support these novel strategies [144].

DRUG AVAILABLE FOR TREATMENT OF ALZHEIMER'S DISEASE

A variety of pharmaceutical medicines are now available to manage symptoms and possibly slow the progression of Alzheimer's disease, marking a substantial evolution in therapeutic choices. Acetylcholinesterase inhibitors, NMDA receptor antagonists, and more recent amyloid-targeting treatments are the main pharmacological classes. A thorough summary of the medications available to treat AD is provided below [145].

Acetylcholinesterase inhibitors

- Donepezil: This medication, which is approved for all stages of AD, inhibits acetylcholinesterase to enhance cognitive performance.
- Rivastigmine: This medicine, which comes in patch or tablet form, works well for mild to moderate AD.
- Galantamine: Used for mild to moderate cases, it improves cholinergic function [146].

NMDA receptor antagonists

- Memantine: Licensed to treat moderate to severe AD, it protects neurons by controlling glutamate activity [147].

Amyloid-lowering therapies

- Aducanumab: The first medication approved by the FDA that targets amyloid plaques and may halt cognitive deterioration.
- Lecanemab: it targets amyloid, was recently approved and showed promise in clinical trials.
- Donanemab: This medication, which is anticipated to be approved, also attempts to lessen the amyloid burden [148].

Other drugs

- Suvorexant: Used to treat sleeplessness in AD patients.
- Brexpiprazole: Approved for agitation related to AD.

Although these medications alleviate symptoms and may halt the progression of the disease, the quest for a permanent cure is still ongoing, underscoring the complexity of AD and the necessity of continuous research into new treatments [149, 150].

CONCLUSION

It is concluded that Alzheimer's disease is a severe neurological problem that affects mental thought, behaviours and memories. Recent advances in pharmacological innovation offer a new option for changing the course of the disease. The key pathogenic pathway for Alzheimer's disease to be counteracted by the emerging therapy, such as tau aggregation inhibitors, neuroinflammatory modulators, inhibition of monoclonal antibodies that target amyloid-β, and mitochondrial protectants. Specific and accurate treatment can also be achieved by using gene and RNA-based medication in combination with the assessment of biomarkers that are used for the diagnosis and prediction of Alzheimer's disease. Because of there are still issues with safety profile, distribution profiles and efficient blood-brain barrier penetration, these advancements can significantly improve the treatment against Alzheimer's disease. Future studies should concentrate on combining lifestyle improvement and medication innovation to achieve complete disease management.

REFERENCES

[1] Scheltens P, De Strooper B, Kivipelto M, *et al.* Alzheimer's disease. Lancet 2021; 397(10284): 1577-90.
[http://dx.doi.org/10.1016/S0140-6736(20)32205-4] [PMID: 33667416]

[2] Finder VH. Alzheimer's disease: a general introduction and pathomechanism. J Alzheimers Dis 2010; 22(s3) (Suppl. 3): S5-S19.
[http://dx.doi.org/10.3233/JAD-2010-100975] [PMID: 20858960]

[3] Uwishema O, Mahmoud A, Sun J, *et al.* Is Alzheimer's disease an infectious neurological disease? A review of the literature. Brain Behav 2022; 12(8)e2728
[http://dx.doi.org/10.1002/brb3.2728] [PMID: 35879909]

[4] Sachdeva P, Ahmad F. In silico Characterization of Predominant Genes Involved in Early Onset of Alzheimer's Disease. J Neurobehav Sci 2021; 8(3): 179-90.
[http://dx.doi.org/10.4103/jnbs.jnbs_34_21]

[5] Simrén J, Leuzy A, Karikari TK, *et al.* The diagnostic and prognostic capabilities of plasma biomarkers in Alzheimer's disease. Alzheimers Dement 2021; 17(7): 1145-56.
[http://dx.doi.org/10.1002/alz.12283] [PMID: 33491853]

[6] Sirkis DW, Bonham LW, Johnson TP, La Joie R, Yokoyama JS. Dissecting the clinical heterogeneity of early-onset Alzheimer's disease. Mol Psychiatry 2022; 27(6): 2674-88.
[http://dx.doi.org/10.1038/s41380-022-01531-9] [PMID: 35393555]

[7] Kamer AR, Craig RG, Niederman R, Fortea J, de Leon MJ. Periodontal disease as a possible cause for Alzheimer's disease. Periodontol 2000 2020; 83(1): 242-71.
[http://dx.doi.org/10.1111/prd.12327] [PMID: 32385876]

[8] Sims R, Hill M, Williams J. The multiplex model of the genetics of Alzheimer's disease. Nat Neurosci 2020; 23(3): 311-22.
[http://dx.doi.org/10.1038/s41593-020-0599-5] [PMID: 32112059]

[9] Piaceri I, Nacmias B, Sorbi S. Genetics of familial and sporadic Alzheimer s disease. Front Biosci (Elite Ed) 2013; E5(1): 167-77.
[http://dx.doi.org/10.2741/E605] [PMID: 23276979]

[10] Khezri MR, Mohebalizadeh M, Ghasemnejad-Berenji M. Therapeutic potential of ADAM10 modulation in Alzheimer's disease: a review of the current evidence. Cell Commun Signal 2023; 21(1): 60.
[http://dx.doi.org/10.1186/s12964-023-01072-w] [PMID: 36918870]

[11] Oliveira Monteiro e Pereira de Almeida MP, Valle Pedroso R, Mantellatto Grigoli M, Vicente Silva T, Manzine PR, Cominetti MR. ADAM10 as a biomarker for Alzheimer's disease: A systematic review. Rev Neurol (Paris) 2024; 180(1-2): 1-11.
[http://dx.doi.org/10.1016/j.neurol.2023.04.002] [PMID: 37460331]

[12] Gabriele RMC. Investigating the function of the Alzheimer's disease risk gene BIN1 and its role in Alzheimer's disease [PhD dissertation]. London: University College London, Institute of Neurology; 2021.

[13] Perdigão CB. The impact of the genetic risk factor BIN1 to Alzheimer's disease development [dissertation]. Oeiras, Portugal: Instituto de Tecnologia Química e Biológica António Xavier, Universidade Nova de Lisboa; 2023.

[14] Ma J, Wang Z, Chen S, *et al.* EphA1 activation induces neuropathological changes in a mouse model of Parkinson's disease through the CXCL12/CXCR4 signaling pathway. Mol Neurobiol 2021; 58(3): 913-25.
[http://dx.doi.org/10.1007/s12035-020-02122-x] [PMID: 33057926]

[15] Eskandari-Sedighi G, Jung J, Macauley MS. CD33 isoforms in microglia and Alzheimer's disease: Friend and foe. Mol Aspects Med 2023; 90: 101111.
[http://dx.doi.org/10.1016/j.mam.2022.101111] [PMID: 35940942]

[16] Cardoso R. Role of TOMM40'523–APOE Haplotypes in Alzheimer's Disease Etiology–from Clinics to Mitochondria [dissertation]. Lisbon, Portugal: Faculdade de Ciências Médicas, Universidade NOVA de Lisboa; 2023.

[17] Alvarez J, Alvarez-Illera P, Santo-Domingo J, Fonteriz RI, Montero M. Modeling Alzheimer's disease in Caenorhabditis elegans. Biomedicines 2022; 10(2): 288.
[http://dx.doi.org/10.3390/biomedicines10020288] [PMID: 35203497]

[18] Kaladhar DS, Yarla NS, Anusha N. Functional analysis and molecular docking studies of medicinal compounds for AchE and BchE in Alzheimer's disease and type 2 diabetes mellitus. Aging Dis 2013; 4(4): 186-200.
[PMID: 23936743]

[19] Andrade-Guerrero J, Santiago-Balmaseda A, Jeronimo-Aguilar P, *et al.* Alzheimer's disease: an updated overview of its genetics. Int J Mol Sci 2023; 24(4): 3754.
[http://dx.doi.org/10.3390/ijms24043754] [PMID: 36835161]

[20] Murphy MP, LeVine H III. Alzheimer's disease and the amyloid-β peptide. J Alzheimers Dis 2010; 19(1): 311-23.
[http://dx.doi.org/10.3233/JAD-2010-1221] [PMID: 20061647]

[21] Suzuki N, Hatta T, Ito M, Kusakabe K. Anti-amyloid-β antibodies and anti-tau therapies for Alzheimer's disease: recent advances and perspectives. Chem Pharm Bull (Tokyo) 2024; 72(7): 602-9.
[http://dx.doi.org/10.1248/cpb.c24-00069] [PMID: 38945936]

[22] Meyer-Luehmann M, Spires-Jones TL, Prada C, *et al.* Rapid appearance and local toxicity of amyloid-β plaques in a mouse model of Alzheimer's disease. Nature 2008; 451(7179): 720-4.

[http://dx.doi.org/10.1038/nature06616] [PMID: 18256671]

[23] Lanoiselée HM, Nicolas G, Wallon D, *et al.* APP, PSEN1, and PSEN2 mutations in early-onset Alzheimer disease: A genetic screening study of familial and sporadic cases. PLoS Med 2017; 14(3)e1002270
[http://dx.doi.org/10.1371/journal.pmed.1002270] [PMID: 28350801]

[24] Li P, Marshall L, Oh G, *et al.* Epigenetic dysregulation of enhancers in neurons is associated with Alzheimer's disease pathology and cognitive symptoms. Nat Commun 2019; 10(1): 2246.
[http://dx.doi.org/10.1038/s41467-019-10101-7] [PMID: 31113950]

[25] Chen T, Pan F, Huang Q, *et al.* Metabolic phenotyping reveals an emerging role of ammonia abnormality in Alzheimer's disease. Nat Commun 2024; 15(1): 3796.
[http://dx.doi.org/10.1038/s41467-024-47897-y] [PMID: 38714706]

[26] Petit D, Fernández SG, Zoltowska KM, *et al.* Aβ profiles generated by Alzheimer's disease causing PSEN1 variants determine the pathogenicity of the mutation and predict age at disease onset. Mol Psychiatry 2022; 27(6): 2821-32.
[http://dx.doi.org/10.1038/s41380-022-01518-6] [PMID: 35365805]

[27] Liu X, Zhao X, Zeng X, *et al.* β-Arrestin1 regulates γ-secretase complex assembly and modulates amyloid-β pathology. Cell Res 2013; 23(3): 351-65.
[http://dx.doi.org/10.1038/cr.2012.167] [PMID: 23208420]

[28] Kim M, Suh J, Romano D, *et al.* Potential late-onset Alzheimer's disease-associated mutations in the ADAM10 gene attenuate α-secretase activity. Hum Mol Genet 2009; 18(20): 3987-96.
[http://dx.doi.org/10.1093/hmg/ddp323] [PMID: 19608551]

[29] Hernández F, Barreda EG, Fuster-Matanzo A, Goñi-Oliver P, Lucas JJ, Avila J. The role of GSK3 in Alzheimer disease. Brain Res Bull 2009; 80(4-5): 248-50.
[http://dx.doi.org/10.1016/j.brainresbull.2009.05.017] [PMID: 19477245]

[30] Penke B, Bogár F, Fülöp L. β-Amyloid and the pathomechanisms of Alzheimer's disease: a comprehensive view. Molecules 2017; 22(10): 1692.
[http://dx.doi.org/10.3390/molecules22101692] [PMID: 28994715]

[31] Cha MY, Han SH, Son SM, *et al.* Mitochondria-specific accumulation of amyloid β induces mitochondrial dysfunction leading to apoptotic cell death. PLoS One 2012; 7(4)e34929
[http://dx.doi.org/10.1371/journal.pone.0034929] [PMID: 22514691]

[32] Hardy J. The discovery of Alzheimer-causing mutations in the APP gene and the formulation of the "amyloid cascade hypothesis". FEBS J 2017; 284(7): 1040-4.
[http://dx.doi.org/10.1111/febs.14004] [PMID: 28054745]

[33] Tam JHK, Pasternak SH. Amyloid and Alzheimer's disease: inside and out. Can J Neurol Sci 2012; 39(3): 286-98.
[http://dx.doi.org/10.1017/S0317167100013408] [PMID: 22547507]

[34] Stancu IC, Vasconcelos B, Terwel D, Dewachter I. Models of β-amyloid induced Tau-pathology: the long and "folded" road to understand the mechanism. Mol Neurodegener 2014; 9(1): 51.
[http://dx.doi.org/10.1186/1750-1326-9-51] [PMID: 25407337]

[35] Levin OS, Vasenina EE. Twenty-five years of the amyloid hypothesis of alzheimer disease: advances, failures and new perspectives. Zh Nevrol Psikhiatr Im S S Korsakova 2016; 116(6. Vyp. 2): 3-9.
[http://dx.doi.org/10.17116/jnevro2016116623-9] [PMID: 27723704]

[36] Zhang Y, Zhao J. TFEB participates in the Aβ-induced pathogenesis of Alzheimer's disease by regulating the autophagy-lysosome pathway. DNA Cell Biol 2015; 34(11): 661-8.
[http://dx.doi.org/10.1089/dna.2014.2738] [PMID: 26368054]

[37] Zhang Y, Zhang C, Liu J, Shao H. Alzheimer's disease: Potential pathogenesis and imaging findings. VIEW 2023; 4(5)20230025
[http://dx.doi.org/10.1002/VIW.20230025]

[38] Levin OS, Vasenina EE. 25 Years of the Amyloid Hypothesis of the Origin of Alzheimer's Disease: Advances, Failures, and New Perspectives. Neurosci Behav Physiol 2017; 47(9): 1065-70. [http://dx.doi.org/10.1007/s11055-017-0513-0]

[39] Karran E, Mercken M, Strooper BD. The amyloid cascade hypothesis for Alzheimer's disease: an appraisal for the development of therapeutics. Nat Rev Drug Discov 2011; 10(9): 698-712. [http://dx.doi.org/10.1038/nrd3505] [PMID: 21852788]

[40] Moloney CM, Lowe VJ, Murray ME. Visualization of neurofibrillary tangle maturity in Alzheimer's disease: A clinicopathologic perspective for biomarker research. Alzheimers Dement 2021; 17(9): 1554-74. [http://dx.doi.org/10.1002/alz.12321] [PMID: 33797838]

[41] Alonso A, Li B, Grundke-Iqbal I, Iqbal K. Mechanism of tau-induced neurodegeneration in Alzheimer disease and related tauopathies. Curr Alzheimer Res 2008; 5(4): 375-84. [http://dx.doi.org/10.2174/156720508785132307] [PMID: 18690834]

[42] Wang J, Liu H, Cao Y. tau protein and Alzheimer's disease. Chinese J Tissue Eng Res 2020; 24(17): 2775.

[43] Acosta DM, Mancinelli C, Bracken C, Eliezer D. Post-translational modifications within tau paired helical filament nucleating motifs perturb microtubule interactions and oligomer formation. J Biol Chem 2022; 298(1)101442 [http://dx.doi.org/10.1016/j.jbc.2021.101442] [PMID: 34838590]

[44] Ohm DT, Peterson C, Lobrovich R, *et al.* Degeneration of the locus coeruleus is a common feature of tauopathies and distinct from TDP-43 proteinopathies in the frontotemporal lobar degeneration spectrum. Acta Neuropathol 2020; 140(5): 675-93. [http://dx.doi.org/10.1007/s00401-020-02210-1] [PMID: 32804255]

[45] Novoa C, Salazar P, Cisternas P, *et al.* Inflammation context in Alzheimer's disease, a relationship intricate to define. Biol Res 2022; 55(1): 39. [http://dx.doi.org/10.1186/s40659-022-00404-3] [PMID: 36550479]

[46] Congdon EE, Sigurdsson EM. Tau-targeting therapies for Alzheimer disease. Nat Rev Neurol 2018; 14(7): 399-415. [http://dx.doi.org/10.1038/s41582-018-0013-z] [PMID: 29895964]

[47] Murtishaw AS, Heaney CF, Bolton MM, Belmonte KCD, Langhardt MA, Kinney JW. Intermittent streptozotocin administration induces behavioral and pathological features relevant to Alzheimer's disease and vascular dementia. Neuropharmacology 2018; 137: 164-77. [http://dx.doi.org/10.1016/j.neuropharm.2018.04.021] [PMID: 29738850]

[48] Passaro AP, Lebos AL, Yao Y, Stice SL. Immune response in neurological pathology: emerging role of central and peripheral immune crosstalk. Front Immunol 2021; 12676621 [http://dx.doi.org/10.3389/fimmu.2021.676621] [PMID: 34177918]

[49] Wenzel TJ, Desjarlais JD, Mousseau DD. Human brain organoids containing microglia that have arisen innately adapt to a β-amyloid challenge better than those in which microglia are integrated by co-culture. Stem Cell Res Ther 2024; 15(1): 258. [http://dx.doi.org/10.1186/s13287-024-03876-0] [PMID: 39135132]

[50] Wang M, Zhang H, Liang J, Huang J, Chen N. Exercise suppresses neuroinflammation for alleviating Alzheimer's disease. J Neuroinflammation 2023; 20(1): 76. [http://dx.doi.org/10.1186/s12974-023-02753-6] [PMID: 36935511]

[51] Krasemann S, Madore C, Cialic R, *et al.* The TREM2-APOE pathway drives the transcriptional phenotype of dysfunctional microglia in neurodegenerative diseases. Immunity 2017; 47(3): 566-581.e9. [http://dx.doi.org/10.1016/j.immuni.2017.08.008] [PMID: 28930663]

[52] Saroja SR, Gorbachev K, Julia TCW, Goate AM, Pereira AC. Astrocyte-secreted glypican-4 drives

APOE4-dependent tau hyperphosphorylation. Proc Natl Acad Sci USA 2022; 119(34)e2108870119 [http://dx.doi.org/10.1073/pnas.2108870119] [PMID: 35969759]

[53] Litvinchuk A, Wan YW, Swartzlander DB, *et al.* Complement C3aR inactivation attenuates tau pathology and reverses an immune network deregulated in tauopathy models and Alzheimer's disease. Neuron 2018; 100(6): 1337-1353.e5. [http://dx.doi.org/10.1016/j.neuron.2018.10.031] [PMID: 30415998]

[54] Galea I. The blood–brain barrier in systemic infection and inflammation. Cell Mol Immunol 2021; 18(11): 2489-501. [http://dx.doi.org/10.1038/s41423-021-00757-x] [PMID: 34594000]

[55] Pizzino G, Irrera N, Cucinotta M, *et al.* Oxidative stress: harms and benefits for human health. Oxid Med Cell Longev 2017; 2017(1)8416763 [http://dx.doi.org/10.1155/2017/8416763] [PMID: 28819546]

[56] Misrani A, Tabassum S, Yang L. Mitochondrial dysfunction and oxidative stress in Alzheimer's disease. Front Aging Neurosci 2021; 13617588 [http://dx.doi.org/10.3389/fnagi.2021.617588] [PMID: 33679375]

[57] Singh A, Kukreti R, Saso L, Kukreti S. Oxidative stress: a key modulator in neurodegenerative diseases. Molecules 2019; 24(8): 1583. [http://dx.doi.org/10.3390/molecules24081583] [PMID: 31013638]

[58] Averchuk AS, Ryazanova MV, Rozanova NA, *et al.* Neurogenesis in Neurogenic Niches of the Brain in Experimental Alzheimer's Disease at the Presymptomatic Stage of Neurodegeneration Development. Cell Tissue Biol 2023; 17(6): 619-26. [http://dx.doi.org/10.1134/S1990519X23060020]

[59] Shahpasand K, Uemura I, Saito T, *et al.* Regulation of mitochondrial transport and inter-microtubule spacing by tau phosphorylation at the sites hyperphosphorylated in Alzheimer's disease. J Neurosci 2012; 32(7): 2430-41. [http://dx.doi.org/10.1523/JNEUROSCI.5927-11.2012] [PMID: 22396417]

[60] Du F, Yu Q, Kanaan NM, Yan SS. Mitochondrial oxidative stress contributes to the pathological aggregation and accumulation of tau oligomers in Alzheimer's disease. Hum Mol Genet 2022; 31(15): 2498-507. [http://dx.doi.org/10.1093/hmg/ddab363] [PMID: 35165721]

[61] Roy RG, Mandal PK, Maroon JC. Oxidative stress occurs prior to amyloid Aβ plaque formation and tau phosphorylation in Alzheimer's disease: Role of glutathione and metal ions. ACS Chem Neurosci 2023; 14(17): 2944-54. [http://dx.doi.org/10.1021/acschemneuro.3c00486] [PMID: 37561556]

[62] Dubey S, Singh E. Antioxidants: an approach for restricting oxidative stress induced neurodegeneration in Alzheimer's disease. Inflammopharmacology 2023; 31(2): 717-30. [http://dx.doi.org/10.1007/s10787-023-01173-5] [PMID: 36933175]

[63] Kuller LH, Lopez OL, Mackey RH, *et al.* Subclinical cardiovascular disease and death, dementia, and coronary heart disease in patients 80+ years. J Am Coll Cardiol 2016; 67(9): 1013-22. [http://dx.doi.org/10.1016/j.jacc.2015.12.034] [PMID: 26940919]

[64] Rajeev V, Fann DY, Dinh QN, *et al.* Pathophysiology of blood brain barrier dysfunction during chronic cerebral hypoperfusion in vascular cognitive impairment. Theranostics 2022; 12(4): 1639-58. [http://dx.doi.org/10.7150/thno.68304] [PMID: 35198062]

[65] Gireud-Goss M, Mack AF, McCullough LD, Urayama A. Cerebral amyloid angiopathy and blood-brain barrier dysfunction. Neuroscientist 2021; 27(6): 668-84. [http://dx.doi.org/10.1177/1073858420954811] [PMID: 33238806]

[66] Cechetto DF, Hachinski V, Whitehead SN. Vascular risk factors and Alzheimer's disease. Expert Rev Neurother 2008; 8(5): 743-50.

[http://dx.doi.org/10.1586/14737175.8.5.743] [PMID: 18457531]

[67] Mei M, Liu M, Mei Y, Zhao J, Li Y. Sphingolipid metabolism in brain insulin resistance and neurological diseases. Front Endocrinol (Lausanne) 2023; 141243132 [http://dx.doi.org/10.3389/fendo.2023.1243132] [PMID: 37867511]

[68] Karisetty BC, Bhatnagar A, Armour EM, Beaver M, Zhang H, Elefant F. Amyloid-β peptide impact on synaptic function and neuroepigenetic gene control reveal new therapeutic strategies for Alzheimer's disease. Front Mol Neurosci 2020; 13: 577622. [http://dx.doi.org/10.3389/fnmol.2020.577622] [PMID: 33304239]

[69] Huber N, Korhonen S, Hoffmann D, *et al.* Deficient neurotransmitter systems and synaptic function in frontotemporal lobar degeneration—Insights into disease mechanisms and current therapeutic approaches. Mol Psychiatry 2022; 27(3): 1300-9. [http://dx.doi.org/10.1038/s41380-021-01384-8] [PMID: 34799692]

[70] Gautam D, Naik UP, Naik MU, Yadav SK, Chaurasia RN, Dash D. Glutamate Receptor Dysregulation and Platelet Glutamate Dynamics in Alzheimer's and Parkinson's Diseases: Insights into Current Medications. Biomolecules 2023; 13(11): 1609. [http://dx.doi.org/10.3390/biom13111609] [PMID: 38002291]

[71] Kim JH, Chang IB, Kim YH, Min CY, Yoo DM, Choi HG. The association of pre-existing diagnoses of Alzheimer's disease and Parkinson's disease and coronavirus disease 2019 infection, severity and mortality: results from the Korean National Health Insurance Database. Front Aging Neurosci 2022; 14: 821235. [http://dx.doi.org/10.3389/fnagi.2022.821235] [PMID: 35317529]

[72] Seo D, Holtzman DM. Current understanding of the Alzheimer's disease-associated microbiome and therapeutic strategies. Exp Mol Med 2024; 56(1): 86-94. [http://dx.doi.org/10.1038/s12276-023-01146-2] [PMID: 38172602]

[73] Meng G, Zhong X, Mei H. A systematic investigation into aging related genes in brain and their relationship with Alzheimer's disease. PLoS One 2016; 11(3)e0150624. [http://dx.doi.org/10.1371/journal.pone.0150624] [PMID: 26937969]

[74] Tam KY, Ju Y. Pathological mechanisms and therapeutic strategies for Alzheimer's disease. Neural Regen Res 2022; 17(3): 543-9. [http://dx.doi.org/10.4103/1673-5374.320970] [PMID: 34380884]

[75] Ramesh M, Govindaraju T. Multipronged diagnostic and therapeutic strategies for Alzheimer's disease. Chem Sci (Camb) 2022; 13(46): 13657-89. [http://dx.doi.org/10.1039/D2SC03932J] [PMID: 36544728]

[76] Cao S, Fisher DW, Yu T, Dong H. The link between chronic pain and Alzheimer's disease. J Neuroinflammation 2019; 16(1): 204. [http://dx.doi.org/10.1186/s12974-019-1608-z] [PMID: 31694670]

[77] Cummings J, Gold M, Mintun M, *et al.* Key considerations for combination therapy in Alzheimer's clinical trials: Perspectives from an expert advisory board convened by the Alzheimer's drug discovery foundation. J Prev Alzheimers Dis 2025; 12(1)100001. [http://dx.doi.org/10.1016/j.tjpad.2024.100001] [PMID: 39800453]

[78] Yeh SJ, Chung MH, Chen BS. Investigating pathogenetic mechanisms of Alzheimer's disease by systems biology approaches for drug discovery. Int J Mol Sci 2021; 22(20): 11280. [http://dx.doi.org/10.3390/ijms222011280] [PMID: 34681938]

[79] Wong LR, Wong P, Ho PCL. Metabolic profiling of female tg2576 mouse brains provides novel evidence supporting intranasal low-dose pioglitazone for long-term treatment at an early stage of alzheimer's disease. Biomedicines 2020; 8(12): 589. [http://dx.doi.org/10.3390/biomedicines8120589] [PMID: 33317213]

[80] Brownell M, Sehar U, Mukherjee U, Reddy PH. Creating Cultural and Lifestyle Awareness About

Dementia and Co-morbidities. J Alzheimers Dis Rep 2024; 8(1): 747-64.
[http://dx.doi.org/10.3233/ADR-240043] [PMID: 38746643]

[81] Song Q, Li J, Li T, Li HW. Nanomaterials that Aid in the Diagnosis and Treatment of Alzheimer's Disease, Resolving Blood–Brain Barrier Crossing Ability. Adv Sci (Weinh) 2024; 11(38)2403473.
[http://dx.doi.org/10.1002/advs.202403473] [PMID: 39101248]

[82] Naser SS, Singh D, Preetam S, *et al.* Posterity of nanoscience as lipid nanosystems for Alzheimer's disease regression. Mater Today Bio 2023; 21: 100701.
[http://dx.doi.org/10.1016/j.mtbio.2023.100701] [PMID: 37415846]

[83] Meng Q, Wang A, Hua H, *et al.* Intranasal delivery of Huperzine A to the brain using lactoferrin-conjugated N-trimethylated chitosan surface-modified PLGA nanoparticles for treatment of Alzheimer's disease. Int J Nanomedicine 2018; 13: 705-18.
[http://dx.doi.org/10.2147/IJN.S151474] [PMID: 29440896]

[84] Mathew A, Fukuda T, Nagaoka Y, *et al.* Curcumin loaded-PLGA nanoparticles conjugated with Tet-1 peptide for potential use in Alzheimer's disease. PLoS One 2012; 7(3)e32616
[http://dx.doi.org/10.1371/journal.pone.0032616] [PMID: 22403681]

[85] Huang D, Wang Q, Cao Y, *et al.* Multiscale NIR-II Imaging-Guided Brain-Targeted Drug Delivery Using Engineered Cell Membrane Nanoformulation for Alzheimer's Disease Therapy. ACS Nano 2023; 17(5): 5033-46.
[http://dx.doi.org/10.1021/acsnano.2c12840] [PMID: 36867454]

[86] Hu B, Dai F, Fan Z, Ma G, Tang Q, Zhang X. Nanotheranostics: Congo Red/Rutin-MNPs with Enhanced Magnetic Resonance Imaging and H_2O_2-Responsive Therapy of Alzheimer's Disease in APPswe/PS1dE9 Transgenic Mice. Adv Mater 2015; 27(37): 5499-505.
[http://dx.doi.org/10.1002/adma.201502227] [PMID: 26270904]

[87] Wang C, Song X, Li P, *et al.* Multifunctional Nanocarrier for Synergistic Treatment of Alzheimer's Disease by Inhibiting β-Amyloid Aggregation and Scavenging Reactive Oxygen Species. ACS Appl Mater Interfaces 2024; 16(21): 27127-38.
[http://dx.doi.org/10.1021/acsami.4c02825] [PMID: 38747495]

[88] Li WB, Xu LL, Wang SL, *et al.* Co-Assembled Nanoparticles toward Multi-Target Combinational Therapy of Alzheimer's Disease by Making Full Use of Molecular Recognition and Self-Assembly. Adv Mater 2024; 36(28)2401918
[http://dx.doi.org/10.1002/adma.202401918] [PMID: 38662940]

[89] Roghani AK, Garcia RI, Roghani A, *et al.* Treating Alzheimer's disease using nanoparticle-mediated drug delivery strategies/systems. Ageing Res Rev 2024; 97: 102291.
[http://dx.doi.org/10.1016/j.arr.2024.102291] [PMID: 38614367]

[90] Ke J, Yu C, Li S, *et al.* Combining Multifunctional Delivery System with Blood–Brain Barrier Reversible Opening Strategy for the Enhanced Treatment of Alzheimer's Disease. Adv Healthc Mater 2024; 13(8)2302939
[http://dx.doi.org/10.1002/adhm.202302939] [PMID: 38117094]

[91] Xie A, Cheng G, Wu J, *et al.* Highly BBB-permeable nanomedicine reverses neuroapoptosis and neuroinflammation to treat Alzheimer's disease. Biomaterials 2025; 312122749
[http://dx.doi.org/10.1016/j.biomaterials.2024.122749] [PMID: 39121725]

[92] Kumari Y, Raj K, Singh PK. Promising Nano-Carriers-Based Targeted Drug Delivery Approaches for the Effective Treatment of Alzheimer's Disease. Bentham Science Publishers 2023; pp. 181-204.

[93] Han G, Bai K, Yang X, *et al.* "Drug-Carrier" Synergy Therapy for Amyloid- *β* Clearance and Inhibition of Tau Phosphorylation *via* Biomimetic Lipid Nanocomposite Assembly. Adv Sci (Weinh) 2022; 9(14)2106072.
[http://dx.doi.org/10.1002/advs.202106072] [PMID: 35307993]

[94] Jiao Y, Yang L, Wang R, *et al.* Drug Delivery Across the Blood–Brain Barrier: A New Strategy for

the Treatment of Neurological Diseases. Pharmaceutics 2024; 16(12): 1611. [http://dx.doi.org/10.3390/pharmaceutics16121611] [PMID: 39771589]

[95] Ding X, Hu Y, Wang Z, *et al.* Enhanced Blood-Brain Barrier Penetrability of BACE1 SiRNA-Loaded Prussian Blue Nanocomplexes for Alzheimer's Disease Synergy Therapy Authorea 2024. [http://dx.doi.org/10.22541/au.171023610.07919917/v1]

[96] Wang C, Ren J. Ultrasound blood–brain barrier opening: A new era of treatment for Alzheimer's disease? Aging Med (Milton) 2024; 7(6): 673-5. [http://dx.doi.org/10.1002/agm2.12371] [PMID: 39777099]

[97] Inamdar A, Gurupadayya B, Halagali P, Nandakumar S, Pathak R, Singh H, *et al.* Cutting-Edge Strategies for Overcoming Therapeutic Barriers in Alzheimer's Disease. Curr Pharm Des 2025; 31(8): 598-618. [PMID: 39492772]

[98] Daraban BS, Popa AS, Stan MS. Latest Perspectives on Alzheimer's Disease Treatment: The Role of Blood-Brain Barrier and Antioxidant-Based Drug Delivery Systems. Molecules 2024; 29(17): 4056. [http://dx.doi.org/10.3390/molecules29174056] [PMID: 39274904]

[99] El Abiad E, Al-Kuwari A, Al-Aani U, Al Jaidah Y, Châari A. Navigating the Alzheimer's Biomarker Landscape: A Comprehensive Analysis of Fluid-Based Diagnostics. Cells 2024; 13(22): 1901. [http://dx.doi.org/10.3390/cells13221901] [PMID: 39594648]

[100] Panwar D, Sharma P, Sharma S, Malik M, Kaur J. Biomarkers for Alzheimer's disease: early diagnosis. In: Kumar A, Kumar S, Gayathri N, Srivel R, D C, eds. Deep generative models for integrative analysis of Alzheimer's biomarkers. Hershey (PA): IGI Global Scientific Publishing; 2025. p. 51–94. [http://dx.doi.org/10.4018/979-8-3693-6442-0.ch003]

[101] Therriault J, Janelidze S, Benedet AL, *et al.* Diagnosis of Alzheimer's disease using plasma biomarkers adjusted to clinical probability. Nature Aging 2024; 4(11): 1529-37. [http://dx.doi.org/10.1038/s43587-024-00731-y] [PMID: 39533113]

[102] VandeVrede L, Schindler SE. Clinical use of biomarkers in the era of Alzheimer's disease treatments. Alzheimers Dement. 2025 Jan; 21(1): e14201. [PMID: 39740074]

[103] Dong Y, Song X, Wang X, Wang S, He Z. The early diagnosis of Alzheimer's disease: Blood-based panel biomarker discovery by proteomics and metabolomics. CNS Neurosci Ther 2024; 30(11)e70060 [http://dx.doi.org/10.1111/cns.70060] [PMID: 39572036]

[104] Vecchio I, Colica C. New Research on Biomarkers in Alzheimer's Continuum. Rev Recent Clin Trials 2025; 20(3): 211-231. [PMID: 39865820]

[105] Hroudová J, Fišar Z. Alzheimer's disease approaches - Focusing on pathology, biomarkers and clinical trial candidates. Prog Neuropsychopharmacol Biol Psychiatry 2024; 134: 111069. [http://dx.doi.org/10.1016/j.pnpbp.2024.111069] [PMID: 38917881]

[106] Pascoal TA, Aguzzoli CS, Lussier FZ, *et al.* Insights into the use of biomarkers in clinical trials in Alzheimer's disease. EBioMedicine 2024; 108: 105322. [http://dx.doi.org/10.1016/j.ebiom.2024.105322] [PMID: 39366844]

[107] Khalid M, Khan F, Broussard JN, Barman A. Graph-Based Biomarker Discovery and Interpretation for Alzheimer's Disease 2024.http://arxiv.org/pdf/2411.18796

[108] Li K, Gao Y, Liu M, Chen Y. Advances in Alzheimer's Disease Biomarkers. Curr Alzheimer Res 2024; 21(11): 791-803. [http://dx.doi.org/10.2174/0115672050366767241223050957] [PMID: 39757626]

[109] Wang Q, Chen S, Wang J, Shang H, Chen X. Advancements in Pharmacological Treatment of Alzheimer's Disease: The Advent of Disease-Modifying Therapies (DMTs). Brain Sci 2024; 14(10):

990.
[http://dx.doi.org/10.3390/brainsci14100990] [PMID: 39452004]

[110] Huang LK, Kuan Y, Lin HW, Hu CJ. Clinical trials of new drugs for Alzheimer disease: a 2020–2023 update. J Biomed Sci 2023; 30(1): 83.
[PMID: 37784171]

[111] Cho E. Approved and late clinical trial monoclonal antibodies for the treatment of Alzheimer's disease. J Stud Res. 2024 Sep 27; 5(2).

[112] Boxer AL, Sperling R. Accelerating Alzheimer's therapeutic development: The past and future of clinical trials. Cell 2023; 186(22): 4757-72.
[http://dx.doi.org/10.1016/j.cell.2023.09.023] [PMID: 37848035]

[113] Waite LM. New and emerging drug therapies for Alzheimer disease. Aust Prescr 2024; 47(3): 75-9.
[http://dx.doi.org/10.18773/austprescr.2024.021] [PMID: 38962384]

[114] Cummings J, Ritter A, Zhong K. Clinical trials for disease-modifying therapies in Alzheimer's disease: a primer, lessons learned, and a blueprint for the future. J Alzheimers Dis 2018; 64(s1): S3-S22.
[http://dx.doi.org/10.3233/JAD-179901] [PMID: 29562511]

[115] Mirzaei M, Ahmadi N, Bagheri Fahraji B, *et al.* A randomized clinical trial evaluating Hydralazine's efficacy in early-stage Alzheimer's disease: The EHSAN Study. Sci Rep. 2024 Nov 21; 14(1): 28837.

[116] Weninger S, Irizarry MC, Fleisher AS, *et al.* Alzheimer's disease drug development in an evolving therapeutic landscape. Alzheimers Dement (N Y) 2024; 10(4)e70015
[http://dx.doi.org/10.1002/trc2.70015] [PMID: 39748838]

[117] Xiao D, Zhang C. Current therapeutics for Alzheimer's disease and clinical trials Explor Neurosci 2024; 3: 255–271.
[http://dx.doi.org/10.37349/en.2024.00048]

[118] Benussi A, Borroni B. Brain Stimulation in Alzheimer's Disease Trials. J Alzheimers Dis 2024; 101(s1): S545-65.
[http://dx.doi.org/10.3233/JAD-230535] [PMID: 39422933]

[119] Mahase E. FDA approves controversial Alzheimer's drug despite uncertainty over effectiveness. BMJ 2021; 373: n1462.https://www.bmj.com/content/373/bmj.n1462
[http://dx.doi.org/10.1136/bmj.n1462] [PMID: 34103308]

[120] Morant AV, Vestergaard HT, Lassen AB, Navikas V. US, EU, and Japanese Regulatory Guidelines for Development of Drugs for Treatment of Alzheimer's Disease: Implications for Global Drug Development. Clin Transl Sci 2020; 13(4): 652-64.
[http://dx.doi.org/10.1111/cts.12755] [PMID: 32043310]

[121] Glymour MM, Mor V. A large pragmatic trial is the right solution for testing anti-amyloid therapies for Alzheimer's disease. J Am Geriatr Soc 2022; 70(5): 1595-8.
[http://dx.doi.org/10.1111/jgs.17695] [PMID: 35128639]

[122] Angioni D, Cummings J, Lansdall CJ, *et al.* Clinical Meaningfulness in Alzheimer's Disease Clinical Trials. A Report from the EU-US CTAD Task Force. J Prev Alzheimers Dis. 2024; 11(5): 1219–1227.

[123] Wu W, Ji Y, Wang Z, *et al.* The FDA-approved anti-amyloid-β monoclonal antibodies for the treatment of Alzheimer's disease: a systematic review and meta-analysis of randomized controlled trials. Eur J Med Res 2023; 28(1): 544.
[http://dx.doi.org/10.1186/s40001-023-01512-w] [PMID: 38017568]

[124] Geldmacher DS. Treatment of Alzheimer Disease. Continuum (Minneap Minn) 2024; 30(6): 1823-44.
[http://dx.doi.org/10.1212/CON.0000000000001503] [PMID: 39620846]

[125] Acuña-Villaorduña A, Baranda JC, Boehmer J, Fashoyin-Aje L, Gore SD. Equitable access to clinical trials: how do we achieve it? Am Soc Clin Oncol Educ Book 2023; 43(43)e389838.

[http://dx.doi.org/10.1200/EDBK_389838] [PMID: 37146264]

[126] Kumar V, Roy K. Recent progress in the treatment strategies for Alzheimer's disease. In: Roy K, ed. Computational modeling of drugs against Alzheimer's disease. New York (NY): Humana; 2023. p. 1–35.
[http://dx.doi.org/10.1007/978-1-0716-3311-3_1]

[127] Brabin DRD, Jayashree M, Jagannathan SK, Kalangi RR. AI-Enhanced Drug Discovery for Alzheimer's. Advances in medical technologies and clinical practice book series 2024;101–13.

[128] González E, Varas P, González-Naranjo P, et al. AI-driven de-novo design and development of non-toxic DYRK1A inhibitors. ChemRxiv. [Preprint]. 2025 Feb 21.
[http://dx.doi.org/10.26434/chemrxiv-2024-rcx7n]

[129] Dominguez-Gortaire J, Ruiz A, Porto-Pazos AB, Rodriguez-Yanez S, Cedrón F. Alzheimer's Disease: Exploring Pathophysiological Hypotheses and the Role of Machine Learning in Drug Discovery. Int J Mol Sci 2025; 26(3): 1004.
[http://dx.doi.org/10.3390/ijms26031004] [PMID: 39940772]

[130] Cheng F, Hou Y, Zhang P, *et al.* Alzheimer's disease genetics and real-world patient data fuel target and drug discovery in European and African American Ancestries using AI/ML approaches. Alzheimers Dement 2024; 20(S1)e087887
[http://dx.doi.org/10.1002/alz.087887]

[131] Kale M, Wankhede N, Pawar R, *et al.* AI-driven innovations in Alzheimer's disease: Integrating early diagnosis, personalized treatment, and prognostic modelling. Ageing Res Rev 2024; 101: 102497.
[http://dx.doi.org/10.1016/j.arr.2024.102497] [PMID: 39293530]

[132] Gupta P. A Review Article on Current Therapeutic Strategies and Future Directions in the Treatment of Alzheimer's Disease. Afr J Biomed Res 2024; 47(4S): 6233-45.
[http://dx.doi.org/10.53555/AJBR.v27i4S.4758]

[133] Fan X, Chen H, He W, *et al.* Integrating genetic and immune profiles for personalized immunotherapy in Alzheimer's disease. Front Med (Lausanne). 2025 Jun 2; 12: 1603553.
[http://dx.doi.org/10.3389/fmed.2025.1603553]

[134] Patwekar M, Patwekar F, Khan S, Sharma R, Kumar D. Navigating the Alzheimer's Treatment Landscape: Unraveling Amyloid-beta Complexities and Pioneering Precision Medicine Approaches. Curr Top Med Chem 2024; 24(19): 1665-82.
[http://dx.doi.org/10.2174/0115680266295495240415114919] [PMID: 38644708]

[135] Mayo P, Pascual J, Crisman E, *et al.* Innovative pathological network-based multitarget approaches for Alzheimer's disease treatment. Med Res Rev. 2024 Nov; 44(6): 2367–2419.
[http://dx.doi.org/10.1002/med.22045]

[136] Dumore NG. Emerging therapies for Alzheimer's disease: The promise of disease-modifying drugs exploring recent breakthroughs in drugs aimed at slowing or halting the progression of Alzheimer's disease. 2024;12:01–4.

[137] Li Y, Serras CP, Blumenfeld J, *et al.* Data-driven discovery of cell-type-directed network-correcting combination therapy for Alzheimer's disease. Cell. 2025 Oct 2; 188(20): 5516–5534.e18.

[138] Umar M, Rehman Y, Ambreen S, *et al.* Innovative approaches to Alzheimer's therapy: Harnessing the power of heterocycles, oxidative stress management, and nanomaterial drug delivery system. Ageing Res Rev 2024; 97102298
[http://dx.doi.org/10.1016/j.arr.2024.102298] [PMID: 38604453]

[139] Cummings JL, Osse AML, Kinney JW. Alzheimer's Disease: Novel Targets and Investigational Drugs for Disease Modification. Drugs 2023; 83(15): 1387-408.
[http://dx.doi.org/10.1007/s40265-023-01938-w] [PMID: 37728864]

[140] Kivipelto M, Mangialasche F, Ngandu T. Lifestyle interventions to prevent cognitive impairment, dementia and Alzheimer disease. Nat Rev Neurol 2018; 14(11): 653-66.

[http://dx.doi.org/10.1038/s41582-018-0070-3] [PMID: 30291317]

[141] Niazi SK, Magoola M, Mariam Z. Innovative Therapeutic Strategies in Alzheimer's Disease: A Synergistic Approach to Neurodegenerative Disorders. Pharmaceuticals (Basel) 2024; 17(6): 741. [http://dx.doi.org/10.3390/ph17060741] [PMID: 38931409]

[142] Hlávka JP, Kinoshita AT, Jeyasingh D, Huang C, Mirsafian L, Jacobson M. Emerging Alzheimer's disease treatment paradigms: A late-stage clinical trial review. Alzheimers Dement (N Y) 2024; 10(4)e70022
[http://dx.doi.org/10.1002/trc2.70022] [PMID: 39748848]

[143] Salarpour S, Barani M, Pardakhty A, Khatami M, Pal Singh Chauhan N. The application of exosomes and Exosome-nanoparticle in treating brain disorders. J Mol Liq 2022; 350: 118549. [http://dx.doi.org/10.1016/j.molliq.2022.118549]

[144] Zhong Y. Cutting Edge Research on Drug Treatment Strategies and Emerging Therapies for Alzheimer's Disease. Highlights in Science, Engineering and Technology 2023; 74: 719-26. [http://dx.doi.org/10.54097/meadgd26]

[145] Vishwas S, Gulati M, Babu MR, *et al.* The overview of drugs used in Alzheimer's disease and their molecular targets. In: Kumar D, Patil VM, Wu D, Thorat N, eds. Deciphering drug targets for Alzheimer's disease. Singapore: Springer; 2023. p. 569–601. [http://dx.doi.org/10.1007/978-981-99-2657-2_15]

[146] Cacabelos R, Martínez-Iglesias O, Cacabelos N, Carrera I, Corzo L, Naidoo V. Therapeutic Options in Alzheimer's Disease: From Classic Acetylcholinesterase Inhibitors to Multi-Target Drugs with Pleiotropic Activity. Life (Basel) 2024; 14(12): 1555. [http://dx.doi.org/10.3390/life14121555] [PMID: 39768263]

[147] Jia J, Yuan Y, Zhang Z. Treatment of Alzheimer's disease based on new drugs. Theoretical and Natural Science. 2023 Dec 20; 20(1): 180–186. [http://dx.doi.org/10.54254/2753-8818/20/20230703]

[148] Ahmad F, Karan A, Sharma R, *et al.* Evolving therapeutic interventions for the management and treatment of Alzheimer's disease. Ageing Res Rev 2024; 95102229 [http://dx.doi.org/10.1016/j.arr.2024.102229] [PMID: 38364913]

[149] Liu X. Research progress and outlook of drugs for the treatment for Alzheimer's disease. Medscien. 2025; 1(3): 1. [http://dx.doi.org/10.61173/4z762m76]

[150] Seifert R. Drugs for the treatment of Alzheimer's disease. In: Basic knowledge of pharmacology. Cham: Springer; 2019. p. 287–291. [http://dx.doi.org/10.1007/978-3-030-18899-3_30]

CHAPTER 11

Recent Advances in Drug Development for the Management of Alzheimer's Disease

Amit Upadhyay[1,*], Abhishek Kumar Singh[1], Nandani Jayaswal[1], Pooja Jaiswal[1], Shashi Kant Singh[1] and **Shashank Tewari[2]**

[1] *Faculty of Pharmaceutical Sciences, Mahayogi Gorakhnath University, Gorakhpur, Uttar Pradesh 273007, India*

[2] *Department of Pharmaceutical Chemistry, Manipal College of Pharmaceutical Sciences, Manipal Academy of Higher Education (MAHE), Manipal, Karnataka 576104, India*

Abstract: Alzheimer's disease is a long-term, neurodegenerative disease of the brain, which is a major cause of dementia affecting the elderly population. It is also characterised by behavioural dysfunction and cognitive impairment. Alzheimer's disease is a complex disease involving multiple genetic and environmental risk factors. Despite significant advancements in the study of Alzheimer's disease, many aspects of the disease are still poorly understood. Although a reduction in acetylcholine and the formation of amyloid-beta plaques are characteristic signs of Alzheimer's disease, the precise causes and mechanisms underlying the illness are still unknown. Significant reduction of neurotransmitters, such as acetylcholine and butyrylcholine, and aggregation of β-amyloid protein in the neocortex of the brain are strongly linked to the etiology of Alzheimer's disease. Although there is currently no treatment for Alzheimer's, both pharmaceutical and non-pharmacological therapy options are being thoroughly studied. Recent research has introduced promising new drugs designed to address the root causes of AD. Acetylcholinesterase inhibitors (*e.g.*, donepezil) and NMDA receptor antagonists (*e.g.*, memantine) are used for symptomatic relief. Donepezil was the first cholinesterase inhibitor drug approved for the symptomatic treatment of Alzheimer's disease. Later, other drugs, such as rivastigmine and galantamine, also got FDA approval. Newly FDA-approved drugs, such as Aducanumab and Lecanemab, have been effective in reducing amyloid plaques. Many hybrid molecules targeting two or more relevant sites of disease have been developed and tested. These hybrid compounds showed antiinflammatory and amyloid aggregation inhibition. Additionally, many global clinical trials are underway testing effective therapies for Alzheimer's disease. Current therapies can control symptoms, but they cannot halt or reverse the disease's course. Researchers are now focusing on target the molecular mechanisms that underlie Alzheimer's disease. This chapter examines the most recent therapeutic approaches targeted at amyloid plaques, tau protein tangles, neurotransmitter imbalances, anneuro inflammation.

* **Corresponding author Amit Upadhyay:** Faculty of Pharmaceutical Sciences, Mahayogi Gorakhnath University, Gorakhpur, Uttar Pradesh 273007, India; E-mail: amit.upadhyay189@gmail.com

Prashant Kumar Dhakad, Raghav Mishra, Shweta Sharma, Sudhanshu Mishra & Nishant Gaur (Eds.)

Keywords: Acetylcholine, Alzheimer's, β-amyloid, Dementia, Donepezil, FDA, Galantamine, NMDA, Rivastigmine.

INTRODUCTION

Alzheimer's Disease (AD) is the most common type of neurodegenerative disorder, marked by a steady deterioration of memory and cognitive function. Effective treatments for this deadly condition remain elusive despite a great deal of research since its discovery in 1906 [1]. The pathophysiology of AD includes dysfunction in cholinergic neurotransmission, aggregation of Aβ protein, accumulation and phosphorylation of tau, and involvement of inflammatory and oxidative pathways [2]. These findings contributed to clinical trial design, therapeutic target development, and current knowledge of AD etiology. There is a global need for effective treatments, as illustrated by the extensive pipeline of medications that have been studied but have not yet demonstrated disease-modifying effects. The FDA has granted approval for only four small-molecule drugs for the management of AD: galantamine, donepezil, and rivastigmine (all cholinesterase inhibitors) and memantine, which is an *N*-methyl-D-aspartate receptor inhibitor [3].

Significant reduction of enzymes, such as acetylcholine, in the neocortex of the brain is strongly associated with AD. Preventing the acetylcholine degradation by acetylcholinesterase enzyme in the brain can improve cholinergic neurotransmission, which is a treatment approach for this disease [4]. In 1993, the FDA approved Tacrine (Cognex) as the first acetylcholinesterase inhibitor (AChEI) to treat Alzheimer's disease (Fig. **1**). However, Tacrine's toxicity to the liver led to its removal from the market in 2013 [5], and Donepezil (**2**), rivastigmine (**3**), and galantamine (**4**) are examples of second-generation AChEIs that are more selective. As first-line medications for AD, they demonstrated better pharmacokinetic profiles or fewer side effects. Furthermore, AChEI, such as donepezil (**2**) and galantamine (**4**), may result in unexpected effects when used in combination with other neurologic medications, metal chelators, or antioxidants [6, 7]. *N*-Methyl-D-Aspartate (NMDA) receptors are involved in glutamate signaling in the brain. Over-activation of NMDA receptors may result in damage to neurons. This takes place in several neurological illnesses, including AD. Memantine (**5**) is an FDA-approved antagonist of the NMDA receptor, which is used to treat moderate to severe AD stages [8]. It regulates the transfer of glutamate and is effective in enhancing patients' behavior, everyday life skills, and cognitive performance. For patients with moderate to severe AD, Namzaric (**6**, fixed dose combination of memantine (**5**) extended release and donepezil (**4**) offers an additional therapeutic alternative [9].

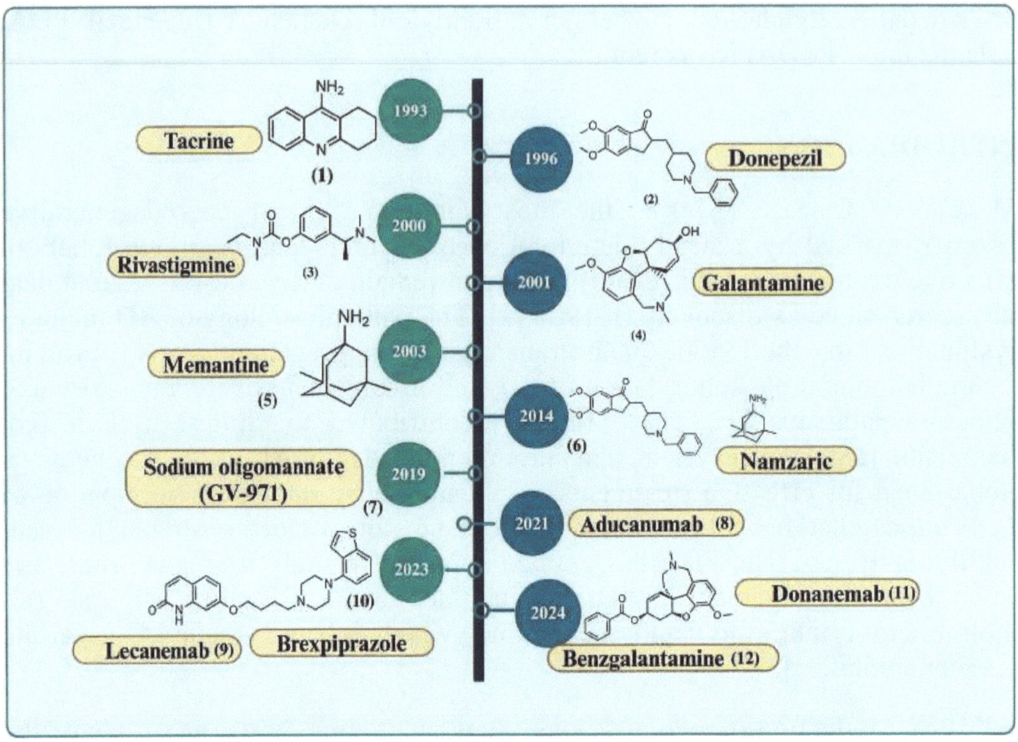

Fig. (1). Timeline of FDA/China authorised drugs for Alzheimer's disease.

Sodium oligomannate (**7**, GV-971) was provisionally approved in 2019 [10]. GV-971 is derived from a marine algae. It was hypothesized that sodium oligomannate (**7**, GV-971) would prevent AD by preventing neuroinflammation induced due to dysbiosis of the gut and inhibiting Aβ fibril production [11]. To further investigate its efficacy and safety, phase four clinical trials (NCT05058040 and NCT05181475) are underway, with an anticipated ongoing until 2025.

The monoclonal antibodies, such as aducanumab (**8**) and lecanemab (**9**), that target Aβ exhibited promising results. Aducanumab (**8**) won FDA approval in 2021. Aducanumab (**8**) targets both insoluble fibrils and soluble oligomers [12]. More soluble Aβ aggregates (oligomers and protofibrils) were bound by lecanemab (**9**) [13]. Lecanemab (**9**) obtained traditional approval in 2023 [14]. Despite the risks and costs associated with amyloid-related imaging abnormalities, all these monoclonal antibody treatments have been shown to effectively remove Aβ plaque and prevent cognitive decline. Brexpiprazole (**10**) targets norepinephrine, dopamine, and serotonin receptors and is frequently used for schizophrenia and depression. It has been shown to lessen agitation in AD patients [15].

Recently, donanemab (**11**) was approved for early-stage AD symptoms [16]. Donanemab, a humanized antibody, reduces Aβ plaques in AD patients [17]. In July 2024, the FDA authorized benzgalantamine (**12**), which is a prodrug of galantamine (**4**), for mild-to-moderate dementia in individuals with AD. Benzagalantamine avoids many of the gastrointestinal side effects linked to peripheral cholinesterase inhibition since it is a prodrug that is inert as it moves through the stomach [18].

Unfortunately, these medications focus on relieving cognitive and memory damage rather than treating diseases. Development of AD is a lengthy, expensive, and resource-intensive process. Because of their high failure rate, drugs that were initially created for other conditions—such as cancer, heart disease, mental health issues, diabetes, and other neurodegenerative diseases—are increasingly being repurposed. In order to ensure safety, evaluate long-term efficacy, and comprehend pharmacological mechanisms, more research is required. This chapter examines the most recent therapeutic approaches targeted at the management of AD.

CURRENT THERAPEUTIC STRATEGIES

AD treatments fall into two groups: symptomatic and disease-modifying medications [19]. Symptomatic treatments seek to alleviate symptoms, but they do not affect the underlying development of the disease. Disease-modifying therapies try to reduce or stop the progression of the disease by addressing Alzheimer's underlying biological mechanisms. Symptomatic treatments include AChEI (*e.g.*, Galantamine, Rivastigmine, Donepezil) that boost memory and NMDA receptor antagonists (Memantine) for psychological assistance in the late stages. Antidepressants (*e.g.*, citalopram) and antipsychotic medications (*e.g.*, olanzapine) are used to treat behavioral symptoms such as anxiety and agitation. Disease-modifying therapies, such as Lecanemab and Aducanumab, target amyloid plaques and delay early-stage disease progression. Ongoing research is looking into therapeutics that target tau proteins, inflammation, and neuroprotection to slow disease development further.

ACETYLCHOLINESTERASE INHIBITORS

Brain function depends on the balance of several neurotransmitter systems, including noradrenaline, dopamine, acetylcholine (ACh), gamma-aminobutyric acid, glutamate and serotonin [20]. Among these, the cholinergic neurotransmission is a major factor in both adult-onset dementia diseases and AD's compromised cognitive performance. Choline acetyltransferase (ChAT) converts acetyl-coenzyme A and choline into acetylcholine (ACh), which is transported into synaptic vesicles by the vesicular acetylcholine transporter (VAChT) (Fig. **2**) [21]. ACh is released upon arrival of neural signal and attaches itself to nicotinic acetylcholine receptors (nAChRs) and muscarinic receptors (mAChRs) on the post-synaptic membrane. Following its degradation into choline by acetylcholinesterase (AChE), ACh is reabsorbed by presynaptic cholinergic neurons [22]. The reduction in ChAT activity and/or increased catalytic activity of AChE causes a reduction in the synthesis, release and concentration of ACh [23].

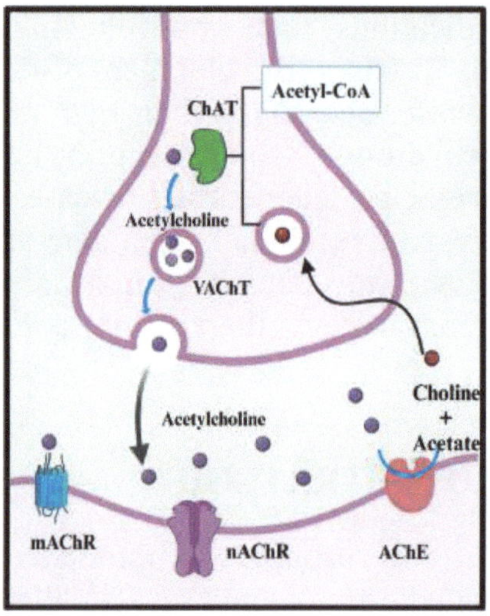

Fig. (2). Synthesis and degradation of acetylcholine.

Currently, improving cholinergic neurotransmission remains a key strategy for treating the behavioral and cognitive symptoms of AD. AChE is primarily responsible for the hydrolysis of acetylcholine, whereas butyrylcholinesterase (BChE) acts as a secondary mediator [24]. Although blood and neural synapses include AChE, BuChE is the primary ChE in the neurons, tangles, neuritic plaques, liver and glia [25]. These enzymes differ in their substrate: ACh and butyrylcholine (BuCh), respectively [4].

Fig. (**3**) depicts the similarities and differences between human AChE and BChE enzymes based on three-dimensional structures. The Peripheral Anionic Site (PAS) near the entrance of BChE and AChE enzymes plays a crucial function in substrate entry. AChE's PAS comprises amino acids, such as tryptophan (trp), aspartic acid (Asp), tyrosine (tyr) and serine (Ser), while BChE's contains glutamine (Gln), aspartic acid (Asp), asparagine (Asn), tyrosine (Tyr), and serine (Ser) in the same area. Both enzymes have acyl and choline-binding sites near the catalytic active gorge (esteratic region) [26]. AChE's acyl and choline-binding sites contain tryptophan, tyrosine, and phenylalanine, while BChE contains leucine, alanine, and valine. Compared to AChE, BChE has a deeper gorge. BChE can interact with significantly more substrates and inhibitors than AChE [27, 28]. The esteratic region of both enzymes is located deep in the gorge and contains serine, glutamic acid, and histidine [29].

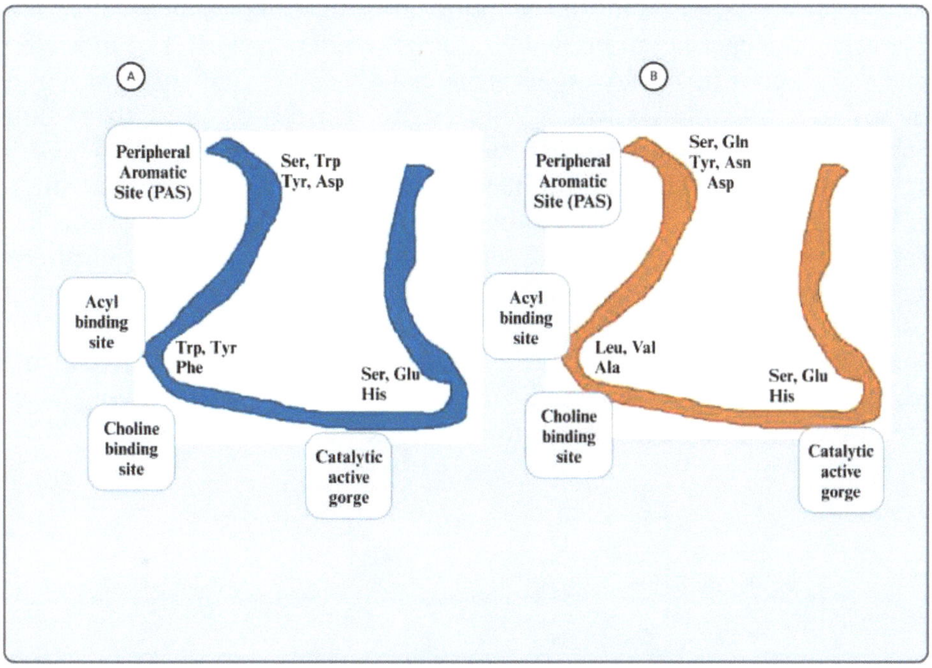

Fig. (3). (A) Acetylcholinesterase and **(B)** butyrylcholinesterase binding sites.

Catalytic activity requires three amino acids, specifically serine, glutamic acid, and histidine. In the anionic pocket, the carbonyl oxygen of the acetyl group, in the anionic pocket of ACh, forms a hydrogen bond with the -OH in serine to bind and recognize the substrate. Water molecule hydrolyses the tetrahedral intermediate, resulting in the release of choline and acetic acid [30].

Cholinesterase inhibitors (ChEI) were the first medications approved for treating symptoms of AD in the US and Europe. AChE and BuChE inhibitors function by increasing acetylcholine (ACh) and butyrylcholine (BuCh) concentrations at the synaptic level, thereby enhancing ACh neurotransmission and facilitating the recovery of cognitive functions in patients. ChEIs are grouped into two types based on their mode of action and structure: (i) molecules that form covalent bonds with the enzyme, such as carbamates and organophosphates, and (ii) molecules with reversible bonds, such as amine groups [31].

Tacrine and its Derivatives

Tacrine (**1**, Cognex®) inhibits both AChE and BuChE reversibly (Fig. **1**). In 1993, the US FDA approved tacrine to treat dementias associated with AD. However, tacrine was removed from the market in 2013 owing to its hepatotoxicity [20]. Following its withdrawal, researchers aimed to reduce the adverse effects of tacrine and develop alternative derivatives. The discovery of bistacrine analogues linked by an alkylene chain was the first successful attempt to synthesize a possible AChEI. Compared to tacrine, some of these dimeric compounds exhibited higher selectivity for AChE and efficacy. The series' most potent inhibitor of AChE was the heptylene-linked bis-(6-chloro)-tacrine (**13**), which had a potency > 3000 times that of tacrine (Fig. **4**) [32]. Several tacrine compounds were developed that effectively inhibit AChE. A group of cinnamates/benzoates/phenylacetates- tacrine hybrids has received significant interest. The most effective drug in this series **14,** not only inhibited AChE but also prevented Aβ aggregation (IC_{50} = 51.81 nM) [33]. Tacrine-benzofuran hybrids **15** inhibited AChE and Aβ aggregation, with efficiency varying based on the size of the linker and substituent groups of each major component [34].

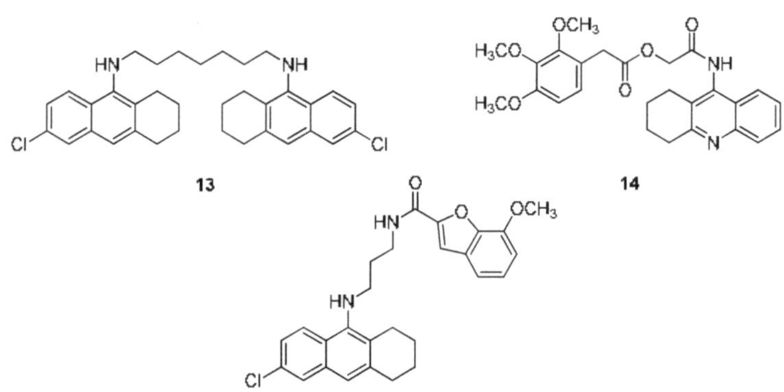

Fig. (4). Tacrine derivatives for the management of AD.

Donepezil and its analogues

In 1996, the FDA authorized donepezil (**2**) as the second medicine for the management of AD (Fig. **1**). It was approved in 2010 (23 mg/day) [35]. Donepezil (Aricept®) is a reversible and highly selective AchEI that increases ACh bioavailability in the synaptic cleft [36]. A library of 2-phenoxy-indan-1-one derivatives containing an alkylamine side chain has been designed to target BuChE and AChE while also exhibiting antioxidant action [37]. AP2238 interacted with AChE and prevented Aβ aggregation [38]. Later, a number of hybrid compounds, for example **16**, were created that included a pharmacophoric fragment of AP2238 and donepezil (Fig. **5**) [39]. Like donepezil, most of them showed strong AChEI action and prevention of Aβ aggregation.

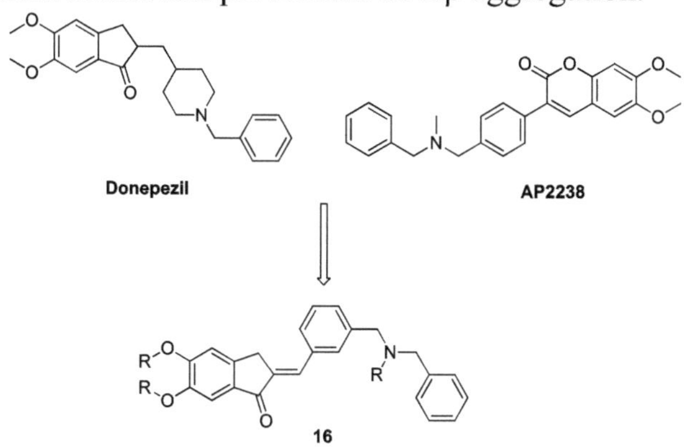

Fig. (5). Donepezil-AP2238 hybrid.

Combining the *N*-benzyl-piperidine subunit of donepezil with the hydroxy-piperidine fragment of AChEI, LASSBio-767, *via* an acylhydrazone linker resulted in a new library of *N*-benzyl-piperidine-aryl-acylhydrazone hybrid derivatives (Fig. **6**) [40]. The hybrids of **17** inhibited AChE, similar to donepezil, and have anti-inflammatory properties to prevent Aβ production.

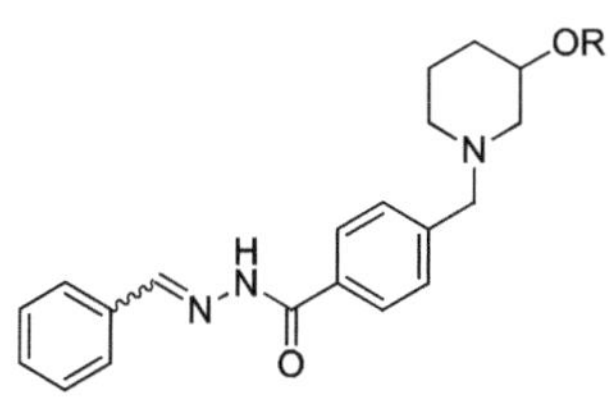

Fig. (6). N-benzyl-piperidine-aryl-acylhydrazone hybrid.

Two novel hybrids containing l-alpha-lipoic acid and donepezil's indanone-piperidine moiety have been produced. One compound had a modest inhibitory effect on AChE but a stronger inhibitory effect on BuChE [41]. Additionally, it had a higher antioxidant activity than alpha-lipoic acid.

Rivastigmine and its analogues

The FDA approved Exelon®, an oral formulation of rivastigmine (**3**), as an irreversible carbamate-selective AChEI and BuChEI, for the management of mild-to-moderate AD in 2000 (Fig. **1**) [42]. Rivastigmine has a reduced protein binding affinity, selective activity, and minimal medication interactions with other drugs. Rivastigmine inhibits the G1 isoform of AChE with greater efficiency than the G4 isoform [43]. G1 is a key enzyme in the hydrolysis of ACh at cholinergic synapses and rises with Alzheimer's disease progression [44].

Substituted 1,2,3,4-tetrahydroquinolin-6 (or-7)-ylcarbamates are a key focus among rivastigmine derivatives. The most potent chemical in this series, compound 18, inhibited AChE (acetylcholinesterase) with an IC_{50} value of 100 nM, making it roughly 10 times more effective than rivastigmine (IC_{50} value of 1030 nM) in the rat brain (Fig. **7**) [45]. Bio-oxidizable prodrugs are a viable technique for creating selective central AChE inhibitors (AChE-Is)because they readily pass the blood-brain barrier (BBB) and activate the cholinergic system. The most potent compound in this series, compound 19, demonstrated a high degree of selectivity: it was inactive against AChE in the peripheral system (IC_{50} > 1nM) but potently inhibited AChE in the brain (IC_{50} = 20 nM) [46].

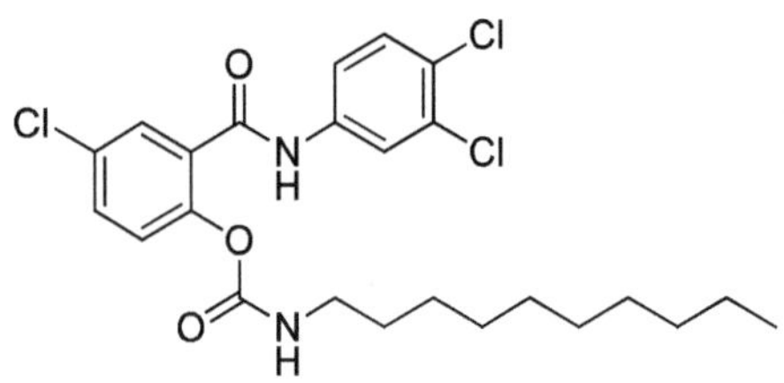

18

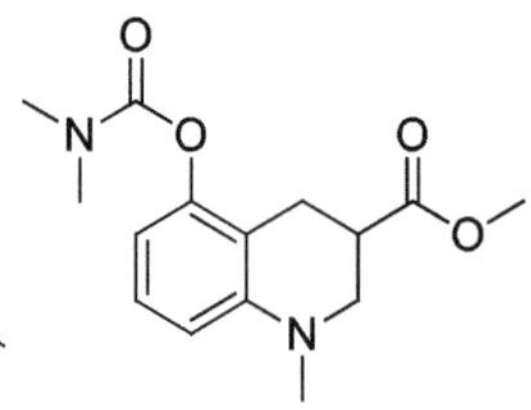

19

Fig. (7). Derivatives of rivastigmine.

Galantamine and its analogues

Galantamine (**4**, Razadyne®), which is an allosteric modulator of nicotinic cholinergic receptors and a selective reversible AchEI, was released in the US in the year 2000 for the management of AD (Fig. **1**) [47]. Galantamine is suitable for AD treatment because it has little effect on peripheral tissues, while selectively inhibiting AChE in the CNS. Prodrug of galantamine, Memogain® **(12)**, was developed (Fig. **1**). The bioavailability of memogain was >15 times higher in the brain than the same amount of galantamine. Memogain, which gets cleaved to galantamine, produces greater cognitive enhancement than the same doses of galantamine without severe side effects [48].

A library of indole-galantamine analogues **19** with potent AChE inhibition was discovered, with the most potent molecule having an IC_{50} of 11 nM (Fig. **8**) [49].

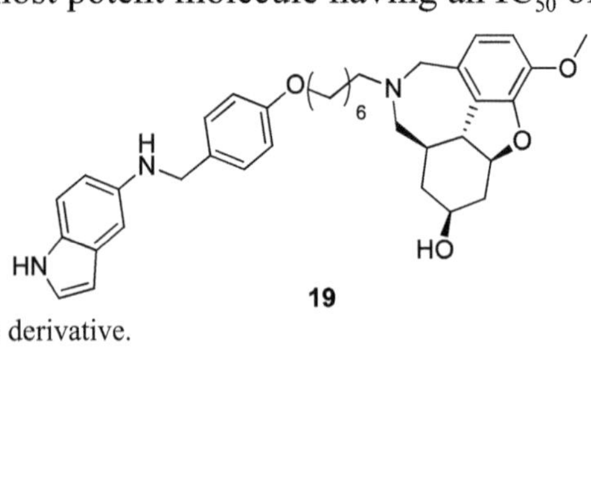

19

Fig. (8). Galantamine-indole derivative.

NMDA receptor antagonist

The nervous system processes and stores information through synaptic transmission.

Synapses are neural connections where presynaptic neurons activate neurotransmitter receptors in postsynaptic neurons [50]. The N-methyl-D-aspartate receptors (NMDARs) play an important role in synaptic transmission, influencing learning and memory [51]. Synaptic NMDARs protect neurons from both internal and extrinsic causes that cause harm. However, NMDARs' overstimulation outside the synapse can reduce mitochondrial membrane potential and lead to cell death [52]. NMDAR inhibitors are promising candidates for the treatment of neurological illnesses, such as AD. Glutamate is a key neurotransmitter that facilitates excitatory synaptic neurotransmission [53]. Excessive glutamatergic signaling can lead to excitotoxicity, which damages nerve cells and causes neurological damage [54]. NMDARs have higher calcium ion permeability than other glutamate receptors. This shows that toxicity is primarily caused by excessive Ca^{2+} entry through NMDARs [55].

NMDA receptors contain multiple agonist and allosteric binding sites, which complicates their pharmacology. An NMDA antagonist can block NMDA receptors by acting on several binding sites, such as the strychnine-insensitive glycine site (glycine B), polyamine site (NR2B specific), phencyclidine site (cationic channel) and competitive primary transmitter site [56]. Developing NMDA receptor inhibitors that prevent pathological activation while allowing normal function is crucial [57]. Only a few NMDA receptor antagonists, such as memantine (**5**) for AD, have proven effective in clinical trials (Fig. **1**). However, there are also downsides to these triumphs, including side effects that can hamper their use. Several NMDA receptor antagonists have failed clinical trials due to unpleasant side effects and pharmacokinetic problems [58].

Anti-Amyloid Therapies

AD is a neurodegenerative condition with two major pathologies: neurofibrillary tangles (intracellular) and β-amyloid plaques (extracellular). Accumulation of β-amyloid leads to neurodegeneration, including neuronal damage, inflammation, synapse loss, gliosis and death [59]. Anti-amyloid treatments, which target extracellular deposits of β-amyloid plaques as the primary neuropathological characteristic of AD, have received significant academic attention in recent decades [60]. Anti-amyloid medicines are the first disease-modifying medications for AD. The FDA authorized aducanumab, the first anti-amyloid drug, in 2021.

However, it was removed from the US market in 2024 [61]. Lecanemab was approved in the US in 2023, while donanemab, a third anti-amyloid treatment, was approved in early 2024 [59].

Anti-amyloid medicines primarily reduce β-amyloid levels by activating microglia, leading to phagocytosis and cell disintegration [62]. Anti-β-amyloid antibodies can be used in active and passive immunotherapies, as well as monoclonal antibody therapy. Monoclonal antibody aducanumab was the first anti-amyloid medication that selectively targets soluble and insoluble β-amyloid [63]. Although aducanumab did not show significant clinical improvement, the FDA approved it due to its ability to remove amyloid from the brain [64].

Lecanemab, which is a humanized IgG1 antibody, targets soluble β-amyloid protofibrils [65]. Additionally, it inhibits fibrinogen's capacity to bind to β-amyloid, potentially contributing to disease progression [66]. The FDA granted an urgent approval to Lecanemab in January 2023 for its capacity to remove amyloid, followed by conventional approval in July 2023 based on the CLARITY-Alzheimer's study results [67].

In June 2024, the FDA authorized donanemab, a humanized monoclonal antibody, as a third anti-amyloid treatment option [68]. This anti-amyloid therapy focuses on a unique pyroglutamine alteration seen solely in brain amyloid plaques, contrary to earlier approaches [69]. Donanemab interacts with the N-terminal pyroglutamate (N3pG) epitope of amyloid-beta. This interaction enables microglial cells to clear amyloid plaques via phagocytosis. Donanemab causes immune cells to recognize and eliminate amyloid, which accumulates in the brains of AD patients. Donanemab is easier to use than lecanemab, and therapy can be terminated after significant amyloid is eliminated [70]. Table **1** lists drugs approved by the FDA.

Table 1. List of drugs for the management of AD approved by the FDA.

S. No.	Name	Class	Chemical structure	Action	References
1.	Donepezil	Acetylcholinesterase (AChE) inhibitor.	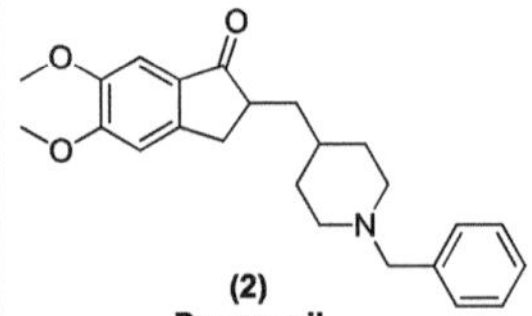 (2) Donepezil	Donepezil reduces Alzheimer's disease cognitive and behavioral symptoms, including apathy, aggressiveness, and confusion, by suppressing AChE.	[36]

(Table 1) cont.....

S. No.	Name	Class	Chemical structure	Action	References
2.	Rivastigmine	AChE inhibitor.	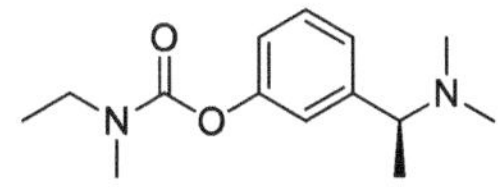 (3) **Rivastigmine**	Rivastigmine inhibits both AChE and butyrylcholinesterase, inhibiting ACh hydrolysis and resulting in an elevated concentration of acetylcholine at cholinergic synapses.	[43]
3.	Galantamine	AChE inhibitor.	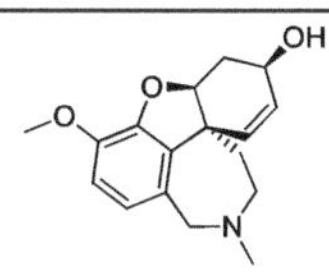 (4) **Galantamine**	Galantamine inhibits AChE, thereby increasing the concentration of ACh.	[47]
4.	Memantine	NMDA receptor antagonist.	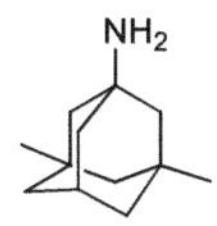 (5) **Memantine**	Memantine inhibits the action of glutamate on the NMDA receptor, avoiding the hyperexcitability of neurons.	[8]
5.	Aducanumab	Monoclonal IgG1antibody to treat mild symptoms of AD.	NA	Selectively targets various types of soluble and insoluble aggregates of β-amyloid.	[63]
6.	Lecanemab	humanized IgG1 antibody, targets soluble β-amyloid protofibril	NA	Reduces αβ-amyloid plaques.	[66]
7.	Donanemab	Monoclonal IgG1 antibody for the early stages of AD.	NA	Donanemab acts by stimulating microglial clearance of existing αβ-plaques.	[69]

EMERGING DRUG TARGETS

Recent drug development strategies for AD focus on neuroinflammation modulation and neuroprotection, expanding beyond amyloid-targeting therapies to address complex pathology. This text discusses the most promising emerging drug targets for AD.

Amyloid-β Pathway Modulators

The amyloid-β pathway is central to AD pathophysiology, with biochemical alterations being promising targets for disease-modifying therapies. Basic research and human genetic studies show that Aβ pathway dyshomeostasis plays a role in AD pathophysiology, highlighting the interplay of different Aβ forms with other AD-related processes [71].

α-Secretase inducers and AD

With emphasis on α-secretase activation as a potentially effective treatment method, finding natural substances that might increase this cleavage is necessary for AD. Targeting nuclear retinoic acid receptor b, ADAM10 expression can be increased, leading to further non-amyloidogenic Amyloid precursor protein APP processing. ADAM10 and ADAM17 cleave a variety of substrates and are not APP specific. In AD brains, the usage of ADAM10 to raise α-secretase levels has been linked to carcinogenesis and the development of tumours. Consequently, researchers have avoided using ADAM10 in their AD study. Therefore, more investigation is required to comprehend the toxicity brought on by ADAM10's impacts on AD neurons. α-Secretase activators have been previously identified in modified compounds based on curcumin (Table **2**). These substances consist of curcumin-isoleucine, curcumin-phenylalanine and curcumin-valine, which are amino acid conjugates that have been shown to raise ADAM10 immunoreactivity and stimulate a-secretase activity [72]. This resulted in the identification of naturally occurring substances that induce α-secretase, including ginsenoside and those present in *Ginkgo biloba* leaves, which were shown to control nonamyloidogenic APP-processing pathways. Similarly, medicinal herb *Tripterygium wilfordii*, which is frequently used to treat inflammatory illnesses, is the source of triptolide, another naturally occurring chemical. Triptolide furthermore prevents both in vitro and in vivo development of βAPP-cleaving enzyme-1 (BACE1) and processing of amyloidogenic APP. Triptolide was shown to lower Aβ40 levels; however, all of these naturally occurring substances have been shown to lower Aβ42 levels. It has been discovered that curcumin and ginkgo biloba not only lower Aβ42 levels but also Aβ40 and Aβ25-35 levels [73]. Accordingly, α-secretase activators typically show beneficial effects, according to a recent study. Additionally, recent studies have demonstrated that α-secretase activators, including ADAM10, exhibit tumour growth in AD brains. It will take more investigation to comprehend the fundamental functions of these substances, as well as the positive and negative effects they have on AD patients.

Table 2. α-Secretase activators and their effect on ADAM10 and BACE1.

S. No.	Drug	Study design	Gene/Protein	Result	References
1.	Curcumin	*In-vivo* and *in-vitro*	BACE1/α-secretase, ADAM10,	Reduce Aβ42 and Aβ40	[74]
2.	*Ginkgo biloba*	*In-vivo*	ADAM10,	Reduce Aβ42 and Aβ25-35	[73]
3.	Ginsenoside	*In-vivo*	ADAM10, BACE1	Decrease Aβ42	[75]
4.	Phenylananine-Curcumin Valine-Curcumin Isoleucine-Curcumin	*In-vitro*	ADAM10	Decrease Aβ42	[72]
5.	Triptolide	*In-vivo*	ADAM10	Decrease Aβ40	[76]

β-Secretase inhibitors and AD

Inhibiting β-site APP cleaving enzyme 1 (BACE1), the enzyme that is essential for the production of Aβ peptides, seems to be the most appealing strategy to stop the synthesis of Aβ oligomers. Oxidative stress increases activity and expression of BACE1, which causes neuronal death and dysfunction in AD [77].

BACE-1 and γ-secretase cleave APP at the N-terminus to produce Aβ peptides. BACE1 inhibition has emerged as a key therapeutic target for anti-amyloid treatment. Asiaticoside, tryptoline, and curcumin were discovered as potential BACE1 inhibitors. Curcumin's capacity to stop aggregation of Aβ and fibril formation was confirmed by both in vitro and in vivo investigations. Additionally, curcumin decreased Aβ (Aβ40/Aβ42) in AD mice *via* α-secretase induction and BACE1 inhibition [78]. Aspartic acid residues Asp32 and Asp228, in the catalytic sites, were discovered to bind to BACE1 by tryptoline, which prevented the cleavage of APP-β [79]. A highly polyphenolic compound, Asiaticoside, decreased Aβ42 levels in AD neurons, preventing amyloidogenesis. Derivatives of fluoro-oxazines, such as CF3, were created by Hilpert *et. al.* to inhibit BACE1. Oral dosages (1 mg/kg) replaced BACE1 inhibitors in rats with oxazine, which markedly decreased Aβ40 and Aβ42 levels [80]. Strong small-molecule BACE1 inhibitors of the third generation have been created more recently. Strong Aβ reduction and adequate blood–brain barrier (BBB) permeability were demonstrated by these inhibitors in vitro. The potential of β-secretase inhibitors to lower Aβ in AD is another area of research. MK-8931, AZD3293, E2609 (all undergoing Phases II/III in human clinical studies) and HPP854 (Phase I in human clinical studies) are β-secretase inhibitors that are currently being studied. Additionally, MK-8931 inhibited the amyloid precursor protein's cleavage *via* binding to its conserved extracellular domain. All things considered, β-secretase inhibitors are encouraging candidates for lowering Aβ levels in AD. Nevertheless, these substances had no effect on the cognitive abilities of elderly people or AD patients. These β-secretase inhibitors also caused negative side effects in AD patients. Therefore, more study is required to determine how these drugs normally function as well as the reasons behind their negative effects in AD patients.

Gamma-Secretase inhibitors and AD

Gamma-secretases are a class of intramembrane-cleaving proteases that are widely expressed and implicated in numerous physiological processes linked to AD. Presenilin (PSEN1 and PSEN2) is encoded by genes on chromosomes 14 and 21 due to changes on those chromosomes. γ-secretase activity, which is triggered by PSEN1 and -2, has been linked to the process of Aβ production in AD [80]. Studies examining the γ-secretase complex in the course of AD have shown that PSENs play a crucial role in preserving the γ-secretase complex. APH1 and nicastrin are two other subunits that have been found to be crucial cofactors for PSEN [81]. γ-Secretase enzymes create a complex that results in the cleavage of APP. Inhibiting these enzymes redirects the amyloidogenic route to a non-amyloidogenic pathway, resulting in decreased Aβ generation. The last ten years have seen a lot of research on γ-secretase inhibition. A variety of γ-secretase inhibitors (GSI) and modulators (GSM) classes have either decreased Aβ peptide levels or altered the composition of Aβ. Some GSIs and GSMs have had serious side effects or inadequate BBB penetration, which has limited their capacity to lower Aβ [82]. *N*-[*N*-(3,5-difluorophe nacetyl)-L-alanyl- -S-phenylglycinetbutylester (DAPT) has been identified as a promising therapeutic target that reduces γ-secretase activity in AD. In cellular tests, DAPT was orally active and had a moderate inhibitory concentration for Aβ (20 nM). The arylsulfonamide series SCH697466 was evaluated as a γ-secretase inhibitor. SCH697466 did not lower Aβ levels, despite producing enough notch signaling. Next, researchers worked to create notch-sparing arylsulfonamide series inhibitors, like begacestat and avagacestat. The way that avagacestat revealed notch-related issues was comparable to that of semagacestat. The γ-secretase activity of the notch-sparing drug ELND006 was examined in a recent clinical investigation. However, liver damage that was not mechanism-based led to the study's termination [83]. In contrast, it has been discovered that γ-secretase inhibitors CHF-5074 and NIC5-15 lower Aβ levels in AD patients. While the Phase II research of CHF-5074 was stopped due to side effects, the NIC5-15 Phase II trial is still continuing.

Tau Phosphorylation

Tau, discovered by Kirschner and colleagues, was identified as a key component of neurofibrillary tangles, a hallmark of AD. The accumulation of phosphorylated and aggregated Tau has been used for tau therapy, but the causal relationship remains debated. Fetal Tau and AD brain Tau have similar phosphorylation levels [84]. In AD, abnormal hyperphosphorylation of tau leads to toxic aggregates, neurofibrillary tangles (NFTs), and neuronal dysfunction.

1. GSK3β inhibitors

GSK3 was found as a regulatory threonine/serine protein kinase in 1980. It is expressed by two genes: GSK3α (chromosome 19) and GSK3β (chromosome 2). GSK3α and GSK3β have different molecular masses: 51 kDa and 47 kDa, respectively. Although these two isoforms differ at the C-terminal and N-terminal regions, they are very similar in the kinase domain [85]. Numerous diseases, including neurological disorders (*e.g.*, AD), cancer and heart disease, have been linked to GSK3β activation [86]. Since GSK3β is known to produce tau phosphorylation and increase Aβ deposits in AD brains, a number of studies have focused on identifying its function in AD pathogenesis in recent years. Due to the enhanced activity of GSK3β, tau phosphorylation was markedly boosted in primary cortical neuronal cultures [85]. Two factors have been suggested to be the primary causes of GSK3b activation: (i) oxidative stress and (ii) the interaction between GSK3b and the Aβ peptide. Tau phosphorylation results from a hyperactive GSK3β enzymatic state. On the other hand, new research suggests that elevated GSK3β activity in AD also causes Aβ production via acting on γ-secretase through PES1 mutations. Researchers looked at the involvement of GSK3β in AD pathogenesis, and it was discovered that the inhibition of GSK3β decreased the AD pathogenesis. Further studies employing AD models discovered that the cognitive deterioration was lessened. In order to lessen the symptoms of AD, researchers looked into medicinal drugs that could inhibit GSK3β due to the toxicity of Aβ.. Numerous GSK3β inhibitors have been found to be able to lower tau hyperphosphorylation and Aβ levels in both neuronal and non-neuronal cells. Furthermore, it was discovered that GSK3β inhibitors improved adult hippocampus neurogenesis both *in vitro* and *in vivo.* Because metal chelators help keep metal ions in balance, the metal ion hypothesis suggests that one of the causes of AD is abnormal metal homeostasis with an Aβ imbalance. GSK3β inhibitors have been shown in recent lithium experiments to have neuroprotective effects in AD and a number of other disorders [85].

CDK5 (Cyclin-Dependent Kinase 5) inhibitors and AD

According to reports, the pathophysiology of AD is closely linked to CDK5, a distinct member of the CDKs. CDK5 plays a crucial role in neural development, including migration, differentiation, synaptic processes, and memory consolidation. Nevertheless, CDK5 becomes hyperactive in response to pathogenic stimuli, which results in the abnormal hyperphosphorylation of tau NFTs and other CDK5-like APP substrates, leading to neurodegenerative disorders like AD.CDK5 inhibitors inhibit CDK5 kinase activity, preventing phosphorylation of the tau protein (Table **3**).

Table 3. List of CDK5 inhibitors.

S. No.	Drug	Protein/Gene	Result	References
1.	Hymenialdisine	GSK3	Decrease Tau–GSK3	[87]
2.	N6 -Isopentenyladenine	CDK5	Decrease CDK5	[88]
3.	Olomoucine	CDK5	Decrease CDK5	[89]
4.	(R)-Roscovitine	CDK5	Decrease CDK5	
5.	Purvalanol A	CDK5	Decrease CDK5 Amloyd beta	[90]
6.	Purvalanol B	CDK5	Decrease CDK5 Amloyd beta	

Neuroinflammation

The role of microglia in neuroinflammation

During neurodevelopment, monocyte precursor cells give rise to the neural microglia, which are macrophages. In order to maintain homeostasis and immunological defense, microglia are essential. In addition, they are necessary for neurogenesis, brain circuit development, and neuronal health [91]. In both diseased and healthy settings, microglia dynamically alter, changing from ramified (at rest) to amoeboid (active) forms. Under typical physiological circumstances, microglia patrol the surrounding environment of the brain to maintain immunological surveillance and neuronal survival. It has been discovered that Aβ aggregates cause microglia to accumulate morphological and molecular changes that result in a broad range of activation. Certain receptors, chemokines, cytokines, and other components are overexpressed in activated microglia known as Disease-Associated Microglia (DAM) [92]. The microglia have a large number of purinergic receptors, including P2Y12 and P2Y6, which bind to extracellular nucleotides generated by damaged cells. Activated microglia, therefore, have a greater capability for phagocytic activity [93]. The gene expression involved in the downstream inflammatory response is regulated by these components in combination. Preclinical and clinical studies have demonstrated the significant role of activated microglia in the progression of AD. Additionally, researchers have found Aβ plaque-associated microglia in the brains of AD patients. Aβ plaque-associated microglia have been shown to exhibit high immunoreactivity towards activation markers (such as COX2). Cytokines (including IL-1, TNFα, IL-1β, and MIP-1α) and chemokine receptors (including CCR3, CCR5) [3]. There have been instances of chemotaxis and astrocyte recruitment around Aβ plaque caused by MCP-1.

Additionally, reactive nitrogen species are produced by Aβ-induced microglia activation through the stimulation of Nitric Oxide Synthase (iNOS) and ROS *via* the NADPH oxidase system. These dangerous substances exacerbate neurotoxicity, impede the clearance of Aβ, and encourage amyloidogenesis. Nevertheless, a number of investigations have not demonstrated the neuroprotective effect of activated microglia. By increasing the number of activated microglia, overexpression of IL-1β has been demonstrated to decrease an Aβ plaque load in the APP/PS1 mouse model. IL-6 overexpression has also been demonstrated to aid APP transgenic mice in preventing the buildup of Aβ plaque through gliosis and microglial activation. As a result, research on the neuroprotective or neurotoxic effects of microglia-mediated inflammation varies. This may be influenced by the degree, timing, and duration of microglia activation.

Astrocytes in neuroinflammation

The most prevalent glial cells in the central nervous system, astrocytes, are produced from neural stem cells (NSCs) [94]. Astrocytes sustain homeostasis and supply growth and metabolic components to neurons due to their physical similarity to microglia. Astrocytes are also necessary for synaptic plasticity and synapse formation. These cells are also important for the control of the balance of extracellular ions, fluid, and free radicals at the BBB [95]. Astrocytes are a crucial part of tripartite synapses, which involve bidirectional communication between neurons and astrocytes in addition to inter-neuronal transmission, and are necessary for maintaining homeostasis and neural survival [96]. The morphological and functional changes of astrocytes, sometimes referred to as reactive astrocytes, under pathological conditions are characterized by cell expansion and release of neurotoxic chemicals. Vimentin and Glial Fibrillary Acidic Protein (GFAP), two important indicators that are overexpressed in reactive astrocytes, are responsible for cell hypertrophy. Studies using animal models and post-mortem examinations have shown that reactive astrocytes are present near Aβ plaque in the setting of AD [97, 98]. When GFAP and vimentin are decreased in the animal model of AD (APP/PS1), the Aβ plaque load increases, and the astrocytic response reduces [99]. The role of reactive astrocytes in neuroprotection may be plausible. This lessens collateral damage from neurotoxic Aβ species by penetrating the Aβ plaque and encircling it, creating a physical barrier known as a glial scar. Additionally, *in situ* and *in vitro* studies have also shown that reactive astrocytes employ phagocytosis to eradicate deposits of Aβ [100, 101]. Furthermore, the research indicates that improving astrocytes' phagocytic capacity may help to restore immunological response, neural circuit dysfunction, and Aβ-mediated illness. However, a number of studies have shown that reactive astrocytes have a hazardous role. The TLR/RAGE-dependent detection of Aβ aggregates by astrocytes, which are microglia, leads to downstream target gene activation and subsequent factor production. Aβ causes increased release of cytokines (*e.g.*, TNF-α), chemokines (like MIP1-α) and neurotoxins [102]. Increased Aβ burden and toxicity are the results of excessive neurotoxic factor production, which also modifies the homeostasis of astrocyte APP processing. Additionally, neuronal death, cognitive decline, and synaptic degeneration have all been connected to inflammatory mediators. In AD, signaling—more especially, IL-1-mediated signaling—is recognized to have a detrimental role. According to the study, blocking IL-1 signaling improves cognitive deficiency, reduces tau and Aβ pathology, and controls inflammatory response in a transgenic mouse model. The crucial role astrocytes play in neuroinflammation has been shown by the evidence that is now accessible. The focus of current research is on astrocyte-mediated neuroinflammatory pathways to halt AD's progression.

RECENT DRUG DEVELOPMENT STRATEGIES

AD medication development is a rapidly evolving sector, moving away from conventional methods and toward more Biologics and Monoclonal Antibodies, small-molecule drugs, and Gene Therapy Methods. The novel approaches to improve clinical success, safety, and efficacy are currently the main focus of study. The most recent approaches to medication development in AD research are listed below.

Monoclonal Antibodies and Biologics

In addition to drugs for managing cognitive symptoms like NMDA antagonists and acetylcholinesterase inhibitors, the FDA has approved three monoclonal antibodies targeting amyloid beta: aducanumab, lecanemab and donanemab [103].

Amyloid Beta Target

Aducanumab, a human immunoglobulin 1 (IgG1) mab, can specifically target aggregated Aβ. However, it cannot target the monomers of Aβ. Anti-Aβ mAbs have shown great promise in treating AD in animal models. By stimulating microglia and preventing Aβ aggregation, these drugs can enhance cognitive abilities and reduce brain damage. It has been demonstrated that aducanumab efficiently eliminates Aβ from mice's brains. In transgenic mice's brains, aducanumab binds with parenchymal Aβ and inhibits both soluble and insoluble Aβ. When aducanumab was administered weekly for four to ten-month-old an animal model of AD (tgAPPPS1-21 mice), it was evident that it improved cell viability and phagocytosis and decreased Aβ toxicity.

Another humanized IgG1 monoclonal antibody, Lecanemab, has shown a strong reduction of fibrillar amyloid in early AD. It targets soluble aggregated Aβ species, or protofibrils [104]. Eric *et. al.* investigated the double-blind, randomized, placebo-controlled research, called "The lecanemab study 201 core" [105]. Eight hundred fifty-six individuals were randomly assigned to either a placebo or one of five dosing regimens. It was found that Aβ decreased dose-dependently with lecanemab, whereas the ratio of plasma Aβ42/40 was improved, and plasma tau was reduced. It has been shown to lower amyloid indicators in the early stages of AD, which has logically slowed the decline in cognitive and functional impairment [106].

Donanemab (11) was recently approved for the treatment of AD symptoms in early stages [16]. Donanemab, a humanized antibody, reduces Aβ plaques in AD patients. Donanemab seems to have cleared up a significant quantity of plaque. The anti-Aβ antibodies tested so far are diverse, focusing on various Aβ peptide domains and different stages of aggregation [69]. While some target the monomers, oligomers, or fibrillar types of Aβ, others are selective for plaques. Some antibodies also attach to normal Aβ, and the composition and shape of Aβ in the brain vary dramatically. Donanemab's targeting of AβpE, which is Aβ found only in cerebral Aβ plaques and is not present in biofluids, is a crucial feature that suggests it is plaque-specific. Determining AβpE's involvement in pathogenesis and its potential as a target for AD treatment requires an understanding of where and how it forms [107].

Gantenerumab is a human Aβ immunoglobulin antibody that works by dissociating peptide aggregates and clearing fibrillar Aβ to help the brain get rid of amyloid plaques [108]. The phagocytosis of aggregated Aβ by microglia mediated by the Fc gamma receptor is facilitated by the human immunoglobulin G1 backbone [109]. Within Aβ plaques, gantenerumab binds to Aβ fibrils in AD patients' brains [108, 110]. Gantenerumab has the strongest affinity for soluble oligomers and aggregated forms of Aβ, such as fibrils and plaques. Evidence from in vitro experiments indicates that binding to oligomers takes place, counteracting the toxicity of oligomers [111]. According to the results presented above, gantenerumab appears to have two significant impacts on aggregated Aβ: it inhibits the growth of aggregations and it triggers the phagocytosis of microglia. Gantenerumab was chosen due to its exceptional capacity to bind to the core amino acids of the Aβ peptide as well as its N-terminal regions. Gantenerumab was distinguished from other anti-amyloid monoclonal antibodies that were being developed at the time by this feature, as they all bound to the same area of the Aβ sequence. Greater binding stability, which is anticipated to be crucial for therapeutic impact, is conferred by gantenerumab's capacity to bind to *N*-terminal regions of Aβ and core parts of fibrillar Aβ [108].

Another humanized monoclonal IgG1 antibody is solanezumab, which binds to the mid-domain epitope of the Aβ, which is present in oligomers and fibrils, targeting the soluble monomers and oligomers. In laboratory experiments, solanezumab's binding to the monomer inhibits the development of amyloid plaque. Preclinical research suggests that solanezumab improves Aβ removal from the brain *via* the formation of antibody-Aβ peptide complexes and their transfer to plasma. In AD models, this pathway improves cognitive performance and reduces Aβ load [112, 113].

Tau Target

The Microtubule-Associated Protein Tau (MAPT) gene, which is located on chromosome 17, encodes the tau protein, which is composed of 352-441 amino acids and exists in six distinct isoforms. Tau is present in the nucleus, cell membrane, axons, dendrites, and synapses of neurons. Tau protein is also found in interstitial fluid (ISF) and can enter the cerebrospinal fluid (CSF) and blood flow. The primary function of the Tau protein is to help microtubules in neuronal axons assemble and stabilize. In addition, tau is involved in motor function, myelination, neurogenesis, memory and learning among other biological processes. Tau's equilibrium between phosphorylation and dephosphorylation controls its attachment to microtubules. The conformational changes in tau result in the formation of paired helical and straight filaments of abnormally phosphorylated tau and tau aggregates, whereas tau is unfolded and phosphorylated in physiological conditions. Also, excessive hyperphosphorylation causes tau to disengage from the microtubules. These clusters have the potential to degenerate neurons and glial cells, which can result in a range of clinical cognitive, behavioral, and motor symptoms that fall under the broad categories of neurodegenerative diseases known as "tauopathies." Primary and secondary tauopathies are the two categories of tauopathies. The basic underlying neurodegenerative mechanism in primary tauopathies is caused by aberrant tau [114]. There are some monoclonal antibodies that target the tau protein to treat AD.

Bepranemab (UCB0107) is a humanized monoclonal IgG4 antibody that targets the core region of tau at amino acids 235-246 [115, 116]. Antibodies targeting the midregion of tau are more effective in preventing pathogenic and aggregated tau from spreading across cells [117]. According to preclinical research, bepranemab has promise in treating tau disease. Targeting an area close to the microtubule-binding domain appears to be a promising method for successfully preventing the spread of pathogenic tau. Notably, it was able to prevent the development of tau pathology in transgenic mice. Additionally, it prevented tau disease from spreading to other parts of the brain [118, 119].

Semorinemab (RO705705), a humanized antibody (IgG4), targets extracellular tau. In humans, it can attach itself to all six isoforms of tau protein. Additionally, in co-culture settings with microglia, it protects neurons against the tau oligomers-induced neurotoxicity. In preclinical research, a murine form of semorinemab demonstrated encouraging results by lowering tau-associated toxicity in a cell culture and preventing the buildup of tau in a mouse model. However, there have been difficulties transferring these preclinical achievements to the clinical environment. A phase 1 study showed a good safety profile and dose-dependent target engagement, which led to additional clinical research in early AD [120].

Tilavonemab, an IgG4 monoclonal antibody, targets the aggregated, extracellular form of pathogenic tau. The mode of action of tilavonemab sets it apart from other anti-tau antibodies; it does not require absorption into neurons. Preclinical studies using the mouse form of this antibody demonstrated its capacity to prevent primary neurons from absorbing tau generated from AD. The parental antibody produced significant results in P301S tau-transgenic mice, such as decreased insoluble tau, brain atrophy and microgliosis. These results highlight the unique characteristics of the preclinical models of the murine form of tilavonemab, highlighting its potential as a viable anti-tau treatment option with a novel mode of action [121].

Gosuranemab (BIIB092) is another IgG4 antibody targeting tau's extracellular *N*-terminal region (residue areas) [118]. The addition of exogenous tau was found to increase Aβ synthesis and cause increased neuronal activity in human cortical neurons in preclinical studies. In other models using human tau transgenic mice, this pattern was consistently seen. The antibody demonstrated robust binding properties to a variety of tau forms, including insoluble tau, fibrils, and monomers [118, 122].

Gene Therapy Approaches

The interaction between amyloid-β and other proteins accelerates the progression of AD [123]. Furthermore, AD's pathophysiology is complex, and a single medication may not produce a positive therapeutic result. Sujun Jiang *et. al.* described a gene therapy to modify microglia activity and intervene in Aβ anabolism. Biomimetic nanovesicles deliver the β-site amyloid precursor protein cleaving enzyme-1 (BACE1), TREM2 plasmid (pTREM2) and siRNA (siBACE1) gene drugs to AD lesions. Angiopep-2 peptides and exosome homing improve drug accumulation. Upregulating TREM2 expression can reprogramme microglia from a pro-inflammatory M1 to an anti-inflammatory M2 phenotype, allowing them to phagocytose Aβ and repair nerves. Furthermore, by inhibiting the BACE1 gene, siRNA lowers the creation of Aβ plaques at their origin. According to the *in vivo* study, TSEL can improve the cognitive impairment of APP/PS1 mice, lowering the buildup of Aβ, and halting the recurrence of neuroinflammation through the synergistic action of two gene medications [124]. Recent developments in neurotrophin delivery aided by gene therapy have produced unexpected outcomes, such as enhanced cognitive function and trophic responses of cholinergic neurons. NGF is a member of the neurotrophin family. Although NGF was initially identified as a neurotrophin, it may have several functions. Recent findings indicate that decreasing NGF in PC12 (pheochromocytoma) cells triggers Aβ generation, connecting NGF and Aβ [125]. Cholinergic neurons in the cerebral cortex and hippocampus have the ability to release the neurotransmitter ACh, which enhances synaptic efficiency and controls cortical circuit activity. These cholinergic neurons severely deteriorate in AD conditions, which leads to memory impairments. Consequently, NGF-based treatment strategies may lessen cholinergic neuron damage in AD situations. AD therapy involves raising NGF levels to modulate the cholinergic system and compensate for low ACh levels. A lentiviral vector was utilized in an in vitro gene therapy study to assess the therapeutic effect of BDNF gene delivery in different animals. Research findings indicate encouraging effects, such as improving cognitive learning, reversing synapse damage, and controlling cell signaling. Furthermore, no negative effects were found. The AD study focuses on targeting BACE-1 (β-site APP cleaving enzyme 1 or β-secretase), an aspartyl protease linked to the aberrant formation of Aβ plaques, which is a hallmark of its pathogenesis. Accordingly, lentiviral expression of anti-BACE-1 small interfering RNA (siRNA) was evaluated in a scientific study. This study shows encouraging results, including a 50% reduction in BACE-1 expression, a 50% reduction in amyloid formation, and a considerable reduction in neuronal death [126]. Furthermore, the study focused on using peripheral APN gene therapy to treat the pathology associated with AD in 5xFAD mice in a novel and manageable therapeutic manner. Laam Ng-Roy Chun created and transformed adeno-associated virus (AAV2/8) encoding the mouse mutant APN gene (APNC39S) into the liver of 5xFAD mice, resulting in only low-molecular-weight trimeric APN (APNTri). Because the overexpressed APNTri was able to pass the BBB, a single injection of AAV2/8-APNC39S in the liver raised cerebral and circulatory APN levels. Both soluble and fibrillar Aβ were reduced in the brains of 5xFAD mice when APNTri was overexpressed. Treatment with AAV2/8-APNTri inhibited microglial NLRP3-inflammasome activation, which decreased Aβ-induced IL-1β and IL-18 production. AAV-APNTri-treated 5xFAD mice showed a considerable improvement in memory capabilities, together with a decrease in dystrophic neurites. These results show that overexpressing trimeric APN through peripheral gene delivery may be a viable treatment for AD [127].

PRECLINICAL STUDIES

Preclinical trial studies of AD are found in various data collected in vitro models, which are less profound than in vivo models. Alternative strategies for AD-related pathological conditions could include investigating unexplored models, which could be future targets for anti-AD drugs and help address current limitations in AD models [128]. The aim of preclinical studies is to increase the options for the treatment and therapeutic targets for AD. Table **4** summarizes some preclinical studies of AD.

CHALLENGES IN DRUG DEVELOPMENT FOR AD

AD is a challenging disease owing to its complex nature, model limitations, and high clinical trial failure rates. Developing drugs and techniques for delivering bioactive substances to the CNS is challenging [137].

High failure rate in preclinical trial

A study of clinical trial failures for AD, from 2004 to present, revealed a 2.0% success rate in recent drug development. This highlights the urgent need for improved drug development in AD [138]. Animal models of AD are essential for determining the efficacy and toxicity before humans are exposed, showing limited tau accumulation, cell death, and cognitive changes in genetically engineered animals [139]. Several anti-amyloid drugs, such as monoclonal antibodies (mAbs), β-secretase inhibitors, intravenous immunoglobulins and γ-secretase inhibitors have not performed well during Phase III investigations in patients with AD [140]. Flurizan, a treatment for AD, failed in phase III due to inadequate potency, low brain penetration, and a lack of target modulation, leaving approximately 2,500 patients with ineffective treatment [141].

Table 4. Recent advancements in preclinical research in drug development for AD.

Sr. No	Target	Drug/Therapy	Preclinical finding	Model use	Reference
1.	Amyloid-beta	solanezumab, verubecestat, and semagacestat	Preclinical research suggests that shorter Aβ forms improve glutamate neurotransmission in humans.	Patients with MCI and a positive Aβ burden.	[129]
		Donanemab (LY3002813)	Donanemab effectively reduced amyloid plaques for up to 72 weeks, including in patients with treatment-emergent antidrug antibodies. The most commonly reported side effect was imaging problems.	61 AD patients with mild-to-moderate dementia tested positive for amyloid plaque.	[130]
		solanezumab	Solanezumab did not slow cognitive deterioration compared to placebo over 240 weeks.	spline model	[131]
		Semagacestat	Semagacestat therapy lowers Aβ levels in plasma, cerebrospinal fluid, and brain in both humans and animals. The most common side effect is the inhibition of Notch cleavage in the peripheral region.	Transgenic mouse	[132]
2.	Tau-Targeting	Antibodies (gosuranemab)	The research uses PBPK-PD modeling to address challenges in the preclinical and clinical study of anti-tau antibody treatments for AD.	The physiologically-based pharmacokinetic (PBPK) model is applicable to mice, rats, monkeys, and humans.	[133]
		Tideglusib, Dl-3-n-butylphthalide, Lithium, Nilotinib	Hyperphosphorylation of Tau, but its clinical application remains unclear.	Transgenic rodent models	[134]

(Table 4) cont.....

Sr. No	Target	Drug/Therapy	Preclinical finding	Model use	Reference
3.	Gene thrapeutics	CRISPR-Cas9	The study investigates the action of SORL1 in AD, revealing early endosome enlargement and altered localization of APP.	anti-Human CD44, Mouse model, and anti-Human CD184 antibodies.	[135]
		antisense oligonucleotides	The study demonstrates that purchased ASOs effectively decrease the level of tau protein and mRNA in the brain, and *in vivo*, decrease phosphorylated tau and gliosis in mice, supporting their broad use by researchers.	Male C57BL/6 mice	[136]

Blood-Brain Barrier (BBB) Penetration

Drug transport into CSF is influenced by drug delivery across the choroid plexus, while drug penetration into the ISF is measured by therapeutic antibodies' distribution in brain tissue [142]. The BBB presents a substantial hurdle to drug delivery to the CNS, and many techniques have been developed to address this issue. Disrupting the BBB for drug delivery has been proposed, despite the potential neurotoxicity of endogenous substances entering the brain from the blood [143].

Biomarkers for Early Diagnosis and Treatment Efficacy

This section examines the problems faced by clinical needs, scientific and technological breakthroughs, regulatory environments, and commercial prospects for determining the use of diagnostic biomarkers for AD. The difficulties in identifying biomarkers cause delayed and uncoordinated clinical application [144]. The AFE-T test was used to investigate memory performance in preclinical AD stages 1 and 2, including immediate recall, learning curves, long-term forgetting rates, and relearning curves. The findings indicate that apparent learning issues could be precursors to persistent memory impairments [145]. Biomarkers are being used to define the preclinical state of these conditions, potentially enabling the development of novel treatments for AD [146]. A multicenter study discovered that cerebrospinal fluid biomarkers, including A42, T-tau, and P-tau, are capable of identifying developing AD, but with less accuracy than single-center studies, emphasizing the necessity for uniformity in methods of analysis and clinical practices [147].

FUTURE DIRECTIONS

Drug development for AD is rapidly changing, focusing on targeting Amyloid-β Pathway Modulators, Tau Protein Aggregation Inhibitors, Neuroinflammation Modulators, Neuroprotection, and Synaptic Regeneration. Experimental studies demonstrate that memantine in conjunction with an AChE inhibitor has a good safety profile and improves the impact of cholinergic treatment in elderly women with AD. The German Competence Network on Dementias is studying whether combining galantamine with memantine therapy can stop or decrease the progression of AD [148]. Direct assessments of therapeutic targets in human studies can assist in determining the success of treatments in treating amyloid plaques, β-amyloid, tau, CNS levels and other biomolecules involved in disease progression. This information helps in determining the duration, magnitude, and required modulation for positive clinical outcomes. Additionally, it helps identify critical Aβ forms and the best targets for therapeutic intervention [149]. AI/ML-driven discovery of the management of CNS diseases has grown significantly due to their success in various fields. Future applications will focus on personalized medicine, patient subtyping, disease driver identification, drug response prediction, autonomous drug design, and disease-specific BBB permeability testing. Despite data and algorithm constraints, ongoing developments will enable more effective CNS drug development [150]. Alzheimer's Disease (AD) treatments aim to relieve cognitive symptoms and slow its progression. The first effective drug for AD treatment was tacrine in 1970, but it was recalled from the market owing to its side effects and liver toxicity. Cholinesterase inhibitors, such as Rivastigmine, Donepezil and Galantamine, are currently used, but can cause adverse effects. Memantine, which is an NMDA receptor antagonist, is used for the treatment of moderate-to-severe AD. In 2021, the FDA approved aducanumab, the first drug to address AD's underlying biology. Memantine's ability to bind with soluble oligomers and Aβ fibrils enhances its therapeutic effectiveness [151].

CONCLUSION

Alzheimer's is a neurological illness that causes memory and cognitive deterioration. Despite substantial studies, effective treatments continue to be elusive. The FDA has granted approval for four small-molecule medications: donepezil, galantamine, rivastigmine, and memantine. Second-generation AChEIs have shown better pharmacokinetic profiles and fewer side effects. Memantine enhances cognitive performance. In 2019, China approved sodium oligomannate. Donanemab and benzgalantamine are humanized antibodies, focusing on cognitive and memory damage. More study is needed to guarantee the long-term efficacy and safety. AD is treated with symptomatic and disease-modifying therapies. Symptomatic treatments aim to relieve symptoms, while disease-modifying therapies aim to reduce or stop the disease's progression. Symptomatic treatments include cholinesterase inhibitors, antidepressants, and antipsychotic medications. Disease-modifying therapies target amyloid plaques and delay early-stage disease progression. Research is ongoing to target tau proteins, inflammation, and neuroprotection to slow disease development further. Cholinergic neurotransmission has a crucial role in Alzheimer's disease and adult-onset dementia. Recent drug development advancements offer promising future therapies, focusing on multi-target approaches, personalized medicine, AI-driven discovery, and novel drug delivery, enhancing effectiveness in Alzheimer's disease treatment. Personalized medicine, artificial intelligence, advanced drug delivery systems, and disease-modifying therapies are revolutionizing drug discovery, early diagnosis, and clinical trial optimization. They are a crucial component of the healthcare innovations of the future due to their versatility and ongoing study, particularly in the fields of precision and individualized treatment.

CONFLICT OF INTEREST

The writers state the absence of any conflicts of interest.

ACKNOWLEDGEMENTS

The Authors would like to thank Bentham Science.

REFERENCES

[1] Piemontese L. Historic recurrences in medicinal chemistry: nature-inspired structures as a new opportunity for novel multi-target anti-Alzheimer's drugs. Neural Regen Res. 2023 Dec; 18(12): 2671–2672.
[http://dx.doi.org/10.4103/1673-5374.373685]

[2] Scheltens P, De Strooper B, Kivipelto M, *et al.* Alzheimer's disease. Lancet 2021; 397(10284): 1577-90.
[http://dx.doi.org/10.1016/S0140-6736(20)32205-4] [PMID: 33667416]

[3] Galimberti D, Scarpini E. Old and new acetylcholinesterase inhibitors for Alzheimer's disease. Expert Opin Investig Drugs. 2016 Oct 1; 25(10): 1181–1187.
[http://dx.doi.org/10.1080/13543784.2016.1216972]

[4] Sharma K. Cholinesterase inhibitors as Alzheimer's therapeutics (Review). Mol Med Rep. 2019 Aug; 20(2): 1479–1487.
[http://dx.doi.org/10.3892/mmr.2019.10374]

[5] Simunkova M, Alwasel SH, Alhazza IM, *et al.* Management of oxidative stress and other pathologies in Alzheimer's disease. Arch Toxicol. 2019 Sep; 93(9): 2491–2513.
[http://dx.doi.org/10.1007/s00204-019-02538-y]

[6] Amat-ur-Rasool H, Ahmed M, Hasnain S, Carter WG. Anti-cholinesterase combination drug therapy as a potential treatment for Alzheimer's disease. Brain Sci. 2021 Feb 2; 11(2): 184. [http://dx.doi.org/10.3390/brainsci11020184]

[7] Kabir MT, Uddin MS, Mamun AA, Jeandet P, Aleya L, Mansouri RA. Combination drug therapy for the management of Alzheimer's disease. Int J Mol Sci. 2020 May 6; 21(9): 3272. [http://dx.doi.org/10.3390/ijms21093272]

[8] Pardo-Moreno T, González-Acedo A, Rivas-Domínguez A, García-Morales V, García-Cozar FJ, Ramos-Rodríguez JJ,. Therapeutic approach to Alzheimer's disease: current treatments and new perspectives. Pharmaceutics. 2022 May 25; 14(6): 1117. [http://dx.doi.org/10.3390/pharmaceutics14061117]

[9] Greig SL. Memantine ER/Donepezil: a review in Alzheimer's disease. CNS Drugs. 2015 Nov; 29(11): 963–970.

[10] Syed YY. Sodium oligomannate: first approval. Drugs. 2020 Mar; 80(4): 441–444. [http://dx.doi.org/10.1007/s40265-020-01268-1]

[11] Wang X, Sun G, Feng T, *et al.* Sodium oligomannate therapeutically remodels gut microbiota and suppresses gut bacterial amino acids-shaped neuroinflammation to inhibit Alzheimer's disease progression. Cell Res. 2019 Oct; 29(10): 787–803. [http://dx.doi.org/10.1038/s41422-019-0216-x]

[12] Sevigny J, Chiao P, Bussière T, *et al.* The antibody aducanumab reduces Aβ plaques in Alzheimer's disease. Nature. 2016 Sep 1; 537(7618): 50–56. [http://dx.doi.org/10.1038/nature19323]

[13] Lannfelt L, Söderberg L, Laudon H, Sahlin C, Johannesson M, Nygren Pl. P4-704: BAN2401 shows stronger binding to soluble aggregated amyloid-beta species than aducanumab. Alzheimers Dement. 2019 Jul; 15(7S): P1601–P1602. [http://dx.doi.org/10.1016/j.jalz.2019.09.068]

[14] Karran E, De Strooper B. The amyloid hypothesis in Alzheimer disease: new insights from new therapeutics. Nat Rev Drug Discov. 2022 Apr; 21(4): 306–318. [http://dx.doi.org/10.1038/s41573-022-00391-w]

[15] Ballard CJJn. Brexpiprazole for the treatment of agitation and aggression in alzheimer disease 2023. [http://dx.doi.org/10.1001/jamaneurol.2023.3967]

[16] Kang C. Donanemab: First Approval. Drugs 2024; 84(10): 1313-8. [http://dx.doi.org/10.1007/s40265-024-02087-4] [PMID: 39237715]

[17] Ballard C. Brexpiprazole for the treatment of agitation and aggression in Alzheimer disease. JAMA Neurol. 2023 Dec 1; 80(12): 1272–1273.

[18] Baakman AC, 't Hart E, Kay DG, *et al.* First in human study with a prodrug of galantamine: Improved benefit-risk ratio? Alzheimers Dement (N Y) 2016; 2(1): 13-22. [http://dx.doi.org/10.1016/j.trci.2015.12.003] [PMID: 29067291]

[19] Galimberti D, Scarpini E. Treatment of Alzheimer's disease: symptomatic and disease-modifying approaches. Curr Aging Sci 2010; 3(1): 46-56. [http://dx.doi.org/10.2174/1874609811003010046] [PMID: 20298170]

[20] Watkins PB, Zimmerman HJ, Knapp MJ, Gracon SI, Lewis KW. Hepatotoxic effects of tacrine administration in patients with Alzheimer's disease. JAMA 1994; 271(13): 992-8. [http://dx.doi.org/10.1001/jama.1994.03510370044030] [PMID: 8139084]

[21] Bekdash RA. The cholinergic system, the adrenergic system and the neuropathology of Alzheimer's disease. Int J Mol Sci. 2021 Jan 28; 22(3): 1273. [http://dx.doi.org/10.3390/ijms22031273]

[22] Stanciu GD, Luca A, Rusu RN, Bild V, Beschea Chiriac SI, Solcan C. Alzheimer's disease

pharmacotherapy in relation to cholinergic system involvement. Biomolecules. 2020 Jan; 10(1): 40. [http://dx.doi.org/10.3390/biom10010040]

[23] Berry AS, Harrison TM. New perspectives on the basal forebrain cholinergic system in Alzheimer's disease. Neurosci Biobehav Rev. 2023 Jul; 150: 105192. [http://dx.doi.org/10.1016/j.neubiorev.2023.105192]

[24] Nordberg A, Ballard C, Bullock R, Darreh-Shori T, Somogyi M. A review of butyrylcholinesterase as a therapeutic target in the treatment of Alzheimer's disease. Prim Care Companion CNS Disord 2013; 15(2). [http://dx.doi.org/10.4088/PCC.12r01412] [PMID: 23930233]

[25] DeTure MA, Dickson DW. The neuropathological diagnosis of Alzheimer's disease. Mol Neurodegeneration. 2019 Aug 2; 14(1): 32. [http://dx.doi.org/10.1186/s13024-019-0333-5]

[26] Kareem RT, Abedinifar F, Mahmood EA, Ebadi AG, Rajabi F, Vessally E. The recent development of donepezil structure-based hybrids as potential multifunctional anti-Alzheimer's agents: highlights from 2010 to 2020. RSC Adv. 2021 Sep 14; 11(49): 30781–30797. [http://dx.doi.org/10.1039/D1RA03718H]

[27] Wang W, Chen XF, Zhang Y, Ran Y, Jin L, Li S. Improved affinity: a customized fluorescent probe for the rapid detection of butyrylcholinesterase. Chemosensors. 2024 May 24; 12(6): 100. [http://dx.doi.org/10.3390/chemosensors12060100]

[28] Atatreh N, Al Rawashdah S, Al Neyadi SS, Abuhamdah SM, Ghattas MA. Discovery of new butyrylcholinesterase inhibitors via structure-based virtual screening. J Enzyme Inhib Med Chem. 2019 Dec; 34(1): 1373–1379. [http://dx.doi.org/10.1080/14756366.2019.1644329]

[29] Peitzika SC, Pontiki EA. A review on recent approaches on molecular docking studies of novel compounds targeting acetylcholinesterase in Alzheimer disease. Molecules. 2023 Jan 21; 28(3): 1084. [http://dx.doi.org/10.3390/molecules28031084]

[30] Vecchio I, Sorrentino L, Paoletti A, Marra R, Arbitrio M. The state of the art on acetylcholinesterase inhibitors in the treatment of Alzheimer's disease. J Cent Nerv Syst Dis. 2021 Jul 7; 13: 11795735211029113.

[31] Giacobini E. Invited Review Cholinesterase inhibitors for Alzheimer's disease therapy: from tacrine to future applications. Neurochem Int 1998; 32(5-6): 413-9. [http://dx.doi.org/10.1016/S0197-0186(97)00124-1] [PMID: 9676739]

[32] Pang Y-P, Brimijoin S. THA analogs useful as cholinesterase inhibitors. Google Patents 1999.

[33] Zhang C, Du QY, Chen LD, *et al.* Design, synthesis and evaluation of novel tacrine-multialkoxybenzene hybrids as multi-targeted compounds against Alzheimer's disease. Eur J Med Chem 2016; 116: 200-9. [http://dx.doi.org/10.1016/j.ejmech.2016.03.077] [PMID: 27061983]

[34] Fancellu G, Chand K, Tomás D, *et al.* Novel tacrine–benzofuran hybrids as potential multi-target drug candidates for the treatment of Alzheimer's Disease. J Enzyme Inhib Med Chem 2020; 35(1): 211-26. [http://dx.doi.org/10.1080/14756366.2019.1689237] [PMID: 31760822]

[35] English C. Donepezil 23 mg: Is it more advantageous compared to the original? Ment Health Clin 2012; 1(11): 272-3. [http://dx.doi.org/10.9740/mhc.n106739]

[36] Szeto JY, Lewis SJ. Current Treatment Options for Alzheimer's Disease and Parkinson's Disease Dementia. Curr Neuropharmacol 2016; 14(4): 326-38. [http://dx.doi.org/10.2174/1570159X14666151208112754] [PMID: 26644155]

[37] Shen Y, Sheng R, Zhang J, He Q, Yang B, Hu Y. 2-Phenoxy-indan-1-one derivatives as acetylcholinesterase inhibitors: A study on the importance of modifications at the side chain on the

activity. Bioorg Med Chem 2008; 16(16): 7646-53.
[http://dx.doi.org/10.1016/j.bmc.2008.07.014] [PMID: 18662884]

[38] Piazzi L, Rampa A, Bisi A, *et al.* 3-(4-[[Benzyl(methyl)amino]methyl]phenyl)-6,7-dimethox--2H-2-chromenone (AP2238) inhibits both acetylcholinesterase and acetylcholinesterase-induced β-amyloid aggregation: a dual function lead for Alzheimer's disease therapy. J Med Chem 2003; 46(12): 2279-82.
[http://dx.doi.org/10.1021/jm0340602] [PMID: 12773032]

[39] Rizzo S, Bartolini M, Ceccarini L, *et al.* Targeting Alzheimer's disease: Novel indanone hybrids bearing a pharmacophoric fragment of AP2238. Bioorg Med Chem 2010; 18(5): 1749-60.
[http://dx.doi.org/10.1016/j.bmc.2010.01.071] [PMID: 20171894]

[40] Ojha S, Mishra S. Nanostructured lipid carriers for targeting central nervous system: Recent advancements. Micro Nanosystems 2023; 82-91.
[http://dx.doi.org/10.2174/1876402915666230518121949]

[41] Terra BS, da Silva PHC, Tramarin A, *et al.* Two novel donepezil-lipoic acid hybrids: synthesis, anticholinesterase and antioxidant activities and theoretical studies. J Braz Chem Soc. 2018 Apr; 29(4): 738–747.

[42] Grossberg GT, Desai AK. Management of Alzheimer's disease. J Gerontol A Biol Sci Med Sci. 2003 Apr; 58(4): M331–M353.
[http://dx.doi.org/10.1093/gerona/58.4.M331]

[43] Desai AK, Grossberg GT. Rivastigmine for Alzheimer's disease. Expert Rev Neurother 2005; 5(5): 563-80.
[http://dx.doi.org/10.1586/14737175.5.5.563] [PMID: 16162080]

[44] Eldufani J, Blaise G. The role of acetylcholinesterase inhibitors such as neostigmine and rivastigmine on chronic pain and cognitive function in aging: A review of recent clinical applications. Alzheimer's Dement (N Y). 2019; 5(1): 175–183.
[http://dx.doi.org/10.1016/j.trci.2019.03.004]

[45] Roy K, Srinivas K, Khanna I, Pillarisetti S. Substituted 1, 2, 3, 4-tetrahydroquinolin-7-yl carbamates their preparation and use thereof as acetylcholinesterase (AChE) inhibitors for the treatment of Alzheimer's and other neurodegenerative disease. US Patent US8946261B2. 2015 Feb 3.

[46] Bohn P, Gourand F, Papamicaël C, *et al.* Dihydroquinoline Carbamate Derivatives as "Bio-oxidizable" Prodrugs for Brain Delivery of Acetylcholinesterase Inhibitors: [^{11} C] Radiosynthesis and Biological Evaluation. ACS Chem Neurosci 2015; 6(5): 737-44.
[http://dx.doi.org/10.1021/cn5003539] [PMID: 25695305]

[47] Amenta F, Parnetti L, Gallai V, Wallin A. Treatment of cognitive dysfunction associated with Alzheimer's disease with cholinergic precursors. Ineffective treatments or inappropriate approaches? Mech Ageing Dev 2001; 122(16): 2025-40.
[http://dx.doi.org/10.1016/S0047-6374(01)00310-4] [PMID: 11589920]

[48] Maelicke A, Hoeffle-Maas A, Ludwig J, *et al.* Memogain is a galantamine pro-drug having dramatically reduced adverse effects and enhanced efficacy. J Mol Neurosci 2010; 40(1-2): 135-7.
[http://dx.doi.org/10.1007/s12031-009-9269-5] [PMID: 19669943]

[49] Atanasova M, Stavrakov G, Philipova I, Zheleva D, Yordanov N, Doytchinova I. Galantamine derivatives with indole moiety: Docking, design, synthesis and acetylcholinesterase inhibitory activity. Bioorg Med Chem 2015; 23(17): 5382-9.
[http://dx.doi.org/10.1016/j.bmc.2015.07.058] [PMID: 26260334]

[50] Bennett MVL, Zukin RS. Electrical coupling and neuronal synchronization in the Mammalian brain. Neuron 2004; 41(4): 495-511.
[http://dx.doi.org/10.1016/S0896-6273(04)00043-1] [PMID: 14980200]

[51] Chen HSV, Lipton SA. The chemical biology of clinically tolerated NMDA receptor antagonists. J

Neurochem. 2006 Jun; 97(6): 1611–1626.
[http://dx.doi.org/10.1111/j.1471-4159.2006.03991.x]

[52] Zhang Y, Li P, Feng J, Wu M. Dysfunction of NMDA receptors in Alzheimer's disease. Neurol Sci. 2014 Jul; 37(7): 1039–1047.
[http://dx.doi.org/10.1007/s10072-016-2546-5]

[53] Dingledine R, Borges K, Bowie D, Traynelis SF. The glutamate receptor ion channels. Pharmacol Rev 1999; 51(1): 7-61.
[http://dx.doi.org/10.1016/S0031-6997(24)01394-2] [PMID: 10049997]

[54] Rothman SM, Olney JW. Glutamate and the pathophysiology of hypoxic-ischemic brain damage. Ann Neurol. 1986 Feb; 19(2): 105–111.
[http://dx.doi.org/10.1002/ana.410190202]

[55] Tan C, Wang SSH, de Nola G, Kaeser PS. Rebuilding essential active zone functions within a synapse. Neuron 2022; 110(9): 1498-1515.e8.
[http://dx.doi.org/10.1016/j.neuron.2022.01.026] [PMID: 35176221]

[56] Parsons CG. NMDA receptors as targets for drug action in neuropathic pain. Eur J Pharmacol 2001; 429(1-3): 71-8.
[http://dx.doi.org/10.1016/S0014-2999(01)01307-3] [PMID: 11698028]

[57] Hallett PJ, Standaert DG. Rationale for and use of NMDA receptor antagonists in Parkinson's disease. Pharmacol Ther 2004; 102(2): 155-74.
[http://dx.doi.org/10.1016/j.pharmthera.2004.04.001] [PMID: 15163596]

[58] Egunlusi AO, Joubert J. NMDA Receptor Antagonists: Emerging Insights into Molecular Mechanisms and Clinical Applications in Neurological Disorders. Pharmaceuticals (Basel) 2024; 17(5): 639.
[http://dx.doi.org/10.3390/ph17050639] [PMID: 38794209]

[59] Fedele E. Anti-Amyloid Therapies for Alzheimer's Disease and the Amyloid Cascade Hypothesis 2023; 24(19): 14499.

[60] Ameen TB, Kashif SN, Abbas SMI, Shakaib B, Qazi MR. Unraveling Alzheimer's: the promise of aducanumab, lecanemab, and donanemab. Egypt J Neurol Psychiatry Neurosurg. 2024 Jun 4; 60(1): 72.
[http://dx.doi.org/10.1186/s41983-024-00845-5]

[61] Dyer O. Aduhelm: Biogen abandons Alzheimer's drug after controversial approval left it unfunded by Medicare. BMJ. 2024 Feb 2; 384: q281.
[http://dx.doi.org/10.1136/bmj.q281]

[62] Cummings J, Osse AML, Cammann D, Powell J, Chen J. Anti-Amyloid Monoclonal Antibodies for the Treatment of Alzheimer's Disease. BioDrugs 2024; 38(1): 5-22.
[http://dx.doi.org/10.1007/s40259-023-00633-2] [PMID: 37955845]

[63] Rech MA, Carpenter CR, Aggarwal NT, Hwang U. Anti-Amyloid Therapies for Alzheimer's Disease and Amyloid-Related Imaging Abnormalities: Implications for the Emergency Medicine Clinician. Ann Emerg Med 2025; 85(6): 526-36.
[http://dx.doi.org/10.1016/j.annemergmed.2024.12.002] [PMID: 39818674]

[64] Bradshaw AC, Georges J. Anti-Amyloid Therapies for Alzheimer's Disease: An Alzheimer Europe Position Paper and Call to Action. J Prev Alzheimers Dis 2024; 11(2): 265-73.
[http://dx.doi.org/10.14283/jpad.2024.37] [PMID: 38374732]

[65] Söderberg L, Johannesson M, Nygren P, *et al.* Lecanemab, Aducanumab, and Gantenerumab — Binding Profiles to Different Forms of Amyloid-Beta Might Explain Efficacy and Side Effects in Clinical Trials for Alzheimer's Disease. Neurotherapeutics 2023; 20(1): 195-206.
[http://dx.doi.org/10.1007/s13311-022-01308-6] [PMID: 36253511]

[66] Singh PK, Simões-Pires EN, Chen ZL, *et al.*, Lecanemab blocks the effects of the Aβ/fibrinogen complex on blood clots and synapse toxicity in organotypic culture. 2024. 121(17): p. e2314450121.

[67] Dyer O. Donanemab: FDA experts recommend approval of Alzheimer's drug. BMJ. 2024 Jun 11; 385: q1295.

[68] Dyer O. Donanemab: FDA experts recommend approval of Alzheimer's drug. British Medical Journal Publishing Group 2024.

[69] Rashad A, Rasool A, Shaheryar M, Sarfraz A, Sarfraz Z, Robles-Velasco K. Donanemab for Alzheimer's disease: a systematic review of clinical trials. Healthcare (Basel). 2023 Jan; 11(1): 32.

[70] Song T, Wang Y, Silverglate BD, Grossberg GT. Pharmacokinetic evaluation of donanemab for the treatment of Alzheimer's. Expert Opin Drug Metab Toxicol. 2024 Jun 1; 20(6): 411–417. [http://dx.doi.org/10.1080/17425255.2024.2357637]

[71] Hampel H, Hardy J, Blennow K, Chen C, Perry G, Bezprozvanny I. The amyloid-β pathway in Alzheimer's disease. Mol Psychiatry. 2021 Oct; 26(10): 5481–5503. [http://dx.doi.org/10.1038/s41380-021-01249-0]

[72] Narasingapa RB, Jargaval MR, Pullabhatla S, Htoo HH, Rao JKS, Hernandez JF. Activation of α-secretase by curcumin-aminoacid conjugates. Biochem Biophys Res Commun. 2012 Aug 10; 424(4): 691–696.
[http://dx.doi.org/10.1016/j.bbrc.2012.07.010]

[73] Shi C, Wu F, Xu J, Zou J. Bilobalide regulates soluble amyloid precursor protein release via phosphatidyl inositol 3 kinase-dependent pathway. Neurochem Int. 2011 Aug; 59(1): 59–64. [http://dx.doi.org/10.1016/j.neuint.2011.03.028]

[74] Liu K, Shen L, Wang J, *et al.* The preventative role of curcumin on the lung inflammatory response induced by cardiopulmonary bypass in rats. J Surg Res. 2012 May 1; 174(1): 73–82 [http://dx.doi.org/10.1016/j.jss.2010.12.002]

[75] Zhang X, Wang J, Xing Y, *et al.,* Effects of ginsenoside Rgl or 17β-estradiol on a cognitively impaired, ovariectomized rat model of Alzheimer's disease. 2012. 220: p. 191-200.

[76] Wang Q, Xiao B, Cui S, *et al.* Triptolide treatment reduces Alzheimer's disease (AD)-like pathology through inhibition of BACE1 in a transgenic mouse model of AD. Dis Model Mech. 2014 Dec; 7(12): 1385–1395.

[77] Tabner, B.J., J. Mayes, and D.J.I.J.o.A.s.D. Allsop, Hypothesis: Soluble Aβ Oligomers in Association with Redox-Active Metal Ions Are the Optimal Generators of Reactive Oxygen Species in Alzheimer′ s Disease. 2011; 2011(1): 546380.

[78] Cole GM, Teter B, Frautschy SA. Neuroprotective effects of curcumin. In: Aggarwal BB, Surh YJ, Shishodia S, eds. The Molecular Targets and Therapeutic Uses of Curcumin in Health and Disease. Adv Exp Med Biol. Boston (MA): Springer; 2007. p. 197–212.
[http://dx.doi.org/10.1007/978-0-387-46401-5_8]

[79] Oehlrich D, Prokopcova H, Gijsen HJM. The evolution of amidine-based brain penetrant BACE1 inhibitors. Bioorg Med Chem Lett. 2014 May 1; 24(9): 2033–2045.
[http://dx.doi.org/10.1016/j.bmcl.2014.03.025]

[80] Selkoe, D.J., D.J.A.r.o.p. Schenk, and toxicology, Alzheimer's disease: molecular understanding predicts amyloid-based therapeutics. 2003. 43(1): 545-584.

[81] De Strooper B. Aph-1, Pen-2, and Nicastrin with Presenilin generate an active γ-secretase complex. Neuron. 2003 Apr 24; 38(1): 9–12.
[http://dx.doi.org/10.1016/S0896-6273(03)00205-8]

[82] Li YM, Xu M, Lai MT, *et al.* Photoactivated γ-secretase inhibitors directed to the active site covalently label presenilin 1. Nature. 2000 Jun 8; 405(6787): 689–694.
[http://dx.doi.org/10.1038/35015085]

[83] Probst G, Xu YZ, Gunawardane J, *et al.* Discovery of (R)-4-Cyclopropyl-7, 8-difluoro-5-(4-(trifluoromethyl) phenylsulfonyl)-4, 5-dihydro-1 H-pyrazolo [4, 3-c] quinoline (ELND006) and

(R)-4-Cyclopropyl-8-fluoro-5-(6-(trifluoromethyl) pyridin-3-ylsulfonyl)-4, 5-dihydro-2 H-pyrazolo [4, 3-c] quinoline (ELND007): Metabolically Stable γ-Secretase Inhibitors that Selectively Inhibit the Production of Amyloid-β over Notch. J Med Chem. 2013; 56(13): 5261–5274.

[84] Wegmann S, Biernat J, Mandelkow E. A current view on Tau protein phosphorylation in Alzheimer's disease. Curr Opin Neurobiol. 2021 Aug; 69: 131–138. [http://dx.doi.org/10.1016/j.conb.2021.03.003]

[85] Reddy PH. Amyloid beta-induced glycogen synthase kinase 3β phosphorylated VDAC1 in Alzheimer's disease: implications for synaptic dysfunction and neuronal damage. Biochim Biophys Acta Mol Basis Dis. 2013 Dec; 1832(12): 1913–1921.

[86] Kremer A, Louis JV, Jaworski T, Van Leuven F. GSK3 and Alzheimer's disease: Facts and fiction… Front Mol Neurosci. 2011 Aug 26; 4: 17. [http://dx.doi.org/10.3389/fnmol.2011.00017]

[87] Meijer L, Thunnissen AM, White AW, et al. Inhibition of cyclin-dependent kinases, GSK-3β and CK1 by hymenialdisine, a marine sponge constituent. Chem Biol. 2000 Jan; 7(1): 51–63.

[88] Ward M, Evans D, Perkinton MS, Richardson JC. Screening methods using fyn in combination with tau protein. Google Patents. US9442119B2; 2016 Sep 13.

[89] RT Abraham, M. Acquarone, A. Andersen, *et al.* Cellular effects of olomoucine, an inhibitor of cyclin-dependent kinases. Biol Cell. 1995; 83(2–3): 105–120. [http://dx.doi.org/10.1016/0248-4900(96)81298-6]

[90] Cicenas J, Valius M. The CDK inhibitors in cancer research and therapy. J Cancer Res Clin Oncol. 2011 Oct; 137(10): 1409–1418. [http://dx.doi.org/10.1007/s00432-011-1039-4]

[91] Ginhoux F, Lim S, Hoeffel G, Low D, Huber T. Origin and differentiation of microglia. Front Cell Neurosci. 2013 Apr 17; 7: 45. [http://dx.doi.org/10.3389/fncel.2013.00045]

[92] Wolf SA, Boddeke HWGM, Kettenmann H. Microglia in physiology and disease. Annu Rev Physiol. 2017 Feb 10; 79: 619–643. [http://dx.doi.org/10.1146/annurev-physiol-022516-034406]

[93] Heneka MT, Kummer MP, Stutz A, *et al.* NLRP3 is activated in Alzheimer's disease and contributes to pathology in APP/PS1 mice. Nature. 2013 Jan 31; 493(7434): 674–678.

[94] Kriegstein A, Alvarez-Buylla A. The glial nature of embryonic and adult neural stem cells. Annu Rev Neurosci. 2009; 32: 149–184. [http://dx.doi.org/10.1146/annurev.neuro.051508.135600]

[95] Sofroniew MV. Molecular dissection of reactive astrogliosis and glial scar formation. Trends Neurosci. 2009 Dec; 32(12): 638–647. [http://dx.doi.org/10.1016/j.tins.2009.08.002]

[96] Farhy-Tselnicker I, Allen NJ. Astrocytes, neurons, synapses: a tripartite view on cortical circuit development. Neural Dev. 2018 May 1; 13(1): 7. [http://dx.doi.org/10.1186/s13064-018-0104-y]

[97] Kato S, Gondo T, Hoshii Y, Takahashi M, Yamada M, Ishihara T. Confocal observation of senile plaques in Alzheimer's disease: Senile plaque morphology and relationship between senile plaques and astrocytes. Pathol Int. 1998 May; 48(5): 332–340. [http://dx.doi.org/10.1111/j.1440-1827.1998.tb03915.x]

[98] Wegiel J, Wang KC, Imaki H, Rubenstein R, Wronska A, Osuchowski M. The role of microglial cells and astrocytes in fibrillar plaque evolution in transgenic APPSW mice. Neurobiol Aging. 2001 Jan; 22(1): 49–61. [http://dx.doi.org/10.1016/S0197-4580(00)00181-0]

[99] Kraft AW, Hu X, Yoon H, et al. Attenuating astrocyte activation accelerates plaque pathogenesis in APP/PS1 mice. FASEB J. 2013 Jan; 27(1): 187–198.

[100] Jin, J.-J., *et al.,* Toll-like receptor 4-dependent upregulation of cytokines in a transgenic mouse model of Alzheimer's disease. 2008. 5: p. 1-10.

[101] Xiao Q, Yan P, Ma X, *et al.* Enhancing astrocytic lysosome biogenesis facilitates Aβ clearance and attenuates amyloid plaque pathogenesis. J Neurosci. 2014 Jul 16; 34(29): 9607–9620. [http://dx.doi.org/10.1523/JNEUROSCI.3788-13.2014]

[102] Gomez-Arboledas A, Davila JC, Sanchez-Mejias E, *et al.* Phagocytic clearance of presynaptic dystrophies by reactive astrocytes in Alzheimer's disease. Glia 2018 Mar; 66(3): 637–653. [http://dx.doi.org/10.1002/glia.23270]

[103] Qi X, Nizamutdinov D, Yi SS, Wu E, Huang JH. Disease Modifying Monoclonal Antibodies and Symptomatic Pharmacological Treatment for Alzheimer's Disease. Biomedicines 2024; 12(11): 2636. [http://dx.doi.org/10.3390/biomedicines12112636] [PMID: 39595200]

[104] Verma A, Rosi SP. A Systemic Review on Aducanumab and Lecanamab in the Development of ARIA in Alzheimer's disease. J Clin Exp Pharmacol 2023; 13: 393.

[105] McDade E, Cummings JL, Dhadda S, *et al.* Lecanemab in patients with early Alzheimer's disease: detailed results on biomarker, cognitive, and clinical effects from the randomized and open-label extension of the phase 2 proof-of-concept study. Alzheimers Res Ther 2022; 14(1): 191. [http://dx.doi.org/10.1186/s13195-022-01124-2] [PMID: 36544184]

[106] Thakkar N, Martis PB, Kutikuppala LVS, Kuchana SK, Mohapatra RK. Lecanemab: A hope in the management of Alzheimer's disease. Brain Circ 2023; 9(3): 194-5. [http://dx.doi.org/10.4103/bc.bc_10_23] [PMID: 38020956]

[107] Alawode DOT, Heslegrave AJ, Fox NC, Zetterberg H. Donanemab removes Alzheimer's plaques: what is special about its target? Lancet Healthy Longev 2021; 2(7): e395-6. [http://dx.doi.org/10.1016/S2666-7568(21)00144-6] [PMID: 36097986]

[108] Bateman RJ, Cummings J, Schobel S, *et al.* Gantenerumab: an anti-amyloid monoclonal antibody with potential disease-modifying effects in early Alzheimer's disease. Alzheimers Res Ther 2022; 14(1): 178.
[http://dx.doi.org/10.1186/s13195-022-01110-8] [PMID: 36447240]

[109] Moussa-Pacha NM, Abdin SM, Omar HA, Alniss H, Al-Tel TH. BACE1 inhibitors: Current status and future directions in treating Alzheimer's disease. Med Res Rev 2020; 40(1): 339-84. [http://dx.doi.org/10.1002/med.21622] [PMID: 31347728]

[110] Ramesh M, Govindaraju T. Multipronged diagnostic and therapeutic strategies for Alzheimer's disease. Chem Sci (Camb) 2022; 13(46): 13657-89. [http://dx.doi.org/10.1039/D2SC03932J] [PMID: 36544728]

[111] Azargoonjahromi A. Immunotherapy in Alzheimer's disease: focusing on the efficacy of gantenerumab on amyloid-β clearance and cognitive decline. J Pharm Pharmacol 2024; 76(9): 1115-31.
[http://dx.doi.org/10.1093/jpp/rgae066] [PMID: 38767981]

[112] Yadollahikhales G, Rojas JC. Anti-amyloid immunotherapies for Alzheimer's disease: a 2023 clinical update. Neurotherapeutics 2023; 20(4): 914-31. [http://dx.doi.org/10.1007/s13311-023-01405-0] [PMID: 37490245]

[113] Holdridge KC, Yaari R, Hoban DB, Andersen S, Sims JR. Targeting amyloid β in Alzheimer's disease: Meta-analysis of low-dose solanezumab in Alzheimer's disease with mild dementia studies. Alzheimers Dement 2023; 19(10): 4619-28. [http://dx.doi.org/10.1002/alz.13031] [PMID: 36946603]

[114] Imbimbo B, Balducci C, Ippati S, Watling M. Initial failures of anti-tau antibodies in Alzheimer's

disease are reminiscent of the amyloid-β story. Neural Regen Res 2023; 18(1): 117-8. [http://dx.doi.org/10.4103/1673-5374.340409] [PMID: 35799522]

[115] Song C, Shi J, Zhang P, *et al.* Immunotherapy for Alzheimer's disease: targeting β-amyloid and beyond. Transl Neurodegener 2022; 11(1): 18. [http://dx.doi.org/10.1186/s40035-022-00292-3] [PMID: 35300725]

[116] Chandrakumaran A, Achuthaprasad A, Daniel P, Jacob J. Tau immunotherapy for Alzheimer's (Review Article). Exp Clin Med Georgia. 2022; (2): 82–84. [https://doi.org/10.52340/jecm.2022.02.15]

[117] Panwar A, Khan MI, Kumar R, Kumar R, Rai SK, Kumar A. Emerging Novel therapeutic approaches for the treatment of Alzheimer's disease. Adv Alzheimer Dis 2024; 13(3): 65-94. [http://dx.doi.org/10.4236/aad.2024.133006]

[118] Neațu M, Covaliu A, Ioniță I, Jugurt A, Davidescu EI, Popescu BO. Monoclonal antibody therapy in Alzheimer's disease. Pharmaceutics 2023; 16(1): 60. [http://dx.doi.org/10.3390/pharmaceutics16010060] [PMID: 38258071]

[119] Ji C, Sigurdsson EM. Current status of clinical trials on tau immunotherapies. Drugs 2021; 81(10): 1135-52. [http://dx.doi.org/10.1007/s40265-021-01546-6] [PMID: 34101156]

[120] Teng E, Manser PT, Pickthorn K, *et al.* Safety and efficacy of semorinemab in individuals with prodromal to mild Alzheimer disease: a randomized clinical trial. JAMA Neurol 2022; 79(8): 758-67. [http://dx.doi.org/10.1001/jamaneurol.2022.1375] [PMID: 35696185]

[121] Albert M, Mairet-Coello G, Danis C, *et al.* Prevention of tau seeding and propagation by immunotherapy with a central tau epitope antibody. Brain 2019; 142(6): 1736-50. [http://dx.doi.org/10.1093/brain/awz100] [PMID: 31038156]

[122] Peng Y, Jin H, Xue Y, *et al.* Current and future therapeutic strategies for Alzheimer's disease: an overview of drug development bottlenecks. Front Aging Neurosci 2023; 15: 1206572. [http://dx.doi.org/10.3389/fnagi.2023.1206572] [PMID: 37600514]

[123] Zhang H, Wei W, Zhao M, *et al.* Interaction between Aβ and tau in the pathogenesis of Alzheimer's disease. Int J Biol Sci 2021; 17(9): 2181-92. [http://dx.doi.org/10.7150/ijbs.57078] [PMID: 34239348]

[124] Jiang S, Cai G, Yang Z, *et al.* Biomimetic nanovesicles as a dual gene delivery system for the synergistic gene therapy of Alzheimer's Disease. ACS Nano 2024; 18(18): 11753-68. [http://dx.doi.org/10.1021/acsnano.3c13150] [PMID: 38649866]

[125] Greig NH, Unnisa A, Kamal M. Nanotechnology-based gene therapy as a credible tool in the treatment of Alzheimer's disease. Neural Regen Res 2023; 18(10): 2127-33. [http://dx.doi.org/10.4103/1673-5374.369096] [PMID: 37056119]

[126] Ataei B, Hokmabadi M, Asadi S, *et al.* A review of the advances, insights, and prospects of gene therapy for Alzheimer's disease: A novel target for therapeutic medicine. Gene 2024; 912: 148368. [http://dx.doi.org/10.1016/j.gene.2024.148368] [PMID: 38485038]

[127] Ng RCL, Jian M, Ma OKF, *et al.* Liver-specific adiponectin gene therapy suppresses microglial NLRP3-inflammasome activation for treating Alzheimer's disease. J Neuroinflammation 2024; 21(1): 77. [http://dx.doi.org/10.1186/s12974-024-03066-y] [PMID: 38539253]

[128] Akhtar A, Gupta SM, Dwivedi S, Kumar D, Shaikh MF, Negi A. Preclinical Models for Alzheimer's Disease: Past, Present, and Future Approaches. ACS Omega 2022; 7(51): 47504-17. [http://dx.doi.org/10.1021/acsomega.2c05609] [PMID: 36591205]

[129] Geerts H, Spiros A, Roberts P. Impact of amyloid-beta changes on cognitive outcomes in Alzheimer's disease: analysis of clinical trials using a quantitative systems pharmacology model. Alzheimers Res Ther 2018; 10(1): 14.

[http://dx.doi.org/10.1186/s13195-018-0343-5] [PMID: 29394903]

[130] Lowe SL, Duggan Evans C, Shcherbinin S, *et al.* Donanemab (LY3002813) Phase 1b Study in Alzheimer's Disease: Rapid and Sustained Reduction of Brain Amyloid Measured by Florbetapir F18 Imaging. J Prev Alzheimers Dis 2021; 8(4): 414-24. [http://dx.doi.org/10.14283/jpad.2021.56] [PMID: 34585215]

[131] Sperling RA, Donohue MC, Raman R, *et al.* Trial of Solanezumab in Preclinical Alzheimer's Disease. N Engl J Med 2023; 389(12): 1096-107. [http://dx.doi.org/10.1056/NEJMoa2305032] [PMID: 37458272]

[132] Henley DB, May PC, Dean RA, Siemers ER. Development of semagacestat (LY450139), a functional γ-secretase inhibitor, for the treatment of Alzheimer's disease. Expert Opin Pharmacother 2009; 10(10): 1657-64. [http://dx.doi.org/10.1517/14656560903044982] [PMID: 19527190]

[133] Bloomingdale P, Bumbaca-Yadav D, Sugam J, *et al.* PBPK-PD modeling for the preclinical development and clinical translation of tau antibodies for Alzheimer's disease. Front Pharmacol 2022; 13867457 [http://dx.doi.org/10.3389/fphar.2022.867457] [PMID: 36120380]

[134] Basheer N, Smolek T, Hassan I, *et al.* Does modulation of tau hyperphosphorylation represent a reasonable therapeutic strategy for Alzheimer's disease? From preclinical studies to the clinical trials. Mol Psychiatry 2023; 28(6): 2197-214. [http://dx.doi.org/10.1038/s41380-023-02113-z] [PMID: 37264120]

[135] Knupp A, Mishra S, Martinez R, *et al.* Depletion of the AD Risk Gene SORL1 Selectively Impairs Neuronal Endosomal Traffic Independent of Amyloidogenic APP Processing. Cell Rep 2020; 31(9)107719. [http://dx.doi.org/10.1016/j.celrep.2020.107719] [PMID: 32492427]

[136] Vemula P, Schoch KM, Miller TM. Evaluating the efficacy of purchased antisense oligonucleotides to reduce mouse and human tau *in vivo.* Front Mol Neurosci 2023; 16: 1320182. [http://dx.doi.org/10.3389/fnmol.2023.1320182] [PMID: 38192302]

[137] Passeri E, Elkhoury K, Morsink M, *et al.* Alzheimer's Disease: Treatment Strategies and Their Limitations. Int J Mol Sci 2022; 23(22): 13954. [http://dx.doi.org/10.3390/ijms232213954] [PMID: 36430432]

[138] Kim CK, Lee YR, Ong L, Gold M, Kalali A, Sarkar J. Alzheimer's Disease: Key Insights from Two Decades of Clinical Trial Failures. J Alzheimers Dis 2022; 87(1): 83-100. [http://dx.doi.org/10.3233/JAD-215699] [PMID: 35342092]

[139] Cummings J. Lessons Learned from Alzheimer Disease: Clinical Trials with Negative Outcomes. Clin Transl Sci 2018; 11(2): 147-52. [http://dx.doi.org/10.1111/cts.12491] [PMID: 28767185]

[140] Yiannopoulou KG, Anastasiou AI, Zachariou V, Pelidou SH. Reasons for Failed Trials of Disease-Modifying Treatments for Alzheimer Disease and Their Contribution in Recent Research. Biomedicines 2019; 7(4): 97. [http://dx.doi.org/10.3390/biomedicines7040097] [PMID: 31835422]

[141] Wan HI, Jacobsen JS, Rutkowski JL, Feuerstein GZ. Translational medicine lessons from flurizan's failure in Alzheimer's disease (AD) trial: Implication for future drug discovery and development for AD. Clin Transl Sci 2009; 2(3): 242-7. [http://dx.doi.org/10.1111/j.1752-8062.2009.00121.x] [PMID: 20443898]

[142] Pardridge WM. Treatment of Alzheimer's Disease and Blood–Brain Barrier Drug Delivery. Pharmaceuticals (Basel) 2020; 13(11): 394. [http://dx.doi.org/10.3390/ph13110394] [PMID: 33207605]

[143] Banks WA. Drug delivery to the brain in Alzheimer's disease: Consideration of the blood–brain

barrier. Adv Drug Deliv Rev 2012; 64(7): 629-39.
[http://dx.doi.org/10.1016/j.addr.2011.12.005] [PMID: 22202501]

[144] Frisoni GB, Boccardi M, Barkhof F, *et al.* Strategic roadmap for an early diagnosis of Alzheimer's disease based on biomarkers. Lancet Neurol 2017; 16(8): 661-76.
[http://dx.doi.org/10.1016/S1474-4422(17)30159-X] [PMID: 28721928]

[145] Tort-Merino A, Valech N, Peñaloza C, *et al.* Early Detection of Learning Difficulties when Confronted with Novel Information in Preclinical Alzheimer's Disease Stage 1. J Alzheimers Dis 2017; 58(3): 855-70.
[http://dx.doi.org/10.3233/JAD-161173] [PMID: 28505969]

[146] Fiandaca MS, Mapstone ME, Cheema AK, Federoff HJ. The critical need for defining preclinical biomarkers in Alzheimer's disease. Alzheimers Dement 2014; 10(3S) (Suppl.): S196-212.
[http://dx.doi.org/10.1016/j.jalz.2014.04.015] [PMID: 24924671]

[147] Mattsson N, Zetterberg H, Hansson O, *et al.* CSF biomarkers and incipient Alzheimer disease in patients with mild cognitive impairment. JAMA 2009; 302(4): 385-93.
[http://dx.doi.org/10.1001/jama.2009.1064] [PMID: 19622817]

[148] Schmitt B, Bernhardt T, Moeller HJ, Heuser I, Frölich L. Combination therapy in Alzheimer's disease: a review of current evidence. CNS Drugs. 2004;18(13):827-44.

[149] Bateman RJ, Klunk WE. Measuring target effect of proposed disease-modifying therapies in Alzheimer's disease. Neurotherapeutics. 2008 Jul; 5(3): 381–390.
[http://dx.doi.org/10.1016/j.nurt.2008.05.009]

[150] Vatansever S, Schlessinger A, Wacker D, *et al.* Artificial intelligence and machine learning□aided drug discovery in central nervous system diseases: State□of□the□arts and future directions 2021; 41(3): 1427-1473.
[http://dx.doi.org/10.1002/med.21764]

[151] Riscado M, Baptista B, Sousa FJP. New RNA-based breakthroughs in Alzheimer's disease diagnosis and therapeutics Pharmaceutics 2021; 13(9): 1397.
[http://dx.doi.org/10.3390/pharmaceutics13091397]

CHAPTER 12

Intranasal Drug Delivery in Treating Alzheimer's Disease

Nitin Kumar[1,*], **Harendar Kumar Nivatya**[2], **Anjali Singh**[3], **Sonam**[2], **Lovy Sharma**[4] and **Vishal Singh**[5]

[1] *Department of Pharmacognosy, College of Pharmacy, Shri Venkateshwara University, Gajraula, Uttar Pradesh 244236, India*

[2] *Faculty of Pharmacy, IFTM University, Moradabad, Uttar Pradesh 244102, India*

[3] *Innovative College of Pharmacy, Greater Noida, Uttar Pradesh 201310, India*

[4] *Department of Pharmaceutical Chemistry, Parmarth College of Pharmacy, Hapur, Uttar Pradesh 245101 India*

[5] *Department of Pharmaceutics, NKBR College of Pharmacy, Meerut, Uttar Pradesh 245206, India*

Abstract: Alzheimer's disease is a monogenic brain illness that impairs thinking, remembrance, judgment, and focus, making it difficult for an individual to carry out regular responsibilities. Drug delivery to the brain is necessary for the effective treatment of diseases of the CNS, including epilepsy, schizophrenia, meningitis, migraine, Parkinson's disease, and Alzheimer's disease. However, because the Blood-Brain Barrier (BBB) separates the Central Nervous System (CNS) from the rest of the body, this remains a challenging area to manage. The blood-brain barrier and additional barriers to oral and other routes, such as decreased bioavailability, quick metabolism, rapid excretion, and drug breakdown by enzymes and stomach juices, make it difficult to transfer drugs to the brain. Promising drug delivery strategies for the efficient nasal delivery of anti-Alzheimer medications with enhanced permeability and bioavailability include lipid particle systems, emulsion-based systems, vesicular drug delivery systems, and other nanocarriers. Effective and targeted drug administration via the nasal route is influenced by charge, size, neurotherapeutic type, and formulation excipients. One of the cutting-edge approaches to brain targeting being explored to overcome the limitations of oral and other modes of administration is the nose-to-brain medication delivery device. A promising substitute for administering medications and enhancing the therapy of Alzheimer's disease is the intranasal route, through the olfactory and trigeminal pathways in the nasal cavity. Intranasal delivery provides a direct channel to the brain. Nasal physiology, however, can restrict bioavailability and impede drug absorption. Despite being limited by the particular circumstances of the

[*] **Corresponding author Nitin Kumar:** Department of Pharmacognosy, College of Pharmacy, Shri Venkateshwara University, Gajraula, Uttar Pradesh 244236, India; E-mail: nitinpharma.np@gmail.com

Prashant Kumar Dhakad, Raghav Mishra, Shweta Sharma, Sudhanshu Mishra & Nishant Gaur (Eds.)

nasal canal, intranasal administration has become a viable option. Using nanotechnology-based nano-carriers for intranasal delivery is a novel strategy. By offering increased bioavailability, greater permeability, efficient blood-brain barrier traversal, prolonged retention inside the body, and precise brain targeting, this approach may be able to get past the present restrictions. Therefore, it is necessary to use technical solutions to maximize the physicochemical features of formulations.

Keywords: Alzheimer's disease, Blood-brain barrier, Intranasal administration, Lipid particulate systems, Novel approaches.

INTRODUCTION

People over 65 are typically affected by Alzheimer's Disease (AD). As the number of senior people rises, the disease's prevalence has gone up [1]. Although this has prompted more research into potential cures, the illness still has no proven cure [2, 3]. Studies have tested over 200 therapeutic agents, and since 2003, the FDA has not approved a new medication for the treatment of the condition [4, 5]. The failure of the applied treatments has been the subject of numerous conjectures. The most significant is misinterpreting the etiology of AD and selecting the incorrect primary treatment objectives [6]. According to research, the brain's distinctive tau nodes and amyloid plaques are a result rather than a cause. Alzheimer's disease is the outcome of improper brain trauma repair and neuroinflammation [7]. Cognitive and behavioral impairments are the outcome of this neurological condition. Acetylcholinesterase inhibitor medications, for example, are poorly soluble, have a low bioavailability, and cannot cross the blood-brain barrier, making them ineffective in many cases [8]. The intranasal approach is highlighted by these drawbacks of the existing treatment. This approach appears to be a viable way to transport medications to the brain [9]. Instead of using parenteral or oral methods, recent research looks at the direct intranasal delivery of pharmacological groups to the CNS.

THE USE OF INTRANASAL DRUG ADMINISTRATION IN ALZHEIMER'S DISEASE TREATMENT

Substitute for parenteral and oral methods. Given that the trigeminal and olfactory nerve cells are in close proximity to the peripheral and central nervous systems, intranasal administration is a non-invasive technique that permits medications to pass through the BBB (Blood-Brain Barrier). Drugs can enter the brain directly from the respiratory epithelium and olfactory area [10, 11]. Thus, the focus is on treating neurological problems [12]. Rapid onset of action, avoidance of the presystemic metabolism of the liver and intestines, decreased systemic exposure, direct administration to the brain and cerebrospinal fluid (CSF), simplicity of use, and improved patient compliance are some benefits when compared to alternative drug administration routes. Limitations of this route include weak nasal permeability and mucociliary clearance [13]. Studies demonstrate that drugs administered into the nasal cavity must have a longer residence time to overcome nasal mucociliary clearance [14].

Drugs are transported either extracellularly or intracellularly across the nasal barriers. Endocytosis is the initial stage of intranasal transport to trigeminal ganglion cells and olfactory sensory neurons, in that order [15]. Agents may not reach the central nervous system for up to 24 hours following nasal delivery due to its extreme slowness [16]. After that, intracellular transport takes place to the brainstem and olfactory bulb. Both passive diffusion and receptor-mediated endocytosis can cause transcytosis between the basolateral membrane and the intracellular space. Diffusion *via* the extracellular channel linked to olfaction into the olfactory bulb takes approximately 0.73 to 2.3 hours. The duration of diffusion into the brainstem is roughly 17–56 hours; substances obtained through this channel are transferred to the blood, olfactory mucosa, and ultimately the central nervous system. This method depends on the drug's size and molecular weight and is less efficient than the transcellular route [17]. On the other hand, this technique is quicker and enables the delivery of low molecular weight medications to the central nervous system in minutes [18].

Mucociliary clearance allows nasal medications to be removed from the nasal cavity. The medication that enters the bloodstream can be eliminated via regular clearance processes or could allow the BBB to enter the brain. Ensuring appropriate therapeutic amounts of medication distribution to target brain regions is the difficult part of the application. Targeting the right receptors in the brain is essential for treating CNS conditions like Parkinson's disease, schizophrenia, AD, brain tumors, meningitis, and migraines. Frey introduced this technique in 1989 to treat AD and other CNS illnesses, and it shows promise in providing current medications and compounds for alternative therapies in AD.

Mucociliary clearance enables the removal of nasal medications from the nasal cavity. It is possible to remove the medication that enters the bloodstream using either regular clearance processes or by crossing the blood-brain barrier. Making sure that the target parts of the brain receive therapeutically appropriate quantities of medication delivery is the difficult part of the application. For the treatment of CNS disorders such as Parkinson's disease, schizophrenia, AD, brain tumors, meningitis, and migraine, the drug's location in the brain must target the essential receptors. Frey introduced this technique in 1989 to treat AD and other CNS illnesses, and it shows promise in providing current medications and compounds for alternative therapies in AD [19].

Intranasally Administered Agents in Alzheimer's Treatment

Insulin

There has been research on the use of intranasal insulin in AD. Research has demonstrated that insulin delivered intranasally is quickly disseminated. In the AD mouse model, it reverses learning and memory loss at the cribriform plate level across the brain. Insulin administered intranasally has little effect on peripheral metabolism and penetrates the bloodstream inadequately. Uptake from the cribriform plate into the brain results in the suppression of Protein Kinase-C (PKC). Insulin is thus transported across the blood-brain barrier by cellular pathway inhibitors. These findings imply that insulin can be delivered to the brain efficiently using the intranasal route [20]. The brain's sensitivity to insulin receptors is diminished in AD patients [21, 22]. Insulin has crucial roles in the central nervous system. Receptors for insulin are located in the brain's cerebral cortex, hippocampus, olfactory bulb, hypothalamus, and cerebellum [23]. Insulin signaling plays a role in both synaptic remodeling and synaptogenesis [24]. By altering glucose consumption, acute insulin absorption into the hippocampus enhances spatial memory in a way that is dependent on phosphatidylinositol 3-kinase (PI3K) [25].

Research has shown that insulin, when administered intranasally or perfused in healthy individuals, enhances cognitive performance [26, 27]. In preclinical studies, intranasal insulin administration prior to anesthesia prevented Alzheimer's disease (AD)-like tau hyperphosphorylation in a transgenic mouse model of AD [28]. Tau hyperphosphorylation, exacerbated by increased anesthetic exposure, is a hallmark pathogenic feature of AD [29]. Additionally, older rats exposed to such conditions exhibit significant learning and memory deficits [30, 31]. In cognitively healthy individuals, insulin signaling in the brain enhances memory and demonstrates neuroprotective properties. These findings suggest that increasing brain insulin levels in AD patients could potentially halt or delay the progression of this debilitating condition. A clinical study involving AD patients utilized positron emission tomography (PET) to assess the effects of intranasal insulin. Patients received either a placebo or 20 or 40 IU of insulin intranasally for four months [32]. Outcomes, evaluated using the Alzheimer's Disease Assessment Scale (ADAS), revealed that the insulin-treated group exhibited improved delayed memory and overall cognitive performance compared to the placebo group. Furthermore, a single dose of intranasal insulin significantly enhanced verbal memory in elderly AD patients, as demonstrated by improved test performance in the AD group. These results highlight the potential of intranasal insulin as a therapeutic approach to improve cognitive function and mitigate AD-related impairments. Only women with the apolipoprotein-e4 (APOe4) allele, however, were immune to high insulin dosages among these groups. Another study found that AD patients with APO-e4 were less insulin sensitive. Since 40–65% of those with late-onset AD are APO-e4 carriers, it is hypothesized that intranasal insulin may not be very useful as a treatment [33].

More clinical research is required, despite the encouraging results of the current investigations. Not much is known about the process through which insulin delivered intravenously enters the brain. There are several theories about how insulin is delivered to the brain intranasally, including via the rostral migration flow, the trigeminal nerve pathway, and axon bundles of olfactory receptor cells on the nasal cavity's roof [34]. According to recent research, the bulk flow in the cerebral blood vessels' perivascular region has a role in the quick distribution of chemicals delivered intranasally through the olfactory tract or trigeminal pathway [35]. These studies all provide evidence that insulin delivered intranasally travels through the nasal epithelium. Current research is investigating the effects on cellular mechanisms of brain intake and cognition in mouse models of AD by looking at the regional distribution and mechanism of action of given insulin.

Erythropoietin

Erythropoietin (EPO) is a glycoprotein cytokine that regulates erythropoiesis [36]. Additionally, it performs biological tasks, including neurogenesis and non-hematopoietic neuroprotection [37]. Because of these factors, EPO has started to be utilized as a neurodegenerative treatment. EPO has not produced encouraging outcomes when used in therapeutic settings. This results from EPO's erythropoietic action [38]. Haematocrit and blood viscosity are raised when recombinant human EPO (rh-EPO) is used to treat neurological disorders. Heart conditions, including infarction and stroke, are brought on by this disorder. EPO compounds that lack erythropoietic action but maintain neuroprotective effects against neuronal damage have started to be used as a solution to this issue [39]. A chemically modified non-erythropoietic variant of EPO called neuro-EPO was created. Chinese hamster ovary cells produce the recombinant human glycoprotein known as neuro-EPO, which has a low sialic acid content. The absence of erythropoietic effect and the strongest neuroprotective properties are important features [40]. Because neuro-EPO is quickly broken down in the liver, it is better administered intranasally [41]. Because it is administered intranasally, it enters the brain rapidly and does not promote erythropoiesis following acute therapies. Animal models of stroke have been used to study the neuroprotective properties of Neuro-EPO, such as the enhancement of cognitive function following a stroke. In one study, non-transgenic Alzheimer's mouse models were given Neuro-EPO intranasally at dosages of 62, 125, and 250 μg/kg.

The study's findings showed that giving Neuro-EPO intranasally shielded the rodent from Ab25-35 toxicity. Showed a strong potential defense against hippocampal amyloid poisoning. Aside from a few dosage-dependent modifications, neuro-EPO has a positive impact on memory loss treatment. In most tests, doses like 125 and 250 μg/kg worked well. But the 125 μg/kg dose had a long-lasting effect that lasted for up to 22 days [42]. Synaptic plasticity is directly impacted by erythropoietin, particularly in the hippocampus [43, 44]. EPO has been shown to enhance primary hippocampus neural networks. In other words, EPO increases neuronal plasticity, synaptic connections, and hippocampal-derived memory networks. Erythropoietin protects against diseases caused by inflammation. Neuro-EPO inhibits the release of TNF-α and IL-1β caused by A25-35 poisoning. EPO's neuroprotective properties stem from its capacity to maintain extrinsic cell homeostasis through cytokine release regulation and microglial activation modulation [45].

The affinity of various EPO variants in activating EPO receptors varies. For the low sialic acid version of EPO to activate EPO receptors, higher doses are required. Another significant aspect of Neuro-EPO is that it may be made to be biologically comparable to endogenous EPO without undergoing chemical alteration, which reduces the likelihood of long-term adverse effects. Erythropoietin provides angiogenesis and protection by specifically triggering the production of the neuroglobin protein in injured areas. Maintaining homeostasis helps to restore lost functions in brain regions that have been injured. EPO administered intranasally is presently being tested for stroke therapy [46]. Research on the use of intranasal EPO in neurodegenerative diseases is still in progress.

Exosomes

The brain's macrophages are called microglial cells. In inflammatory illnesses, microglial cells produce interleukin, which contributes to inflammation, such as Alzheimer's. As a result, targeting microglial cells with anti-inflammatory therapies makes logical sense [47]. There is currently no clear plan in place to target brain microglial cells [48]. Certain drugs do not cross the blood-brain barrier, yet they do target inflammatory cells without damaging healthy organs [49 - 51]. Consequently, anti-inflammatory medications have been administered to the brain non-invasively through the intranasal route using exosomes. Curcumin, a naturally occurring polyphenol present in the rhizomes of *Curcuma longa* (turmeric), is one of these anti-inflammatory medications. Curcumin has anti-inflammatory properties like antioxidants, chemopreventives, and antitumor [52 - 54].

Curcumin presents challenges in clinical application due to its limited bioavailability and solubility [55]. Exosomes have the ability to be employed as medication transporters in nanoparticle form. Exosomes are nanoparticles with a size range of 30 to 100 nm that can be released into the extracellular space. After loading curcumin into exosomes at 22°C, the exosomes were centrifuged using a sucrose gradient. As a result, curcumin was encapsulated into exosomes, improving its stability, bioavailability, and solubility *in vitro* [56]. To illustrate the therapeutic application of curcumin with exosomes (Exo-cur), animals in an inflammatory condition generated by lipopolysaccharide (LPS) were employed. Curcumin was given intravenously after being loaded into exosomes.

One hour after Exo-cur, the quantity of curcumin in brain lysates was measured using high-performance liquid chromatography. Curcumin was found in brain lysates for up to twelve hours. This indicates that microglial cells are the precise target of exosomes. There was a subsequent decrease in the number of activated inflammatory microglial cells in the brains of mice administered LPS. Exosomes' rapid and precise transport to the brain makes them a viable option for delivering medications. Research has demonstrated that exosomes target brain microglial cells, but it has not addressed the mechanism underlying this selectivity. Nobody knows. If exosomes enter recipient cells *via* the endosomal pathway, or if T cell-derived exosomes enter microglial cells *via* the endosomal pathway [57]. It's also unknown how deeply exosomes penetrate different parts of the brain and spinal cord and whether or not exosomes target glial cell types like astrocytes. Furthermore, it is unclear if inflammation influences or coincides with the uptake of exosomes by microglial cells in the brain. It is believed that the relevance of intranasal exosome technology to neurological disorders will become evident after these questions are addressed.

Mesenchymal Stem Cells

Alzheimer's disease is caused by a variety of intricate processes rather than a single factor. As a result, strategies to address the disease's intricate pathology are required. Applications of stem cells are anticipated to offer multi-targeted treatments necessary for the disease's resolution [58]. Choosing the right stem cell is the most crucial step before beginning stem cell therapy. Mesenchymal stem cells (MSCs) are easily applied and can be derived from a variety of cell types. It is the most researched kind of stem cell because of these factors [59]. Neuroprotective stem cells with anti-amyloidogenic properties are mesenchymal stem cells [60]. The use of mesenchymal stem cells can be therapeutic for AD. In animal models of AD, the treatment of mesenchymal stem cells showed anti-inflammatory and anti-amyloid qualities, improving memory. As a result, it shows promise in treating a number of neurological conditions [61]. In addition to proliferating endogenous cells and providing neuroprotection through secretory factors, mesenchymal stem cells also have immunomodulatory effects on the cells that cause illness [62]. However, its clinical use is restricted due to implantation concerns. These restrictions include the invasive cell separation procedure, potency loss, short lifespan, and high expense of large-scale expansion [63] and minimal risk of unchecked cell proliferation [64].

Rifampicin

The majority of research to date has looked into ways to eradicate the pathophysiology of neurodegenerative dementia, including cerebral tauopathy and amyloid oligomer buildup, to treat AD [65]. According to clinical research, patients' cognitive levels do not significantly improve as a result of these treatments [66]. As a result of these detrimental effects, rifampicin, an oral antibiotic with minimal side effects and the ability to significantly lower neurotoxic oligomers, was developed as a novel disease prevention strategy [67]. Under specific circumstances, rifampicin has been demonstrated to have preventative effects against AD [68]. Thus, it is necessary to start preventive treatment prior to the onset of memory issues [69]. The presence of AD-type hypometabolism is enough justification to start preventative treatment in patients who do not have memory loss. As a preventative measure, rifampicin has been employed. It was thought that 450 mg per day or more—600 mg per day, for example—was the effective dosage. For older patients, rifampicin medication must be maintained for a minimum of 12 months. If rifampicin is taken consistently for a number of years, it is believed that memory loss can be avoided. Rifampicin given orally lowers tau and amyloid beta (Ab) pathology in mice.

However, prolonged use is not recommended due to its negative effects, which include liver damage. This circumstance has led researchers to look for a safer way to administer rifampicin. Thus, rifampicin's therapeutic efficacy and safety were assessed by giving it to transgenic AD mouse models orally, intranasally, and subcutaneously for a duration of one month. The experiment's findings showed that oral treatment was not as effective at enhancing memory as intranasal or subcutaneous administration. Furthermore, a reduction in neuropathology includes Ab oligomer formation, aberrant phosphorylation of tau and synaptic loss. Pharmacokinetic research indicates that the intranasal route of rifampicin delivery results in the largest amount of rifampicin in the brain. For long-term rifampicin dosing, intranasal administration makes sense because it is simple to use and non-invasive. The possibility for long-term usage of rifampicin is further enhanced by its affordable availability [70].

Numerous approaches to treating AD have not worked. This has led scholars to adopt various methodologies. Unlike parenteral and oral approaches, intranasal treatment delivers directly to the central nervous system. It is a promising non-invasive technique to penetrate the blood-brain barrier since the trigeminal nerves and olfactory nerve cells are directly in contact with the environment and the central nervous system. This makes it possible to treat AD, a neurological condition, by achieving direct medication passage from the olfactory area and respiratory epithelium to the brain. Rapid start of action, avoidance of the liver's and intestines' presystemic metabolism, decreased systemic exposure, direct administration to the brain and CSF, simplicity of use, and improved patient compliance are some benefits when compared to alternative drug administration methods. Mucociliary clearance and inadequate intranasal nasal permeability are the route's drawbacks. Given medications. This approach is used to apply or attempt to apply a large number of agents. Research conducted on animal models has shown that drugs including insulin, erythropoietin, exosomes, MSCs, and rifampicin can be administered intranasally to treat AD. It will be easier to assess the suitability of these medicines for the treatment of AD if the issues around therapeutic dosages and mechanisms of action are clarified.

INTRANASAL ROUTE: NOSE-TO-BRAIN DELIVERY

Current treatments play a significant part in curing CNS diseases and lowering death rates. But for the majority of CNS conditions, there are still unsolved problems and no surefire therapy [71]. The ability of medications to pass through the blood-brain barrier in large enough amounts to reach healing levels is the primary obstacle. An estimated 98% of low molecular weight medications and 100% of macromolecules are thought to be blocked from passing across the barrier, resulting in a markedly reduced CNS bioavailability [72]. For the local delivery of active therapeutics to the brain, a range of techniques are being used, such as catheter infusions, mini-pump-assisted intracranial delivery, intraparenchymal or intracerebroventricular injections, precise ultrasound techniques, or electromagnetic force-field techniques. However, many of these approaches are inappropriate for long-term care and are intrusive, dangerous, and cause neurotoxic consequences at the delivery site. Therefore, the main factor preventing drugs from reaching the brain through systemic circulation is the presence of the BBB [73].

There are a number of strategies being developed to get around the BBB, and nose-to-brain administration is an important one. This method enables quick and direct distribution to the brain by avoiding the blood-brain barrier [74]. The supplied medication enters the brain by evading the blood-brain barrier, which extends its residence time in the active site. Additionally, this method lowers systemic toxicity and limits needless drug exposure. Direct access to the brain is possible through the nasal cavity's olfactory area, which extends to the cranial cavity. The pace of drug transport is influenced by various variables and pharmacokinetic parameters that need to be assessed at the clinical level [75, 76]. Currently, there is a lot of focus on intranasal drug delivery to the brain using delivery devices based on nanotechnology. Over the past few decades, a lot of research has been done on nanosized drug delivery devices as a novel way to get around the low bioavailability of many pharmaceutical medications [77]. The literature has reported successful intranasal medicine delivery to the brain using a variety of methods, including carbon-based nanoformulations, liposomes, microspheres, dendrimers, nanoemulsions, and nanoparticles [78]. For instance, animals given talinolol in nanoemulsions via intranasal administration had greater brain concentrations after 15 minutes compared to those given it intravenously [79]. Numerous commercially available nanoemulsions have been developed for use in biology and medicine [80, 81].

PATHWAY FOR THE INTRANASAL ROUTE OF BRAIN DELIVERY

Delivering the appropriate medication concentrations to the site of action is the main goal of various drug delivery mechanisms [82, 83]. Furthermore, it is possible to reduce both physical clearance and the breakdown of medications through metabolism; Fig. (**1**) provides an overview of drug transport [84, 85]. Rapid drug absorption to the brain is made possible by the extremely permeable nasal epithelium, which has a large surface area, a porous endothelium membrane, a high total blood flow, and the capacity to bypass first-pass metabolism. Numerous therapeutic substances, both small and large, can be administered to the central nervous system (CNS) via the intranasal route. Furthermore, the therapeutic agent does not need to be altered for nasal medication delivery, nor does the drug need to be paired with a carrier. With its strong benefits over alternative drug delivery techniques, nasal drug delivery with multiple channels has long been a major research focus for pharmaceutical and medical device firms [86, 87].

Olfactory Pathway

Drugs can enter olfactory epithelium cells by either passively diffusing via the narrow interstitial space or endocytosing across the cell membrane, or they can be carried by neurons [88]. Most medications applied to the olfactory area are moved between cells extracellularly. Several experimental *in vitro* and *in vivo* findings have demonstrated the clinical significance of P-glycoprotein in intranasal drug administration. Additionally, drug penetration efflux transporter substrata at the nasal barrier have been tested in vitro using RPMI 2650 cells and 3D MucilAir™

nasal models in order to determine their potential nose-to-brain drug delivery pathway [89].

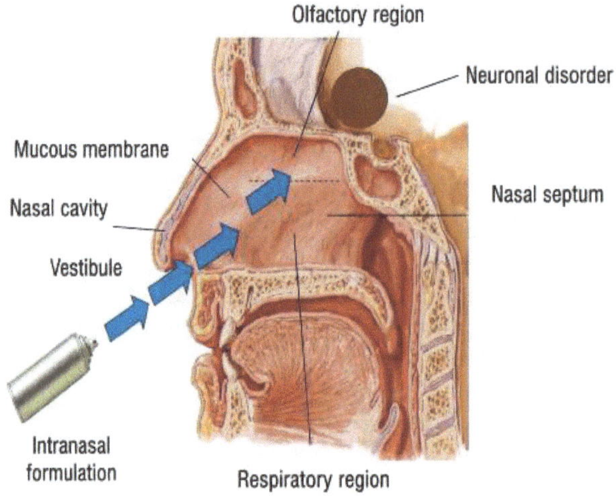

Fig. (1). An example of how drugs are transported from the nose to the brain.

When it comes to intranasal medication delivery to the brain, olfactory neurons play a crucial role [90]. Through the intracellular axonal channel, drugs enter the olfactory bulb and are then disseminated throughout the brain (Fig. **2**) [91]. Since the human olfactory axon has a diameter of roughly 0.1 to 0.7 µm, chemicals or nanoparticles with diameters in this range can be easily supplied by this channel. Axonal transport is slower than drug transport via the epithelial route. Olfactory pathways carry drugs via both intracellular and extracellular routes. Hydrophilic medications are transported via the paracellular pathway, whereas the majority of lipophilic pharmaceuticals are carried by passive diffusion. The manner by which medications are absorbed is significantly influenced by their molecular weight and lipophilicity. For medications with extremely high lipophilicity, the transcellular mechanism is particularly pertinent.

AD is Treated with a Novel Drug Delivery Method

An innovative drug delivery system (NDDS) is a cutting-edge technique for administering therapeutic drugs that outperforms conventional drug delivery methods in terms of efficacy, safety, and targeting. Particularly for difficult situations like infections, cancer, and neurological disorders, NDDS emphasizes enhanced bioavailability, site-specific targeting, and controlled release [92]. The following promising NDDS are presently being researched or developed:

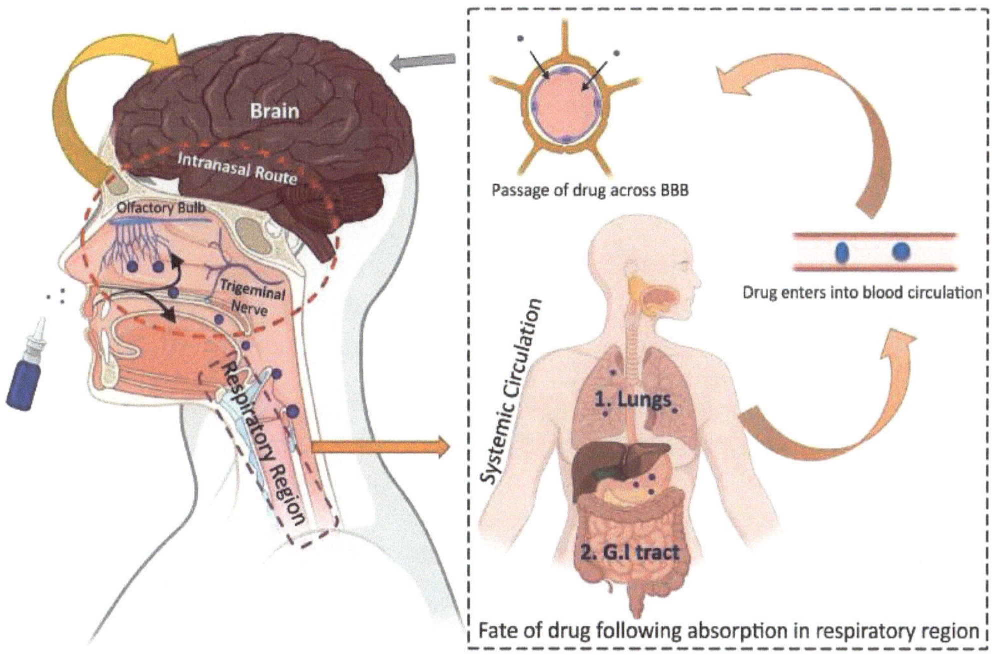

Fig. (2). A graphic depicting the delivery system's course following passage through the olfactory, trigeminal, and respiratory intranasal pathways.

Nanoparticles

Colloidal solid particles, the nanoparticles range in size from 1 to 1000 nm. Only the tiniest nanoparticles in the 1–100 nm size range can get through the mucosal membrane by paracellular transport via tight junctions. This kind of delivery mechanism is perfect for nasal vaccinations. For instance, miR-132 maintains neuron survival and regulates neuron structure. Additionally, it improves the microenvironment of hematoma lesions, reduces cell death following cerebral bleeding, and exhibits protective effects following cerebral ischemia. In a mouse model with AD, the intranasal administration of Wheat Germ Agglutinin (WGA), nanoparticles (NPs), and miR132 results in a substantial upregulation of synaptic protein expression levels [93].

Nanoparticles based on Polymers

Chitosan is produced when the crustacean cell's chitin is deacetylated. It is employed in intranasal distribution due to its bioadhesive qualities, which prolong the drug's nasal residence period by preventing nasal evacuation [94]. For instance, both the particle size and the polydispersity index were tuned in the formulation of the Memantine-HCl cross-linked chitosan nanoparticles. When the concentrations of tripolyphosphates (TPP) and chitosan are 2:1, the stirring time is 4 hours, and the stirring speed is 1000 rpm, the nanoparticle size was determined to be 129 nm. When used intranasally, memantine-loaded chitosan nanoparticles are probably going to be a novel way to treat Alzheimer's disease. Another polymer used to deliver nanoparticles *via* the intranasal route is poly (lactic-co-glycolic acid) (PLGA). The best example is the intranasal delivery of olanzapine-loaded PLGA nanoparticles, which improves medication absorption [95].

Nanoparticles Made of Lipids

This system is a versatile carrier system for brain targeting and offers numerous advantages due to its lipid nature. This system includes Solid Lipid Nanoparticles (SLNs) and Nanostructured Lipid Carriers (NLCs). In albino Wistar rats, for instance, pharmacokinetic and biodistribution studies revealed that donepezil-SLNs given intranasally had an $AUC_0{-}\infty$ in the brain that was 2.26 times better than DPL-Sol given intranasally and nearly 2.61 times higher than donepezil-Sol given intravenously [94].

Liposomes

These are phospholipid vesicles with lipid bilayers enclosing one or more aqueous regions. Depending on their solubility in fat and water, drugs are either coated in a lipid layer or a water layer [96]. Small and big molecules with a wide range of hydrophilicity and pKa values are efficiently encapsulated by this system [97, 98]. For instance, a study using a mouse model of AD found that administering quercetin liposomes through the nose decreases hippocampal degeneration in cortical and cholinergic neurons.

Microemulsions

Intranasal microemulsions are a useful tool for brain targeting. The hybrid system of clonazepam as a microemulsion formulation with a mucoadhesive component offers a longer duration of effect and a quicker onset of action [99].

Nanogels

These suspensions or solutions are extremely viscous. There are several reasons why nanogels are better than other systems.

a. It may cause the postnasal drip to become thinner.
b. It prevents the formulation from leaking anteriorly.
c. It has an emollient and calming effect.

To improve insulin transfer to the brain, the intranasal route is used to administer the insulin-loaded nanogels.

In-situ gelling system

Before being administered, they are fluid, and then they can change from a liquid to a gel by a variety of external factors, such as temperature, pH, ion changes, magnetic fields, or biological environments [100, 101]. Pharmacokinetic studies, for instance, showed that when compared to rivastigmine solution given by the same route, intranasal (IN) and intravenous (IV) NLCs had a significantly greater area under the curve (AUC) and t-half. Additionally, biodistribution showed that IV NLCs (457.24 ± 38.4112 ng/mL) and IN NLCs (736.42 ± 34.54 ng/mL) possess the capacity to cross the blood-brain barrier, and their brain concentrations are 4.6 and 5.3 times greater than those of IV and IN solutions, respectively.

Hydrogels

These are three-dimensional (3D) networks that are cross-linked and compatible with biological systems, and they can absorb large amounts of water [102, 103]. For instance, the rabbit was used in the pharmacokinetics investigation. When compared to donepezil oral pills, the optimized hydrogel was found to improve the area under the curve (AUC) and average peak medication concentration by 46% and 39%, respectively, via the intranasal method.

Nanoemulsions (NEs)

Both nanoscale globules and the nasal mucosa can absorb these lipophilic complexes. NEs come in two varieties: water-in-oil (w/o) emulsions and oil-in-water (o/w) emulsions. The primary purposes of oil-in-water (o/w) emulsions are to encapsulate lipophilic medications, shield them from enzyme breakdown, increase their solubility in liquid medium, control drug release, and boost bioavailability [104, 105]. Rivastigmine-loaded nanoemulsions, for example, in vitro drug release and ex vivo diffusion tests showed significantly greater drug concentrations in the brain than the solution. Studies on nasal ciliotoxicity in goat nasal mucosa demonstrated that the improved formulation did not cause nasal ciliotoxicity. All of it pertains to NDDS for Alzheimer's disease, and Table **1** lists a few delivery methods, their examples, and a description of how they work.

Table 1. Several cutting-edge approaches to AD treatment.

Sr. No.	Delivery system	Experimental Data	Drug Examples
1.	Nanoemulsions	Studies on nasal ciliotoxicity in goat nasal mucosa have shown that the improved formulation did not cause nasal ciliotoxicity [106].	Rivastigmine (Cholinesterase inhibitor)
2.	Nanoemulsions	Studies on drug diffusion in vitro and drug penetration in vivo revealed that donepezil penetrates the body very well via the intranasal route, in contrast with other methods. Polymers are a useful strategy for enhancing medication bio-adhesion and nasal mucosal penetration, which eventually raises donepezil's bioavailability [107].	Donepezil (Cholinesterase inhibitor)
3.	Nanoparticles	Neuron survival and morphological homeostasis are maintained by miR-132. Additionally, it improves the microenvironment of hematoma lesions, reduces cell death following cerebral bleeding, and exhibits protective effects following cerebral ischemia. Agglutinin from wheat germ (WGA)-nanoparticles (NPs)-miR132 markedly elevated synaptic protein expression levels in AD mouse models when given intranasally [108].	miR-132
4.	Cationic nanostructured lipid carriers (NLCs) in situ gelling	Pharmacokinetic tests revealed a considerably greater area under the curve (AUC) and t-half for the sustained release of intranasal (IN) and intravenous (IV) NLCs when compared to rivastigmine solution administered by the same route. The results of biodistribution showed that IV NLCs (457.24 ± 38.4112 ng/mL) and IN NLCs (736.42 ± 34.54 ng/mL) could penetrate the blood-brain barrier, with brain concentrations 4.6 and 5.3 times higher than IV and IN solutions, respectively [109].	Rivastigmine
5.	Solid lipid nanoparticles (SLNs)	Studies on pharmacokinetics and biodistribution in albino Wistar rats revealed that donepezil-SLNs given intranasally had an AUC_0–∞ in the brain that was 2.26 times superior to intranasal DPL-Sol and around 2.61 times superior to donepezil-Sol given intravenously [110].	Donepezil
6.	Lipidic nano-emulsion based on hyaluronic acid that co-encapsulates two polyphenols	The hyaluronic acid-based nano-emulsion had a zeta potential of −23.9 ± 1.7 and a particle size of 115.2 ± 0.15. The mucoadhesive nano-emulsion was deemed safe for the nasal mucosa and enhanced the levels in the brain (about seven and nine times, respectively, in AUC 0–7 h for curcumin and resveratrol) [111].	Polyphenols (Curcumin-resveratrol)

(Table 1) cont.....

Sr. No.	Delivery system	Experimental Data	Drug Examples
7.	*In situ*-gelling system	By extending the medication's nasal retention period and examining the impact of many factors on formulation creation, including temperature, pH, drug content, viscosity, and others, the formulation was developed to increase the nasal bioavailability of the medication [112].	Donepezil
8.	Chitosan nanoparticle	Memantine-HCl cross-linked chitosan nanoparticles were prepared with the polydispersibility index and particle size in mind. At a 2:1 ratio of chitosan to tripolyphosphates (TPP), the nanoparticle size was found to be 129 nm, with 4 hours of stirring time and 1000 rpm of stirring speed. When used intranasally, memantine-loaded chitosan nanoparticles are probably going to be a novel way to treat Alzheimer's disease [113].	Memantine hydrochloride (NMDA receptor antagonist)
9.	Liposomes	Quercetin liposomes administered by the nasal route have been shown to prevent AD. A mouse model study of AD revealed the loss of cholinergic and cortical neurons in the hippocampus [114].	Quercetin
10.	Exosomes	Naturally, extracellular vesicles at the nanoscale can pass through the BBB and can be modified to deliver medications, siRNA, or CRISPR constructs that target tau or amyloid proteins [115].	BACE1-targeting siRNA or quercetin supplied by exosomes.
11.	Intranasal Drug Delivery	A non-invasive method that uses the olfactory and trigeminal pathways to pass the blood-brain barrier. Reduces systemic side effects and delivers drugs directly to the central nervous system [116].	Neuropeptides (such as oxytocin or BDNF analogs) or intranasal insulin.

According to our research and a number of review articles, it can be concluded that delivery methods such as liposomes, exosomes, intranasal delivery, and nanoparticle-based delivery are currently available and may be followed in the future by gene therapy vectors, microneedle patches, 3D-printed drug delivery systems, artificial intelligence-assisted delivery design, stimuli-responsive nanocarriers, and more.

Penetration enhancers to improve drug delivery through the intranasal route

Some different approaches for the improvement of medication bioavailability while creating intranasal formulations have been discussed. Adding excipients to enhance drug penetration and absorption is one method that makes this goal possible. According to the pharmacopeia, an excipient is any material that is purposefully added to the dosage form and is distinct from the active pharmaceutical ingredients. These are the agents that increase the absorption rate by enhancing the permeability of the biological membrane rather than the solubility. Increment is referred to as an absorption enhancer; for this reason, these compounds are occasionally more precisely referred to as permeation enhancers [117 - 119]. The effectiveness of active medicinal constituents, proteins, and peptides with inadequate permeation has been investigated in relation to penetration enhancers. To increase the ability of active pharmaceutical components to pass through biological barriers, the functional excipients are added in the formulation, called penetration enhancers [120].

A perfect penetration enhancer should be pharmacologically inert at the concentration employed, have a precise, appropriate, and predictable impact, and work well with a smaller quantity. Additionally, they should not have an unpleasant taste or smell, should not damage the membrane, and should be compatible with medication and with additional additives. It should also be in touch with the mucosal membrane of the nasal cavity long enough to prolong its optimal effectiveness. Penetration enhancers' modes of action are not fully known. Nonetheless, it is believed that they work by causing one or both of the following outcomes:

Physicochemical effects: By changing the drug's solubility, partition coefficient, or weak interactions, certain penetration enhancers can change a drug's physicochemical characteristics. Low concentrations are where these mechanisms are seen.

Membrane effects: A wide variety of penetration enhancers act on the nasal membrane's surface; while the majority of these enhancers are safe, it is necessary to ascertain their extent and effects. Most of the time, penetration enhancers have a temporary effect with no irreversible effects [121, 122].

Cyclodextrins

Cyclodextrins have garnered attention in recent years from a variety of application domains, with the pharmaceutical industry showing particular interest. Penetration enhancers are present in the majority of intranasal formulations to improve the nasal penetration, aqueous solubility, and absorption of lipophilic drugs [123]. When enzymes break down starch, they create cyclic oligosaccharides called cyclodextrins [124]. Even though they were first identified in 1891, Ono Pharmaceutical Co. did not use them in a pharmaceutical product until 1976 [125, 126]. Six, seven, and eight glucose units combine to generate the most prevalent natural cyclodextrins, α-, β-, and γ-cyclodextrins. Derivatives of cyclodextrins are produced for the substitution of hydroxide groups rather than the desired functional groups. Only a small number of the more than 1,500 distinct cyclodextrin derivatives that have been identified recently are accessible as pharmaceutical excipients due to the high cost of toxicological investigation. Methyl, hydroxypropyl, and sulfobutyl ether derivatives are frequently used in pharmaceutical goods to increase their solubility and inclusion potential over native cyclodextrins [127].

For nasal medication administration, cyclodextrins serve as solubilizers and permeability enhancers. Cyclodextrins increase the apparent solubility of poorly water-soluble compounds by interacting with them. Nowadays, the exact mechanism by which cyclodextrins improve medication penetration is still unclear and depends on certain circumstances. Methylated β-cyclodextrin has been shown to improve transmucosal absorption by increasing the transcellular and paracellular motions of peptidic medicines. Additionally, tetradecyl-β-maltosi-e-cyclodextrin has been demonstrated to function by facilitating the opening of tight cell junctions. The USFDA (United States Food and Drug Administration) lists which excipients are natural and are derivatives of cyclodextrins and their derivatives as approved excipients. The route of administration affects the toxicity of cyclodextrins. They are benign and well-tolerated when taken orally, but they become dangerous when given intravenously because they have the potential to crystallize. Methylated β-cyclodextrin causes less local nasal irritation than benzalkonium chloride [128]. Intranasal penetration of calcitonin, insulin, glucagon, and recombinant granulocyte colony-stimulating factor has been improved by cyclodextrins. Due to its low toxicity and ability to dissolve drugs, KLEPTOSE® HPB (hydroxypropyl-β-cyclodextrin), a commercial cyclodextrin product, is suggested as an appealing excipient for intranasal drug administration [129].

Chitosan

Fungal cell walls contain chitosan, a cationic linear polymer that is produced by deacetylating chitin, which is the second most common polymer in nature after cellulose, insect exoskeletons, and crustacean shells (crabs and shrimp) [130, 131]. Being biocompatible, biodegradable, nontoxic, inexpensive, and typically nonallergenic, chitosan and its derivatives are referred to as versatile materials due to their wide range of applications. In addition, they possess exceptional physical characteristics, such as a wide surface area, porosity, conductivity, and ease of malleability in a variety of forms. Additionally, because of their antibacterial properties, they have been included in antifungal, anticancer, and antiulcer properties.

Chitosan not only improves medication absorption, but it also stabilizes the formulation. There have been reports of improved absorption of calcitonin, insulin, and buserelin through the intestinal and nasal epithelia [132, 133]. Quaternary chitosan derivatives known as N-trimethyl chitosan chlorides (TMC) have improved intestinal permeability, water solubility, and, at a broad pH range, improved absorption. The TMC polymer is classified as TMC-20%, TMC-40%, and TMC-60% based on the level of methylation.

Chitosan may be utilized in intranasal formulations to transport proteins, according to research. The mucoadhesive qualities of chitosan are probably owing to an ionic interaction between the negatively charged sialic acid residues in mucus and the positively charged amino groups in chitosan. It has been demonstrated that chitosan's potent cationic properties promote contact with the mucosa's sialic acid, releasing the nasal epithelial cells' tight junctions. Chitosan-containing intranasal formulations may improve drug time residence in tissues and cells, demonstrating sustained drug release. They may also improve drug bioavailability and decrease the need for repeated dose administration. It has also been discovered to improve medication penetration via the mucosa without compromising biological membrane integrity. By employing salmon calcitonin in rats, the effectiveness of chitosan as a permeation enhancer was verified. Rats treated with 1% free amine had a greater absolute bioavailability of calcitonin than rats treated with 5% dimethylated β-cyclodextrins. Free amines and chitosan salts have been shown to cause moderate nasal discomfort. These findings suggest that chitosan is a safe and efficient supplement that can be administered intranasally [134, 135].

Surfactants

A polar asymmetric structure is formed by the amphiphilic molecules known as surfactants, which have properties like a lipophobic head and a lipophilic tail. They work by reducing the surface tension between the air and water phases; surfactants can increase a substance's miscibility. Surfactants can have simple or polymeric chemical structures, be soluble or insoluble, and be natural or manufactured. They are categorized as anionic, cationic, non-ionic, and zwitterionic surfactants based on the charge of the molecules [136].

The most popular excipients for enhancing medication permeability through the mucosa in intranasal formulations are surfactants. By altering the nasal membrane's permeability, the addition of surfactant to an intranasal formulation may improve medication absorption. According to reports, a number of surfactants enhance the nasal absorption of drugs, greatly boosting their bioavailability. Biological membranes can be changed or even dissolved by certain surfactants. Furthermore, the aerosol nasal formulation has made extensive use of surfactants. These compounds aid in the formulation's atomization in addition to lowering surface water tension. The most popular excipients for enhancing medication permeability through the mucosa in intranasal formulations are surfactants [137]. By altering the nasal membrane's permeability, the addition of surfactant to an intranasal formulation may improve medication absorption. According to reports, a number of surfactants enhance the nasal absorption of drugs, greatly boosting their bioavailability. Biological membranes can be changed or even dissolved by certain surfactants [138].

Furthermore, the aerosol nasal formulation has made extensive use of surfactants. These compounds aid in the formulation's atomization in addition to lowering surface water tension. Surfactants are frequently used in intranasal spray formulations to provide an effective dose (by droplet size and nebulization), as the production of a uniform nebulization depends on the formulation and the device [139]. Because of their ability to enhance intranasal medication absorption, surfactants have consequently been widely used in the formulation. When added to intranasal formulations as an enhancer, surfactants may result in ciliated activity, nasal discomfort, and epithelium toxicity. The majority of chemically produced surfactants, on the other hand, are biosurfactants, which are surfactants made from a broad range of chemicals by microorganisms. They have special qualities like biodegradability, low toxicity, and usefulness in harsh environments [140, 141].

Salt & derivatives of bile acid

In order to enhance medication penetration through the blood-brain barrier and mucosa, including the buccal, pulmonary, ophthalmic, nasal, and oral, bile acids are utilized in drug formulation. They increase the apical and basolateral fluidity and boost the chemical and enzymatic action of medications by opening tight junctions, influencing mucolytic activity, and making hydrophobic pharmaceuticals more soluble [142, 143]. These qualities form the basis of their penetration-enhancing mechanism. The main component of bile is bile acids, which are amphiphilic molecules made by the liver from cholesterol. They function as a biological surfactant in the colon, helping to solubilize and absorb lipids from food fats as well as endogenous soluble compounds, including cholesterol, fatty acids, and hydrophobic medications. Despite some studies indicating that they may cause nasal irritation when used in concentrations greater than 0.3% [144]. They play a role in digestion as well as other vital physiological processes, including the regulation of glucose homeostasis, the regulation of lipid metabolism, the modulation of immunomodulatory and inflammatory processes, and the antibacterial activity in the small intestine [145].

Tauro Dihydro Fusidate Sodium (DFS)

The search for more structurally comparable compounds that may increase mucosal permeability, show improved protein compatibility, and reduce associated toxicity and irritation has been sparked by the restricted ability of bile salts to boost penetration through the mucosa [146]. In 1962, LEO Pharmaceuticals (Denmark) created the sodium salt of fusidic acid by fermenting the fungus Fusidium coccineum [147]. TDHFS was frequently utilized to increase the intranasal penetration of protein and peptide medications. Using TDHFS as a penetration enhancer was found to increase the intranasal bioavailability of peptide and protein medications, including insulin, calcitonin, human growth hormone, and octreotide. According to reports, TDHFS has a slight impact on ciliary movement and a mild effect on nasal morphology. Hermens et al. used TDHFS to investigate the impact of the human nasal mucosa's ciliary movement frequency in vitro as a supplement [148]. It has been demonstrated that when the concentration of TDHFS rises, ciliary movement falls. These side effects subsided when TDHFS was used at doses of 0.1% and 0.2%; however, ciliostasis is induced at concentrations of 0.3% and above.

Phospholipids

Phospholipids are surfactant substances that have been investigated as nasal penetration enhancers and are present in both plant and animal cells. These compounds are lipids with a hydrophilic head, a hydrophobic tail, and phosphorus. Alcohol is connected to both the hydrophilic head and the hydrophobic tail. They can be divided into glycerophospholipids and sphingomyelins based on the chemical makeup of the alcohol. Glycerol makes up the backbone of glycerophospholipids. Fatty acids of different lengths and saturation levels esterify the glycerol backbone at places sn-1 and sn-2, while phosphoric acid esterifies the sn-3 position. Different glycerophospholipids, including phosphatidylcholine, phosphatidylethanolamine, phosphatidylglycerol, phosphatidylinositol, or phosphatidylserine, are produced when the head group varies. Because of their tiny size and high ability to absorb both hydrophobic and hydrophilic medications, phospholipids—substances that can self-assemble in an aqueous medium and form colloidal structures like liposomes, emulsions, or lipid microspheres—have been the subject of much research. Phospholipids also have emulsifying and moisturizing qualities. They discovered that the intranasal delivery of this excipient was quick, irritant-free, and had a bioavailability that was 8.3% more than that of subcutaneous injection. Additionally, phospholipid-based nanoparticles have been created in recent years as a means of overcoming barriers in the nose. The first carrier was created by using phospholipid-based vesicles that were altered with ethanol and ethylene glycol for intranasal delivery, which targets the systemic circulation and brain [149].

Intranasal drug delivery devices

Devices for intranasal administration are yet another effective tactic that will be used in the therapeutic use of intranasal medications. Keep in mind that the olfactory occupies 10% less of the nasal cavity and is situated on the superior aspect. Additionally, the motile respiratory regions are known for their quick mucus clearance. According to biology, in order to produce significant clinical applications, the maximum dose must be administered to the appropriate location. The olfactory region receives less than 3% of the dose from traditional spray pumps, which typically only reach the anterior and lateral parts of the nasal cavity [150].

Many patients of all ages cannot use other options, such as nasal sprays, because they require the patients to arrange their bodies precisely. All of the previously mentioned data were collected by qualified experts using model individuals that were ideally positioned in a controlled setting. To see findings translate from labs into clinics, it will be necessary to recreate this effective delivery, specifically to the olfactory region. Thankfully, there are gadgets that can help with this problem. Not even they are a panacea. For the majority of therapies, the patient must be compliant and conscious during the device insertion and release process. To date, a variety of devices have been created and investigated, all in preclinical settings. Electronic nebulizers, like ViaNase™, come first. Studies have shown a notable improvement in cognition when using this device to provide insulin for the treatment of AD [151, 152].

This is an excellent illustration of the unrealized clinical potential benefits of using this method of administering insulin. Since there is little proof that these devices genuinely improve distribution to the olfactory region and do not release a substantial amount into the lungs, where additional pain or damage exists, human testing of these devices is still lacking. Many powdered devices are offered by various pharmaceutical companies [153]. The advantages of powdered formulations include improved nasal residence and, in some cases, increased stability. One example of a powdered product designed for intranasal nose-to-brain transfer is the Opt-Powder™ produced by the Optinose company. Compared to liquids, this device delivered over six times as much powder to the upper nasal cavity, demonstrating another advantage of powders. This could just be because the device works better with powders, but some devices can effectively transport both powders and liquids, like the Precision Olfactory Delivery (POD®) device from Impel Neuro Pharma [154]. Both of these devices are powered by the patient's simultaneous mouth exhalation. Many patients with diminished pulmonary or cognitive function may find it difficult to utilize the device appropriately, even though this pushes the soft palate to close and reduces the amount of medication that could inadvertently enter the lungs. We have described the many gels, nasal coatings, additives, and devices that can be utilized to raise the brain AUC of medications that are delivered intranasally. Many seem to be able to facilitate this delivery route and assist in overcoming the challenges that the tissue of the nasal cavity naturally presents [155 - 158].

TRADITIONAL METHODS OF TREATING ALZHEIMER'S DISEASE

AChE inhibitors (Acetylcholinesterase inhibitors) are among the medications that effectively lessen AD symptoms. They work by preventing the cholinesterase enzyme from degrading ACh (acetylcholine), thereby raising its level and prolonging the duration of the neurotransmitter's action. They can be categorized as either reversible (donepezil, rivastigmine, or galantamine) or irreversible (organophosphorus insecticides or organophosphorus nerve agents/gases) based on how they work. While the irreversible ones frequently have deleterious effects, the reversible ones have therapeutic uses, primarily in treating AD, because they can be competitive or not. These medications are used to treat conditions that impair language, cognition, and memory [159]. The European Medicines Agency and the Food and Drug Administration have granted licenses for the use of donepezil, rivastigmine, and memantine, an N-methyl-D-aspartate receptor antagonist, in the treatment of AD.

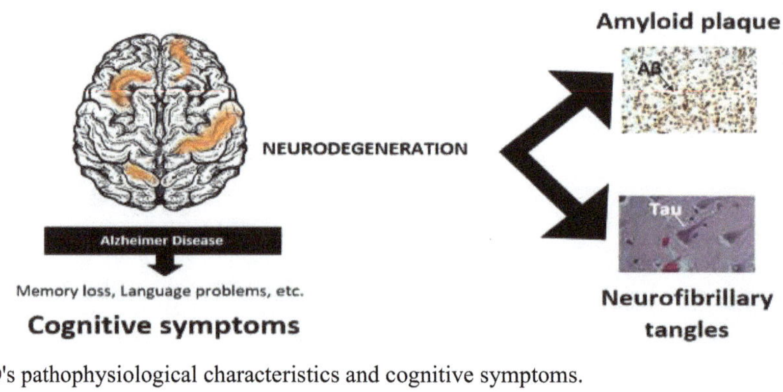

Fig. (3). AD's pathophysiological characteristics and cognitive symptoms.

As a whole, oral administration is the primary method used in current therapy for the pathologies linked to neurodegenerative disorders. However, there are many barriers against this mode of delivery, particularly when discussing neurodegenerative illnesses, because many molecules, particularly those with a high molecular weight, have difficulty crossing the blood-brain barrier [160].

Furthermore, oral medications may have their bioavailability significantly decreased by first-pass digestion, resulting in only a small quantity of the medication entering the bloodstream. Because of this, medications taken orally may also take a long time to get to the brain (Fig. **3**). The binding of medications to plasma proteins, which reduces their capacity to pass the blood-brain barrier, is another issue with oral delivery. Given that plasma proteins bind medicines, those that attach to plasma proteins can keep them in the vascular area for a specific amount of time, even though they cannot flow through capillaries. The only portion that can pass across membranes and interact with drug receptors and have a pharmacological effect is the fraction that is not linked to plasma proteins. One crucial factor in the quantity of drug available to breach the blood-brain barrier is the reversible link between the drug-free fraction and the plasma proteins, which varies depending on the concentration of the free fraction and the binding to the plasma proteins. If that bond is irreversible, the free fraction's concentration will remain constant; therefore, a low concentration will have negligible pharmacological effects [161]. Some benefits and drawbacks of administering medications intranasally are mentioned in Table **2**.

Table 2. Benefits and limitations of intranasal drug delivery.

Benefit	Restriction
Non-invasive	Only applicable to potent drugs
Possibility of self-administration	Small volumes (25–200 μL)
Direct medication administration to the central nervous system	Short retention time
Prevents first-pass metabolism	The absorption of drugs is influenced by nasal secretion.
A reasonable-sized absorption area (160 cm^2)	Enzymatic degradation
Low risk of infections	Low nasal epithelial pH
Rapid absorption	Active mucociliary clearance

CONCLUSION

AD is a neurodegenerative condition that is difficult to treat. The majority of medications cannot enter the central nervous system due to the BBB. Numerous compounds have been investigated; however, their transport to the central nervous system remains ineffective. It is feasible to cross the blood-brain barrier and enter the brain through the olfactory or trigeminal pathways when administered intranasally. But it also has drawbacks, which is why this method of drug delivery for the treatment of AD is still in the preclinical stage, because of toxicity and safety concerns. The formulations' extended interaction with the nasal mucosa may result in tissue damage, ciliotoxicity, irritation, or epithelial toxicity. Create the ideal conditions for microbial development. In order to reach particular brain regions of interest and prevent a number of negative effects, the use of substrates like albumin and cyclodextrins as permeation enhancers, nanoparticle delivery systems, and co-administration with vasoconstrictors may undoubtedly have major advantages. A definite nose-to-brain connection is not evident in several studies that use the IN route.

However, they could exhibit decreased bloodstream clearance. It increased brain bioavailability, which suggests that it was accomplished in conjunction with a different intranasal route via the innervated olfactory pathway (trigeminal nerve and olfactory bulb). It is more difficult to decipher the mechanisms underlying the shown benefits when the pathway approaches that apply to each study are not clearly defined. Indeed, the enhancements might be the result of the medications seeping from the epithelial cells into the bloodstream or CSF, or even through the trigeminal nerve or innervated olfactory bulbs. Drugs and other neuroprotective compounds can be transported and administered in the brain using nanoparticles, which have been seen as a potential method. But the clinical trials have not yet been conducted to assess efficacy and possible toxicological consequences. Insulin, glutathione, and other high-potential medications have been characterized as promising for the intranasal route. To fully understand dosage and safety following intranasal medication administration, more long-term research in humans and animals is still required.

REFERENCES

[1] Kingston A, Comas-Herrera A, Jagger C. Forecasting the care needs of the older population in England over the next 20 years: estimates from the Population Ageing and Care Simulation (PACSim) modelling study. Lancet Public Health 2018; 3(9): e447-55. [http://dx.doi.org/10.1016/S2468-2667(18)30118-X] [PMID: 30174210]

[2] Scheltens P, Blennow K, Breteler MMB, *et al.* Alzheimer's disease. Lancet 2016; 388(10043): 505-17. [http://dx.doi.org/10.1016/S0140-6736(15)01124-1] [PMID: 26921134]

[3] Cummings J, Lee G, Ritter A, Sabbagh M, Zhong K. Alzheimer's disease drug development pipeline: 2019. Alzheimers Dement (N Y) 2019; 5(1): 272-93. [http://dx.doi.org/10.1016/j.trci.2019.05.008] [PMID: 31334330]

[4] Anderson RM, Hadjichrysanthou C, Evans S, Wong MM. Why do so many clinical trials of therapies for Alzheimer's disease fail? Lancet 2017; 390(10110): 2327-9. [http://dx.doi.org/10.1016/S0140-6736(17)32399-1] [PMID: 29185425]

[5] Atri A. Current and future treatments in Alzheimer's disease Semin Neurol 2019; 39(2): 227-240. [http://dx.doi.org/10.1055/s-0039-1678581]

[6] Gauthier S, Albert M, Fox N, *et al.* Why has therapy development for dementia failed in the last two decades? Alzheimers Dement 2016; 12(1): 60-4. [http://dx.doi.org/10.1016/j.jalz.2015.12.003] [PMID: 26710325]

[7] Lehrer S, Rheinstein PH. Nasal steroids as a possible treatment for Alzheimer's disease. Discov Med 2017; 24(132): 147-52.

[PMID: 29272691]

[8] Yin F, Sancheti H, Patil I, Cadenas E. Energy metabolism and inflammation in brain aging and Alzheimer's disease. Free Radic Biol Med 2016; 100: 108-22. [http://dx.doi.org/10.1016/j.freeradbiomed.2016.04.200] [PMID: 27154981]

[9] Sood S, Jain K, Gowthamarajan K. Intranasal therapeutic strategies for management of Alzheimer's disease. J Drug Target 2014; 22(4): 279-94. [http://dx.doi.org/10.3109/1061186X.2013.876644] [PMID: 24404923]

[10] Illum L. Nasal drug delivery—possibilities, problems and solutions. J Control Release 2003; 87(1-3): 187-98. [http://dx.doi.org/10.1016/S0168-3659(02)00363-2] [PMID: 12618035]

[11] Frey WH, Liu J, Chen X, *et al.* Delivery of 125I-NGF to the brain via the olfactory route. Drug Deliv 1997; 4(2): 87-92. [http://dx.doi.org/10.3109/10717549709051878]

[12] Reger MA, Watson GS, Frey WH II, *et al.* Effects of intranasal insulin on cognition in memory-impaired older adults: Modulation by APOE genotype. Neurobiol Aging 2006; 27(3): 451-8. [http://dx.doi.org/10.1016/j.neurobiolaging.2005.03.016] [PMID: 15964100]

[13] Illum L. Transport of drugs from the nasal cavity to the central nervous system. Eur J Pharm Sci 2000; 11(1): 1-18. [http://dx.doi.org/10.1016/S0928-0987(00)00087-7] [PMID: 10913748]

[14] Ugwoke MI, Verbeke N, Kinget R. The biopharmaceutical aspects of nasal mucoadhesive drug delivery. J Pharm Pharmacol 2001; 53(1): 3-21. [http://dx.doi.org/10.1211/0022357011775145] [PMID: 11206189]

[15] Lochhead JJ, Thorne RG. Intranasal delivery of biologics to the central nervous system. Adv Drug Deliv Rev 2012; 64(7): 614-28. [http://dx.doi.org/10.1016/j.addr.2011.11.002] [PMID: 22119441]

[16] Kristensson K, Olsson Y. Uptake of exogenous proteins in mouse olfactory cells. Acta Neuropathol 1971; 19(2): 145-54. [http://dx.doi.org/10.1007/BF00688493] [PMID: 5126829]

[17] Dalpiaz A, Contado C, Vighi E, Tosi G, Leo E. Preparation and characterization of particulate drug delivery systems for the brain targeting InHandbook of particulate drug delivery. American Scientific Publishers 2008; pp. 1-26.

[18] Chou KJ, Donovan MD. Lidocaine distribution into the CNS following nasal and arterial delivery: a comparison of local sampling and microdialysis techniques. Int J Pharm 1998; 171(1): 53-61. [http://dx.doi.org/10.1016/S0378-5173(98)00170-7]

[19] Salameh TS, Bullock KM, Hujoel IA, *et al.* Central nervous system delivery of intranasal insulin: mechanisms of uptake and effects on cognition. J Alzheimers Dis 2015; 47(3): 715-28. [http://dx.doi.org/10.3233/JAD-150307] [PMID: 26401706]

[20] Rivera EJ, Goldin A, Fulmer N, Tavares R, Wands JR, de la Monte SM. Insulin and insulin-like growth factor expression and function deteriorate with progression of Alzheimer's disease: Link to brain reductions in acetylcholine. J Alzheimers Dis 2005; 8(3): 247-68. [http://dx.doi.org/10.3233/JAD-2005-8304] [PMID: 16340083]

[21] Talbot K, Wang HY, Kazi H, *et al.* Demonstrated brain insulin resistance in Alzheimer's disease patients is associated with IGF-1 resistance, IRS-1 dysregulation, and cognitive decline. J Clin Invest 2012; 122(4): 1316-38. [http://dx.doi.org/10.1172/JCI59903] [PMID: 22476197]

[22] Plum L, Schubert M, Brüning JC. The role of insulin receptor signaling in the brain. Trends Endocrinol Metab 2005; 16(2): 59-65. [http://dx.doi.org/10.1016/j.tem.2005.01.008] [PMID: 15734146]

[23] Abbott MA, Wells DG, Fallon JR. The insulin receptor tyrosine kinase substrate p58/53 and the insulin receptor are components of CNS synapses. J Neurosci 1999; 19(17): 7300-8. [http://dx.doi.org/10.1523/JNEUROSCI.19-17-07300.1999] [PMID: 10460236]

[24] McNay EC, Ong CT, McCrimmon RJ, Cresswell J, Bogan JS, Sherwin RS. Hippocampal memory processes are modulated by insulin and high-fat-induced insulin resistance. Neurobiol Learn Mem 2010; 93(4): 546-53. [http://dx.doi.org/10.1016/j.nlm.2010.02.002] [PMID: 20176121]

[25] Benedict C, Hallschmid M, Hatke A, *et al.* Intranasal insulin improves memory in humans. Psychoneuroendocrinology 2004; 29(10): 1326-34. [http://dx.doi.org/10.1016/j.psyneuen.2004.04.003] [PMID: 15288712]

[26] Kern W, Peters A, Fruehwald-Schultes B, Deininger E, Born J, Fehm HL. Improving influence of insulin on cognitive functions in humans. Neuroendocrinology 2001; 74(4): 270-80. [http://dx.doi.org/10.1159/000054694] [PMID: 11598383]

[27] Chen Y, Run X, Liang Z, *et al.* Intranasal insulin prevents anesthesia-induced hyperphosphorylation of tau in 3xTg-AD mice. Front Aging Neurosci 2014; 6: 100. [http://dx.doi.org/10.3389/fnagi.2014.00100] [PMID: 24910612]

[28] Dong Y, Wu X, Xu Z, Zhang Y, Xie Z. Anesthetic isoflurane increases phosphorylated tau levels mediated by caspase activation and Aβ generation. PLoS One 2012; 7(6)e39386 [http://dx.doi.org/10.1371/journal.pone.0039386] [PMID: 22745746]

[29] Shen X, Dong Y, Xu Z, *et al.* Selective anesthesia-induced neuroinflammation in developing mouse brain and cognitive impairment. Anesthesiology 2013; 118(3): 502-15. [http://dx.doi.org/10.1097/ALN.0b013e3182834d77] [PMID: 23314110]

[30] Bianchi SL, Tran T, Liu C, *et al.* Brain and behavior changes in 12-month-old Tg2576 and nontransgenic mice exposed to anesthetics. Neurobiol Aging 2008; 29(7): 1002-10. [http://dx.doi.org/10.1016/j.neurobiolaging.2007.02.009] [PMID: 17346857]

[31] Claxton A, Baker LD, Wilkinson CW, *et al.* Sex and ApoE genotype differences in treatment response to two doses of intranasal insulin in adults with mild cognitive impairment or Alzheimer's disease. J Alzheimers Dis 2013; 35(4): 789-97. [http://dx.doi.org/10.3233/JAD-122308] [PMID: 23507773]

[32] Rosenbloom MH, Barclay TR, Pyle M, *et al.* A single-dose pilot trial of intranasal rapid-acting insulin in apolipoprotein E4 carriers with mild-moderate Alzheimer's disease. CNS Drugs 2014; 28(12): 1185-9. [http://dx.doi.org/10.1007/s40263-014-0214-y] [PMID: 25373630]

[33] Lochhead JJ, Thorne RG. Intranasal delivery of biologics to the central nervous system. Adv Drug Deliv Rev 2012; 64(7): 614-28. [http://dx.doi.org/10.1016/j.addr.2011.11.002] [PMID: 22119441]

[34] Lochhead JJ, Wolak DJ, Pizzo ME, Thorne RG. Rapid transport within cerebral perivascular spaces underlies widespread tracer distribution in the brain after intranasal administration. J Cereb Blood Flow Metab 2015; 35(3): 371-81. [http://dx.doi.org/10.1038/jcbfm.2014.215] [PMID: 25492117]

[35] Sun J, Martin JM, Vanderpoel V, Sumbria RK. The promises and challenges of erythropoietin for treatment of Alzheimer's disease. Neuromolecular Med 2019; 21(1): 12-24. [http://dx.doi.org/10.1007/s12017-019-08524-y] [PMID: 30656553]

[36] Rama R, Garzón F, Rodríguez-Cruz Y, Maurice T, García-Rodríguez JC. Neuroprotective effect of Neuro-EPO in neurodegenerative diseases: "Alea jacta est". Neural Regen Res 2019; 14(9): 1519-21. [http://dx.doi.org/10.4103/1673-5374.255968] [PMID: 31089047]

[37] Ehrenreich H, Weissenborn K, Prange H, *et al.* Recombinant human erythropoietin in the treatment of acute ischemic stroke. Stroke 2009; 40(12): e647-56.

[http://dx.doi.org/10.1161/STROKEAHA.109.564872] [PMID: 19834012]

[38] Leist M, Ghezzi P, Grasso G, *et al.* Derivatives of erythropoietin that are tissue protective but not erythropoietic. Science 2004; 305(5681): 239-42. [http://dx.doi.org/10.1126/science.1098313] [PMID: 15247477]

[39] Rodríguez Cruz Y, Yuneidys Mengana T, Muñoz Cernuda A, *et al.* Treatment with nasal neuro-EPO improves the neurological, cognitive, and histological state in a gerbil model of focal ischemia. ScientificWorldJournal 2010; 10(1): 2288-300. [http://dx.doi.org/10.1100/tsw.2010.215] [PMID: 21103798]

[40] García-Rodríguez JC, Sosa Teste I. The nasal route as a potential pathway for delivery of erythropoietin in the treatment of acute ischemic stroke in humans. ScientificWorldJournal 2009; 9(1): 970-81. [http://dx.doi.org/10.1100/tsw.2009.103] [PMID: 19768354]

[41] Maurice T, Mustafa MH, Desrumaux C, *et al.* Intranasal formulation of erythropoietin (EPO) showed potent protective activity against amyloid toxicity in the $A\beta_{25-35}$ non-transgenic mouse model of Alzheimer's disease. J Psychopharmacol 2013; 27(11): 1044-57. [http://dx.doi.org/10.1177/0269881113494939] [PMID: 23813967]

[42] Adamcio B, Sargin D, Stradomska A, *et al.* Erythropoietin enhances hippocampal long-term potentiation and memory. BMC Biol 2008; 6(1): 37. [http://dx.doi.org/10.1186/1741-7007-6-37] [PMID: 18782446]

[43] Kamal A, Al Shaibani T, Ramakers G. Erythropoietin decreases the excitatory neurotransmitter release probability and enhances synaptic plasticity in mice hippocampal slices. Brain Res 2011; 1410: 33-7. [http://dx.doi.org/10.1016/j.brainres.2011.06.059] [PMID: 21803333]

[44] Maiese K, Li F, Chong ZZ. Erythropoietin in the brain: can the promise to protect be fulfilled? Trends Pharmacol Sci 2004; 25(11): 577-83. [http://dx.doi.org/10.1016/j.tips.2004.09.006] [PMID: 15491780]

[45] Fletcher L, Kohli S, Sprague SM, *et al.* Intranasal delivery of erythropoietin plus insulin-like growth factor–I for acute neuroprotection in stroke. J Neurosurg 2009; 111(1): 164-70. [http://dx.doi.org/10.3171/2009.2.JNS081199] [PMID: 19284235]

[46] Lakhal S, Wood MJA. Intranasal exosomes for treatment of neuroinflammation? Prospects and limitations. Mol Ther 2011; 19(10): 1754-6. [http://dx.doi.org/10.1038/mt.2011.198] [PMID: 21964306]

[47] Zhuang X, Xiang X, Grizzle W, *et al.* Treatment of brain inflammatory diseases by delivering exosome encapsulated anti-inflammatory drugs from the nasal region to the brain. Mol Ther 2011; 19(10): 1769-79. [http://dx.doi.org/10.1038/mt.2011.164] [PMID: 21915101]

[48] Potschka H. Targeting the brain--surmounting or bypassing the blood-brain barrier. Handb Exp Pharmacol 2010; 197(197): 411-31. [http://dx.doi.org/10.1007/978-3-642-00477-3_14] [PMID: 20217538]

[49] Gabathuler R. Blood-brain barrier transport of drugs for the treatment of brain diseases. CNS Neurol Disord Drug Targets 2009; 8(3): 195-204. [http://dx.doi.org/10.2174/187152709788680652] [PMID: 19601817]

[50] Bidros DS, Vogelbaum MA. Novel drug delivery strategies in neuro-oncology. Neurotherapeutics 2009; 6(3): 539-46. [http://dx.doi.org/10.1016/j.nurt.2009.04.004] [PMID: 19560743]

[51] Ravindran J, Prasad S, Aggarwal BB. Curcumin and cancer cells: how many ways can curry kill tumor cells selectively? AAPS J 2009; 11(3): 495-510. [http://dx.doi.org/10.1208/s12248-009-9128-x] [PMID: 19590964]

[52] Anand P, Thomas SG, Kunnumakkara AB, *et al.* Biological activities of curcumin and its analogues

(Congeners) made by man and Mother Nature. Biochem Pharmacol 2008; 76(11): 1590-611. [http://dx.doi.org/10.1016/j.bcp.2008.08.008] [PMID: 18775680]

[53] Aggarwal BB, Harikumar KB. Potential therapeutic effects of curcumin, the anti-inflammatory agent, against neurodegenerative, cardiovascular, pulmonary, metabolic, autoimmune and neoplastic diseases. Int J Biochem Cell Biol 2009; 41(1): 40-59. [http://dx.doi.org/10.1016/j.biocel.2008.06.010] [PMID: 18662800]

[54] Anand P, Kunnumakkara AB, Newman RA, Aggarwal BB. Bioavailability of curcumin: problems and promises. Mol Pharm 2007; 4(6): 807-18. [http://dx.doi.org/10.1021/mp700113r] [PMID: 17999464]

[55] Sun D, Zhuang X, Xiang X, *et al.* A novel nanoparticle drug delivery system: the anti-inflammatory activity of curcumin is enhanced when encapsulated in exosomes. Mol Ther 2010; 18(9): 1606-14. [http://dx.doi.org/10.1038/mt.2010.105] [PMID: 20571541]

[56] Record M, Subra C, Silvente-Poirot S, Poirot M. Exosomes as intercellular signalosomes and pharmacological effectors. Biochem Pharmacol 2011; 81(10): 1171-82. [http://dx.doi.org/10.1016/j.bcp.2011.02.011] [PMID: 21371441]

[57] Santamaria G, Brandi E, Vitola PL, *et al.* Intranasal delivery of mesenchymal stem cell secretome repairs the brain of Alzheimer's mice. Cell Death Differ 2021; 28(1): 203-18. [http://dx.doi.org/10.1038/s41418-020-0592-2] [PMID: 32704089]

[58] Duncan T, Valenzuela M. Alzheimer's disease, dementia, and stem cell therapy. Stem Cell Res Ther 2017; 8(1): 111. [http://dx.doi.org/10.1186/s13287-017-0567-5] [PMID: 28494803]

[59] Fan X, Sun D, Tang X, Cai Y, Yin ZQ, Xu H. Stem-cell challenges in the treatment of Alzheimer's disease: a long way from bench to bedside. Med Res Rev 2014; 34(5): 957-78. [http://dx.doi.org/10.1002/med.21309] [PMID: 24500883]

[60] Laroni A, Novi G, Kerlero de Rosbo N, Uccelli A. Towards clinical application of mesenchymal stem cells for treatment of neurological diseases of the central nervous system. J Neuroimmune Pharmacol 2013; 8(5): 1062-76. [http://dx.doi.org/10.1007/s11481-013-9456-6] [PMID: 23579931]

[61] Bali P, Lahiri D, Banik A, Nehru B, Anand A. Potential for stem cells therapy in Alzheimer's disease: do neurotrophic factors play critical role? Curr Alzheimer Res 2017; 14(2): 208-20. [http://dx.doi.org/10.2174/1567205013666160314145347] [PMID: 26971940]

[62] Abbasi-Malati Z, Roushandeh AM, Kuwahara Y, Roudkenar MH. Mesenchymal stem cells on horizon: a new arsenal of therapeutic agents. Stem Cell Rev 2018; 14(4): 484-99. [http://dx.doi.org/10.1007/s12015-018-9817-x] [PMID: 29687338]

[63] Teixeira FG, Carvalho MM, Sousa N, Salgado AJ. Mesenchymal stem cells secretome: a new paradigm for central nervous system regeneration? Cell Mol Life Sci 2013; 70(20): 3871-82. [http://dx.doi.org/10.1007/s00018-013-1290-8] [PMID: 23456256]

[64] Forloni G, Artuso V, La Vitola P, Balducci C. Oligomeropathies and pathogenesis of Alzheimer and Parkinson's diseases. Mov Disord 2016; 31(6): 771-81. [http://dx.doi.org/10.1002/mds.26624] [PMID: 27030592]

[65] Benilova I, Karran E, De Strooper B. The toxic Aβ oligomer and Alzheimer's disease: an emperor in need of clothes. Nat Neurosci 2012; 15(3): 349-57. [http://dx.doi.org/10.1038/nn.3028] [PMID: 22286176]

[66] Imbimbo BP, Giardina GA. γ-secretase inhibitors and modulators for the treatment of Alzheimer's disease: disappointments and hopes. Curr Top Med Chem 2011; 11(12): 1555-70. [http://dx.doi.org/10.2174/156802611795860942] [PMID: 21510832]

[67] Umeda T, Ono K, Sakai A, *et al.* Rifampicin is a candidate preventive medicine against amyloid-β and tau oligomers. Brain 2016; 139(5): 1568-86.

[http://dx.doi.org/10.1093/brain/aww042] [PMID: 27020329]

[68] Iizuka T, Morimoto K, Sasaki Y, *et al.* Preventive effect of rifampicin on Alzheimer disease needs at least 450 mg daily for 1 year: An FDG-PET follow-up study. Dement Geriatr Cogn Disord Extra 2017; 7(2): 204-14.
[http://dx.doi.org/10.1159/000477343] [PMID: 28690634]

[69] Jack CR Jr, Knopman DS, Jagust WJ, *et al.* Hypothetical model of dynamic biomarkers of the Alzheimer's pathological cascade. Lancet Neurol 2010; 9(1): 119-28.
[http://dx.doi.org/10.1016/S1474-4422(09)70299-6] [PMID: 20083042]

[70] Umeda T, Tanaka A, Sakai A, Yamamoto A, Sakane T, Tomiyama T. Intranasal rifampicin for Alzheimer's disease prevention. Alzheimers Dement (N Y) 2018; 4(1): 304-13.
[http://dx.doi.org/10.1016/j.trci.2018.06.012] [PMID: 30094330]

[71] Martins PP, Smyth HDC, Cui Z. Strategies to facilitate or block nose-to-brain drug delivery. Int J Pharm 2019; 570118635
[http://dx.doi.org/10.1016/j.ijpharm.2019.118635] [PMID: 31445062]

[72] Pardridge WM. The blood-brain barrier: Bottleneck in brain drug development. NeuroRx 2005; 2(1): 3-14.
[http://dx.doi.org/10.1602/neurorx.2.1.3] [PMID: 15717053]

[73] Gadhave D, Choudhury H, Kokare C. Neutropenia and leukopenia protective intranasal olanzapine-loaded lipid-based nanocarriers engineered for brain delivery. Appl Nanosci 2019; 9(2): 151-68.
[http://dx.doi.org/10.1007/s13204-018-0909-3]

[74] Pardeshi CV, Belgamwar VS. Direct nose to brain drug delivery *via* integrated nerve pathways bypassing the blood–brain barrier: an excellent platform for brain targeting. Expert Opin Drug Deliv 2013; 10(7): 957-72.
[http://dx.doi.org/10.1517/17425247.2013.790887] [PMID: 23586809]

[75] Hada N, Netzer WJ, Belhassan F, Wennogle LP, Gizurarson S. Nose-to-brain transport of imatinib mesylate: A pharmacokinetic evaluation. Eur J Pharm Sci 2017; 102: 46-54.
[http://dx.doi.org/10.1016/j.ejps.2017.02.032] [PMID: 28238945]

[76] Bonaccorso A, Gigliobianco M, Pellitteri R, *et al.* Optimization of curcumin nanocrystals as a promising strategy for nose-to-brain delivery application. Pharmaceutics 2020; 12(5): 476.
[http://dx.doi.org/10.3390/pharmaceutics12050476] [PMID: 32456163]

[77] Fahmy U, Badr-Eldin S, Ahmed O, *et al.* RETRACTED: Intranasal Niosomal In Situ Gel as a Promising Approach for Enhancing Flibanserin Bioavailability and Brain Delivery: In Vitro Optimization and Ex Vivo/In Vivo Evaluation. Pharmaceutics 2020; 12(6): 485.
[http://dx.doi.org/10.3390/pharmaceutics12060485] [PMID: 32471119]

[78] Bourganis V, Kammona O, Alexopoulos A, Kiparissides C. Recent advances in carrier mediated nose-to-brain delivery of pharmaceutics. Eur J Pharm Biopharm 2018; 128: 337-62.
[http://dx.doi.org/10.1016/j.ejpb.2018.05.009] [PMID: 29733950]

[79] Shingaki T, Hidalgo IJ, Furubayashi T, *et al.* Nasal delivery of P-gp substrates to the brain through the nose-brain pathway. Drug Metab Pharmacokinet 2011; 26(3): 248-55.
[http://dx.doi.org/10.2133/dmpk.DMPK-10-RG-108] [PMID: 21317541]

[80] Choudhury H, Gorain B, Chatterjee B, Mandal UK, Sengupta P, Tekade RK. Pharmacokinetic and pharmacodynamic features of nanoemulsion following oral, intravenous, topical, and nasal routes. Curr Pharm Des 2017; 23(17): 2504-31.
[http://dx.doi.org/10.2174/1381612822666161201143600] [PMID: 27908273]

[81] Illum L. Transport of drugs from the nasal cavity to the central nervous system. Eur J Pharm Sci 2000; 11(1): 1-18.
[http://dx.doi.org/10.1016/S0928-0987(00)00087-7] [PMID: 10913748]

[82] Illum L. Is nose-to-brain transport of drugs in man a reality? J Pharm Pharmacol 2004; 56(1): 3-17.

[http://dx.doi.org/10.1211/0022357022539] [PMID: 14979996]

[83] Bonferoni MC, Rossi S, Sandri G, *et al.* Nanoemulsions for "nose-to-brain" drug delivery. Pharmaceutics 2019; 11(2): 84.
[http://dx.doi.org/10.3390/pharmaceutics11020084] [PMID: 30781585]

[84] Rassu G, Soddu E, Cossu M, *et al.* Solid microparticles based on chitosan or methyl-β-cyclodextrin: A first formulative approach to increase the nose-to-brain transport of deferoxamine mesylate. J Control Release 2015; 201: 68-77.
[http://dx.doi.org/10.1016/j.jconrel.2015.01.025] [PMID: 25620068]

[85] Mistry A, Stolnik S, Illum L. Nose-to-brain delivery: investigation of the transport of nanoparticles with different surface characteristics and sizes in excised porcine olfactory epithelium. Mol Pharm 2015; 12(8): 2755-66.
[http://dx.doi.org/10.1021/acs.molpharmaceut.5b00088] [PMID: 25997083]

[86] Quintana DS, Guastella AJ, Westlye LT, Andreassen OA. The promise and pitfalls of intranasally administering psychopharmacological agents for the treatment of psychiatric disorders. Mol Psychiatry 2016; 21(1): 29-38.
[http://dx.doi.org/10.1038/mp.2015.166] [PMID: 26552590]

[87] Wu H, Hu K, Jiang X. From nose to brain: understanding transport capacity and transport rate of drugs. Expert Opin Drug Deliv 2008; 5(10): 1159-68.
[http://dx.doi.org/10.1517/17425247.5.10.1159] [PMID: 18817519]

[88] Agrawal M, Saraf S, Saraf S, *et al.* Nose-to-brain drug delivery: An update on clinical challenges and progress towards approval of anti-Alzheimer drugs. J Control Release 2018; 281: 139-77.
[http://dx.doi.org/10.1016/j.jconrel.2018.05.011] [PMID: 29772289]

[89] Bors LA, Bajza Á, Mándoki M, *et al.* Modulation of nose-to-brain delivery of a P-glycoprotein (MDR1) substrate model drug (quinidine) in rats. Brain Res Bull 2020; 160: 65-73.
[http://dx.doi.org/10.1016/j.brainresbull.2020.04.012] [PMID: 32344126]

[90] Mittal D, Ali A, Md S, Baboota S, Sahni JK, Ali J. Insights into direct nose to brain delivery: current status and future perspective. Drug Deliv 2014; 21(2): 75-86.
[http://dx.doi.org/10.3109/10717544.2013.838713] [PMID: 24102636]

[91] Casettari L, Illum L. Chitosan in nasal delivery systems for therapeutic drugs. J Control Release 2014; 190: 189-200.
[http://dx.doi.org/10.1016/j.jconrel.2014.05.003] [PMID: 24818769]

[92] Feng Y, He H, Li F, Lu Y, Qi J, Wu W. An update on the role of nanovehicles in nose-to-brain drug delivery. Drug Discov Today 2018; 23(5): 1079-88.
[http://dx.doi.org/10.1016/j.drudis.2018.01.005] [PMID: 29330120]

[93] Kozlovskaya L, Abou-Kaoud M, Stepensky D. Quantitative analysis of drug delivery to the brain via nasal route. J Control Release 2014; 189: 133-40.
[http://dx.doi.org/10.1016/j.jconrel.2014.06.053] [PMID: 24997277]

[94] Soane RJ, Hinchcliffe M, Davis SS, Illum L. Clearance characteristics of chitosan based formulations in the sheep nasal cavity. Int J Pharm 2001; 217(1-2): 183-91.
[http://dx.doi.org/10.1016/S0378-5173(01)00602-0] [PMID: 11292554]

[95] Seju U, Kumar A, Sawant KK. Development and evaluation of olanzapine-loaded PLGA nanoparticles for nose-to-brain delivery: *in vitro* and *in vivo* studies Acta Biomaterialia 2011; 7(12): 4169-4176.
[http://dx.doi.org/10.1016/j.actbio.2011.07.025]

[96] Rai A, Jain A, Jain A, *et al.* Targeted SLNs for management of HIV-1 associated dementia. Drug Dev Ind Pharm 2015; 41(8): 1321-7.
[http://dx.doi.org/10.3109/03639045.2014.948453] [PMID: 25113430]

[97] Ganeshpurkar A, Ganeshpurkar A, Agnihotri A, Pandey V, Vishwakarma N, Bansal D, Dubey N. Chondroitin sulfate surface engineered docetaxel-loaded liposomes for tumor targeting: Design,

development, and characterization. InProceedings of All India Seminar on Biomedical Engineering 2012 (AISOBE 2012) 2012 Nov 2 (pp. 77-82). India: Springer India.
[http://dx.doi.org/10.1007/978-81-322-0970-6_9]

[98] Alsarra IA, Hamed AY, Alanazi FK. Acyclovir liposomes for intranasal systemic delivery: development and pharmacokinetics evaluation. Drug Deliv 2008; 15(5): 313-21.
[http://dx.doi.org/10.1080/10717540802035251] [PMID: 18763162]

[99] Bansal D, Yadav K, Pandey V, Ganeshpurkar A, Agnihotri A, Dubey N. Lactobionic acid coupled liposomes: an innovative strategy for targeting hepatocellular carcinoma. Drug Deliv 2016; 23(1): 140-6.
[http://dx.doi.org/10.3109/10717544.2014.907373] [PMID: 24786484]

[100] Vyas TK, Babbar AK, Sharma RK, Singh S, Misra A. Intranasal mucoadhesive microemulsions of clonazepam: preliminary studies on brain targeting. J Pharm Sci 2006; 95(3): 570-80.
[http://dx.doi.org/10.1002/jps.20480] [PMID: 16419051]

[101] Aderibigbe B, Buyana B. Alginate in wound dressings. Pharmaceutics 2018; 10(2): 42.
[http://dx.doi.org/10.3390/pharmaceutics10020042] [PMID: 29614804]

[102] Sosnik A, Seremeta K. Polymeric hydrogels as technology platform for drug delivery applications. Gels 2017; 3(3): 25.
[http://dx.doi.org/10.3390/gels3030025] [PMID: 30920522]

[103] Al Harthi S, Alavi SE, Radwan MA, El Khatib MM, AlSarra IA. Nasal delivery of donepezil HCl-loaded hydrogels for the treatment of Alzheimer's disease. Sci Rep 2019; 9(1): 9563.
[http://dx.doi.org/10.1038/s41598-019-46032-y] [PMID: 31266990]

[104] Carvalho FC, Bruschi ML, Evangelista RC, Gremião MPD. Mucoadhesive drug delivery systems. Braz J Pharm Sci 2010; 46(1): 1-17.
[http://dx.doi.org/10.1590/S1984-82502010000100002]

[105] Bahadur S, Pardhi DM, Rautio J, Rosenholm JM, Pathak K. Intranasal nanoemulsions for direct nose-to-brain delivery of actives for CNS disorders. Pharmaceutics 2020; 12(12): 1230.
[http://dx.doi.org/10.3390/pharmaceutics12121230] [PMID: 33352959]

[106] Sessa M, Balestrieri ML, Ferrari G, *et al.* Bioavailability of encapsulated resveratrol into nanoemulsion-based delivery systems. Food Chem 2014; 147: 42-50.
[http://dx.doi.org/10.1016/j.foodchem.2013.09.088] [PMID: 24206683]

[107] Haider MF, Khan S, Gaba B, *et al.* Optimization of rivastigmine nanoemulsion for enhanced brain delivery: in-vivo and toxicity evaluation. J Mol Liq 2018; 255: 384-96.
[http://dx.doi.org/10.1016/j.molliq.2018.01.123]

[108] Espinoza LC, Silva-Abreu M, Clares B, *et al.* Formulation strategies to improve nose-to-brain delivery of donepezil. Pharmaceutics 2019; 11(2): 64.
[http://dx.doi.org/10.3390/pharmaceutics11020064] [PMID: 30717264]

[109] Su Y, Sun B, Gao X, *et al.* Intranasal delivery of targeted nanoparticles loaded with miR-132 to brain for the treatment of neurodegenerative diseases. Front Pharmacol 2020; 11: 1165.
[http://dx.doi.org/10.3389/fphar.2020.01165] [PMID: 32848773]

[110] Wavikar P, Pai R, Vavia P. Nose to brain delivery of rivastigmine by in situ gelling cationic nanostructured lipid carriers: enhanced brain distribution and pharmacodynamics. J Pharm Sci 2017; 106(12): 3613-22.
[http://dx.doi.org/10.1016/j.xphs.2017.08.024] [PMID: 28923321]

[111] Yasir M, Sara UVS. Solid lipid nanoparticles for nose to brain delivery of haloperidol: in vitro drug release and pharmacokinetics evaluation. Acta Pharm Sin B 2014; 4(6): 454-63.
[http://dx.doi.org/10.1016/j.apsb.2014.10.005] [PMID: 26579417]

[112] Nasr M. Development of an optimized hyaluronic acid-based lipidic nanoemulsion co-encapsulating two polyphenols for nose to brain delivery. Drug Deliv 2016; 23(4): 1444-52.

[http://dx.doi.org/10.3109/10717544.2015.1092619] [PMID: 26401600]

[113] Trejo-Lopez JA, Yachnis AT, Prokop S. Neuropathology of Alzheimer's disease. Neurotherapeutics 2022; 19(1): 173-85.
[http://dx.doi.org/10.1007/s13311-021-01146-y] [PMID: 34729690]

[114] Ruby JJ, Pandey VP. Chitosan nanoparticles as a nasal drug delivery for memantine hydrochloride. Int J Pharm Pharm Sci. 2015 Jan; 7(1): 34–37. https://journals.innovareacademics.in/index.php/ijpps/article/view/3459

[115] Phachonpai W, Wattanathorn J, Muchimapura S, Tong-Un T, Preechagoon D. Neuroprotective effect of quercetin encapsulated liposomes: A novel therapeutic strategy against Alzheimer's disease. Am J Appl Sci. 2010; 7(4): 480–485.
[http://dx.doi.org/10.3844/ajassp.2010.480.485]

[116] Chattopadhyay S, Das S, Sarma KN. Nose-to-brain drug delivery: An update to the alternative path to successful targeted antimigraine drugs. International Journal of Applied Pharmaceutics 2021; 13(2): 67-75.
[http://dx.doi.org/10.22159/ijap.2021v13i2.40404]

[117] Kim D-D. In Vitro Cellular Models for Nasal Drug Absorption Studies. 1st ed. California, USA 2008; pp. 216-34.
[http://dx.doi.org/10.1007/978-0-387-74901-3_9]

[118] Aungst BJ. Absorption enhancers: applications and advances. AAPS J 2012; 14(1): 10-8.
[http://dx.doi.org/10.1208/s12248-011-9307-4] [PMID: 22105442]

[119] Villafuerte Robles L. Los excipientes y su funcionalidad en productos farmacéuticos sólidos. Rev Mex Cienc Farm 2011; 42: 18-36.

[120] Ghadiri M, Young PM, Traini D. Strategies to Enhance Drug Absorption via Nasal and Pulmonary Routes. Pharmaceutics 2019; 11(3): 113.
[http://dx.doi.org/10.3390/pharmaceutics11030113] [PMID: 30861990]

[121] Behl CR, Pimplaskar HK, Sileno AP, et al. Optimization of systemic nasal drug delivery with pharmaceutical excipients. Adv Drug Deliv Rev. 1998 Jan 5; 29(1–2): 117–133.
[http://dx.doi.org/10.1016/S0169-409X(97)00064-1]

[122] Na L, Mao S, Wang J, Sun W. Comparison of different absorption enhancers on the intranasal absorption of isosorbide dinitrate in rats. Int J Pharm 2010; 397(1-2): 59-66.
[http://dx.doi.org/10.1016/j.ijpharm.2010.06.048] [PMID: 20599486]

[123] Leyva E, Moctezuma E, Leyva R, Oros S. Estudio de los complejos de inclusión de ácido nalidíxico y ácido oxolínico con ciclodextrinas. Rev Soc Quím Méx 2004; 48: 189.

[124] Connors KA. The Stability of Cyclodextrin Complexes in Solution. Chem Rev 1997; 97(5): 1325-58.
[http://dx.doi.org/10.1021/cr960371r] [PMID: 11851454]

[125] Brewster ME, Loftsson T. Cyclodextrins as pharmaceutical solubilizers. Adv Drug Deliv Rev 2007; 59(7): 645-66.
[http://dx.doi.org/10.1016/j.addr.2007.05.012] [PMID: 17601630]

[126] Villiers A [On the fermentation of starch by the action of butyric ferment].. Comptes rendus de l'Académie des sciences Sci 1891;112:536–8.

[127] Loftsson T, Duchêne D. Cyclodextrins and their pharmaceutical applications. Int J Pharm 2007; 329(1-2): 1-11.
[http://dx.doi.org/10.1016/j.ijpharm.2006.10.044] [PMID: 17137734]

[128] Popr M. Synthesis of cyclodextrin derivatives for practical applications [PhD thesis]. Prague: Charles University; 2016.

[129] Muankaew C, Loftsson T. Cyclodextrin-Based Formulations: A Non-Invasive Platform for Targeted Drug Delivery. Basic Clin Pharmacol Toxicol 2018; 122(1): 46-55.

[http://dx.doi.org/10.1111/bcpt.12917] [PMID: 29024354]

[130] Merkus F, Verhoef JC, Marttin E, Romeijn SG, Hermens WA, Schipper NG. Cyclodextrins in nasal drug delivery. Adv Drug Deliv Rev 1999; 36(1): 41-57.
[http://dx.doi.org/10.1016/S0169-409X(98)00054-4] [PMID: 10837708]

[131] Cheung R, Ng T, Wong J, Chan W. Chitosan: An Update on Potential Biomedical and Pharmaceutical Applications. Mar Drugs 2015; 13(8): 5156-86.
[http://dx.doi.org/10.3390/md13085156] [PMID: 26287217]

[132] Schipper NGM, Vårum KM, Stenberg P, Ocklind G, Lennernäs H, Artursson P. Chitosans as absorption enhancers of poorly absorbable drugs. Eur J Pharm Sci 1999; 8(4): 335-43.
[http://dx.doi.org/10.1016/S0928-0987(99)00032-9] [PMID: 10425384]

[133] Dodane V, Vilivalam VD. Pharmaceutical applications of chitosan. Pharm Sci Technol Today 1998; 1(6): 246-53.
[http://dx.doi.org/10.1016/S1461-5347(98)00059-5]

[134] Bansal V, Sharma P, Sharma N, Pal O, Malviya R. Applications of Chitosan and Chitosan Derivatives in Drug Delivery. Adv Biol Res 2011; 5: 28-37.

[135] Shariatinia Z. Pharmaceutical applications of chitosan. Adv Colloid Interface Sci 2019; 263: 131-94.
[http://dx.doi.org/10.1016/j.cis.2018.11.008] [PMID: 30530176]

[136] Illum L, Jabbal-Gill I, Hinchcliffe M, Fisher AN, Davis SS. Chitosan as a novel nasal delivery system for vaccines. Adv Drug Deliv Rev 2001; 51(1-3): 81-96.
[http://dx.doi.org/10.1016/S0169-409X(01)00171-5] [PMID: 11516781]

[137] Aspden TJ, Mason JDT, Jones NS, Lowe J, Skaugrud Ø, Illum L. Chitosan as a nasal delivery system: the effect of chitosan solutions on in vitro and in vivo mucociliary transport rates in human turbinates and volunteers. J Pharm Sci 1997; 86(4): 509-13.
[http://dx.doi.org/10.1021/js960182o] [PMID: 9109057]

[138] Suhail M, Janakiraman AK, Khan A, Naeem A, Badshah SF. Surfactants and their role in Pharmaceutical Product Development: An Overview. J Pharm Pharm 2019; 6: 72-82.

[139] Rosen MJ. Characteristic Features of Surfactants. In: Surfactants and Interfacial Phenomena. 3rd ed. Hoboken (NJ): John Wiley & Sons; 2004. p. 1–33.

[140] Chhajed S, Sangale S, Barhate SD. Advantageous Nasal Drug Delivery System: A Review. Int J Pharm Sci Res. 2011; 2(6): 1322–1330.
[http://dx.doi.org/10.1002/9781118228920.ch1]

[141] Kulkarni VS, Shaw C. Essential Chemistry for Formulators of Semisolid and Liquid Dosages. 1st ed., London, United Kingdom: Academic Press 1999.

[142] De S, Malik S, Ghosh A, Saha R, Saha B. A review on natural surfactants. RSC Advances 2015; 5(81): 65757-67.
[http://dx.doi.org/10.1039/C5RA11101C]

[143] Saravanan V, Vijayakuma S. Biosurfactants-Types, Sources and Applications. Research Journal of Microbiology 2015; 10(5): 181-92.
[http://dx.doi.org/10.3923/jm.2015.181.192]

[144] Pavlović N, Goločorbin-Kon S, Ðanić M, *et al.* Bile Acids and Their Derivatives as Potential Modifiers of Drug Release and Pharmacokinetic Profiles. Front Pharmacol 2018; 9: 1283.
[http://dx.doi.org/10.3389/fphar.2018.01283] [PMID: 30467479]

[145] Moghimipour E, Ameri A, Handali S. Absorption-Enhancing Effects of Bile Salts. Molecules 2015; 20(8): 14451-73.
[http://dx.doi.org/10.3390/molecules200814451] [PMID: 26266402]

[146] Moiseev RV, Morrison PWJ, Steele F, Khutoryanskiy VV. Penetration Enhancers in Ocular Drug Delivery. Pharmaceutics 2019; 11(7): 321.

[http://dx.doi.org/10.3390/pharmaceutics11070321] [PMID: 31324063]

[147] Faustino C, Serafim C, Rijo P, Reis CP. Bile acids and bile acid derivatives: use in drug delivery systems and as therapeutic agents. Expert Opin Drug Deliv 2016; 13(8): 1133-48. [http://dx.doi.org/10.1080/17425247.2016.1178233] [PMID: 27102882]

[148] Stojančević M, Pavlović N, Goločorbin-Kon S, Mikov M. Application of bile acids in drug formulation and delivery. Front Life Sci 2013; 7(3-4): 112-22. [http://dx.doi.org/10.1080/21553769.2013.879925]

[149] Lee WA, Lu HFL, Maffuid PW, *et al.* The synthesis, characterization and biological testing of a novel class of mucosal permeation enhancers. J Control Release 1992; 22(3): 223-37. [http://dx.doi.org/10.1016/0168-3659(92)90097-B]

[150] Longenecker JP, Moses AC, Flier JS, Silver RD, Carey MC, Dubovi EJ. Effects of sodium taurodihydrofusidate on nasal absorption of insulin in sheep. J Pharm Sci 1987; 76(5): 351-5. [http://dx.doi.org/10.1002/jps.2600760502] [PMID: 3309255]

[151] Hermens WAJJ, Hooymans PM, Verhoef JC, Merkus FWHM. Effects of absorption enhancers on human nasal tissue ciliary movement in vitro. Pharm Res 1990; 7(2): 144-6. [http://dx.doi.org/10.1023/A:1015872617511] [PMID: 2408030]

[152] Drescher S, van Hoogevest P. The Phospholipid Research Center: Current Research in Phospholipids and Their Use in Drug Delivery. Pharmaceutics 2020; 12(12): 1235. [http://dx.doi.org/10.3390/pharmaceutics12121235] [PMID: 33353254]

[153] Djupesland PG, Skretting A. Nasal deposition and clearance in man: comparison of a bidirectional powder device and a traditional liquid spray pump. J Aerosol Med Pulm Drug Deliv 2012; 25(5): 280-9. [http://dx.doi.org/10.1089/jamp.2011.0924] [PMID: 22251061]

[154] Reger MA, Watson GS, Green PS, *et al.* Intranasal insulin improves cognition and modulates β-amyloid in early AD. Neurology 2008; 70(6): 440-8. [http://dx.doi.org/10.1212/01.WNL.0000265401.62434.36] [PMID: 17942819]

[155] Craft S, Baker LD, Montine TJ, *et al.* Intranasal insulin therapy for Alzheimer disease and amnestic mild cognitive impairment: a pilot clinical trial. Arch Neurol 2012; 69(1): 29-38. [http://dx.doi.org/10.1001/archneurol.2011.233] [PMID: 21911655]

[156] Djupesland PG. Nasal drug delivery devices: characteristics and performance in a clinical perspective—a review. Drug Deliv Transl Res 2013; 3(1): 42-62. [http://dx.doi.org/10.1007/s13346-012-0108-9] [PMID: 23316447]

[157] Hoekman JD, Ho RJY. Enhanced analgesic responses after preferential delivery of morphine and fentanyl to the olfactory epithelium in rats. Anesth Analg 2011; 113(3): 641-51. [http://dx.doi.org/10.1213/ANE.0b013e3182239b8c] [PMID: 21709146]

[158] Xi J, Zhang Z, Si X. Improving intranasal delivery of neurological nanomedicine to the olfactory region using magnetophoretic guidance of microsphere carriers. Int J Nanomedicine 2015; 10: 1211-22. [http://dx.doi.org/10.2147/IJN.S77520] [PMID: 25709443]

[159] Colović MB, Krstić DZ, Lazarević-Pašti TD, Bondžić AM, Vasić VM. Acetylcholinesterase inhibitors: pharmacology and toxicology. Curr Neuropharmacol 2013; 11(3): 315-35. [http://dx.doi.org/10.2174/1570159X11311030006] [PMID: 24179466]

[160] Lindup WE, Orme MC. Clinical pharmacology: plasma protein binding of drugs. BMJ 1981; 282(6259): 212-4. [http://dx.doi.org/10.1136/bmj.282.6259.212] [PMID: 6779954]

[161] Fonseca LC, Lopes JA, Vieira J, *et al.* Intranasal drug delivery for treatment of Alzheimer's disease. Drug Deliv Transl Res 2021; 11(2): 411-25. [http://dx.doi.org/10.1007/s13346-021-00940-7] [PMID: 33638130]

CHAPTER 13

Case Studies and Patents in Alzheimer's Disease Research

Deepak Kumar[1], **Piyush Anand**[1], **Juhi Tiwari**[1] and **Shashi Kant Singh**[1,*]

[1] *Faculty of Pharmaceutical Sciences, Mahayogi Gorakhnath University, Gorakhpur, Uttar Pradesh 273007, India*

Abstract: Alzheimer's disease is a progressive neurodegenerative disorder that affects cognition, memory, and neuronal loss. Treatment strategies remain largely underdeveloped, despite numerous studies. The chapter reviews the most recent case studies and patents to further novel therapeutic approaches in the treatment of AD. Case studies can depict the potential to exemplify the patient's response in the course of such therapy, where personal treatment choices are considered in conjunction with the efficacy of newly introduced therapies. Patents are the indicators of innovative drug discovery, biomarkers, and neuroprotective strategies. We focus on patents covering Multi-Target Directed Ligands (MTDLs), monoclonal antibodies, and small molecules targeting amyloid-beta (Aβ) and tau proteins. We also scan patents centered around cholinesterase inhibitors, neuroinflammation modulators, and techniques in gene therapy. Case studies that involve success stories of drug development in clinical settings, along with personalized medicine approaches, are also examined. New biotechnological interventions like drug delivery systems *via* nanotechnology and AI-based diagnostic tools are emerging to transform AD treatment. Furthermore, regulatory policies and intellectual property rights also create an atmosphere conducive to speeding or slowing the commercialization of innovative products for AD. Within view, this chapter gives an integrated view of current patents with a case study of the management of AD concerning their real-world applications, as well as current discoveries in its management.

Keywords: Alzheimer's disease, Amyloid-beta, Artificial intelligence, Biomarkers, Case studies, Cholinesterase inhibitors, Drug discovery, Gene therapy, Neurodegeneration, Neuroinflammation, Patents, Personalized medicine, Tau protein.

[*] **Corresponding author Shashi Kant Singh:** Faculty of Pharmaceutical Sciences, Mahayogi Gorakhnath University, Gorakhpur, Uttar Pradesh 273007, India; E-mail: principal.fps@mgug.ac.in

Prashant Kumar Dhakad, Raghav Mishra, Shweta Sharma, Sudhanshu Mishra & Nishant Gaur (Eds.)

INTRODUCTION

The global incidence of dementia, particularly Alzheimer's disease, has seen a significant rise over the last thirty years. This concerning trend underscores the urgent requirement for effective intervention strategies. While various therapeutic agents have been developed to alleviate symptoms, delay onset, or even arrest or reverse the progression of this challenging condition, results to date have yielded limited success. Consequently, research and development efforts focused on this pressing issue have been expanding in response to the high demand for effective solutions. However, the low efficacy outcomes reported in numerous clinical trials have highlighted the need for coordinated and collaborative initiatives [1]. These collaborative efforts are designed to enhance knowledge sharing and promote mutual learning among diverse stakeholders, including researchers, clinicians, and organizations engaged in addressing this intricate challenge. In this context, we will examine patents and scientific publications to pinpoint the most promising constructs and strategies currently underpinning therapeutic approaches. This analysis is part of a broader strategic and forward-looking examination aimed at identifying the most effective avenues for combating dementia. Dementia itself is a syndrome composed of various conditions, all marked by a progressive decline in cognitive function, primarily affecting older adults. Among these conditions, Alzheimer's disease is the most common and recognized cause, further illustrating the complexities of this syndrome [2]. The likelihood of developing Alzheimer's disease increases sharply after the age of 60, and the association between aging and the disease's onset is concerning. For example, global life expectancy was approximately 66.4 years in 1980, whereas it now averages around 71 years. This increase in life expectancy serves as a significant risk factor for the growing prevalence of Alzheimer's disease, as it also reduces the risk of mortality from other, more prevalent causes. This situation highlights the critical need for continued research and innovative approaches to better understand Alzheimer's disease and to develop effective strategies to mitigate its substantial impact on individuals and society as a whole [3].

UNDERSTANDING AD

As the world's population ages, the prevalence of chronic neurodegenerative diseases such as Alzheimer's disease becomes an urgent health concern for society. Early and accurate diagnosis is essential for timely and effective intervention; however, most conventional diagnostic methods have limited effectiveness in early diagnosis [4]. To find effective biomarkers for early diagnosis of Alzheimer's disease, researchers investigate the applications of patents and publications in Alzheimer's disease research. In this entry, two case studies are presented: (1) the application analysis of Alzheimer's disease patents based on a database; and (2) a general introduction to the Alzheimer's disease research literature, followed by an example showing how to analyze the co-occurrence of keywords in Alzheimer's disease research. Together with the previous chapter, we hope it will provide a good basis for patent- and literature-based research on Alzheimer's disease. Fig. (**1**) depicts the features analyzed in the post-mortem brain for the treatment of AD with the help of different available information after case studies [5, 6].

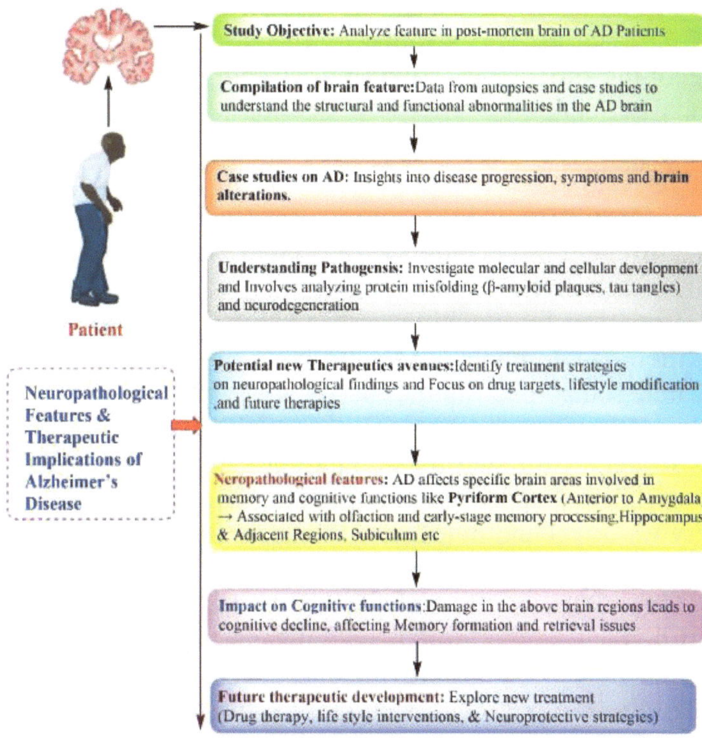

Fig. (1). Depicts a treatment strategy for AD through a case.

Alzheimer's disease is the most common form of dementia. Pathological features include the formation of senile plaques and neurofibrillary tangles in the brain. To diagnose Alzheimer's disease, one must identify two types of basic lesions, which are amyloid deposits and neurofibrillary tangles. Amyloid deposits are mainly amyloid βA4 protein, and this defect is particularly useful in diagnosing Alzheimer's disease histopathologically [7, 8]. However, this is usually only done in autopsies. Alzheimer's disease is not just a natural decline in cognitive function but involves multiple cognitive dysfunctions. As a neurodegenerative disease, the most prominent clinical manifestation is a decline in cognitive function. In terms of progress, Alzheimer's disease usually has a largely insidious onset. Postmortem examinations of the brains of patients with late-onset Alzheimer's disease reveal the presence of both amyloid plaques and neurofibrillary tangles [9, 10].

GLOBAL PREVALENCE AND SOCIO-ECONOMIC IMPACT

Alzheimer's Disease (AD) has been widely recognized as the most prevalent and common form of dementia afflicting the elderly population. Its prevalence is not just a static figure; rather, it is steadily increasing as a direct consequence of the world's burgeoning older population and demographic changes. This condition presents global effects that are both economically substantial and socially significant; the total financial cost associated with Alzheimer's has been estimated to be quite significant and is expected to reach an even higher mark shortly due to the continuous rise in cases [11]. The increased financial burden of long-term care services is significantly and closely associated with decreased well-being and mental health of the caregiver, which ultimately complicates the situation even further. Therefore, the far-reaching costs of this debilitating disease extend not only to the afflicted individuals but also to families and society at large, creating a ripple effect that can impact numerous lives beyond just those diagnosed [12]. Unfortunately, at present, the only available alternatives to traditional management strategies utilized for alleviating symptoms of Alzheimer's disease are a limited number of approved pharmacological drugs. There is an urgent and clear necessity for novel and innovative therapies to be developed, and significant emphasis must be placed on intensified research efforts right now to forge new avenues for the treatment of Alzheimer's disease [13]. Moreover, the rising prevalence of dementia on a global scale means that pharmaceutical companies are facing an absolute need to innovate and develop new products to effectively treat dementia. This development is crucial for them to sustain and maintain their competitive market position while addressing an ever-growing public health concern [14].

CHALLENGES IN AD RESEARCH AND TREATMENT

AD is a major neurodegenerative disease, with a continuous increase in the occurrence of the aging population. Nevertheless, a group of promising drug candidates in clinical studies has failed to show significant effects on patients, which underlies the outdated methodology to characterize drug candidates and patient populations and recruit patients [15]. It takes time and space to translate basic research into clinical or pharmaceutical practice. The latest developments in translational achievements in disease-relevant research are necessary for the strategic shift in Alzheimer's drug development, such as setting suitable drug candidate screening criteria, distinguishing characteristics of different patient subgroups, and discussing prevention strategies. This aids in establishing the therapeutic benefits of multicomponent drugs used in chronic diseases like AD, where different pathophysiologic mechanisms operate in different individuals [16, 17]. Characterization of certain therapeutic indications can indicate favorable properties of a drug candidate. This may highlight new therapeutic uses of old drugs or drugs that can be candidates for investigation as a starting point toward innovative therapy [18]. The patent search and textual data corresponding to a novel utilization may provide the possibility to screen drugs that could be involved in personalized or tailored medicine approaches for cost-effective solutions for therapy and prevention [19]. Overall, these investigations support the assumption that the knowledge of existing information may be important to develop new patents in therapeutic approaches for pathological conditions of interest. Current models of dementia care are based on a poorly integrated systems-level understanding of this complex disease, which causes many researchers to rely on incomplete models due to a lack of unbiased large-scale molecular measurements. Our analysis suggests that bioinformatic analysis tools and freely accessible tools can be a useful platform for integration [20, 21].

CASE STUDIES ON ALZHEIMER'S DISEASE

The term Alzheimer's disease is referenced in millions of patents, articles, books, and dissertations that are diligently analyzed by researchers spanning various disciplines from all corners of the globe. There are over 650,000 dedicated scientists who are actively registered and involved in the mission to combat Alzheimer's disease. Despite this impressive number of participants in the fight against this condition, the number of approved medications remains frustratingly stagnant. Nonetheless, it is also important to acknowledge that there are notable successes in the field [22]. These successes are not necessarily tied solely to the chemical properties or nature of the new drugs that are being developed for therapeutic purposes. Rather, they are significantly connected to the fact that the populations affected by Alzheimer's disease are increasingly living longer, which in turn leads to the emergence of new problems and challenges [23]. This demographic shift paves the way for the identification and discovery of new genetic factors and variants that may have a propensity either for or against the development of Alzheimer's disease itself. Table **1** describes recent case studies with their key findings, limitations, and treatment approach for the treatment of AD [24, 25].

Table 1. Describes recent case studies for the treatment of AD.

S.No	Case Study Type	Years	Key Findings	Limitations	Treatment Approach	References
1.	Palliative & End-of-Life Care	2021	Advance Care Planning, Hospice Care Benefit	Communication Barriers, Variability in Access to Care, Limited Research on AD-Specific Palliative Care, EHR Screening Challenges	1. Reducing symptom distress 2. Minimizing burdensome treatments 3. Enhancing caregiver support through shared decision-making and specialized palliative care	[26]
2.	Non-Pharmacological Therapy	2016	Improved Mood & Social Reintegration, Stabilized Cognitive Decline, Effectiveness of Outdoor Exposure,	Lack of a Standardized Community Care Model, Challenges in Scaling the Model, and Reliance on MMSE as the Primary Measure	1. Social Engagement 2. Cognitive Stimulation 3. Caregiver and Community Support 4. Multimodal Therapy Approach	[27]
3.	Lifestyle Interventions	2024	Feasibility of Multimodal Lifestyle Interventions, Positive Impact of Lifestyle and Medical Food on Cognitive and Cardiovascular Health, and Higher Adherence in Combined Interventions	Short Study Duration & Small Sample Size, Demographic Limitations and Recruitment Challenges, Challenges in Measuring Lifestyle Changes	1. Dietary Interventions 2. Physical Activity 3. Cognitive Training & Social Engagement 4. Multimodal Approach	[28]
4.	Gene Therapy	2019	The amyloid precursor protein (APP) gene has been implicated in a novel PSEN2 mutation of adenosine replacing cytosine at codon 222	Lack of Family Genetic Testing, Influence of ApoE4 Allele, Small Sample Size, and Novel Mutation	1. Symptomatic Management 2. Targeting Aβ Accumulation 3. Investigational Therapies 4. Family Genetic Counseling	[29]

(Table 1) cont.....

S.No	Case Study Type	Years	Key Findings	Limitations	Treatment Approach	References
5.	Neuromodulation	2018	Severe Volume loss, Amyloid Beta deposition	No Effect parameters, PET process not implemented, Mechanism of cognitive changes s	1. Postoperative Management and Monitoring 2. Targeted Neuroimaging 3. Thermoregulatory Observations	[30]
6.	Sleep & Circadian Rhythm	2021	No Significant Differences in Circadian Parameters, Phase Relationship Changes in AD Patients, Skin Temperature Variability,	Home Environment Influence, Small Sample Size, Lack of Severe AD Cases	1. Circadian Rhythm Stabilization Strategies 2. Environmental and Behavioral Interventions	[31]
7.	Gut-Brain Axis	2025	Cognitive impairment, Probiotics show effectiveness	Variations in outcome measurements, variations in strains and dosages of probiotics treatment, and AD pathological conditions created in animal models	1. long-term use of probiotics on memory impairment and cognitive function should be assessed 2. The safety of each strain or formula is the primary concern 3. Different species respond differently to the same interventions	[32]

(Table 1) cont.....

S.No	Case Study Type	Years	Key Findings	Limitations	Treatment Approach	References
8.	Social & Environmental Factors	2003 to 2011	Physical and Social Environments Impact AD Risk, Urbanization Levels Modify Environmental Effects, Community Centers, and Individual Factors	Disease Misclassification, Confounding Variables, Limited Exposure Time, Lack of Biomarkers and Imaging Data	1. Incorporate biomarkers (*e.g.*, amyloid PET, CSF biomarkers) to improve diagnostic accuracy. 2. Addressing Environmental and Social Factors 3. Encourage lifestyle modifications, including smoking cessation, physical exercise, and cognitive training.	[33]
9.	Longitudinal Studies	2022	Faster QoL Decline in Amyloid-Positive Patients, Faster QoL Decline in Dementia Patients, QoL Improvement in Amyloid-Negative Patients, Importance of Longitudinal Data	Selective Dropout Bias, Use of Different EQ-5D Versions, Limited Sensitivity of EQ-5D for Early AD Stage,	1. Early Detection and Intervention 2. Psychosocial Support for Amyloid-Positive Patients 3. Tailoring QoL Assessment in AD Patients 4. Evaluating Future Disease-Modifying Treatments	[34]
10.	Behavioral & Psychological Interventions	2015	CBT-Based Multicomponent Treatment for Dementia, Potential Alternative to Medication, Empowering Caregivers and Enhancing Well-Being, Potential Cost Savings and Delayed Institutionalization	Long-Term Sustainability of Benefits, Caregiver Burden and Adherence, Comparison with Other Psychosocial Interventions	1. CBT-Based Psychological Intervention 2. Caregiver Involvement and Training 3. Potential Integration into National Dementia Plans 4. Dissemination of Treatment Manual and Workshops	[35]

This chapter aims to present a comprehensive overview of the Alzheimer's research area, delving into various dimensions of this critical subject. The issue at hand is in line with a detailed "what can be patented" paper, seeking to clarify the boundaries and possibilities within the realm of intellectual property in this field. In particular, this work aims to answer a pivotal question: How do the currently approved drugs differ from those that are being offered and subsequently patented? The period in focus studies only drugs that have received approval and patents that have expired or are on the brink of expiration [36]. The main contributions of this study include a thorough identification of the various areas of technology that are of significant interest concerning the universe of Alzheimer's drugs, encompassing both the simple and the more complex compound drugs, as well as the careful selection of relevant economic indicators that reflect scientific interest within this vital area of research [37, 38].

Clinical Case Studies on Early and Late-Onset AD

Alzheimer's disease is a complex condition that is primarily characterized by a notable decline or complete loss of various cognitive functions, including memory, language, problem-solving skills, and additional cognitive abilities. These impairments significantly impact an individual's ability to perform essential everyday activities, leading to a diminished quality of life. As a way to better understand the disorder, clinical case studies have proven to be invaluable, offering essential insights into patients who are afflicted by rare forms of Alzheimer's disease. These detailed case analyses not only provide clues about altered or potentially new underlying pathogenic mechanisms of the disease but also raise the prospect for the development of innovative treatment options that may alleviate the symptoms or slow the progression of this challenging condition. This paper, therefore, aims to provide an extensive compilation of the characteristic features found in the post-mortem brains of individuals who suffer from Alzheimer's disease. The case studies included within this document vividly portray the plight of the patients and also enhance our overall understanding of the underlying pathogenesis associated with Alzheimer's disease [39]. Furthermore, these studies suggest possible avenues for the development of new therapeutics that could one day revolutionize treatment protocols. The neuropathological features that typify Alzheimer's disease tend to be demonstrable in regions of the brain that are critically involved in cognitive functions, reflecting a typically intense level of metabolic activity. Particular areas of interest include the pyriform cortex, which is located anterior to the amygdala, the hippocampus, and its adjacent regions, as well as the subiculum and the neocortical areas found on the mesial aspect of the temporal lobe. Each of these regions plays a crucial role in memory formation and retrieval, highlighting the profound impact that Alzheimer's disease has on the cognitive capabilities of affected individuals [40, 41].

Genetic and Environmental Influences in AD Progression

Genetic and environmental factors both play a significant role in the complex development and progression of the disease. The advancement in our understanding of these critical elements has been remarkably rapid through the intensive use of detailed case studies that are focused on potential treatments for this devastating ailment that affects millions worldwide [42]. Two separate and distinct model case studies are also described in detail, one specifically for Alzheimer's disease and the other for stroke, to bring these common afflictions into sharp focus and offer deeper insights into their mechanisms [43]. Furthermore, patents associated with these studies have been meticulously classified according to their general type: discovery, diagnosis/prognosis, drug development, treatment, or use, providing a structured approach to understanding the innovation landscape in this vital area of healthcare. This categorization not only aids researchers but also helps in guiding future investigations and therapeutic strategies [44, 45].

Alzheimer's disease is a large and growing health problem with three hallmarks of histologic lesions: β-amyloid plaques, neurofibrillary tangles within neurons, and neuron loss leading to brain atrophy. We have examined trends in the genetic and environmental etiologies of diseases in the research literature by looking at what terms are used, who profits through patent ideas, where the ideas come from, especially in terms of the type of organization and country, and which stage of Alzheimer's disease is tackled. Of the patents finally analyzed, the most obvious observation is that of patient care; the use of genetic tests for established known genes for Alzheimer's disease and other tests is the focus for a large majority of the patents. This is highly correlated with the trend in the literature. Unlike what is reported, this study did see a significant lag in patents, as only a small fraction of the articles examined mention patent applications. Inventors appear to be the most active in producing and commercializing Alzheimer's disease patents [46].

Case Reports on Current Therapeutic Approaches

Alzheimer's Disease (AD) stands as a complex, multifactorial disorder that challenges both researchers and clinicians alike. It is becoming increasingly clear that any prospective therapy aimed at treating AD would likely need to address a wide and varied range of underlying pathologies. In recent years, we have witnessed the emergence of several new and promising therapeutic approaches that could potentially alter the course of this debilitating disease. However, due to their infrequency and the open-label design utilized in some instances, case reports generally do not result in immediate modifications to clinical practice.

These reports are often one of the initial steps in the protracted process of drug development, which is why they typically receive limited attention from the broader medical community. In this special issue, we strive to provide a comprehensive overview of multiple therapeutic strategies under investigation. Our goal is to inspire further exploration of the broad spectrum of therapeutic options that hold the potential to lead to effective treatment alternatives for patients afflicted with Alzheimer's disease [47]. The proliferation of advertisements and promotional materials has made it relatively easy to access case reports depicting various treatments from all corners of the globe, and we are sincerely appreciative of the extensive big data approach that has been adopted in the realm of research concerning AD. Nevertheless, in this special issue, we intentionally selected a range of more innovative studies that combine substantial evidence supporting the proposed mechanisms of action. We extend our heartfelt gratitude to the authors who have dedicated their time and expertise to contribute to this body of work. The resultant limited number of publications leads us to believe that the concept of individualized therapy remains underappreciated within the medical field. By summarizing the case reports presented in this special issue, we hope to encourage more researchers and clinicians to seek, report, and engage in discussions about the numerous underrecognized potential treatment options that exist for patients suffering from the relentless progression of Alzheimer's disease. Ultimately, our aspiration is that these efforts will promote a richer dialogue within the scientific community about the diverse therapeutic avenues worth exploring for better patient outcomes in the future [48, 49].

Non-Pharmacological Interventions: A Case-Based Perspective

Connecting dementia prevention strategies with their origin in real life has the potential to promote the application of evidence-based interventions in a person-relevant manner. In this narrative review, we discuss evidence on the effects and the underlying mechanisms of three promising, but heterogeneous non-pharmacological intervention strategies, namely, cognitive training, sleep-EE--aided memory intervention, and self-efficacy training [50, 51]. We show and discuss how these single-strategy findings are related to and thus embedded in the multicomponent mode of action of two currently ongoing dementia prevention trials. Overall, our narrative review indicates the feasibility and utility of person-centered research in the field of cognitive science and dementia [52, 53].

Relatives' reports on persons with dementia frequently show that instrumental activities of daily life contribute most to restrictions in the lives of both the person with dementia and the relative. To succeed in maintaining patients' capabilities for as long as possible, profound knowledge of how the elderly brain may contribute to successful functioning in demanding tasks is needed. However, evidence indicates that established and future cognitive intervention trials frequently do not take the trial participants' complex and multifaceted living contexts into account to a satisfying degree [54]. To consider the multifactorial nature of elderly cognitive aging and to explicitly include everyday life realizations of evidence-based findings, person-relevant research in the field of adaptive cognitive interventions seems necessary [55]. Based on our recent experimental special issue, this narrative review showed that real-life conditions, as models of dementia-relevant organizational framework conditions, may promote transfer effects also among cognitively very fit participants. Our review specifically focuses on dementia prevention strategies, including cognitive training, sleep-EEG-aided memory intervention, and self-efficacy training, addressing evidence and potential transfer toward individual activity stairs within models of successful cognitive aging [56, 57].

DRUG DEVELOPMENT AND CLINICAL TRIALS

The current hypothesis concerning the main cause of Alzheimer's disease is an abnormal aggregation in the brain of the amyloid beta protein. Through several research studies, several agents that would

act to resolve the disorder have been identified and are being tested in clinical trials [58, 59]. These translational studies have resulted in inventions that can be protected by patents. Large enterprises may hold patents for different inventions, but with a common possible usefulness to them in the treatment of Alzheimer's disease. This analysis takes European patents present in the Patent Information System and includes the derivatives in these patents of European applications filed under the Patent Cooperation Treaty [60, 61].

The current prospect of drug development and clinical trials in Alzheimer's disease centers around agents that will remove the existing plaque, agents that may prevent the formation of more plaque, and agents that will mop up the amyloid protein. Since the common interest of the free market is to develop and commercialize agents as quickly and efficiently as possible to replace their costs, an essential protection for the business is offered by exclusive industrial property rights. When the enterprise exceeds an equilibrium point of basic research, patents begin to appear. At this point, we begin to see the investment in patents associated with Alzheimer's disease research [62, 63].

FDA-Approved Medications for AD: A Critical Analysis

Cholinesterase inhibitors represent one of the limited classes of medications that have gained official approval for the treatment of patients diagnosed with Alzheimer's disease. In this chapter, we take a closer look at the clinical drug development history surrounding cholinesterase inhibitors, culminating in some intriguing and noteworthy conclusions. It is essential to highlight that, as of now, there exists no definitive clinical trial evidence that unequivocally demonstrates the efficacy of any of the cholinesterase inhibitors that are currently available on the market [64, 65]. In particular, none of these drugs has managed to showcase even a modest benefit when it comes to the cognitive subscale of the Alzheimer's Disease Assessment Scale, especially once the confounding influence of practice, as evidenced by the performance of the placebo group, is taken into consideration and subsequently removed from the overall analysis. The present data available on cholinesterase inhibitors fail to exhibit any convincing evidence of disease-modifying actions. The implications stemming from these empirical findings will be examined in detail, particularly about future clinical research endeavors involving cholinesterase inhibitors as they pertain to the complex landscape of Alzheimer's disease [66, 67]. The clear responsibility to ensure the quality, efficacy, and safety of pharmaceutical drugs has, at times, necessitated the approval of less-than-ideal compounds for the treatment of moderate-to-severe Alzheimer's disease. Indeed, the approval of every drug for the treatment of Alzheimer's disease rests solely on the results of a single phase III clinical trial [68]. Except for selegiline, which is a selective monoamine oxidase B inhibitor, and the non-steroidal anti-inflammatory drugs, none of the approved compounds show any disease-modifying effect, and the non-steroidal anti-inflammatory drugs have a large number of potential side effects [69, 70].

Experimental Therapies and Ongoing Clinical Trials

To significantly accelerate the discovery, development, and ultimately the delivery of effective therapies for Alzheimer's disease, the role of intellectual property—which is determined and protected through patents—plays a crucial role in the translation of both academic and industry research into viable commercial products. This chapter thoroughly addresses the most recent advancements concerning experimental protein or RNA-based amyloid β therapies and their associated patents, as well as the substantial number of ongoing clinical trials that focus on these experimental therapeutic approaches [71]. Most applications that are likely to contain therapeutic proteins relevant to this field have already been filed and granted, indicating a robust framework for innovation. The research also comprehensively identified a range of significant inventions that relate specifically to therapies for Alzheimer's disease. Additionally, many ongoing commercial clinical trials, which are crucial for validating these innovations, were identified along with potential patent families that could arise from disclosures made during these clinical trials, as well as potential new patent filings that may result from recently published scientific papers in the field [72, 73]. The research conducted an extensive identification of patents or application documents that specifically consist of the targeted keywords "Alzheimer's disease" and "therapy," which encompass a wide range of relevant topics, including amyloid proteins, RNA therapies, and various other experimental therapeutic candidates that are currently being explored [74, 75]. The main aim of this chapter is to thoroughly present the comprehensive number of patents that were either filed or granted during the significant period between the years 2000 and 2018. Additionally, it seeks to highlight the major patent assignees and applicants that were active in this particular timeframe, alongside detailing the number, phase, and status of ongoing clinical trials related to Alzheimer's disease [76]. This detailed process of counting, categorizing, and further counting was systematically conducted with the helpful aid of specialized software designed to facilitate such research endeavors [77, 78].

Multi-Target Approaches for AD Management

Alzheimer's Disease (AD) is a multifactorial pathology with many molecular players involved. Thus, knocking down only one specific target has so far resulted in manifest failure. However, the multitarget paradigm may be the cornerstone that leads to a successful cure for AD. The multitarget paradigm is attracting considerable attention as a promising strategy for the development of multitarget drugs for AD. These multitarget drugs may offer hope for preventing, retarding, or curing complex neurodegenerative disorders such as AD due to their synergistic or additive effects at the molecular, biochemical, and pharmacological levels [79, 80]. To maximize this paradigm, a multitarget approach tackling the two major histopathological features of the disease, the production of amyloid-β peptides and the neurofibrillary tangle formation, is required. In this paper, an updated review of recent patents, together with selected milestones in the field of multitarget-directed ligands as a strategy for the development of effective medications and therapeutic agents against AD, will be presented [81, 82]. Patents indexed in the comprehensive database spanning the years from 2006 to 2016 were meticulously searched. The specific criteria employed were focused on identifying active patents that are owned by international companies. The review presented herein offers a thorough summary of the active patents concerning the treatment of Alzheimer's Disease (AD) within prominent international companies located in two major developed nations: the United States and Germany [83]. Our extensive review showcases not only the most promising directives but also highlights the active patents in the two corresponding countries, which are crucial for the further development of effective drugs that modify the course of AD. Additionally, a brief mini-review of selected significant milestones was thoroughly considered in our analysis. This particular term refers specifically to patents concerning either a new active substance or the innovative use of an existing active substance. This exploration into patents aims to provide insight into current trends and advancements in the fight against Alzheimer's disease [84, 85].

INTELLECTUAL PROPERTY AND PATENT TRENDS IN ALZHEIMER'S DISEASE

More than 72,000 patents in the field of Alzheimer's Disease (AD) research have been published and granted worldwide; 63% are in force at present. Technology transfer in the field reached its peak of activity in 2010, for patent applications as well as for the number of assignees involved in the registration of these patents [86, 87]. Worldwide, 28 countries stand out as leaders in patent activity in AD research. These countries are China, the United States, France, Japan, and Spain. In countries like China, the number of patents is increasing aggressively, while in European countries, it is more stable [88, 89]. Although the European Union leads the project in terms of federated AD research, the number of patents in force in this territory is less than 4%. China and the United States accumulate more than 60% of the patents in force [90, 91].

The most represented areas in AD research among academic and commercial institutions include the treatment of ailments by altering genes, clinical tests on the diagnosis of blood, and the diagnosis of genetic diseases [92, 93]. Technologies such as these are already commercially available, and the application of different approved drugs to a variety of neurodegenerative disorders is being used. There is a large patent landscape with a high level of competition, with glatiramer acetate, donepezil, mitoxantrone, and iron-based MRI contrast agents playing important roles [94]. Although with lower intensity, the number of patented technologies such as radiotherapy, neurostimulation, electrical stimulation, or deep brain stimulation is still increasing. These results confirm that the relevance of AD is not given only by the increase in its incidence, but also by economic and social factors that are increasing commercial competition [95, 96].

Overview of Patent Filings in AD Research

Alzheimer's Disease (AD) represents the most significant cause of cognitive impairment in later stages of life. Unfortunately, there has yet to be a developed disease-modifying therapy that effectively alters the trajectory of this debilitating condition. While case studies have historically constituted an essential aspect of clinical medicine, it is noteworthy that only a limited number of studies have genuinely focused on therapeutic interventions or medications specifically for addressing the complexities of this disease [97, 98]. To effectively bridge the existing gap between clinical practice and the ongoing endeavors within neuroscience research, patents are increasingly being sought for various inventions that may hold promise, albeit sometimes with only preliminary supportive evidence available. As a result, conducting an intensive and comprehensive review that centers on these critical aspects bears significant merit, as it serves to direct ongoing translational research efforts while simultaneously offering additional inspiration for future investigations. Table **2** describes the current status of patents with their patent number, title, inventor, description, and patent year related to AD treatments [99, 100].

Table 2. Describes the current status of patents related to the treatment of AD.

S.No	Patent Number	Title	Inventor	Description	Year	Status	References
1.	US4663318A	AD Treatment with Galantamine	Synaptech Inc	"An effective Alzheimer's disease cognitive-enhancing dosage of galantamine or a pharmaceutically acceptable acid addition salt thereof is administered to mammals, including humans, as part of a treatment approach for Alzheimer's disease and related dementias. Additionally, a radioactively labeled molecule version could be used as a diagnostic test for Alzheimer's disease".	1987	Expired	[101]
2.	US4897388A	Method of Treating AD using Vitamin D_2 and D_3	Hartmut H. Malluche, Versailles, Ky.	"To cure Alzheimer's disease, the current invention involves giving a pharmaceutically acceptable, physiologically active vitamin D3 or D2 substance to the individual who needs it. Parenterally, topically, or orally, these components are applied. The way a substance is administered, its activity, the patient's unique state, and the illness stage all affect how much of a substance should be taken".	1990	Expired	[102]
3.	US4758430A	Method of treatment of Alzheimer's disease using phytic acid	Robert Sabin	"Giving a patient a dose of a compound chosen from the group that includes phytic acid, phytate salt, an isomer or hydrolysate of phytic acid or a phytate salt, or a mixture of any combination of these, is one of the steps in the current invention's Alzheimer's disease treatment method. The oral route is the recommended administration method."	1988	Expired	[103]

(Table 2) cont.....

S.No	Patent Number	Title	Inventor	Description	Year	Status	References
4.	US4419365A	Method of treating Alzheimer's disease	Donald R. McLachlan	"When compared to the untreated group of eleven patients, the six deferoxamine-treated individuals showed a significantly smaller deterioration in brain function as assessed by each of the three criteria used within a comparable test interval".	1983	Expired	[104]
5.	US20050112543A1	Method of screening for drugs useful in treating Alzheimer's disease	General Hospital Corp	"The methods for finding potential pharmaceutical compounds to treat and/or prevent Alzheimer's disease and/or similar pathological conditions are the subject of the invention".	2005	Abandoned	[105]
6.	US5705401A	Method of assaying for Alzheimer's disease	University of Melbourne	"The current invention relates to a technique for diagnosing Alzheimer's disease in humans by measuring the relative amounts of one or more forms of amyloid precursor protein (APP) or the enzyme that produces them in the blood. It also relates to a treatment strategy that involves adjusting the way heparin, divalent cations, and/or trivalent cations interact with APP".	1998	Expired	[106]
7.	US4666829A	Polypeptide marker for Alzheimer's disease and its use for diagnosis	University of California	"The highly pure form of Alzheimer's Amyloid Polypeptide (AAP), a new polypeptide that is extracted from amyloid plaques in Alzheimer's disease patients, is supplied. The polypeptide's amino acid sequence is as follows: These are H2N-AS- -ALA-GLU-PHE-ARG-HIS-ASP- -SE- -GLY-TYR-GLN-VAL-HIS-GLN- -LY- -LEU-VAL-PHE-PHE-ALA-GLU-ASP- +TR -VAL-GLY-SER-ASN-LYS-COOH. The polypeptide or its fragments could be utilized to create antibodies that are helpful in an Alzheimer's disease diagnostic test. Additionally, helpful for diagnostic purposes are nucleotide probes that match certain regions of the polypeptide".	1987	Expired	[107]
8.	US4912206A	cDNA clone encoding the brain amyloid of Alzheimer's disease	US Department of Health and Human Services	"Using an oligonucleotide probe that matches the first 20 amino acids of the Alzheimer's disease brain amyloid polypeptide, four clones were extracted from an adult human brain cDNA library. This polypeptide's known amino acid sequence was among the 97 amino acids that the sequenced clone's open reading frame coded for. Human thymus and animal brains have been found to have the 3.5 kb messenger RNA. The gene has been localized to human chromosome 21 and has been substantially conserved throughout evolution".	1990	Expired	[108]

(Table 2) cont.....

S.No	Patent Number	Title	Inventor	Description	Year	Status	References
9.	US5223482A	Recombinant Alzheimer's protease inhibitory amyloid protein and method of use	Scios LLC	The DNA sequences that encode proteins linked to Alzheimer's disease that are β-amyloid-related are revealed. A DNA sequence encoding a new protease inhibitor is also included in this document. Recombinant β-amyloid core protein, β-amyloid-related proteins, and synthetic or recombinant immunogenic peptides are all made or constructed using these sequences. Alzheimer's disease can be diagnosed by serum protein or brain fluid using antibodies produced against the recombinant proteins or immunogenic peptides derived from them.	1993	Expired	[109]
10.	US6114133A	Methods for aiding in the diagnosis of Alzheimer's disease by measuring amyloid-β peptide (x-□41)	Elan Pharmaceuticals LLC Eli Lilly and Co	"This invention offers techniques that can help with Alzheimer's disease diagnosis. The techniques entail determining the concentration of amyloid-β peptide (x-□41) in a patient's cerebrospinal fluid. In general, elevated peptide levels are not consistent with an Alzheimer's diagnosis. When combined with other testing, low levels of the peptide can yield a positive diagnosis and are consistent with the condition".	2000	Expired	[110]

To provide practical information for clinical medical investigation, we first summarized patent filings within AD research. We compared the filings from private and public sectors, *i.e.*, from academia and pharmaceutical companies. We found more filings from pharmaceutical industries, which emphasize technology development, and more filings from academia, which emphasize biological mechanisms [111]. This feature was also found in the last half-decade, in the genomic era. Preclinical research was a predominant focus in academia, while clinical-stage research was focused on by pharmaceutical companies. Additionally, we analyzed the disease coverage of these patents and observed a trend to target not only the etiological cause but also its pathological consequences, including the diseased brain. The patents discussed became more numerous and comprehensive over the years, and were increasing. This patent analysis might provide clues on emerging technologies and drug development strategies in industry and academic alliances to combat AD, guiding the direction of preclinical research designs and helping clinicians reach shared decisions with patients and care partners [112, 113].

Key Patents on Amyloid Beta and Tau-Targeting Therapies

A diverse range of innovative therapeutic strategies for individuals who are affected by Alzheimer's disease is currently under rigorous investigation across various scientific platforms. This paper provides a detailed examination of two specific therapeutic strategies: therapies aimed at modifying the aggregation of amyloid-beta, a crucial process in Alzheimer's pathology, and those that focus on directly targeting tau, another key protein that significantly contributes to the disease's progression and symptoms. The comprehensive study evaluates both cutting-edge scientific findings and extensive patent literature to methodically pinpoint the most promising and viable therapeutic approaches [114]. Furthermore, it illustrates the vital connection between academic research and industrial applications, whether these applications manifest in the form of completed market products or those still existing in the early phases of research and development. Through targeted software analysis, a substantial total of 62 patents have been meticulously identified, which underscores the active progress in this field. Additionally, a total of 75 patent families have been carefully assessed and are visually represented graphically, showcasing the diverse innovations being pursued in the quest for effective treatments for Alzheimer's disease [115, 116].

The analysis revealed a strong industrial interest in antibodies and peptides and a preference for preventing the formation of neurofibrillary tangles by stabilizing microtubules. While overall research findings favored amyloid beta as the target for therapeutic intervention or use in diagnostics, the preference for tau appears in earlier stages of research, as shown by comparison with the corresponding patents [117]. The seriousness of the disease has also been reflected in fast-track regulatory procedures for the majority of the patents. The study gives additional weight to the observation of highly interdisciplinary approaches and the existence of collaborations between academia and industry [118, 119].

Nanotechnology and Drug Delivery Patents for AD Treatment

Dendrimers are a significant category of substances employed as drug delivery systems in innovative therapies. They have a well-established presence in scientific literature and have been the subject of extensive analysis in various review articles. Each dendrimer is structured around a central initiator core from which multiple chains radiate. These chains consist of several repeating units, commonly referred to as generations, with specific inventions detailing the number of generations involved. While some chains from earlier generations may remain unchanged, others typically need to be functionalized [120, 121]. The choice of functional groups is influenced by several factors, including the type of molecule to be solubilized and/or delivered. A diverse range of hydrophilic and hydrophobic branch combinations can be synthesized based on the characteristics of the target molecule that requires solubilization in water, as well as considerations related to steric effects, intermolecular interactions, and generation concentration. Each configuration offers unique benefits depending on the parameters analyzed. For instance, hydrophobic interactions may develop at higher generations; the hydrophobic chains can become integrated within the structure, utilizing the entropy of the solvent to disrupt water organization and improve the solvent's interaction with the dendrimer. This process is vital for forming molecular complexes between the dendrimer and hydrophobic substances [122, 123].

Patents on Phytochemicals and Natural Compounds for AD

The patent lists reveal that compounds specific for enzyme inhibitors and antagonist compositions are predominantly concentrated on the BACE1 enzyme inhibitor. Within this comprehensive list of patents, there exists a relatively higher number of natural compounds, phytochemicals, and products that are derived from an array of food sources, diverse plants, and various extracts. These patents further disclose the compound as either a singular compound or as a combination of various compounds, which are utilized for alternative treatment approaches, and preventive health measures, and are also included as components of dietary supplements aimed at combating Alzheimer's disease within their claims. In this particular chapter, we aim to provide more detailed and extensive information on the potent natural compounds sourced from plants, while discussing some that lack patents or are deemed too late for optimizing the chances of securing patents, and we will explore the other compounds that can serve as starting molecules for the new development and potential directions in anti-Alzheimer's disease drug research. The current patent lists found in the Natural Compounds and Phytochemicals for Alzheimer's Disease discovery section are thoroughly reviewed based on the monthly issues of the United States Patent and Trademark Office Official Gazette for the specific time frame from June 2020 to May 2021, thereby capturing the United States perspective on this subject matter. The timing of our search is solely indicative of the access to the available information regarding the issuance of these patents and does not necessarily reflect the importance or the sequential development of the various projects initiated by researchers and industrial laboratories over time. The timeline charts depicting the progression in the discovery of these natural molecules have been meticulously generated from earlier literature reports coupled with data obtained from the Chembio database. These informative charts indicate that the contribution of individual natural compounds sourced from plants, which act as BACE inhibitors, is relatively low or rare, with many of these contributions emerging from the smaller parts of Chencovirus fruits, highlighting the uniqueness of this particular botanical source [124, 125].

Innovations in AD Diagnosis and Biomarkers

Alzheimer's disease biomarkers have emerged as one of the pioneering pharmacodiagnostics utilized to effectively stratify patients participating in clinical trials. Over the past few years, several of the most recent patents related to these biomarkers have garnered significant recognition within esteemed journals in an impressively short period, typically ranging from just 5 to 8 months. In addition to this, various global regulatory authorities have engaged in collaborative initiatives that could be perceived as strategically advancing at a pace that somewhat eclipses the traditional timelines of patent claims. Indeed, the recent collaboration between different regulatory bodies might lead to a more expeditious engagement in the development process. The concerted efforts among these authorities have fostered the creation of robust platforms dedicated to high-performance data analysis, which plays a pivotal role in accelerating drug discovery efforts. Moreover, there has also been considerable attention directed toward enhancing the strategic timeliness of preclinical regulatory review collaborations, alongside the development of observatories with similar objectives to streamline the overall process [126]. A subset of the current discrimination related to Alzheimer's disease includes critical elements such as MTL, Aβ, and phosphorylated tau. Furthermore, the Longitude Prize competition is actively considering innovative, portable solutions for point-of-care diagnostic devices that specifically address Alzheimer's disease and other neurodegenerative biomarkers. This text aims to provide a concise review of some of the most recent advancements in both the development of diagnostic methods and the creation of effective biomarkers for Alzheimer's disease. Subsequently, a series of detailed case studies that were developed in the year 2019, along with some of the notable patents that have already been granted this year, will be described in detail [127, 128]. The documented case studies focus on the significant presence of Aβ1-42 and phosphorylated tau, wherein the tau: S396 P-tau may be increasingly preferred, whether used individually or in combination with other biomarkers. In addition, this text emphasizes the importance of third-generation biomarkers, which are proving to be crucial in clinical research settings and provide insight into the underlying pathology that is associated with these biomarkers. The exploration of these advancements reflects the ongoing efforts to enhance our understanding and diagnostic capabilities concerning Alzheimer's disease and related neurodegenerative disorders [129, 130].

Patent Landscape of Biomarker-based Diagnosis

Alzheimer's disease is recognized as the leading cause of dementia globally, impacting the lives of more than 35 million individuals worldwide. This escalating figure highlights the urgency surrounding the need for effective interventions and approaches to tackle this complex illness. Unfortunately, the only available confirmatory test for diagnosing Alzheimer's disease comes with a range of cumbersome financial constraints and ethical limitations that make it challenging to implement widely. Consequently, the development of minimally invasive and efficient tests aimed at facilitating early diagnosis has emerged as one of the key focal points in the healthcare segment. In this perspective, we took the initiative to construct a comprehensive map of patents that includes innovative tests developed for screening purposes, risk prediction, early diagnosis, and even the monitoring of the progression of Alzheimer's disease over time. Our snapshot unveiled unprecedented data on the current state of the art within this critical area of research, significantly reinforcing our understanding of the innovative activity arising from research projects that are actively engaged in the creation of novel tests tailored specifically for the early diagnosis of Alzheimer's disease. Notably, there is an increasing trend towards new and innovative studies within this field, reflecting the significant interest and investment from the business sector. Despite this growing engagement, successful predictive studies that lead to the development of easy-to-use tests designed for preclinical or asymptomatic stages have proven to be fewer than initially anticipated [131, 132]. Recent advances in medical diagnosis can largely be attributed to the ongoing search for viable alternatives to the traditional models of diagnostic tools that have been in use for quite some time. These conventional tools were often invasive and presented a range of limitations in various aspects, including time consumption, costs involved, and potential side effects on the health of individuals being tested. As a result, these diagnostic approaches frequently favored the monitoring of symptoms rather than focusing on preventing issues arising from the underlying causes. Given these significant drawbacks, it is not surprising that the landscape of medical diagnostics is evolving. In this context, patent offices have witnessed, and continue to witness, a significant rise in the number of applications that are explicitly focused on commercialization efforts in this domain [133]. This particular trend is especially noteworthy when it comes to the field of diagnostic tests aimed at the early determination of biological markers. These markers are crucially important for the diagnosis of neurodegenerative diseases, which are characterized by a multifactorial basis and generally exhibit a slow progression over time. The underlying reason for this heightened interest within the medical community and the patent system is largely due to the significant clinical and economic value associated with the potential for earlier interventions on individual patients. Such timely interventions can greatly improve patient outcomes and help in managing these debilitating conditions more effectively [134, 135].

Artificial Intelligence and Machine Learning in AD Detection

Artificial Intelligence (AI) is rapidly emerging as a revolutionary game changer in a wide array of domains, with a notable impact in medical imaging, particularly within the specialized field of neuroimaging. Over the past few years, there has been tremendous and unprecedented progress in both the development and the application of AI and Machine Learning (ML) methods. These advancements are being utilized for a broad spectrum of applications, ranging from the accurate identification of common neuropathologies to the intricate detection and precise diagnosis of rare ultra-orphan diseases. Beyond merely focusing on image and pattern recognition, cutting-edge AI/ML algorithms possess the powerful ability to identify and quantify minute brain abnormalities that often arise in complex neurodegenerative disorders, demonstrating an impressive and commendable level of classification performance. The capabilities of these AI/ML algorithms extend beyond initial diagnosis; they have the potential to inform, enrich, and augment disease research across a varying range of disciplines, potentially leading to better outcomes in healthcare. Recent advancements in machine learning stand poised to significantly enhance existing methods for the early diagnosis and effective prevention of Alzheimer's disease. These pioneering techniques can likely be effectively utilized to create sophisticated predictive models when they are applied to the available biomarker data collected in real-world clinical settings. Consequently, this application could facilitate the development of effective interventions and targeted treatments, which in turn may lead to a significant reduction in the public health impact of Alzheimer's disease. This chapter will comprehensively focus on discussing the current issues and challenges that are being faced in the development of AI and ML diagnostic tools specifically designed for Alzheimer's disease (AD). These discussions will be based on in-depth analyses of both comorbidity-derived and genetics-derived case studies and patents that contribute to the ongoing research in the field of Alzheimer's disease [136, 137].

Neuroimaging and Blood-based Biomarkers in AD Diagnostics

The role of neuroscience research in biomarkers and diagnostics of Alzheimer's disease and the impact of patenting activities on innovation in the field can be tracked using the information provided by patent archives. We selected patents related to a specific study of Alzheimer's disease to compare technologies and academic output—the use of neuroimaging and omics technologies, and cerebrospinal fluid and blood-based biomarkers for Alzheimer's disease research. We examine three periods of public and private patented output in Alzheimer's disease research to show that universities, especially public research organizations, are the main contributors to Alzheimer's disease research [138, 139]. The regions with the highest patenting impact are those with the best organizations and geographical features. Rates of international cooperation are also affected by the quality of the regional system. Biomarkers and neuroimaging products may be related to different scientific paradigms that completely ignore the advancements of one another. Both have become extremely promising research areas in recent years as part of the integration of genome-wide association studies results with further technologies to advance a paradigm shift in the field of Alzheimer's disease [140]. Scientists have started to visualize gene protein interaction networks as the biomarkers of preclinical stages and heritability for Aβ and tau biomarkers with imaging-driven biomarker methods. Brain signal sensation regions can be compared with gene interaction networks developed using omics data mining based on research linked to attention deficit hyperactivity disorder. Such progress can facilitate the consideration of personalized medicine from both ontological ontogeny and imaging phenotypic diversity. The term omics technology relates to all the biological methods by which biological-process component data, including those from neuroimaging surveillance, are assessed in depth or quantitatively [141, 142].

PATENTS ON NON-PHARMACOLOGICAL APPROACHES

Building on the nuances of the non-pharmacological aspects of research in terms of clinical care, we also show the relevance of non-pharmacological interventions as a field of investigation per se, by analyzing their use in global patent families. One issue that may need further discussion is the question of whether non-pharmacological interventions, such as clinical practice or the development of specific neurorehabilitation programs to treat fragile individuals with Alzheimer's dementia, including community-based interventions, present a type of innovation that society expects from scientific research. Such a scenario further highlights how this research may be seen from different societal perspectives. With increasing longevity, age-related diseases, and progressive brain dysfunction associated with neurodegenerative diseases, such as Alzheimer's dementia, have intertwined with history, health, and civilization [143]. In a globalized world, non-pharmacological approaches, predominantly preventive and after the occurrence of the disease, such as the non-pharmacological interventions analyzed as the object of the patent demand made worldwide, are increasingly sought after by individuals, families, and even political decision-makers. In the European Union, Alzheimer's and other dementias prevention and early diagnosis, and more generally, for aging well, are top public health research priorities due to the rapid aging of the Union's population, and this position is supported by various institutions. Our findings below are interesting not only from a scientific point of view but also for understanding societal investments in innovative non-pharmacological research designed to care for those with such frailty and progressive fragility [144, 145].

Cognitive Training and Digital Therapeutics Patents

Many cognitive training patents have been filed. Some of the earliest patents were limited to video technologies, while more recent patents are directed at the software and combine elements of training, monitoring, and updating. Several patents have been filed either by the company itself or through its assignee. On the other hand, several wearable sensor patents related to monitoring and adapting the content and delivery of a cognitive training regimen based on user performance data have been filed. In addition, several patents aimed at the personalization of cognitive tasks, either through the overall selection of cognitive tasks or specific task difficulty levels, have been published. This is done by dynamically tuning cognitive tasks in embodiment programming according to users of prototype models based on target cognitive areas reached and actual performance. Moreover, a project of patentable interest is being worked on [146]. Very recently, there has been a notable increase in the interest surrounding the development of innovative digital therapeutic interventions aimed at addressing various human conditions and health challenges. The general term digital therapy refers to the strategic use of advanced digital technologies or specialized software applications in the prevention, management, and treatment of a wide range of medical conditions or diseases. The surge in the creation of digital therapeutic solutions is a direct response to the escalating number of patients suffering from chronic diseases, coupled with the rising healthcare costs and associated expenses that burden both individuals and healthcare systems alike [147]. Chronic conditions such as heart disease, depression, and Alzheimer's disease are significant contributors to this trend, leading to the steepest growth in venture capital funding in any digital health category to date. Specific examples in the cognitive health area highlight the implementation of Cognitive Behavioral Therapy techniques, biofeedback mechanisms, telehealth platforms, enhancements in spatial working memory, advancements in upper limb function therapies, innovative speech therapy approaches, and sophisticated facial recognition technologies. In this growing context, one can observe an increasing number of patents emerging in the specialized field of Alzheimer's disease, showcasing the ongoing innovation and attention dedicated to finding effective digital therapeutic solutions for this challenging and impactful condition [148].

Wearable Technology and Remote Monitoring Systems

Development of early in-home tools that allow early identification and monitoring of Alzheimer's Disease (AD) is needed so that interventions can be initiated to improve or preserve cognitive function by mechanisms such as stimulating neuronal growth or slowing neurodegeneration following diagnosis. Generally, the burden of disease includes the measurable expression of neuropathology or associated symptoms such as cognitive impairment. Problems associated with traditional methods for assessing cognition include accessibility, financial cost, and motivation to participate. For data collection, patient engagement ranges from passive measures—defined as continuous real-time data of behavior or cognition retrieved with no input from participants—to the active capture of participant-generated data. Passive measures typically use commercially available systems that collect data continuously, while active measures require active participation from the participants. Currently, the most effective way to capture particular broad-ranging measures of brain health is *via* trained healthcare professionals. These measures are based on symptoms and activities that can be monitored by caregivers, neurologists, neuropsychologists, and general practitioners through clinical history and neuropsychological testing. Unauthorized sharing of personal patient data has prevented patients from benefiting from advances in machine learning and big data analytics. Furthermore, late-stage patient diagnosis is a barrier to patient inclusion in clinical trials or drug efficacy studies that can significantly delay the approval of potential pharmaceutical treatments. The unmet needs in AD early diagnosis are estimated to be that 77% of cases that would need support remain undiagnosed worldwide. There is a need for additional studies and advances in computer technology to study and analyze the underlying basic causes of AD and cognitive deficits. Fortunately, for passive data, consent is often sought from affected family members rather than the patients themselves. Crucially, patient risk can be minimized only when significant changes in cognitive function, alongside significant changes in behavior, are reported to the patient's healthcare professional; data sharing is required for early diagnosis of the condition. Passive patient engagement presents unprecedented opportunities for capturing data and has the potential to transform healthcare and clinical study design, with obvious links to electrical engineering, computer science, and psychology. Wearable technology, in particular, has the potential to play a pivotal role in reducing the burden on traditionally delivered healthcare while increasing the stream of valuable, large-scale patient data suitable for the application of various validation techniques [149, 150].

Precision Medicine and Personalized AD Treatment Strategies

Alzheimer's disease (AD) is one of the most important health care disorders. As the number of patients has increased, treatment for AD has become the second most expensive diagnosis, and it is expected to grow. Despite several efforts to prevent or treat AD, these therapies still have not been validated. The technologies based on personalized medicine for the diagnosis, prognosis, or care of each individual are called precision medicine. The concept of precision medicine is providing more homogeneous groups of individuals affected by similar diseases. Since strategies for the personalized treatment of AD have not been established yet, we reviewed the patent applications and granted patents. Based on the granted patents, there are several relevant methods based on the identification of molecular markers, detecting the plasma amyloid precursor protein level, detecting beta-amyloid protein, or insulin resistance [151]. These strategies could be used to manage potential future AD patents. Dementia and AD/AD Spectrum Disorders are a wide spectrum. SNPs related to APOE, glutathione S-transferase mu-1, methylenetetrahydrofolate reductase, LDL-receptor-related protein 6, and others are all related to the disease. Recent progress in the personalized treatment of AD is defined as the matching of an individual patient's profile with the most effective treatment, and this individual might receive a safer version with the highest chance of success. The concept of precision medicine means providing more homogeneous groups of individuals affected by similar diseases under the same treatment. Although these concepts are not new and considerable efforts have been conducted, the strategies for the personalized treatment of AD have not been established. In this review, we are going to summarize the current strategies, especially the patents disclosed on the issue, providing a suggestion for the future management of the disease [152, 153].

Ethical, Legal, and Commercial Considerations

Government, academic, and industry investigators are separately trying to create property interests in their discoveries using the patent system. Patentees seek to exclude others from commercial use without their consent for the information contained in the covered published patent. These issues operate differently for governments, universities, and private industry. In the case of government, albeit funded by the public through taxation, sharing knowledge as widely as possible is generally paramount [154]. Universities are funded, in part, through government grants or contracts and seek public benefits. They also generate revenue through technology transfer to optimize resources. In contrast, the central concern of industry is to secure private profits. Patent law is the primary legal instrument used to secure private rights for patentees [155, 156]. For patents in biomedical research, a key issue is whether the research raises ethical considerations. With preclinical research, for example, scientists at universities engage in commercial activities through formal and informal collaborations with corporate entities. Patents, of course, are frequently held both by the scientists and the universities [157, 158]. With public funds, does commercialization *via* patents produce some benefit for the public? Frequently, university patents are licensed to established corporate entities for development or other purposes or are assigned directly to these business entities. In the preclinical space, the project manager can identify who is involved in the project, the title, the topic, the impact factor of the journal where the project was published, and the patent application number [159, 160].

Regulatory Challenges in AD Drug Patenting

Patents usually result from substantial research to develop new products or to find better ways to manufacture them. Preclinical research, the first step in developing potential new drugs, generally begins with basic biochemical studies that aim to identify and validate a molecular target. Once a target is identified, researchers can develop chemical compounds that affect the target through high-throughput screening of combinatorial libraries of small molecules [161]. If the work to identify and develop a molecule leads to a compound that appears promising as a new drug, the developer – who may be a commercial entity, government, or educational institution—files a patent application on the molecule. Once all necessary testing data is collected, the developer applies for permission to begin human clinical testing [162, 163].

Most results from government-funded research, including valuable compounds and extensive datasets, are generally not eligible for patents, except for proprietary biomedical information generated during the discovery process. Researchers often share their findings and datasets with non-profit organizations and databases. This vast array of data is subsequently presented at scientific conferences and published in research journals, which allows manufacturers and other researchers to advance their discovery and development efforts. A notable illustration of this collaborative approach is the creation of cholinesterase inhibitors for treating Alzheimer's disease. In 1977, despite notable advancements related to the cholinergic hypothesis, there was a lack of phenotypic markers for assessing the impact of lowering a patient's cholinesterase levels as a treatment method. Throughout the 1980s and 1990s, both Massachusetts and Lilly opted not to seek patents for their initiatives aimed at developing cholinesterase inhibitors [164, 165].

Ethical Aspects of Intellectual Property in AD Research

Asp—Ethical Aspects of Intellectual Property in Alzheimer's Disease Research. The concentration of economic gain associated with some types of invention is in sharp contrast to the moral imperatives associated with diseases that impose a major health burden on developed and developing societies. Patent law has been designed, traditionally, to strike a balance between competing interests when incentives for innovation were of critical importance [166, 167]. Three currently pending patent applications related to the beta-amyloid and related presenilin cleavages have raised fundamental questions regarding where to draw the line when it comes to providing exclusivity to induce researchers to share their discoveries. This paper addresses the applicability of patent law to the activities of scientists in the health field, particularly in the Alzheimer's Disease area, and embraces the positive and constructive role that industry is playing to help promote the translation of creativity into practical benefit for patients [168, 169]. Other frequently voiced concerns regarding patenting are its impact on collaboration, access to products, training, and fostering compliance with ethical guidelines and regulations in the field of research. There exists no conclusive evidence that demonstrates that patents impede research. The exact effects of patents on scientific research are still not clear, as dynamic scientific developments make it hard to provide an empirical framework to evaluate these effects. We have, however, presented counterexamples where access to some inventions has been facilitated, and claimed that having patent protection for intellectual property helps in increasing the prospect of creating scientific knowledge and aids in the dissemination of knowledge [170, 171].

Bridging the Gap Between Patents and Clinical Applications

It has been over a hundred years since Alois Alzheimer first identified the unique neurological features associated with a significant patient's condition. Much time has elapsed since that groundbreaking autopsy was performed. A topic that consistently engages both the media and the public is the continuous exploration of various approaches aimed at finding a cure or improving treatments for the considerable distress caused by this condition. This includes the numerous patents that have emerged over the years. In this analysis, we will undertake a comparative review of the main patented methods, most of which are primarily secured in the United States. Our examination will involve two distinct analytical approaches. The first method focuses on identifying potential interactions through molecular docking, which plays a crucial role in these patented techniques. Subsequently, we will juxtapose these results with analogous chemical compounds documented in the scientific literature and developed by individuals who do not possess patent rights. This comparative analysis seeks to highlight the progress achieved and the ongoing initiatives within the scientific community [172]. Recent therapeutic, diagnostic, and genetic discoveries related to diseases, with a particular emphasis on human genetic disorders, have significantly benefited from a strategic approach that incorporates the use of patents. The identification and understanding of the specific mutations associated with Alzheimer's disease are now expanding from the three principal genes initially identified to encompass a broader range of genes that play critical roles in the physiopathology of the disease [173]. This review aims to provide a comprehensive analysis of the diverse patented inventions associated with Alzheimer's disease, which can be categorized systematically according to the following types of patented technologies: therapeutic patents, diagnostic methods, genetic patents, as well as patents concerning biochemical and animal models [174]. Additionally, the review will include detailed case studies focusing on Alzheimer's disease patents, wherein we will conduct a thorough comparison between the prescribed applications and the commercial uses attributed to the patents, along with an in-depth examination of the similarities or differences regarding the chemical structures of a curated list of compounds and the corresponding compounds that are documented in the relevant patents [175, 176].

FUTURE PERSPECTIVES AND RESEARCH DIRECTIONS

It is now essential to conduct longitudinal studies to identify predictive molecular biomarkers of AD. These studies will be fundamental to determining the stages of AD. Thus, knowledge of the molecular pathways controlling the development of affected brain areas at each stage will be necessary to understand the disease's progression. Furthermore, new experimental approaches such as accurate clinical phenotyping, longitudinal data, and stratification analysis will increase the ability to detect target phenotypes. Massive parallel nucleotide sequencing technologies and study design will allow better identification of genetic variation affecting the expression of a specific target gene across several tissues [177]. The integration of genetic, transcriptomic, epigenomic, proteomic, and metabonomic data with clinical and molecular imaging has the potential to address the question of how and when molecular brain pathology arises longitudinally. These data will help us understand the specific role of brain developmental stages in AD susceptibility. Understanding the role of these crucial developmental stages in the preclinical development of neurodegenerative diseases will help us design new, rational intervention strategies, such that they will be more effective than current therapies [178, 179].

Emerging Technologies in AD Management

The last decade has seen the birth, development, and validation of several technologies that hold promise to monitor and even treat certain symptoms in Alzheimer's disease patients, such as sundowning, wandering, changes in living activities, and socialization. This work provides a better understanding of the general shift toward the development of technologies that can be used with or be of aid to both patients and their caregivers. Case studies also help in obtaining a more complete vision of the new technologies' features and goals. A more accurate overview of collaborative activity in the field of Alzheimer's disease is provided by analyzing the characteristics of the patents filed in the last 10 years. This work can be the first to consider the kind of international collaboration found and trends in technological management [180, 181].

Potential Breakthroughs in Drug Development and Innovation

Innopharma's case is centered around the critical task of conducting foundational research aimed at unraveling the complex mechanisms behind various diseases. This research is essential as it seeks to integrate a comprehensive understanding of the intricacies of disease mechanisms with the advanced technologies that are developed based on this foundational knowledge. By doing so, Innopharma aims to create innovative commercial opportunities that can positively impact healthcare. Their efforts are particularly directed towards neurodegenerative diseases, where they have developed substantial expertise in two primary research priorities [182]. As a direct result of this narrowed focus, InnoPharmaLimits was established in 2005 to provide specialized contractual research solutions tailored to address these areas where there is a significant unmet clinical need. This concentrated effort, combined with a dedicated team composed of some of the most seasoned drug developers in the neurodegenerative landscape, has facilitated the swift development of a robust and loyal customer base. These customers continuously seek out additional custom-built solutions, thereby reinforcing the value of Innopharma's innovative approach and their commitment to advancing the field of neurodegenerative disease research [183]. InnoPharmaL's core capability is in reverse engineering drugs and developing a detailed knowledge of the mechanisms of action of small molecules, using these drugs as tools to advance understanding of disease mechanisms and identifying new drug targets through comprehensive *in vivo* studies to GLP standards where necessary. Its functional heads design, perform, interpret, and report quality-assured solutions *in vitro*, in silico, or by imaging, in particular, with responsibility for projects passing through project management before delivery to the customer. Critical mass is being formed through a mixture of excellence in science, appropriate control systems, and a strong consultancy team and professional infrastructures that support such roles. Through proactive engagement with stakeholders, significant client validation, and some niche marketing wherein InnoPharmaL's expert drug discovery and development scientists are involved with existing and potential customers and as expert speakers and, in varying capacities, mentors [184].

Future of Personalized and Precision Medicine in AD

Alzheimer's Disease (AD) is the most common cause of dementia and will be an increasing medical and economic problem in the upcoming decades. The future of AD treatment and prevention aims at early and very early intervention in the time window before the onset of dementia. The advent of better characterization techniques, such as blood biomarkers or advanced EEG analysis, has the potential to close the remaining time gaps and to allow, in the future, the identification of presymptomatic patients with AD with certainty. Personalized preventive interventions can then be tested for the terminal outcome "onset of AD" in a rather short time. The invention of preventive interventions can come from two different research areas. One is case-based research, while the other is molecular, especially genetic research. Both have existing patent activity [185].

It is the perspective of scientific research that, at some point shortly, therapeutic interventions in the field of Alzheimer's disease (AD) will be offered before the onset of dementia to an increasingly large population of older and very old patients. These types of therapeutic interventions will be based on one of two concepts, namely personalized and precision medicine. In personalized medicine, the term "personalized" originates from the fact that therapeutic interventions can be personalized to patients after considering several patient-specific factors such as environmental impact, individual therapeutic response, *etc.* In contrast, the main task and goal of precision medicine is to ensure that all patients receive treatments that are specifically tailored to that person and thus contribute to a precision or individualized treatment. Similar to personalized medicine, this tailoring of medical treatment should achieve the highest likelihood of therapeutic success. The increasing individualization of medical services and products is expected to accompany – and to be accompanied by – a proportional increase in associated patent activity [186].

CONCLUSION

The discussion on patents revealed that not only is the existing scientific development on the subject instructive, but obtaining patents also has a role as an innovative vector that went unnoticed by the rest of the world. The existence and intensity of patent deposits, often filed by researchers from the main work centers, including universities, is significant. The increase in the number of patents is related to the significant increase in dementia and its treatments, which began in the second half of the past century. Currently, patents represent a distinct source of innovation and expansion of the competitive process, acting as a basis for the exchange of information between companies, governments, and associations. They are important documentation tools, and all patents should be analyzed to expand knowledge in the areas they cover. Regarding the business center, every company seeks to guarantee its commercial interests, and the protection of its property is essential. Furthermore, the detailed description of the invention or discovery concerning an application can provide an important summary of the state of the art of a specific field, and consequently, its respective patenting modality, through the addition of innovation, or a more advanced or more detailed exemplification. The inclusion of academic research work conclusions into patents is part of the valorization of the research conducted in universities and also a path toward revealing the high degree of innovation, relative to the economy's needs. To provide more information about research work, one can investigate the scope of the concepts described in the patents in more depth. However, for this, we need other public information sources, such as databases from life sciences journals, the frequency of seminars held by other entities, and symposia conferences that discuss the issues related to concepts furnished by patents, among others.

AUTHORS' CONTRIBUTIONS

Conceptualization: All the Authors contributed equally to the development of this book's chapter framework and concept. This collaborative effort ensured that a well-rounded perspective was preserved throughout the study.

LIST OF ABBREVIATIONS

AD - Alzheimer's Disease

Aβ - Amyloid-beta

MTDLs - Multi-target directed ligands

AI - Artificial intelligence

APP - Amyloid precursor protein

RNA - Ribonucleic acid

FDA - Food and Drug Administration

BACE1 - Beta-secretase 1

ML - Machine Learning

CNS - Central Nervous System

MRI - Magnetic resonance imaging

ACKNOWLEDGEMENTS

The authors are thankful to Mahayogi Gorakhnath University, Gorakhpur, for providing the necessary facilities to execute this manuscript.

REFERENCES

[1] Li X, Feng X, Sun X, Hou N, Han F, Liu Y. Global, regional, and national burden of Alzheimer's disease and other dementias, 1990–2019. Front Aging Neurosci 2022; 14: 937486. [http://dx.doi.org/10.3389/fnagi.2022.937486] [PMID: 36299608]

[2] Mishra P, Davies DA, Albensi BC. The interaction between NF-κB and estrogen in Alzheimer's disease. Mol Neurobiol 2023; 60(3): 1515-26. [http://dx.doi.org/10.1007/s12035-022-03152-3] [PMID: 36512265]

[3] Amelianchik A, Sweetland-Martin L, Norris EH. The effect of dietary fat consumption on Alzheimer's disease pathogenesis in mouse models. Transl Psychiatry 2022; 12(1): 293. [http://dx.doi.org/10.1038/s41398-022-02067-w] [PMID: 35869065]

[4] Navarro-Gómez N, Valdes-Gonzalez M, Garrido-Suárez BB, Garrido G. Pharmacological inventions for Alzheimer treatment in the United States of America: A revision patent from 2010–2020. J Prev Alzheimers Dis 2023; 10(1): 50-68. [http://dx.doi.org/10.14283/jpad.2023.2] [PMID: 36641610]

[5] Fish PV, Steadman D, Bayle ED, Whiting P. New approaches for the treatment of Alzheimer's disease. Bioorg Med Chem Lett 2019; 29(2): 125-33. [http://dx.doi.org/10.1016/j.bmcl.2018.11.034] [PMID: 30501965]

[6] Gadiya Y, Gribbon P, Hofmann-Apitius M, Zaliani A. Pharmaceutical patent landscaping: A novel approach to understand patents from the drug discovery perspective. Artificial Intelligence in the Life Sciences 2023; 3100069 [http://dx.doi.org/10.1016/j.ailsci.2023.100069]

[7] Liang CS, Li DJ, Yang FC, *et al.* Mortality rates in Alzheimer's disease and non-Alzheimer's dementias: a systematic review and meta-analysis. Lancet Healthy Longev 2021; 2(8): e479-88. [http://dx.doi.org/10.1016/S2666-7568(21)00140-9] [PMID: 36097997]

[8] Tahami Monfared AA, Byrnes MJ, White LA, Zhang Q. Alzheimer's disease: epidemiology and clinical progression. Neurol Ther 2022; 11(2): 553-69. [http://dx.doi.org/10.1007/s40120-022-00338-8] [PMID: 35286590]

[9] Lin Q, Che C, Hu H, Zhao X, Li S. A comprehensive study on early Alzheimer's disease detection through advanced machine learning techniques on MRI data. Academic Journal of Science and Technology 2023; 8(1): 281-5. [http://dx.doi.org/10.54097/ajst.v8i1.14334]

[10] Inoue Y, Shue F, Bu G, Kanekiyo T. Pathophysiology and probable etiology of cerebral small vessel disease in vascular dementia and Alzheimer's disease. Mol Neurodegener 2023; 18(1): 46. [http://dx.doi.org/10.1186/s13024-023-00640-5] [PMID: 37434208]

[11] Mirakhori F, Moafi M, Milanifard M, Asadi Rizi A, Tahernia H. Diagnosis and Treatment Methods in Alzheimer's Patients Based on Modern Techniques: The Orginal Article. J Pharm Negat Results 2022; 13(1): 1889-1907.

[12] Li YB, Fu Q, Guo M, Du Y, Chen Y, Cheng Y. MicroRNAs: pioneering regulators in Alzheimer's disease pathogenesis, diagnosis, and therapy. Transl Psychiatry 2024; 14(1): 367. [http://dx.doi.org/10.1038/s41398-024-03075-8] [PMID: 39256358]

[13] Kesika P, Suganthy N, Sivamaruthi BS, Chaiyasut C. Role of gut-brain axis, gut microbial composition, and probiotic intervention in Alzheimer's disease. Life Sci 2021; 264118627 [http://dx.doi.org/10.1016/j.lfs.2020.118627] [PMID: 33169684]

[14] Masurkar AV, Marsh K, Morgan B, Leitner D, Wisniewski T. Factors Affecting Resilience and Prevention of Alzheimer's Disease and Related Dementias. Ann Neurol 2024; 96(4): 633-49. [http://dx.doi.org/10.1002/ana.27055] [PMID: 39152774]

[15] Biswas M, Rahman A, Kaiser MS, Al Mamun S, Mizan KSE, Islam MS, Mahmud M. *Indoor navigation support system for patients with neurodegenerative diseases.* September 17–19, 2021; Springer. Proceedings 14 2021; Springer. [http://dx.doi.org/10.1007/978-3-030-86993-9_37]

[16] Hu C, Chen C, Dong XP. Impact of COVID-19 pandemic on patients with neurodegenerative diseases. Front Aging Neurosci 2021; 13: 664965. [http://dx.doi.org/10.3389/fnagi.2021.664965] [PMID: 33897410]

[17] Zhang XX, Tian Y, Wang ZT, Ma YH, Tan L, Yu JT. The epidemiology of Alzheimer's disease modifiable risk factors and prevention. J Prev Alzheimers Dis 2021; 8(3): 313-21. [http://dx.doi.org/10.14283/jpad.2021.15] [PMID: 34101789]

[18] McNabney MK, Green AR, Burke M, *et al.* Complexities of care: Common components of models of care in geriatrics. J Am Geriatr Soc 2022; 70(7): 1960-72. [http://dx.doi.org/10.1111/jgs.17811] [PMID: 35485287]

[19] Chowdhury MR, Karamveer K, Tiwary BK, Nampoothiri NK, Erva RR, Deepa VS. Integrated systems pharmacology, molecular docking, and MD simulations investigation elucidating the therapeutic mechanisms of BHD in Alzheimer's disease treatment. Metab Brain Dis 2024; 40(1): 8. [http://dx.doi.org/10.1007/s11011-024-01460-2] [PMID: 39556154]

[20] Eeza MNH, Bashirova N, Zuberi Z, Matysik J, Berry JP, Alia A. An integrated systems-level model of ochratoxin A toxicity in the zebrafish (Danio rerio) embryo based on NMR metabolic profiling. Sci Rep 2022; 12(1): 6341. [http://dx.doi.org/10.1038/s41598-022-09726-4] [PMID: 35428752]

[21] Kallmyer BA, Bass D, Baumgart M, *et al.* Dementia care navigation: Building toward a common definition, key principles, and outcomes. Alzheimers Dement (N Y) 2023; 9(3)e12408 [http://dx.doi.org/10.1002/trc2.12408] [PMID: 37533688]

[22] Possemis N, Verhey F, Prickaerts J, Blokland A, Ramakers I. A proof of concept phase II study with the PDE-4 inhibitor roflumilast in patients with mild cognitive impairment or mild Alzheimer's disease dementia (ROMEMA): study protocol of a double-blind, randomized, placebo-controlled, between-subjects trial. Trials 2024; 25(1): 162. [http://dx.doi.org/10.1186/s13063-024-08001-3] [PMID: 38438923]

[23] Kuroda Y, Fujita K, Sugimoto T, *et al.* Feasibility of a community-adapted multi-domain intervention for dementia prevention among older adults: a research protocol. Arch Public Health 2023; 81(1): 191. [http://dx.doi.org/10.1186/s13690-023-01205-0] [PMID: 37907936]

[24] Berkowitz E, Schultz A, DeStefano LH *et al.* Advancing the Health of the Public in the United States and Globally.A History of the National Academy of Medicine: 50 Years of Transformational Leadership. US: National Academies Press 2023.

[25] Giovannetti A, Susi G, Casti P, *et al.* Deep-MEG: spatiotemporal CNN features and multiband ensemble classification for predicting the early signs of Alzheimer's disease with magnetoencephalography. Neural Comput Appl 2021; 33(21): 14651-67. [http://dx.doi.org/10.1007/s00521-021-06105-4]

[26] Toles M, Kistler C, Lin FC, *et al.* Palliative care for persons with late-stage Alzheimer's and related dementias and their caregivers: protocol for a randomized clinical trial. Trials 2023; 24(1): 606. [http://dx.doi.org/10.1186/s13063-023-07614-4] [PMID: 37743478]

[27] Quail Z, Carter MM, Wei A, Li X. Management of cognitive decline in Alzheimer's disease using a non-pharmacological intervention program. Medicine (Baltimore) 2020; 99(21)e20128 [http://dx.doi.org/10.1097/MD.0000000000020128] [PMID: 32481282]

[28] Thunborg C, Wang R, Rosenberg A, *et al.* Integrating a multimodal lifestyle intervention with medical food in prodromal Alzheimer's disease: the MIND-AD_{mini} randomized controlled trial. Alzheimers Res Ther 2024; 16(1): 118. [http://dx.doi.org/10.1186/s13195-024-01468-x] [PMID: 38812047]

[29] Wells JL, Pasternak SH. Case Report of a 63-Year-Old Patient With Alzheimer Disease and a Novel Presenilin 2 Mutation. Alzheimer Dis Assoc Disord 2019; 33(2): 166-9. [http://dx.doi.org/10.1097/WAD.0000000000000269] [PMID: 30119059]

[30] 2018 Alzheimer's disease facts and figures. Alzheimers Dement 2018; 14(3): 367-429. [http://dx.doi.org/10.1016/j.jalz.2018.02.001]

[31] Lee JH, Friedland R, Whitehouse PJ, Woo JI. Twenty-four-hour rhythms of sleep-wake cycle and temperature in Alzheimer's disease. J Neuropsychiatry Clin Neurosci 2004; 16(2): 192-8. [http://dx.doi.org/10.1176/jnp.16.2.192] [PMID: 15260371]

[32] Siripaopradit Y, Chatsirisakul O, Ariyapaisalkul T, Sereemaspun A. Exploring the gut-brain axis in alzheimer's disease treatment *via* probiotics: evidence from animal studies-a systematic review and meta-analysis. BMC Neurol 2024; 24(1): 481. [http://dx.doi.org/10.1186/s12883-024-03978-5] [PMID: 39695988]

[33] Liu CC, Li CY, Kung SF, *et al.* Association of environmental features and the risk of Alzheimer's dementia in older adults: a nationwide longitudinal case-control study. Int J Environ Res Public Health 2019; 16(16): 2828. [http://dx.doi.org/10.3390/ijerph16162828] [PMID: 31398817]

[34] Mank A, Rijnhart JJM, van Maurik IS, *et al.* A longitudinal study on quality of life along the spectrum of Alzheimer's disease. Alzheimers Res Ther 2022; 14(1): 132. [http://dx.doi.org/10.1186/s13195-022-01075-8] [PMID: 36109800]

[35] Forstmeier S, Maercker A, Savaskan E, Roth T. Cognitive behavioural treatment for mild Alzheimer's patients and their caregivers (CBTAC): study protocol for a randomized controlled trial. Trials 2015; 16(1): 526. [http://dx.doi.org/10.1186/s13063-015-1043-0] [PMID: 26576633]

[36] Bleibel M, El Cheikh A, Sadier NS, Abou-Abbas L. The effect of music therapy on cognitive functions in patients with Alzheimer's disease: a systematic review of randomized controlled trials. Alzheimers Res Ther 2023; 15(1): 65. [http://dx.doi.org/10.1186/s13195-023-01214-9] [PMID: 36973733]

[37] Gillani N, Arslan T. Intelligent sensing technologies for the diagnosis, monitoring, and therapy of Alzheimer's disease: a systematic review. Sensors (Basel) 2021; 21(12): 4249. [http://dx.doi.org/10.3390/s21124249] [PMID: 34205793]

[38] Simrén J, Elmgren A, Blennow K, Zetterberg H. Fluid biomarkers in Alzheimer's disease. Adv Clin Chem 2023; 112: 249-81. [http://dx.doi.org/10.1016/bs.acc.2022.09.006] [PMID: 36642485]

[39] Porsteinsson AP, Isaacson RS, Knox S, Sabbagh MN, Rubino I. Diagnosis of early Alzheimer's disease: clinical practice in 2021. J Prev Alzheimers Dis 2021; 8(3): 371-86. [http://dx.doi.org/10.14283/jpad.2021.23] [PMID: 34101796]

[40] Assunção SS, Sperling RA, Ritchie C, *et al.* Meaningful benefits: a framework to assess disease-modifying therapies in preclinical and early Alzheimer's disease. Alzheimers Res Ther 2022; 14(1): 54. [http://dx.doi.org/10.1186/s13195-022-00984-y] [PMID: 35440022]

[41] Bogdanovic B, Eftimov T, Simjanoska M. In-depth insights into Alzheimer's disease by using

explainable machine learning approach. Sci Rep 2022; 12(1): 6508.
[http://dx.doi.org/10.1038/s41598-022-10202-2] [PMID: 35444165]

[42] Suzgun M, Melas-Kyriazi L, Sarkar SK, Kominers SD, Shieber SM. The Harvard USPTO patent dataset: A large-scale, well-structured, and multipurpose corpus of patent applications. Adv Neural Inf Process Syst 2023; 36: 57908-46.

[43] Ghane M, Ang MC, Cavallucci D, Abdul Kadir R, Ng KW, Sorooshian S. Semantic TRIZ feasibility in technology development, innovation, and production: A systematic review. Heliyon 2024; 10(1)e23775
[http://dx.doi.org/10.1016/j.heliyon.2023.e23775] [PMID: 38226209]

[44] Chen N, Vaseei M, Peivandizadeh A. Forecasting directions, dates, and causes of future technological revolutions concerning the growth of human capital. Discrete Dyn Nat Soc 2022; 2022(1): 2494916.
[http://dx.doi.org/10.1155/2022/2494916]

[45] Rodríguez-Ibánez M, Casánez-Ventura A, Castejón-Mateos F, Cuenca-Jiménez P-M. A review on sentiment analysis from social media platforms. Expert Syst Appl 2023; 223: 119862.
[http://dx.doi.org/10.1016/j.eswa.2023.119862]

[46] Fanlo-Ucar H, Picón-Pagès P, Herrera-Fernández V, ILL-Raga G, Muñoz FJ. The Dual Role of Amyloid Beta-Peptide in Oxidative Stress and Inflammation: Unveiling Their Connections in Alzheimer's Disease Etiopathology. Antioxidants 2024; 13(10): 1208.
[http://dx.doi.org/10.3390/antiox13101208] [PMID: 39456461]

[47] Abdelnour C, Agosta F, Bozzali M, *et al.* Perspectives and challenges in patient stratification in Alzheimer's disease. Alzheimers Res Ther 2022; 14(1): 112.
[http://dx.doi.org/10.1186/s13195-022-01055-y] [PMID: 35964143]

[48] Tan SZK, Zhao RC, Chakrabarti S, Stambler I, Jin K, Lim LW. Interdisciplinary research in Alzheimer's disease and the roles international societies can play. Aging Dis 2021; 12(1): 36-41.
[http://dx.doi.org/10.14336/AD.2020.0602] [PMID: 33532125]

[49] Pless A, Ware D, Saggu S, Rehman H, Morgan J, Wang Q. Understanding neuropsychiatric symptoms in Alzheimer's disease: challenges and advances in diagnosis and treatment. Front Neurosci 2023; 171263771
[http://dx.doi.org/10.3389/fnins.2023.1263771] [PMID: 37732300]

[50] Shi Y, Wu W. Multimodal non-invasive non-pharmacological therapies for chronic pain: mechanisms and progress. BMC Med 2023; 21(1): 372.
[http://dx.doi.org/10.1186/s12916-023-03076-2] [PMID: 37775758]

[51] Eller OC, Willits AB, Young EE, Baumbauer KM. Pharmacological and non-pharmacological therapeutic interventions for the treatment of spinal cord injury-induced pain. Frontiers in Pain Research 2022; 3991736
[http://dx.doi.org/10.3389/fpain.2022.991736] [PMID: 36093389]

[52] Bennett S, Laver K, MacAndrew M, *et al.* Implementation of evidence-based, non-pharmacological interventions addressing behavior and psychological symptoms of dementia: a systematic review focused on implementation strategies. Int Psychogeriatr 2021; 33(9): 947-75.
[http://dx.doi.org/10.1017/S1041610220001702] [PMID: 33190660]

[53] Sikkes SAM, Tang Y, Jutten RJ, *et al.* Toward a theory-based specification of non-pharmacological treatments in aging and dementia: Focused reviews and methodological recommendations. Alzheimers Dement 2021; 17(2): 255-70.
[http://dx.doi.org/10.1002/alz.12188] [PMID: 33215876]

[54] Amato C, Burridge G, Basic D, *et al.* Assistance provided in daily tasks and difficulty experienced by caregivers for people living with dementia. Aust Occup Ther J 2021; 68(3): 236-45.
[http://dx.doi.org/10.1111/1440-1630.12720] [PMID: 33533025]

[55] Jang S, *et al.* Performance-based vs informant-reported instrumental activities of daily living in

predicting dementia 2022.
[http://dx.doi.org/10.1016/j.jamda.2021.09.020]

[56] Hazzan AA, Dauenhauer J, Follansbee P, Hazzan JO, Allen K, Omobepade I. Family caregiver quality of life and the care provided to older people living with dementia: qualitative analyses of caregiver interviews. BMC Geriatr 2022; 22(1): 86.
[http://dx.doi.org/10.1186/s12877-022-02787-0] [PMID: 35101000]

[57] Kim B, Noh GO, Kim K. Behavioural and psychological symptoms of dementia in patients with Alzheimer's disease and family caregiver burden: a path analysis. BMC Geriatr 2021; 21(1): 160.
[http://dx.doi.org/10.1186/s12877-021-02109-w] [PMID: 33663416]

[58] Sehar U, Rawat P, Reddy AP, Kopel J, Reddy PH. Amyloid beta in aging and Alzheimer's disease. Int J Mol Sci 2022; 23(21): 12924.
[http://dx.doi.org/10.3390/ijms232112924] [PMID: 36361714]

[59] Hampel H, Hardy J, Blennow K, *et al.* The amyloid-β pathway in Alzheimer's disease. Mol Psychiatry 2021; 26(10): 5481-503.
[http://dx.doi.org/10.1038/s41380-021-01249-0] [PMID: 34456336]

[60] Azargoonjahromi A. The duality of amyloid-β: its role in normal and Alzheimer's disease states. Mol Brain 2024; 17(1): 44.
[http://dx.doi.org/10.1186/s13041-024-01118-1] [PMID: 39020435]

[61] Ma C, Hong F, Yang S. Amyloidosis in Alzheimer's disease: Pathogeny, etiology, and related therapeutic directions. Molecules 2022; 27(4): 1210.
[http://dx.doi.org/10.3390/molecules27041210] [PMID: 35209007]

[62] Yin X, Xu B, Li J, Wu J. Environmental technology's diminishing marginal returns: a study of green patents and emission reductions in China. Front Environ Sci 2025; 131524824
[http://dx.doi.org/10.3389/fenvs.2025.1524824]

[63] Bagna E, Cotta Ramusino E, Denicolai S. Innovation through patents and intangible assets: effects on growth and profitability of European companies. J Open Innov 2021; 7(4): 220.
[http://dx.doi.org/10.3390/joitmc7040220]

[64] Meziou L, Haubrick K. A Systematic Review of the Effectiveness of Acetylcholinesterase Inhibitors on Cognition for Patients with Alzheimer's Disease Med Biomed Pharm 2024; 5(6).

[65] Truong C, Recto C, Lafont C, Canoui-Poitrine F, Belmin JB, Lafuente-Lafuente C. Effect of cholinesterase inhibitors on mortality in patients with dementia: a systematic review of randomized and nonrandomized trials. Neurology 2022; 99(20): e2313-25.
[http://dx.doi.org/10.1212/WNL.0000000000201161] [PMID: 36096687]

[66] Basheer N, Smolek T, Hassan I, *et al.* Does modulation of tau hyperphosphorylation represent a reasonable therapeutic strategy for Alzheimer's disease? From preclinical studies to the clinical trials. Mol Psychiatry 2023; 28(6): 2197-214.
[http://dx.doi.org/10.1038/s41380-023-02113-z] [PMID: 37264120]

[67] Vaidya S, Guerin AA, Walker LC, Lawrence AJ. Clinical effectiveness of muscarinic receptor-targeted interventions in neuropsychiatric disorders: A systematic review. CNS Drugs 2022; 36(11): 1171-206.
[http://dx.doi.org/10.1007/s40263-022-00964-8] [PMID: 36269510]

[68] Huang LK, Kuan YC, Lin HW, Hu CJ. Clinical trials of new drugs for Alzheimer disease: a 2020–2023 update. J Biomed Sci 2023; 30(1): 83.
[http://dx.doi.org/10.1186/s12929-023-00976-6] [PMID: 37784171]

[69] Reardon S. FDA approves Alzheimer's drug lecanemab amid safety concerns. Nature 2023; 613(7943): 227-8.
[http://dx.doi.org/10.1038/d41586-023-00030-3] [PMID: 36627422]

[70] Dunn B, Stein P, Cavazzoni P. Approval of aducanumab for Alzheimer's disease—the FDA's

perspective. JAMA Intern Med 2021; 181(10): 1276-8.
[http://dx.doi.org/10.1001/jamainternmed.2021.4607] [PMID: 34254984]

[71] Ghazarian AL, Haim T, Sauma S, Katiyar P. National Institute on Aging seed funding enables Alzheimer's disease startups to reach key value inflection points. Alzheimers Dement 2022; 18(2): 348-59.
[http://dx.doi.org/10.1002/alz.12392] [PMID: 34374496]

[72] Jeremic D, Jiménez-Díaz L, Navarro-López JD. Targeting epigenetics: A novel promise for Alzheimer's disease treatment. Ageing Res Rev 2023; 90102003
[http://dx.doi.org/10.1016/j.arr.2023.102003] [PMID: 37422087]

[73] Aisen PS, Bateman RJ, Carrillo M, *et al.* Platform trials to expedite drug development in Alzheimer's disease: a report from the EU/US CTAD Task Force. J Prev Alzheimers Dis 2021; 8(3): 306-12.
[http://dx.doi.org/10.14283/jpad.2021.21] [PMID: 34101788]

[74] Bamakan SMH, Nezhadsistani N, Bodaghi O, Qu Q. Patents and intellectual property assets as non-fungible tokens; key technologies and challenges. Sci Rep 2022; 12(1): 2178.
[http://dx.doi.org/10.1038/s41598-022-05920-6] [PMID: 35140251]

[75] Bamakan SMH, Babaei Bondarti A, Babaei Bondarti P, Qu Q. Blockchain technology forecasting by patent analytics and text mining. Blockchain: Research and Applications 2021; 2(2)100019
[http://dx.doi.org/10.1016/j.bcra.2021.100019]

[76] Kirikkaleli D, Sofuoğlu E, Ojekemi O. Does patents on environmental technologies matter for the ecological footprint in the USA? Evidence from the novel Fourier ARDL approach. Geoscience Frontiers 2023; 14(4)101564
[http://dx.doi.org/10.1016/j.gsf.2023.101564]

[77] Santoni S, Gugliandolo SG, Sponchioni M, Moscatelli D, Colosimo BM. 3D bioprinting: current status and trends—a guide to the literature and industrial practice. Biodes Manuf 2022; 5(1): 14-42.
[http://dx.doi.org/10.1007/s42242-021-00165-0]

[78] Hu F, Qiu L, Xiang Y, *et al.* Spatial network and driving factors of low-carbon patent applications in China from a public health perspective. Front Public Health 2023; 111121860
[http://dx.doi.org/10.3389/fpubh.2023.1121860] [PMID: 36875394]

[79] Puzzo M, De Santo M, Morelli C, Leggio A, Pasqua L. The Advent of Molecular Targeted Therapies Against Cancer. Toward Multi-Targeting Drugs Through Materials Engineering: A Possible Future Scenario. Small Sci 2024; 4(8)2400113
[http://dx.doi.org/10.1002/smsc.202400113] [PMID: 40212559]

[80] Duan XP, Qin BD, Jiao XD, Liu K, Wang Z, Zang YS. New clinical trial design in precision medicine: discovery, development and direction. Signal Transduct Target Ther 2024; 9(1): 57.
[http://dx.doi.org/10.1038/s41392-024-01760-0] [PMID: 38438349]

[81] Patvegar MT, Jarag R, Sathe P. Targets Identification Using Machine Learning: Accelerating The Hunt For Potential Drug Targets. Int J Behav Nutr Phys Act. 2024; 7(2): 488–498.

[82] Abatematteo FS, Niso M, Contino M, Leopoldo M, Abate C. Multi-target directed ligands (MTDLs) binding the σ1 receptor as promising therapeutics: State of the art and perspectives. Int J Mol Sci 2021; 22(12): 6359.
[http://dx.doi.org/10.3390/ijms22126359] [PMID: 34198620]

[83] Bortolami M, Rocco D, Messore A, *et al.* Acetylcholinesterase inhibitors for the treatment of Alzheimer's disease – a patent review (2016–present). Expert Opin Ther Pat 2021; 31(5): 399-420.
[http://dx.doi.org/10.1080/13543776.2021.1874344] [PMID: 33428491]

[84] Muheem A, Jahangir MA, Jaiswal CP, *et al.* Recent patents, regulatory issues, and toxicity of nanoparticles in neuronal disorders. Curr Drug Metab 2021; 22(4): 263-79.
https://www.eurekaselect.com/article/112327
[PMID: 33305703]

[85] Fernández-Bolaños JG, López Ó. Butyrylcholinesterase inhibitors as potential anti-Alzheimer's agents: an updated patent review (2018-present). Expert Opin Ther Pat 2022; 32(8): 913-32. [http://dx.doi.org/10.1080/13543776.2022.2083956] [PMID: 35623095]

[86] Pipes SC. The World's Medicine Chest: How America Achieved Pharmaceutical Supremacy—and How to Keep It. Encounter Books 2025.

[87] Bauer M. Strategic Autonomy and Long-term Innovation Competitiveness: On the Importance of Intellectual Property Rights for the Production of High-Value Medicines in the EU. Brussels: European Centre for International Political Economy (ECIPE); 2022. (Occasional Paper No. 02/2022). https://ecipe.org/publications/intellectual-property-rights-for-medicines-in-eu/

[88] Blum K, Baron D, McLaughlin T, *et al.* Summary document research on RDS anti-addiction modeling: annotated bibliography. Journal of Addiction Psychiatry 2024; 8(1): 1-33. [PMID: 38765881]

[89] Reşit Çakmak , Başaran E, Boğa M, Erdoğan Ö, Çınar E, Çevik Ö. Schiff base derivatives of 4-amino antipyrine as promising molecules: synthesis, structural characterization, and biological activities. Russ J Bioorganic Chem 2022; 48(2): 334-44. [http://dx.doi.org/10.1134/S1068162022020182]

[90] Dubovsky SL, Marshall D. Benzodiazepines remain important therapeutic options in psychiatric practice. Psychother Psychosom 2022; 91(5): 307-34. [http://dx.doi.org/10.1159/000524400] [PMID: 35504267]

[91] Skinner BJ. New cancer drugs in Canada 2012 to 2021: an economic analysis of cost, benefit, availability, and public insurance coverage. Canadian Health Policy 2023; 2023(4): 9492. [http://dx.doi.org/10.54194/IYUA1228]

[92] Au HKE, Isalan M, Mielcarek M. Gene therapy advances: a meta-analysis of AAV usage in clinical settings. Front Med (Lausanne) 2022; 8: 809118. [http://dx.doi.org/10.3389/fmed.2021.809118] [PMID: 35223884]

[93] Wang JY, Doudna JA. CRISPR technology: A decade of genome editing is only the beginning. Science 2023; 379(6629): 643. [http://dx.doi.org/10.1126/science.add8643] [PMID: 36656942]

[94] Marwaha S, Knowles JW, Ashley EA. A guide for the diagnosis of rare and undiagnosed disease: beyond the exome. Genome Med 2022; 14(1): 23. [http://dx.doi.org/10.1186/s13073-022-01026-w] [PMID: 35220969]

[95] Tran L, Xiao JF, Agarwal N, Duex JE, Theodorescu D. Advances in bladder cancer biology and therapy. Nat Rev Cancer 2021; 21(2): 104-21. [http://dx.doi.org/10.1038/s41568-020-00313-1] [PMID: 33268841]

[96] Tabrizi SJ, Estevez-Fraga C, van Roon-Mom WMC, *et al.* Potential disease-modifying therapies for Huntington's disease: lessons learned and future opportunities. Lancet Neurol 2022; 21(7): 645-58. [http://dx.doi.org/10.1016/S1474-4422(22)00121-1] [PMID: 35716694]

[97] Pardo-Moreno T, González-Acedo A, Rivas-Domínguez A, *et al.* Therapeutic approach to Alzheimer's disease: current treatments and new perspectives. Pharmaceutics 2022; 14(6): 1117. [http://dx.doi.org/10.3390/pharmaceutics14061117] [PMID: 35745693]

[98] Passeri E, Elkhoury K, Morsink M, *et al.* Alzheimer's disease: treatment strategies and their limitations. Int J Mol Sci 2022; 23(22): 13954. [http://dx.doi.org/10.3390/ijms232213954] [PMID: 36430432]

[99] Stoiljkovic M, Horvath TL, Hajós M. Therapy for Alzheimer's disease: Missing targets and functional markers? Ageing Res Rev 2021; 68101318 [http://dx.doi.org/10.1016/j.arr.2021.101318] [PMID: 33711510]

[100] Jeremic D, Jiménez-Díaz L, Navarro-López JD. Past, present and future of therapeutic strategies

against amyloid-β peptides in Alzheimer's disease: a systematic review. Ageing Res Rev 2021; 72101496
[http://dx.doi.org/10.1016/j.arr.2021.101496] [PMID: 34687956]

[101] Davis B. Method of treating Alzheimer's disease. Google Patents 1987.

[102] Malluche HH. Method of treating Alzheimer's disease. Google Patents 1990.

[103] Sabin R. Method of treatment of Alzheimer's disease using phytic acid. Google Patents 1988.

[104] McLachlan DR. Method of treating Alzheimer's disease. Google Patents 1983.

[105] Bush A, *et al.* Method of screening for drugs useful in treating Alzheimer's disease. Google Patents 2005.

[106] Masters CL, Bush AI, Beyreuther KT. Method of assaying for Alzheimer's disease. Google Patents 1998.

[107] Glenner GG, Wong CW. Polypeptide marker for Alzheimer's disease and its use for diagnosis. Google Patents 1987.

[108] Goldgaber, D.Y., M. Lerman, and D.C. Gajdusek, cDNA clone encoding brain amyloid of Alzheimer's disease. 1990, Google Patents.

[109] Schilling JW Jr, Ponte PA, Cordell B. Recombinant Alzheimer's protease inhibitory amyloid protein and method of use. Google Patents 1993.

[110] Seubert PA, et al. Methods for aiding in the diagnosis of Alzheimer's disease by measuring amyloid-β peptide (X->41) and tau. Google Patents 2000.

[111] Abraham J. Science, politics and the pharmaceutical industry: Controversy and bias in drug regulation. Routledge 2023.

[112] Kulkov I. The role of artificial intelligence in business transformation: A case of pharmaceutical companies. Technol Soc 2021; 66: 101629.
[http://dx.doi.org/10.1016/j.techsoc.2021.101629]

[113] Destro F, Barolo M. A review on the modernization of pharmaceutical development and manufacturing – Trends, perspectives, and the role of mathematical modeling. Int J Pharm 2022; 620: 121715.
[http://dx.doi.org/10.1016/j.ijpharm.2022.121715] [PMID: 35367580]

[114] Rombouts F, Kusakabe K, Hsiao CC, Gijsen HJM. Small-molecule BACE1 inhibitors: a patent literature review (2011 to 2020). Expert Opin Ther Pat 2021; 31(1): 25-52.
[http://dx.doi.org/10.1080/13543776.2021.1832463] [PMID: 33006491]

[115] Gonzales MM, Garbarino VR, Kautz TF, *et al.* Senolytic therapy in mild Alzheimer's disease: a phase 1 feasibility trial. Nat Med 2023; 29(10): 2481-8.
[http://dx.doi.org/10.1038/s41591-023-02543-w] [PMID: 37679434]

[116] Jakovljevic M, Deceuninck P, Pistollato F, *et al.* Return on investment in science: twenty years of European Commission funded research in Alzheimer's dementia, breast cancer and prostate cancer. Cost Eff Resour Alloc 2024; 22(1): 51.
[http://dx.doi.org/10.1186/s12962-024-00540-5] [PMID: 38880873]

[117] Peltomaa R, Barderas R, Benito-Peña E, Moreno-Bondi MC. Recombinant antibodies and their use for food immunoanalysis. Anal Bioanal Chem 2022; 414(1): 193-217.
[http://dx.doi.org/10.1007/s00216-021-03619-7] [PMID: 34417836]

[118] Wang L, Wang N, Zhang W, *et al.* Therapeutic peptides: current applications and future directions. Signal Transduct Target Ther 2022; 7(1): 48.
[http://dx.doi.org/10.1038/s41392-022-00904-4] [PMID: 35165272]

[119] Wang Z, Wang G, Lu H, *et al.* Development of therapeutic antibodies for the treatment of diseases. Mol Biomed. 2022 Oct 11; 3: 35.

[http://dx.doi.org/10.1186/s43556-022-00100-4]

[120] Wang J, Li B, Qiu L, Qiao X, Yang H. Dendrimer-based drug delivery systems: history, challenges, and latest developments. J Biol Eng 2022; 16(1): 18.
[http://dx.doi.org/10.1186/s13036-022-00298-5] [PMID: 35879774]

[121] Kheraldine H, Rachid O, Habib AM, Al Moustafa AE, Benter IF, Akhtar S. Emerging innate biological properties of nano-drug delivery systems: A focus on PAMAM dendrimers and their clinical potential. Adv Drug Deliv Rev 2021; 178113908
[http://dx.doi.org/10.1016/j.addr.2021.113908] [PMID: 34390777]

[122] Moorthy H, Govindaraju T. Dendrimer architectonics to treat cancer and neurodegenerative diseases with implications in theranostics and personalized medicine. ACS Appl Bio Mater 2021; 4(2): 1115-39.
[http://dx.doi.org/10.1021/acsabm.0c01319] [PMID: 35014470]

[123] Sahu T, Ratre YK, Chauhan S, Bhaskar LVKS, Nair MP, Verma HK. Nanotechnology based drug delivery system: Current strategies and emerging therapeutic potential for medical science. J Drug Deliv Sci Technol 2021; 63: 102487.
[http://dx.doi.org/10.1016/j.jddst.2021.102487]

[124] Dai R, Sun Y, Su R, Gao H. Anti-Alzheimer's disease potential of traditional chinese medicinal herbs as inhibitors of BACE1 and AChE enzymes. Biomed Pharmacother 2022; 154: 113576.
[http://dx.doi.org/10.1016/j.biopha.2022.113576] [PMID: 36007279]

[125] Coimbra JRM, Resende R, Custódio JBA, Salvador JAR, Santos AE. BACE1 inhibitors for Alzheimer's Disease: current challenges and Future perspectives. J Alzheimers Dis 2024; 101(s1): S53-78.
[http://dx.doi.org/10.3233/JAD-240146] [PMID: 38943390]

[126] De Klerk M, *et al.* Incorporating familial risk, lifestyle factors, and pharmacogenomic insights into personalized noncommunicable disease (NCD) reports for healthcare funder beneficiaries participating in the Open Genome Project. Ann Hum Genet 2024.
[PMID: 39470271]

[127] Krishnadas N, Doré V, Groot C, *et al.* Mesial temporal tau in amyloid-β-negative cognitively normal older persons. Alzheimers Res Ther 2022; 14(1): 51.
[http://dx.doi.org/10.1186/s13195-022-00993-x] [PMID: 35395950]

[128] Gonzalez-Ortiz F, Kac PR, Brum WS, Zetterberg H, Blennow K, Karikari TK. Plasma phospho-tau in Alzheimer's disease: towards diagnostic and therapeutic trial applications. Mol Neurodegener 2023; 18(1): 18.
[http://dx.doi.org/10.1186/s13024-023-00605-8] [PMID: 36927491]

[129] Gobom J, Benedet AL, Mattsson-Carlgren N, *et al.* Antibody-free measurement of cerebrospinal fluid tau phosphorylation across the Alzheimer's disease continuum. Mol Neurodegener 2022; 17(1): 81.
[http://dx.doi.org/10.1186/s13024-022-00586-0] [PMID: 36510321]

[130] Hong S, Baek SH, Lai MKP, Arumugam TV, Jo DG. Aging-associated sensory decline and Alzheimer's disease. Mol Neurodegener 2024; 19(1): 93.
[http://dx.doi.org/10.1186/s13024-024-00776-y] [PMID: 39633396]

[131] Javaid SF, Giebel C, Khan MAB, Hashim MJ. Epidemiology of Alzheimer's disease and other dementias: rising global burden and forecasted trends. F1000 Res 2021; 10: 425.
[http://dx.doi.org/10.12688/f1000research.50786.1]

[132] Kerwin D, Abdelnour C, Caramelli P, *et al.* Alzheimer's disease diagnosis and management: Perspectives from around the world. Alzheimers Dement (Amst) 2022; 14(1)e12334
[http://dx.doi.org/10.1002/dad2.12334] [PMID: 35898519]

[133] Hoyt AS. The impact of uncertainty regarding patent eligible subject matter for investment in US medical diagnostic technologies. Wash Lee Law Rev 2022; 79: 397.

[134] Carneiro DC, Fernandez LG, Monteiro-Cunha JP, Benevides RG, Cunha Lima ST. A patent review of the antimicrobial applications of lectins: Perspectives on therapy of infectious diseases. J Appl Microbiol 2022; 132(2): 841-54.
[http://dx.doi.org/10.1111/jam.15263] [PMID: 34416098]

[135] Ingrole RSJ, Azizoglu E, Dul M, Birchall JC, Gill HS, Prausnitz MR. Trends of microneedle technology in the scientific literature, patents, clinical trials and internet activity. Biomaterials 2021; 267: 120491.
[http://dx.doi.org/10.1016/j.biomaterials.2020.120491] [PMID: 33217629]

[136] Uddin KMM, Alam MJ, Jannat-E-Anawar , Uddin MA, Aryal S. A novel approach utilizing machine learning for the early diagnosis of Alzheimer's disease. Biomedical Materials & Devices 2023; 1(2): 882-98.
[http://dx.doi.org/10.1007/s44174-023-00078-9] [PMID: 37363136]

[137] Kavitha C, Mani V, Srividhya SR, Khalaf OI, Tavera Romero CA. Early-stage Alzheimer's disease prediction using machine learning models. Front Public Health 2022; 10: 853294.
[http://dx.doi.org/10.3389/fpubh.2022.853294] [PMID: 35309200]

[138] An X, He J, Bi B, *et al.* The application of artificial intelligence in diagnosis of Alzheimer's disease: a bibliometric analysis. Front Neurol 2024; 151510729
[http://dx.doi.org/10.3389/fneur.2024.1510729] [PMID: 39703357]

[139] An X, He J, Bi B, *et al.* The role of astrocytes in Alzheimer's disease: a bibliometric analysis. Front Aging Neurosci 2024; 161481748
[http://dx.doi.org/10.3389/fnagi.2024.1481748] [PMID: 39665038]

[140] Li S, Zhang Q, Liu J, *et al.* Bibliometric Analysis of Alzheimer's Disease and Depression. Curr Neuropharmacol 2025; 23(1): 98-115.
[http://dx.doi.org/10.2174/1570159X22666240730154834] [PMID: 39092642]

[141] Hu JH, Zhang X, Yang HM, Xu YL, Zhang M, Niu X. A bibliometric analysis of the relationship between traumatic brain injury and Alzheimer's disease (1993-2023). Front Aging Neurosci 2024; 16: 1462132.
[http://dx.doi.org/10.3389/fnagi.2024.1462132] [PMID: 39507204]

[142] Ciurea VA, Covache-Busuioc RA, Mohan AG, Costin HP, Voicu V. Alzheimer's disease: 120 years of research and progress. J Med Life 2023; 16(2): 173-7.
[PMID: 36937482]

[143] Blackman J, Swirski M, Clynes J, Harding S, Leng Y, Coulthard E. Pharmacological and non-pharmacological interventions to enhance sleep in mild cognitive impairment and mild Alzheimer's disease: A systematic review. J Sleep Res 2021; 30(4)e13229
[http://dx.doi.org/10.1111/jsr.13229] [PMID: 33289311]

[144] Li X, Ji M, Zhang H, *et al.* Non-drug therapies for Alzheimer's disease: a review. Neurol Ther 2023; 12(1): 39-72.
[http://dx.doi.org/10.1007/s40120-022-00416-x] [PMID: 36376734]

[145] Moreno N. Pharmacological and Non-Pharmacological Treatments for Cognitive Impairments in Older Adults with Alzheimer's Disease: A Narrative Review 2025.

[146] Menardi A, Rossi S, Koch G, *et al.* Toward noninvasive brain stimulation 2.0 in Alzheimer's disease. Ageing Res Rev 2022; 75101555
[http://dx.doi.org/10.1016/j.arr.2021.101555] [PMID: 34973457]

[147] Ramirez-Gomez J, Dalal S, Devara D, Sharma B, Rodarte D, Kumar S. MicroRNA-based recent research developments in Alzheimer's disease. J Alzheimers Dis 2025; 104(1): 14-31.
[http://dx.doi.org/10.1177/13872877241313397] [PMID: 39894921]

[148] Mecca AP, O'Dell RS, Sharp ES, *et al.* Synaptic density and cognitive performance in Alzheimer's disease: A PET imaging study with [^{11} C]UCB-J. Alzheimers Dement 2022; 18(12): 2527-36.

[http://dx.doi.org/10.1002/alz.12582] [PMID: 35174954]

[149] Watts GF, Gidding SS, Hegele RA, *et al.* International Atherosclerosis Society guidance for implementing best practice in the care of familial hypercholesterolaemia. Nat Rev Cardiol 2023; 20(12): 845-69.
[http://dx.doi.org/10.1038/s41569-023-00892-0] [PMID: 37322181]

[150] Cui F, Blach S, Manzengo Mingiedi C, *et al.* Global reporting of progress towards elimination of hepatitis B and hepatitis C. Lancet Gastroenterol Hepatol 2023; 8(4): 332-42.
[http://dx.doi.org/10.1016/S2468-1253(22)00386-7] [PMID: 36764320]

[151] Liss JL, Seleri Assunção S, Cummings J, *et al.* Practical recommendations for timely, accurate diagnosis of symptomatic Alzheimer's disease (MCI and dementia) in primary care: a review and synthesis. J Intern Med 2021; 290(2): 310-34.
[http://dx.doi.org/10.1111/joim.13244] [PMID: 33458891]

[152] Chandra S, Verma S, Lim WM, Kumar S, Donthu N. Personalization in personalized marketing: Trends and ways forward. Psychol Mark 2022; 39(8): 1529-62.
[http://dx.doi.org/10.1002/mar.21670]

[153] Yang C, Huang X, Li Y, Chen J, Lv Y, Dai S. Prognosis and personalized treatment prediction in *TP53* -mutant hepatocellular carcinoma: an *in silico* strategy towards precision oncology. Brief Bioinform 2021; 22(3)bbaa164
[http://dx.doi.org/10.1093/bib/bbaa164] [PMID: 32789496]

[154] Kontogeorgis GM, Dohrn R, Economou IG, *et al.* Industrial requirements for thermodynamic and transport properties: 2020. Ind Eng Chem Res 2021; 60(13): 4987-5013.
[http://dx.doi.org/10.1021/acs.iecr.0c05356] [PMID: 33840887]

[155] Pascaris AS, Schelly C, Burnham L, Pearce JM. Integrating solar energy with agriculture: Industry perspectives on the market, community, and socio-political dimensions of agrivoltaics. Energy Res Soc Sci 2021; 75102023
[http://dx.doi.org/10.1016/j.erss.2021.102023]

[156] O'Dwyer M, Filieri R, O'Malley L. Establishing successful university–industry collaborations: barriers and enablers deconstructed. J Technol Transf 2023; 48(3): 900-31.
[http://dx.doi.org/10.1007/s10961-022-09932-2]

[157] Emmerich CH, Gamboa LM, Hofmann MCJ, *et al.* Improving target assessment in biomedical research: the GOT-IT recommendations. Nat Rev Drug Discov 2021; 20(1): 64-81.
[http://dx.doi.org/10.1038/s41573-020-0087-3] [PMID: 33199880]

[158] Stahl BC, Eke D. The ethics of ChatGPT – Exploring the ethical issues of an emerging technology. Int J Inf Manage 2024; 74102700
[http://dx.doi.org/10.1016/j.ijinfomgt.2023.102700]

[159] Mukherjee P, Roy S, Ghosh D, Nandi SK. Role of animal models in biomedical research: a review. Lab Anim Res 2022; 38(1): 18.
[http://dx.doi.org/10.1186/s42826-022-00128-1] [PMID: 35778730]

[160] Teare HJA, Prictor M, Kaye J. Reflections on dynamic consent in biomedical research: the story so far. Eur J Hum Genet 2021; 29(4): 649-56.
[http://dx.doi.org/10.1038/s41431-020-00771-z] [PMID: 33249421]

[161] de Freitas FV, Gomes MVM, Winkler I. Benefits and challenges of virtual-reality-based industrial usability testing and design reviews: A patents landscape and literature review. Appl Sci (Basel) 2022; 12(3): 1755.
[http://dx.doi.org/10.3390/app12031755]

[162] Zatsu V, Shine AE, Tharakan JM, *et al.* Revolutionizing the food industry: The transformative power of artificial intelligence-a review. Food Chem X 2024; 24: 101867.
[http://dx.doi.org/10.1016/j.fochx.2024.101867] [PMID: 39431210]

[163] Qamruzzaman M, Karim S. Green energy, green innovation, and political stability led to green growth in OECD nations. Energy Strategy Reviews 2024; 55: 101519. [http://dx.doi.org/10.1016/j.esr.2024.101519]

[164] Festari C, Massa F, Cotta Ramusino M, *et al.* European consensus for the diagnosis of MCI and mild dementia: Preparatory phase. Alzheimers Dement 2023; 19(5): 1729-41. [http://dx.doi.org/10.1002/alz.12798] [PMID: 36209379]

[165] Dubbelman MA, Verrijp M, Terwee CB, *et al.* Determining the minimal important change of everyday functioning in dementia: pursuing clinical meaningfulness. Neurology 2022; 99(9): e954-64. [http://dx.doi.org/10.1212/WNL.0000000000200781] [PMID: 35641309]

[166] González JF, Sánchez-Montero JM. How to address the complexity of multi-targeted drug discovery for Alzheimer's disease?. Taylor & Francis 2024; pp. 1149-52.

[167] Chhabra A, Solanki S, Saravanabawan P, Venkiteswaran A, Nimmathota N, Modi NM. A systematic review of the efficacy and safety of anti-amyloid beta monoclonal antibodies in treatment of Alzheimer's disease. Expert Opin Biol Ther 2024; 24(11): 1261-9. [http://dx.doi.org/10.1080/14712598.2024.2416947] [PMID: 39432414]

[168] Sexton C, Snyder H, Beher D, *et al.* Current directions in tau research: Highlights from Tau 2020. Alzheimers Dement 2022; 18(5): 988-1007. [http://dx.doi.org/10.1002/alz.12452] [PMID: 34581500]

[169] Phan LMT, Hoang TX, Vo TAT, *et al.* Nanomaterial-based Optical and Electrochemical Biosensors for Amyloid beta and Tau: Potential for early diagnosis of Alzheimer's Disease. Expert Rev Mol Diagn 2021; 21(2): 175-93. [http://dx.doi.org/10.1080/14737159.2021.1887732] [PMID: 33560154]

[170] Blind K, Krieger B, Pellens M. The interplay between product innovation, publishing, patenting and developing standards. Res Policy 2022; 51(7)104556. [http://dx.doi.org/10.1016/j.respol.2022.104556]

[171] Bekkers R, Tur EM, Henkel J, van der Vorst T, Driesse M, Contreras JL. Overcoming inefficiencies in patent licensing: A method to assess patent essentiality for technical standards. Res Policy 2022; 51(10)104590. [http://dx.doi.org/10.1016/j.respol.2022.104590]

[172] Thawabteh AM, Ghanem AW, AbuMadi S, *et al.* Recent Advances in Therapeutics for the Treatment of Alzheimer's Disease. Molecules 2024; 29(21): 5131. [http://dx.doi.org/10.3390/molecules29215131] [PMID: 39519769]

[173] Prokopenko D, Morgan SL, Mullin K, *et al.* Whole-genome sequencing reveals new Alzheimer's disease–associated rare variants in loci related to synaptic function and neuronal development. Alzheimers Dement 2021; 17(9): 1509-27. [http://dx.doi.org/10.1002/alz.12319] [PMID: 33797837]

[174] Andrade-Guerrero J, Santiago-Balmaseda A, Jeronimo-Aguilar P, *et al.* Alzheimer's disease: an updated overview of its genetics. Int J Mol Sci 2023; 24(4): 3754. [http://dx.doi.org/10.3390/ijms24043754] [PMID: 36835161]

[175] Andrews SJ, Renton AE, Fulton-Howard B, Podlesny-Drabiniok A, Marcora E, Goate AM. The complex genetic architecture of Alzheimer's disease: novel insights and future directions. EBioMedicine 2023; 90104511 [http://dx.doi.org/10.1016/j.ebiom.2023.104511] [PMID: 36907103]

[176] Gazestani V, Kamath T, Nadaf NM, et al. Early Alzheimer's disease pathology in human cortex involves transient cell states. Cell 2023 Sep 28; 186(20): 4438–4453. [http://dx.doi.org/10.1016/j.cell.2023.08.005]

[177] Li TR, Yang Q, Hu X, Han Y. Biomarkers and tools for predicting Alzheimer's disease in the preclinical stage. Curr Neuropharmacol 2022; 20(4): 713-37.

[http://dx.doi.org/10.2174/1570159X19666210524153901] [PMID: 34030620]

[178] Iturria-Medina Y, Adewale Q, Khan AF, *et al.* Unified epigenomic, transcriptomic, proteomic, and metabolomic taxonomy of Alzheimer's disease progression and heterogeneity. Sci Adv 2022; 8(46)eabo6764
[http://dx.doi.org/10.1126/sciadv.abo6764] [PMID: 36399579]

[179] Wang M, Song W, Ming C, *et al.* Guidelines for bioinformatics of single-cell sequencing data analysis in Alzheimer's disease: review, recommendation, implementation and application. Mol Neurodegener 2022; 17(1): 17.
[http://dx.doi.org/10.1186/s13024-022-00517-z] [PMID: 35236372]

[180] Miranda J, Navarrete C, Noguez J, *et al.* The core components of education 4.0 in higher education: Three case studies in engineering education. Comput Electr Eng 2021; 93107278
[http://dx.doi.org/10.1016/j.compeleceng.2021.107278]

[181] Zhang X, Chen Y, Hu L, Wang Y. The metaverse in education: Definition, framework, features, potential applications, challenges, and future research topics. Front Psychol 2022; 131016300
[http://dx.doi.org/10.3389/fpsyg.2022.1016300] [PMID: 36304866]

[182] Pont C, Ginex T, Griñán-Ferré C, *et al.* From virtual screening hits targeting a cryptic pocket in BACE-1 to a nontoxic brain permeable multitarget anti-Alzheimer lead with disease-modifying and cognition-enhancing effects. Eur J Med Chem 2021; 225113779
[http://dx.doi.org/10.1016/j.ejmech.2021.113779] [PMID: 34418785]

[183] Mitroshina EV, Saviuk M, Vedunova MV. Necroptosis in CNS diseases: Focus on astrocytes. Front Aging Neurosci 2023; 141016053
[http://dx.doi.org/10.3389/fnagi.2022.1016053] [PMID: 36778591]

[184] Vieyra E, Calderón R, Linares R, *et al.* Pharmacologic blockade of nicotinic receptors in the suprachiasmatic nucleus increases ovarian atresia and inhibits follicular growth. J Neuroendocrinol 2024; 36(9)e13421
[http://dx.doi.org/10.1111/jne.13421] [PMID: 38826071]

[185] Laganà V, Bruno F, Altomari N, *et al.* Neuropsychiatric or behavioral and psychological symptoms of dementia (BPSD): focus on prevalence and natural history in Alzheimer's disease and frontotemporal dementia. Front Neurol 2022; 13832199
[http://dx.doi.org/10.3389/fneur.2022.832199] [PMID: 35812082]

[186] van der Schaar J, Visser LNC, Bouwman FH, *et al.* Considerations regarding a diagnosis of Alzheimer's disease before dementia: a systematic review. Alzheimers Res Ther 2022; 14(1): 31.
[http://dx.doi.org/10.1186/s13195-022-00971-3] [PMID: 35144684]

SUBJECT INDEX

A

B

C

D

E

F

G

H

I

L

M

N

O

P

Q

R

S

T

V

Z

www.ingramcontent.com/pod-product-compliance
Lightning Source LLC
LaVergne TN
LVHW070944210926
848482LV00002B/86
* 9 7 9 8 8 9 8 8 1 4 9 7 7 *